MECHANICAL ENGINEERING FOR SUSTAINABLE DEVELOPMENT

State-of-the-Art Research

MECHANICAL ENGINEERING FOR SUSTAINABLE DEVELOPMENT

State-of-the-Art Research

Edited by

C. S. P. Rao, PhD

G. Amba Prasad Rao, PhD

N. Selvaraj, PhD

P. S. C. Bose, PhD

V. P. Chandramohan, PhD

AAP | APPLE ACADEMIC PRESS

Apple Academic Press Inc.
3333 Mistwell Crescent
Oakville, ON L6L 0A2
Canada

Apple Academic Press Inc.
9 Spinnaker Way
Waretown, NJ 08758
USA

Exclusive worldwide distribution by CRC Press, a member of Taylor & Francis Group

International Standard Book Number-13: 978-1-77188-681-9 (Hardcover)
International Standard Book Number-13: 978-1-351-17016-1 (eBook)

Library and Archives Canada Cataloguing in Publication

ISME Conference on Mechanical Engineering (18th : 2017 : Warangal, India)

Mechanical engineering for sustainable development : state-of-the-art research / edited by C.S.P. Rao, PhD, G. Amba Prasad Rao, PhD, N. Selvaraj, PhD, P.S.C. Bose, PhD, V.P. Chandramohan.

This volume is a compilation of selected research papers prepared and presented at the Eighteenth National Conference of the Indian Society of Mechanical Engineers held at NIT Warangal, February 23-25, 2017.

Includes bibliographical references and index.

Issued in print and electronic formats.

ISBN 978-1-77188-681-9 (hardcover).--ISBN 978-1-351-17016-1 (PDF)

1. Mechanical engineering--Congresses. 2. Sustainable engineering--Congresses. 3. Sustainable development--Congresses. I. Rao, C. S. P., editor II. Title.

TJ5.I86 2018 621 C2018-906143-X C2018-906144-8

CIP data on file with US Library of Congress

Apple Academic Press also publishes its books in a variety of electronic formats. Some content that appears in print may not be available in electronic format. For information about Apple Academic Press products, visit our website at **www.appleacademicpress.com** and the CRC Press website at **www.crcpress.com**

ABOUT THE EDITORS

C. S. P. Rao, PhD

C. S. P. Rao, PhD, is presently working as a Professor in the Department of Mechanical Engineering at the National Institute of Technology, Warangal, India. He has more than 30 years of teaching and research experience. He has authored two textbooks and has published around 250 research papers in national and international journals and conference proceedings. He has supervised more than 20 PhD candidates and guided about 100 postgraduate dissertations and about 50 undergraduate projects. Dr. Rao received the Engineer of the Year Award 2008 on the occasion of the 41st Engineers Day Celebrations from the Government of Andhra Pradesh and the Institution of Engineers (India), AP State Center, as well as an award for Andhra Pradesh Scientist 2008 on the occasion of the National Science Day Celebrations from the Andhra Pradesh Council of Science and Technology (APCOST), Govt. of Andhra Pradesh. Dr. Rao was instrumental in establishing several manufacturing engineering related laboratories and has also held several administrative positions at the National Institute of Technology. He has organized and attended technical conferences, seminars, etc. Dr. Rao's main area of expertise is manufacturing engineering. Other areas of interest include CAD/CAM, CIM, modeling and simulation, evolutionary computation, metal cutting and metal-matrix composites, and flexible manufacturing system.

G. Amba Prasad Rao, PhD

G. Amba Prasad Rao, PhD, is a full Professor in the Department of Mechanical Engineering at the National Institute of Technology, Warangal, India. He has more than 26 years of teaching and research experience and has supervised eight PhD candidates and guided 24 undergraduate design projects and 40 postgraduate dissertations. He has published around 60 research papers in national and international journals and conferences. Dr. Rao has been a reviewer for many journals. He has organized several short-term courses for faculty and also organized the prestigious 1st International and 18th National Indian Society of Mechanical Engineers conference. He has handled a good number of sponsored research projects. He has held many administrative positions at the National Institute of Technology, and at present he is also the Professor in Charge of Training and Placement at the Institute. He has attended many technical conferences, seminars, etc., in India and internationally and has given expert talks. His main areas of research include studies on internal combustion engines for improved performance with alternate fuels, emissions control, engine combustion modeling, and propulsion.

N. Selvaraj, PhD

N. Selvaraj, PhD, is currently working as a Professor in the Department of Mechanical Engineering at the National Institute of Technology, Warangal, India. He guided seven PhD students and 28 undergraduate design projects and 56 postgraduate dissertations. He has published around 50 research papers in national and international journals. He is associated with senior faculty in the development of undergraduate and postgraduate laboratories. He also worked at various administrative positions at the National Institute of Technology. Dr. Selvaraj has organized and attended technical conferences and seminars. His areas of interest include composite materials, modeling and simulation, flexible manufacturing systems, computer numerical control technology, machine tools, and pull systems.

P. S. C. Bose, PhD

P. Subhash Chandra Bose, PhD, is an Assistant Professor in the Department of Mechanical Engineering at the National Institute of Technology, Warangal, India. He has done extensive consulting and research in India on various aspects of machining processes to with grants of over $1 million US. He is associated with the Defense Research and Development Organization (DRDO), the Naval Research Board (NRB), and the Research Centre Imarat (RCI). He has written more than 30 articles on various fields of machining processes in national and international journals. He has been teaching manufacturing engineering and computer numerical control and AM technologies for postgraduate and undergraduate students for the past 13 years and has guided 50 undergraduate, 25 postgraduate, and six research students. He has pioneered five workshops and conferences at the national and international levels. He is also actively participating in administrative work at the Institute. Dr. Bose's research interests include fabrication and machining of ceramic composites, machining of super alloys, computer numerical control technologies, optimization techniques, and additive manufacturing.

V. P. Chandramohan, PhD

V. P. Chandramohan, PhD, is an Assistant Professor of Mechanical Engineering at the National Institute of Technology Warangal, India. He has been teaching at the Institute since 2013. His research contributions are in the field of convection and conduction heat transfer, drying and simultaneous solution of heat and mass transfer, and solar energy and alternative fuels. He has 15 years of teaching and research experience. Dr. Chandramohan received his ME degree from Annamalai University, Tamil Nadu, and his PhD degree from the Indian Institute of Technology, Delhi. He received the Dr. S. Sathik prize for first rank in the batch of 2002 of ME students from Annamalai University, Tamil Nadu.

CONTENTS

CONTRIBUTORS

Md. Abid Ali
Department of Mechanical Engineering, Ramachandra College of Engineering, Eluru, India

S. A. Alur
Mechanical Engineering Department, Hirasugar Institute of Technology Nidasoshi, Belagavi, Karnataka, India

T. R. Anil
Department of Mechanical Engineering, Gogte Institute of Technology, Belgaum, Karnataka 591236, India

Aswatha
Department of Mechanical Engineering, Bangalore Institute of Technology, Bengaluru, Karnataka, India

A. T. Autee
Mechanical Engineering Department, Maharashtra Institute of Technology, Aurangabad, Maharashtra 431010, India

S. G. Auti
Mechanical Engineering Department, Maharashtra Institute of Technology, Aurangabad, Maharashtra 431010, India

S. Nagendra Babu
Defence Research and Development Laboratory (DRDL), Hyderabad, Telangana, India

Athota Rathan Babu
Department of Aeronautical Engineering, Institute of Aeronautical Engineering, Hyderabad, Telangana, India

Ashok Bharati
Department of Mechanical Engineering, Maharashtra Institute of Technology, Aurangabad, Maharashtra 431010, India

Anil Kumar Bodukuri
Department of Mechanical Engineering, Kakatiya Institute of Technology & Science, Warangal, Telangana, India

V. Phanindra Bogu
Department of Mechanical Engineering, NIT Warangal, Warangal, Telangana, India

P. Sai Chaitanya
GMR Institute of Technology, Rajam, Srikakulam District, Andhra Pradesh, India

K. V. P. Chakradhar
Department of Mechanical Engineering, Madanapalle Institute of Technology and Science, Madanapalle, Andhra Pradesh, India

R. V. Chalam
National Institute of Technology, Warangal, Telangana 506004, India

Subhash Chandra Bose P.
Department of Mechanical Engineering, National Institute of Technology, Warangal 506004, India

A. N. Chapgaon
Department of Mechanical Engineering, Ashokrao Mane Group of Institute, Vathar, Wadgaon, Kolhapur, Maharashtra, India

Sandip Chattopadhyay
Defence Research and Development Laboratory (DRDL), Hyderabad, Telangana, India

M. Cheeranjivi
Department of Mechanical Engineering, Amrita Sai Institute of Technology, Andhra Pradesh, India

M. Bala Chennaiah
Department of Mechanical Engineering, V. R. Siddhartha Engineering College, Kanuru, Vijayawada, Andhra Pradesh, India

Animesh Chhotray
Robotics Laboratory, Mechanical Engineering Department, NIT, Rourkela, Odisha 769008, India

Neelima Devi Chinta
Department of Mechanical Engineering, JNTUK University College of Engineering, Vizianagaram, Andhra Pradesh, India

Praveen Choudhary
Department of Mechanical Engineering, School of Engineering, Gautam Buddha University, Greater Noida, Uttar Pradesh 201312, India

B. M. Dodamani
Mechanical Engineering Department, Hirasugar Institute of Technology Nidasoshi, Belagavi, Karnataka, India

Himansu S. Dash
Mechanical Engineering Department, VFSTR University, Vadlamudi, Andhra Pradesh, India

Akshay M. Dashwant
Mehanical Engineering Department, Kolhapur Institute of Technology, Kolhapur, Maharashtra 416234, India

Rabisankar Debnath
Production Engineering, National Institute of Technology, Agartala National Institute of Technology, Agartala, Tripura, India

Jaya Krishna Devanuri
Department of Mechanical Engineering, NIT Warangal, Telangana 506004, India

H. M. Dharmadhikari
Mechanical Engineering, Marathwada Institute of Technology, Aurangabad, Maharashtra 431010, India

Vilas G. Dhore
Department of Mechanical Engineering, Veermata Jijabai Technological Institute, Mumbai, Maharashtra 400019, India

B. M. Dodamani
Mechanical Engineering Department, Hirasugar Institute of Technology Nidasoshi, Belagavi, Karnataka, India

K. Eswaraiah
Department of Mechanical Engineering, Kakatiya Institute of Technology & Science, Warangal, Telangana, India

Nagarjunavarma Ganna
Department of Mechanical Engineering, Anurag Engineering College, Kodad, Suryapet, Telangana, India

Neville Anton George
Department of Aerospace Engineering Indian Institute of Space Science and Technology, Trivandrum, Kerala, India

S. A. Goudadi
Department of Mechanical Engineering, Hirasugar Institute of Technology, Nidasoshi, Belagavi, Karnataka 591236, India

D. Govardhan
Department of Aeronautical Engineering, Institute of Aeronautical Engineering, Hyderabad, Telangana, India

B. P. Harsha
Department of Mechanical Engineering, JSSATE, Noida, Uttar Pradesh 201301, India

D. Jaya Krishna
Department of Mechanical Engineering, National Institute of Technology, Warangal, Telangana 506004, India, E-mail: djayakrishna.iitm@gmail.com

Romit M. Kamble
Department of Automobile Engineering, Rajarambapu Institute of Technology, Sakharale, Sangli 415414, India

U. C. Kapale
Department of Mechanical Engineering, S.G. Balekundri Institute of Technology, Belgaum, Karnataka 591236, India

A. J. Keche
Mechanical Engineering Department, Maharashtra Institute of Technology, Aurangabad, Maharashtra 431010, India

Nanda Kishore
Department of Chemical Engineering, Indian Institute of Technology, Guwahati 781039, Assam, India

Suhas C. Kongre
Mechanical Engineering Department ASP Pipri, Wardha, Maharashtra, India

N. Rama Krishna
Department of Mechanical Engineering, KITS, Singapur, Huzurabad, Telangana, India

Lanka Krishnanand
Department of Mechanical Engineering, National Institute of Technology Warangal, Warangal, Telangana

Sai Sarath Kruthiventi
Department of Mechanical Engineering, K L University Guntur, Andhra Pradesh, India

Arun J. Kulangara
Department of Mechanical Engineering, National Institute of Technology, Warangal 506004, India

S. R. Kulkarni
Department of Mechanical Engineering, Hirasugar Institute of Technology, Nidasoshi, Belagavi, Karnataka 591236, India

A. Kumar
Department of Mechanical Engineering, NITW, Warangal, Telangana, India

Priyadarshi Biplab Kumar
Robotics Laboratory, Mechanical Engineering Department, NIT, Rourkela, Odisha 769008, India

K. Ch. Kishor Kumar
Mechanical Engineering Department, Gudlavalleru Engineering College,
Sheshadri Rao Knowledge Village, Gudlavalleru, Andhra Pradesh 521356, India

B. Karuna Kumar
Mechanical Engineering Department, Gudlavalleru Engineering College,
Sheshadri Rao Knowledge Village, Gudlavalleru, Andhra Pradesh 521356, India

Monu Kumar
Department of Mechanical Engineering, DCR University of Science and Technology, Murthal,
Sonipat, Haryana 131039, India

P. Nanda Kumar
Department of Mechanical Engineering, N.B.K.R. Institute of Science and Technology, Vidhyanagar,
Nellore, Andhra Pradesh, India

Y. Ravi Kumar
Department of Mechanical Engineering, NIT Warangal, Warangal, Telangana, India

K. Asit Kumar
Department of Metallurgical and Material Engineering, NIT Warangal, Warangal, Telangana, India

Santhosha Kumari
Mechanical Engineering, CJITS, Warangal

S. G. Kumbhar
Department of Automobile Engineering, Rajarambapu Institute of Technology,Rajaramnagar,
Sakharale, Sangli, Maharashtra, India, E-mail: surajkumar.kumbhar@ritindia.edu

Jagdeep Kshirsagar
Department of Mechanical Engineering, Maharashtra Institute of Technology, Aurangabad,
Maharashtra 431010, India

A. K. Kumbhar
Department of Automobile Engineering, Annasaheb Dange College of Engineering and Technology,
Ashta, Maharashtra, India, E-mail: avadhut.k94@gmail.com

Atul G. Lodhekar
Mechanical Engineering Department, Rajiv Gandhi Institute of Technology, Mumbai,
Maharashtra, India

M. N. Madhu
Department of Mechanical Engineering, NIT Warangal, Warangal, Telangana, India

Y. C. Madhukumar
Siddaganga Institute of Technology, Tumkur, Karnataka, India

D. V. Madhuri
Mechanical Engineering, CJITS, Warangal, India

V. Mahesh
Department of Mechanical Engineering, SR Engineering College, Warangal, Telangana, India

T. U. Mali
Department of Automobile Engineering, Annasaheb Dange College of Engineering and Technology,
Ashta, Maharashtra, India, E-mail: tusharmali2020@gmail.com

G. Mallaiah
Department of Mechanical Engineering, KITS Singapur, Huzurabad, Telangana, India

T. G. Mamatha
Department of Mechanical Engineering, JSSATE, Noida, Uttar Pradesh 201301, India

K. Madhu Murthy
Department of Mechanical Engineering, NIT Warangal, Warangal, Telangana 506004, India

G. Naga Srinivasulu
Department of Mechanical Engineering, National Institute of Technology Warangal, Telangana, India

Jalaiah Nandanavanam
Department of Mechanical Engineering, BITS, Pilani, Hyderabad Campus, Hyderabad, Telangana, India

Govind Nandipati
Department of Mechanical Engineering, RVR and JC College of Engineering, Guntur, Andhra Pradesh, India

Dhanraj Savary Nasan
Department of Mechanical Engineering National Institute of Technology, Tadepalligudem, Andhra Pradesh, India

Yin Kwee Eddie Ng
School of Mechanical and Aerospace Engineering, Nanyang Technological University, Singapore

Jibitesh Kumar Panda
Production Engineering, National Institute of Technology, Agartala National Institute of Technology, Agartala, Tripura, India

Krishna Kant Pandey
Robotics Laboratory, Mechanical Engineering Department, NIT, Rourkela, Odisha 769008, India

Satyendra Kumar Pankaj
Department of Mechanical Engineering, School of Engineering, Gautam Buddha University, Greater Noida, Uttar Pradesh 201312, India

Dayal R. Parhi
Robotics Laboratory, Mechanical Engineering Department, NIT, Rourkela, Odisha 769008, India

K. N. Patil
Department of Mechanical Engineering, K. J. Somaiya College of Engineering, Vidyavihar, Mumbai, Maharashtra 400077, India

Satyajit R. Patil
Automobile Engineering Department, Rajarambapu Institute of Technology, Sakharale, Sangli, Maharashtra 415414, India

Amar Patnaik
Department of Mechanical Engineering, MNIT, Jaipur Rajasthan, India

V. J. Pillewan
Department of Production Engineering, Veermata Jijabai Technological Institute, Matunga, Mumbai, Maharashtra 400019, India

B. M. Preetham
Department of Mechanical Engineering, Sri Venkateshwara College of Engineering, Bengaluru, Karnataka, India

Ram Naresh Rai
Production Engineering, National Institute of Technology, Agartala National Institute of Technology, Agartala, Tripura, India

Parameshwaran Rajagopalan
Department of Mechanical Engineering, BITS, Pilani, Hyderabad Campus, Hyderabad, Telangana, India

Katla Rajendar
Department of Mechanical Engineering, Kakatiya Institute of Technology & Science, Warangal, Telangana, India

Rega Rajendra
Department of Mechanical Engineering, College of Engineering, Osmania University, Hyderabad, Telangana, India

A. Bala Raju
Madanapalle Institute of Technology and Science, Madanapalle, Andhra Pradesh, India

L. Suvarna Raju
Department of Mechanical Engineering, VFSTR University, Vadlamudi, Andhra Pradesh, India

B. S. V. Ramarao
Department of Mechanical Engineering, Aurora's Scientific & Technological Institute, Ghatkesar, Telangana, India

Rahul Ramdas Ramteke
Department of Chemical Engineering, Indian Institute of Technology, Guwahati 781039, Assam, India

C. S. P. Rao
Department of Mechanical Engineering, National Institute of Technology, Warangal 506004, India

G. Amba Prasada Rao
Department of Mechanical Engineering, NIT Warangal, Warangal, Telangana 506004, India

A. Neelakanteswara Rao
Department of Mechanical Engineering, National Institute of Technology Warangal, Warangal, Telangana

K. Prahlada Rao
Department of Mechanical Engineering, J.N.T. University, Anantapur, Andhra Pradesh 515002, India

Ch. Srinivasa Rao
Department AU College of Engineering (A), Visakhapatnam, Andhra Pradesh, India

Venkata Koteswara Rao K.
Department of Mechanical Engineering, NIT Warangal, Warangal, Telangana 506004, India

G. Venkateswara Rao
NIT, Warangal, Telangana, India

W. S. Rathod
Department of Mechanical Engineering, Veermata Jijabai Technological Institute, Mumbai, Maharashtra 400019, India

D. N. Raut
Department of Production Engineering, Veermata Jijabai Technological Institute, Matunga, Mumbai, Maharashtra 400019, India

K. Jagan Mohan Reddy
Department of Mechanical Engineering, National Institute of Technology Warangal, Warangal, Telangana 506004, India

V. Harsha Vardhan Reddy
Department of Aerospace Engineering Indian Institute of Space Science and Technology, Trivandrum, Kerala, India

P. Sailesh
Department of Mechanical Engineering, Methodist College of Engineering, Hyderabad, Telangana, India

V. Sampath
Department of Mechanical Engineering, Kakatiya Institute of Technology & Science, Warangal, Telangana, India

K. Aravind Sankeerth
AUCE(A), Andhra University, Visakhapatnam, Andhra Pradesh, India

G. Ravi Kiran Sastry
Mechanical Engineering, National Institute of Technology, Agartala National Institute of Technology, Agartala, Tripura, India

Suresh M. Sawant
Department of Automobile Engineering, Rajarambapu Institute of Technology, Sakharale, Sangli 415414, India

Ashish Saxena
School of Mechanical and Aerospace Engineering, Nanyang Technological University, Singapore

K. N. Seetharamu
Department of Mechanical Engineering, PES University, Bengaluru, Karnataka, India

N. Selvaraj
Department of Mechanical Engineering, National Institute of Technology, Warangal, Telangana, India

Amit Sharma
Department of Mechanical Engineering, DCR University of Science and Technology, Murthal, Sonipat, Haryana 131039, India

T. Karthikeya Sharma
Department of Mechanical Engineering National Institute of Technology, Tadepalligudem, Andhra Pradesh, India

D. K. Shinde
Department of Production Engineering, Veermata Jijabai Technological Institute, Matunga, Mumbai, Maharashtra 400019, India

S. M. Shinde
Department of Automobile Engineering, Annasaheb Dange College of Engineering and Technology, Ashta, Maharashtra, India, E-mail: shreeshinde6514@gmail.com

Mahantesh. M. Shivashimpi
Mechanical Engineering Department, Hirasugar Institute of Technology Nidasoshi, Belagavi, Karnataka, India

K. Simhadri
GMR Institute of Technology, Rajam, Srikakulam District, Andhra Pradesh, India

R. Sindhu
Department of Mechanical Engineering, NIT Warangal, Warangal, Telangana 506004, India

Ajay Kumar Singh
DRDO, Hyderabad, Telangana, India

Sayyed Siraj
Mechanical Engineering, Marathwada Institute of Technology, Aurangabad, Maharashtra, India

T. V. S. Siva
Sasi Institute of Technology and Engineering Tadepalligudem, West Godavari District, Andhra Pradesh, India

G. Naga Srinivasulu
Department of Mechanical Engineering, NIT Warangal, Warangal, Telangana, India

Rayapati Subbarao
Department of Mechanical Engineering, NITTTR Kolkata, Kolkata, West Bengal, India

Goitom Tesfaye
Department of Mechanical Engineering, College of Engineering, Osmania University, Hyderabad, Telangana, India

S. M. Teli
Department of Automobile Engineering, Annasaheb Dange College of Engineering and Technology, Ashta, Maharashtra, India, E-mail: satishteli1221@gmail.com

Nagaraju Tenali
Mechanical Engineering Department, Gudlavalleru Engineering College, Sheshadri Rao Knowledge Village, Gudlavalleru, Andhra Pradesh 521356, India

Harishchandra Thakur
Department of Mechanical Engineering, School of Engineering, Gautam Buddha University, Greater Noida, Uttar Pradesh 201312, India

Nagaveni Thallapalli
Department of Mechanical Engineering, University College of Technology, Osmania University, Hyderabad, 500007, Hyderabad, India, E-mail: tnagaveni@gmail.com

Narasimha Suri Tinnaluri
Department of Mechanical Engineering, National Institute of Technology, Warangal, Telangana 506004, India

Pramod K. Tiwari
Mechanical Engineering Department, Rajiv Gandhi Institute of Technology, Mumbai, Maharashtra, India

S. R. Todkar
Department of Mechanical Engineering, D. Y. Patil College of Engineering and Technology, Kolhapur, Maharashtra 416006, India

S. N. Topannavar
Mechanical Engineering Department, Hirasugar Institute of Technology Nidasoshi, Belagavi, Karnataka, India

Umashankar
Siddaganga Institute of Technology, Tumkur, Karnataka, India

T. S. Vandali
Department of Mechanical Engineering, Hirasugar Institute of Technology, Nidasoshi, Belagavi, Karnataka 591236, India

P. Varalaxmi
Mechanical Engineering, CJITS, Warangal

Venkateswarlu Velisala
Department of Mechanical Engineering, NIT Warangal, Warangal, Telangana, India

K. Venkatasubbaiah
AUCE(A), Andhra University, Visakhapatnam, Andhra Pradesh, India

S. Vijay
Department Bapatla Engineering College Bapatla, Andhra Pradesh, India

D. R. Yadav
Defence Research and Development Laboratory (DRDL), Hyderabad, Telangana, India

P. M. Zende
Department of Automobile Engineering, Annasaheb Dange College of Engineering and Technology, Ashta, Maharashtra, India, E-mail: pandhrinathzende@gmail.com

ABBREVIATIONS

2D	two-dimensional
3D	three-dimensional
AC	air-conditioner
AC	air-conditioning
AC	alternating current
AD	anaerobic digestion
ADI	alternating direction implicit
ADVISOR	advanced vehicle simulator
AFC	alkaline-based fuel cell
AGVs	automatic-guided vehicles
AI	artificial intelligence
AISI	American Iron and Steel Institute
AM	averaged Mises
ANFIS	adaptive network-based fuzzy inference system
ANN	artificial neural network
ANOVA	analysis of variance
APF	artificial potential field
AR	aspect ratio
ASTM	American society for testing and materials
AWH	air water heater
B&K	Brüel & Kjær
BCP	biphasic calcium phosphate
BD	biodiesel
BF	barrier following
BL	blends
BM	base metal
BP	brake power
BSFC	brake specific fuel consumption
BT	braking torque
bTDC	before top dead center
BTE	brake thermal efficiency
CA	crank angle
CAD	computer-aided design
CAD	crank angle degree

CCC	cylindrical combustion chamber
CFD	computational fluid dynamics
CHB	conventional hydraulic brake
CHB	conventional hydraulic braking system
CI	carbonyl iron
CI	compression ignition
CL	catalyst layer
CMM	coordinate measuring machine
CNC	computer numerical control
CNT	carbon nanotube
COD	chemical oxygen demand
COP	coefficient of performance
CR	compression ratio
CVD	chemical vapor deposition
CVI	chemical vapor infiltration
DARS-TIF	digital analysis of reaction system–transient interactive flamelets model
DC	direct current
DI	direct injection
DMA	dimethylaniline
DMRL	Defense Metallurgical Research Laboratory
DoE	design of experiment
DOE	design of experiments
DOF	degrees of freedom
DS	digested slurry
ECFM-3Z	three-zone extended coherent flame combustion model
ECFM-CLEH	extended coherent flame model-combustion limited by equilibrium enthalpy
ECN	epoxy-clay nanocomposites
EDM	electrical discharge machining
EDS	energy dispersive spectroscopy
EGR	exhaust gas recirculation
EI	electrolytic irons
EL	elongation
ESR	equivalent series resistor
FC	fuel cell
FDM	finite difference method
FE	finite element
FEM	finite element method
FFA	free fatty acid

FFT	fast Fourier transform
FGM	functionally graded materials
FIS	fuzzy inference system
FIV	flow-induced vibration
FLC	fuzzy logic controller
FOD	front obstacle distance
FSP	friction stir processing
FSW	friction stir welding
F-test	Fisher's test
FVM	finite volume method
FYM	farmyard manure
GCI	grid convergence index
GDL	gas diffusion layer
GFRG	grey fuzzy reasoning grade
GRA	grey relational analysis
GRC	grey relational coefficient
GRG	grey relational grade
GUI	graphical user interface
HA	hydroxyapatite
HAZ	heat-affected zone
HCC	hemispherical combustion chamber
HCCI	homogeneous charge compression ignition
HcHCr	high carbon high chromium
HC	hydrocarbon
HFC	hydrogen fluoride hydrocarbon
HIFP	heated inclined finned plate
HRB	Rockwell hardness number
HRR	heat release rate
HSD	honest significant difference
HWFET	highway fuel economy test
HWT	hypersonic wind tunnel
ICEM	integrated computer-aided engineering and manufacturing
IC	internal combustion
ID	industrial dynamics
IEEE	Institute of Electrical and Electronics Engineers
ILSS	interlaminar shear strength
ISFC	indicated specific fuel consumption
ISI	Institute for Scientific Information
IT	injection timing
IUTM	Instron universal testing machine

JIT	Just-In-Time
KH	Kelvin Helmholtz
KOME	karanja oil methyl esters
KSTePS	Karnataka Science and Technology Promoting Society
LDC	lid-driven cavity
LOD	left obstacle distance
LPG	liquefied petroleum gas
LTC	low-temperature combustion
LUA	limited under aging
MAM	methacrylamide
MBAM	N, N′-methylenebisacrylamide
MCFC	molten carbonate-based FC
MEKP	methyl ethyl ketone peroxide
MF	membership function
MGT	mean gas temperature
MHD	magnetohydrodynamics
MH	manual handling
MIG	metal inert gas
MIT	Massachusetts Institute of Technology
MMCs	metal matrix composites
MME	mahua methyl ester
MMT	montmorillonite
MRB	magneto rheological brake
MRF	magneto rheological fluid
MR	magneto rheological
MRR	material removal rate
MSE	mean squared error
MWCNT	multiwall carbon nanotube
MWCNTs	multiwall CNTs
MWNTs	multiwall nanotubes
Ni–MH	nickel–metal hydride
NN	non-negativeness
NOME	neem oil methyl esters
NURBS	nonuniform rational basis spline
OA	obstacle avoidance
OA	orthogonal array
OA	over aging
OFAT	one-factor-at-a-time
OM	optical metallography
PAFC	phosphoric acid-based FC

PA	peak aging
PEMFC	polymer electrolyte membrane-based FC
PEM	proton exchange membrane
PID	proportional integral derivative
PMEDM	powder-mixed electric discharge machining
PME	palm methyl ester
PM	particulate matter
PM	performance measurement
POC	poly(1,8-octanediol–co-citrate)
POME	pongamia oil methyl esters
PPF	poly(propylene fumarate)
RC	resistor capacitor
RC	roller conveyor
ROD	right obstacle distance
RON	research octane number
RP	rapid prototyping
RSAH	regenerative storage air heater
RSM	response surface methodology
RT	Rayleigh–Taylor
S/N	signal to noise
SAIF	Sophisticated Analytical Instrument Facility
SCM	supply chain management
SCPM	supply chain performance measurement
SC	supply chain
SD	synchronous dynamic
SD	system dynamics
SEM-EDS	scanning electron microscopy-energy dispersive spectroscopy
SEM	scanning electron microscopy
SENB	single edge notched beam
SFC	specific fuel consumption
SiCp	sillicon carbide particles
SI	International System of Units
SI	spark ignition
SMAC	semi-implicit marker and cell
SME	small manufacturing enterprises
SN	sum normalization
SOC	state of charge
SOFC	solid oxide-based FC
SOI	start of injection

SPH	smooth particle hydrodynamics
SQL	structured query language
SR	surface roughness
SST	shear stress transport
SUV	sport-utility vehicle
SVD	singular value decomposition
SVM	support vector machine
SZ	stir zone
TA	turning angle
TEMED	tetramethylethylenediammine
TEM	transmission electron microscopy
TG-DTA	thermogravimetric/differential thermal analysis
TIF	transient interactive flamelet
TMT	thermo mechanical treated
TO	topology optimization
TQM	Total Quality Management
TS	target seeking
TS	tensile strength
TWR	tool wear rate
UA	under aging
UBHC	unburnt hydrocarbon
UDDS	urban dynamometer driving schedule
UHC	unburned hydrocarbon
UP	unsaturated polyester
UTM	universal testing machine
UTS	ultimate tensile strength
VARS	vapor absorption refrigeration system
VCC	vapor compression cycle
VCR	variable compression ratio
VC	vapor compression
VHN	Vickers Hardness Number
VREP	virtual robot experimentation platform
WC	tungsten carbide
WZ	weld zone
XRD	X-ray diffraction
YS	yield strength

PREFACE

The faculty of mechanical engineering is a versatile branch among the various disciplines of engineering. It works hand in hand with interdisciplinary subjects such as electronics, computer science, automatic controls, environment, management, and so forth, in addition to its own core subjects. These are embedded into each other. The subject can especially be seen in practice in the field of automotive engineering where it has undergone a multitude of developments to meet customer demand and satisfaction in terms of safety, automation, economics, and so forth. Of late, the field of mechanical engineering has seen substantial growth and interest to attain sustainable development.

The book is a compilation of selected research papers prepared and presented at the First International and Eighteenth National Conference of the Indian Society of Mechanical Engineers held at NIT Warangal February 23–25, 2017. The theme of the conference was "Advances in Mechanical Engineering: Enabling Sustainable Development." The broad objective of bringing out such a compendium is to provide some insight into the topics related to mechanical engineering and the state-of-the-art of work that is being carried out across the world by budding researchers and scientists with the help of eminent experts.

Broadly, the theme is subdivided into three divisions: machine design, materials and manufacturing, and thermal engineering. The papers presented are both experimental and numerical (computational/simulation) in nature. The area of thermal engineering broadly covers the use of alternate fuels such as biodiesel, etc., with an objective of reducing the burden on petroleum reserves and the environment. Also, the researchers are focusing on making use of a split injection strategy with the aim of mitigating harmful emissions from existing engines with conventional fuels. The conference also witnessed works related to combustion so as to achieve sustained supersonic combustion and to obtain and maintain hot air for such types of combustion. The use of new combustion strategies such as homogeneous charge compression ignition (HCCI) is being explored in depth. These days, fuel cells are emerging as a viable alternate power source either fully or in hybrid vehicles.

Nowadays, many researchers specialized in heat transfer are working in the area of numerical heat transfer with applications for cooling of electronic equipment. The works deal with the different configurations of sources of heat in order to dissipate or maximize the heat transfer with reduced energy consumption. Research includes reports on the use of nanofluids to enhance the rate of heat transfer.

Materials technology is growing at a faster pace, such as with nanomaterials, composite materials, functional-graded materials, in addition to developments in the use of optimization tools in cutting and forming processes. Materials characterization has become an important tool in assessing proper materials for application-oriented jobs. Unconventional machining processes, such as electro discharge machining (EDM), are emerging in addition to friction stir welding and forming processes. There is also new development in design engineering, especially of new products, use of new tools for assessing proper designs, rapid prototyping tools, and so forth.

The purpose of this book is, therefore, to present new technologies and the development scenario, and not to give any indication about the direction that should be given to the research in this complex and multidisciplinary challenging field.

The editors take this opportunity to thank all the chapter contributors for sharing their works. The editors are also grateful to Apple Academic Press, especially Mr. Ashish Kumar, Mr. Rakesh Kumar, and Ms. Sandra Sickels, for extending their cooperation in bringing out this compendium. They are also thankful for the cooperation extended by authorities of NIT Warangal and governing body of ISME for their support.

PART I
Machine Design (MD)

CHAPTER 1

PROTOTYPE OF A COLLAPSIBLE TROLLEY: A NEW DESIGN METHODOLOGY

S. R. KULKARNI[1,*], T. S. VANDALI[1,2], and S. A. GOUDADI[1,3]

[1]*Department of Mechanical Engineering, Hirasugar Institute of Technology, Nidasoshi, Belagavi, Karnataka 591236, India*

[2]*tsvandali.mech@hsit.ac.in*

[3]*sagoudadi.mech@hsit.ac.in*

[*]*Corresponding author. E-mail: srkulkarni.mech@hsit.ac.in*

ABSTRACT

In the present market scenario, everyone is focusing upon better quality products. Today's market is not the product driven but the customer driven. The manufactures should always produce the custom-made products. The specifications are defined by the customers itself. Once a particular customer is dissatisfied he/she will not buy the products from the same company again. Therefore, the manufacturers must include the innovative ideas of changing the product features. It may include reducing the cost or changing the features of the product. At the same time, the design must be simple. Many times the trolleys manufactured in small scale or medium scale industries may not be as per the standard designing methods. There is variation in designing methodology, which has been already done by the various designers. This chapter focuses on the new design methodology.

1.1 INTRODUCTION

As per the survey, more number of accidents occurs due to more number of trolleys being connected to a single tractor along with more load and it becomes difficult for the driver to control the tractor and its attached trolleys. In our project, we have designed a collapsible trolley that can be adjusted to various sizes. Therefore, to eliminate all the trolley-related problems and to reduce the cost, we have designed a simple trolley so that the work of two trolleys can be done by a single trolley. With this kind of design, it becomes easy for the drivers to drive the tractor and the trolley to the long distance safely.

1.2 LITERATURE SURVEY

In the year 2015, Indrasen Karogal and Beshah Ayalew have compared torque distribution management strategies for vehicle stability and control of vehicles for independently driven wheels. Yaw rate and lateral acceleration criteria were considered for the study.[1] In the year 2014, Ms. Kshitija A. Bhat and Prof. Harish V. Katore have redesigned the trolley because of several disadvantages associated with its original design. The trolley failures occur because of excess material usage and self-weight. In dynamic loading conditions, the leaf springs and axle fails. All these problems could be overcome by simulating in finite element (FE) model.[2] Recently, in the year 2016, Shinde has performed the stress analysis on an actual tractor trolley chassis structure consisting of C section beams design application of 6 t. The material of the structure is mild steel with 248 MPa of yield strength. Design optimization is done by keeping the material and dimension constant and using sensitive analysis reduces the weight. As raw material required is less, cost of chassis ultimately reduces.[3] Bansal and Kumar have redesigned trolleys considering the static load conditions. AutoCAD model is prepared. When the cross section of the axle was reduced, there was considerable reduction in the weight of the trolley. The dimensions for the new axle were much changed from the regular trolleys which are already available in the market. The design optimization was done.[4] Acharya has analyzed the failure of the axle occurring at the root of the spline of the rear axle. The time period considered was 600–1000 h and the probability of failure of the component after the replacement was checked again. The analysis has been made from the point of view of mechanics and certain simple feasible measures have been proposed for preventing the failure.[5] In the year 2013, Manasa and Reddy replaced rectangular cross section with circular section. Static analysis is done to determine von Mises stress, equivalent elastic strain, maximum shear stress, and total deformation. Finally, it was concluded that 20% of the weight has been reduced after the replacement.[6]

In 1789, Nicolas J. Cugnot invented the first tractor trolley. He thought of attaching the regular trolley to a tractor which could help tractor to utilize it to pull the loads. This invention helped to carry heavy loads from one place to another.

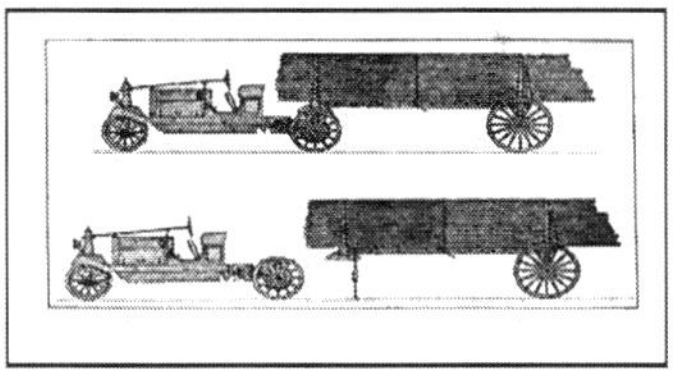

FIGURE 1.1 Nicolas J. Cugnot's first tractor trolley.

FIGURE 1.2 Graff and Hipple model.

In 1884, Graff and Hipple came up with the idea of the mechanism of dumping a trolley automatically with the help of a hydraulic cylinder. They successfully developed this dumping trolley which proved useful in dumping the heavy loads easily which could have been difficult and time-consuming task to do it manually.

FIGURE 1.3 First Lowboy trolley.

Mathew F, in 1945, worked on first Lowboy trolley. These trolleys provided the ability to carry legal loads up to 12 ft. tall.

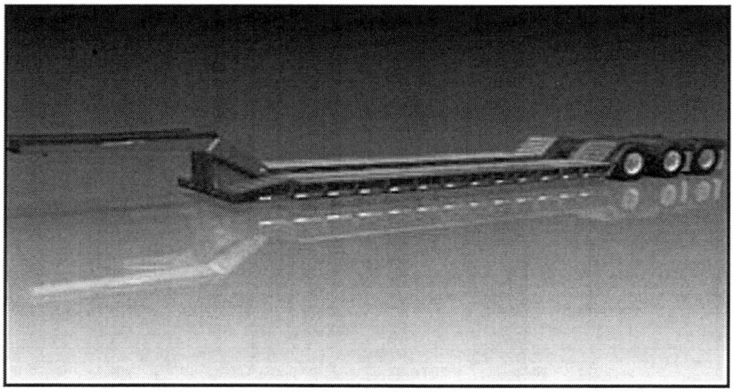

FIGURE 1.4 Kishan equipment's commercial trolleys.

Kishan equipment is an Indian company to remarkably contribute commercial trolleys in our country. It made the tractor trolleys easily available to the Indian market.

1.3 CONSTRUCTION AND WORKING PRINCIPLE

Two trolley parts are connected with sliding contact pairs. One of the trolley parts is supported by two axles with two wheels each. Another trolley part is supported by one axle with two wheels. Two trolleys are free to slide into each other. At the time of the trolley to be loaded, it is expanded by using hydraulic force provided by hydraulic cylinders arranged so conveniently under the trolleys. Whenever the trolley is empty, the trolley can be compressed back into smaller size by operating the hydraulic cylinders. The hydraulic cylinders can be operated by hydraulic pumps in the tractor and do not need separate operating system. The wheels are mounted on the axle. The wheels are fixed to the axle, with bearings or bushings provided at the mounting points where the axle is supported. The axle maintains the position of the wheels relative to each other and the vehicle body. Dead axle does not transmit power.

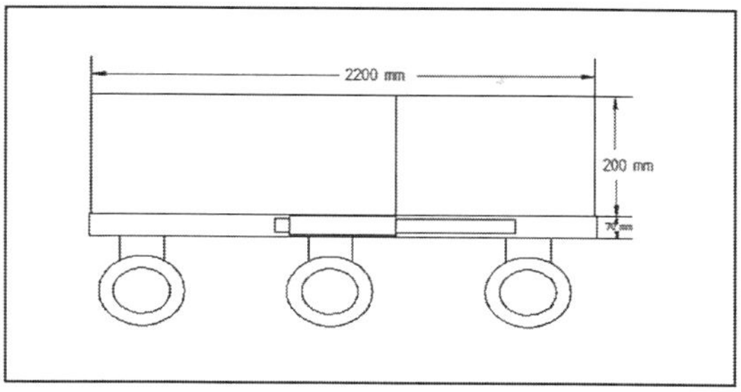

FIGURE 1.5 Side view of the collapsible trolley.

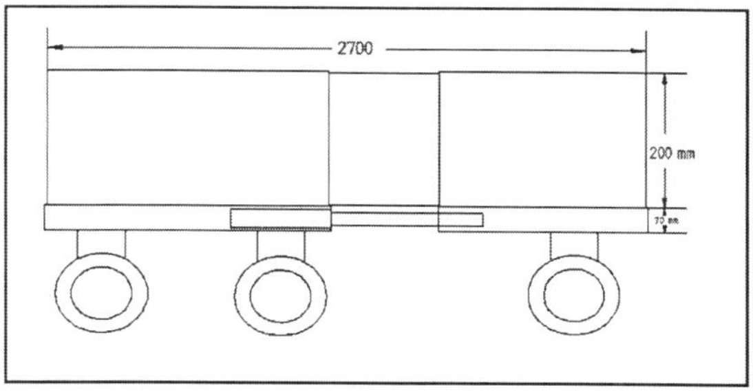

FIGURE 1.6 Side view of expanded trolley.

1.4 MANUFACTURING OF PARTS INVOLVED

1.4.1 TURNING

Turning is a machining operation which depends on divisibility property of the material. The Lathe machine is used for turning operation. The workpiece is to be fixed in between the two dead centers. Since the power is supplied to the lathe machine, the work piece in the chuck rotates and the tool fixed in the tool post travels to the required length. Nowadays CNC machines can also do turning operations. In our project, we have turned the rod joining wheels with the axle.

1.4.2 DRILLING

In order to produce a hole of the required diameter, the drilling machine is used. There are many standard drills available in the market.

1.4.3 CUTTING

Cutting is the process of making parts of the material using a sharp cutting tool. We have done the cutting of sheet metal using a hacksaw.

1.4.4 WELDING

Welding is the process of joining together two or more metal parts by heating the surfaces to the point of melting with or without filler material and pressure. It depends on the fusibility property of the material. We have used welding process to join all the steel plates.

1.4.5 THREADING

Threading is the operation of forming the threads on the outer or inner side of the shaft. We have threaded the shaft which bears the wheels to fasten the spindle nut.

1.5 DESIGN CALCULATIONS

1.5.1 DESIGN OF AXLE

1.5.1.1 STATIC LOAD ANALYSIS

Consider real size model of the trolley. Initially, the trolley is loaded fully and static load on the axle is considered for calculations. The total capacity of the trolley is 180 kN. We have three axles. Consider the load on one axle to be 60 kN. However, weight of trolley and the axle assembly itself is 13 kN. Therefore, we consider the gross weight of the axle to be 73 kN. The trolley chassis is supported by the two points over the axle where the total load is divided equally as shown in the load diagram. Figure 1.9 shows the two points supports.

These two points act as the concentrated loads and transfer the total load on the axle. The trolley is connected to the tractor by hook joint. The trolley is provided with landing legs toward the front of the trailer that supports the trailer when it is not supported by the tractor.

1.5.2 BENDING MOMENT AND SHEAR FORCE ON AXLE

R_A and R_B show the reactions provided by the tires of the trolley. Load 75 kg shows its own weight of axle and (3650+3650)=7300 kg, that is, 73,000 N is load of goods on the trolley for which it is designed and this load is divided equally.

In equilibrium condition,

total upward loads must be equal to the total downward loads.

That is, $R_A + R_B = 36,500 + 750 + 36,500 = 73,750$ N,

Reactions R_A and R_B are as follows.

Taking moment about R_A

$$R_B \times 1310 = 36,500 \times 1155 + 750 \times 655 + 36,500 \times 155$$

$$R_B \times 1310 = 48,306,250$$

$$R_B = 48,306,250 / 1310 = 36,875 \text{ N}$$

Therefore $R_A = 36,875$ N

Calculations for Shear Forces acting on beam:

Considering Right hand side the downward loads are taken as positive and upward loads are taken as negative

$$\text{SFB} = -R_B = -\ 36,875 \text{ N}$$

$$\text{SFE (up to E)} = -36,875 \text{ N}$$

$$\text{SFE (at E)} = -36,875 + 36,500 = -375 \text{ N}$$

$$\text{SFD (up to D)} = -375 \text{ N}$$

$$\text{SFD (at D)} = -375 + 750 = 375 \text{ N}$$

$$\text{SFC (up to C)} = 375 \text{ N}$$

$$\text{SFC (at C)} = 375 + 36,500 = 36,875 \text{ N}$$

$$\text{SFA} = 36,875 \text{ N}.$$

1.5.3 BENDING MOMENTS CALCULATIONS

From figure 7, taking downward moments as positive and upward moments as negative as well as it is noticed that the axle is symmetrically loaded. In addition, it is also observed that the shear force transforms symbol from negative to positive at position D. Hence at point D, the maximum bending moment occurs.

$$\text{Mmax} = M_D = R_B \times 655 - 36{,}500 \times 500 = 5{,}903{,}125\ \text{N - mm}$$

$$M_B = R_B \times 0 = 0$$

$$M_E = R_B \times 155 - 36{,}500 \times 0 = 5{,}715{,}625\ \text{N - mm}$$

$$M_C = R_B \times 1155 - 36{,}500 \times 1000 - 750 \times 500$$
$$= 5{,}715{,}625\ \text{N - mm}$$

$$M_A = R_B \times 1310 - 36{,}500 \times 1155 - 750 \times 655 - 36{,}500 \times 155 = 0$$

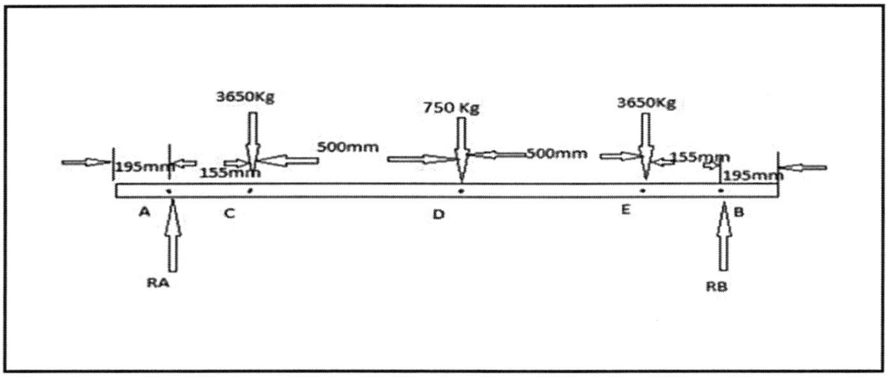

FIGURE 1.7 Trolley chassis main frame.

1.5.4 EXISTING DESIGN CALCULATIONS

$$\text{Volume of hollow portion} = (90^2 - 75^2) \times 1100$$
$$= 2{,}722{,}500\ \text{mm}^3$$

$$\text{Volume of solid portion} = 75^2 \times 800 = 4{,}500{,}000\ \text{mm}^3$$

$$\text{Total volume} = 2{,}722{,}500 + 4{,}500{,}000 = 7{,}222{,}500\ \text{mm}^3$$
$$= 0.0072225\ \text{m}^3$$

$$\text{Mass} = \text{density} \times \text{volume} = 7800 \times 0.0072225 = 56.4\ \text{kg}$$

The general bending equation is given by $M/I = \sigma/y$

Now, to find the bending stress induced in the axle due to the maximum bending moment of 5,903,125 N-mm by bending equation.

$$\text{Mmax} / I_{min} = \sigma / \text{ymax}$$

For cross section of existing axle, ymax = 90/2 = 45 mm

And M.O.I = bd3/12 for rectangular or square sections

Therefore, $I = 75 \times 75^3 /12$

That is, I = 2,636,719mm^4 for solid portion

and $I = 90{\times}90^3 /12 - 75{\times}75^3 /12$

that is, I = 2,830, 782mm^4 for hollow portion

Thus, I_{min} = 2,636, 719mm^4

Therefore, from bending equation

5,903,125/2,636,719 = σ /4

That is bending stress σ = 100.75 N / mm^2 = 100.75MPa,

which is less than the allowable bending stress 430 MPa.

Hence, the design of the axle is safe and its weight is 56.4 kg.

1.5.5 SHEAR FORCE AND BENDING MOMENT

The above figure shows the free body diagram, shear force, and bending moment diagrams. The bending moment is maximum at point D, the sign change also occurs at the same point. Bending moment at point C and E will remain the same but at point A and B, it will be zero.

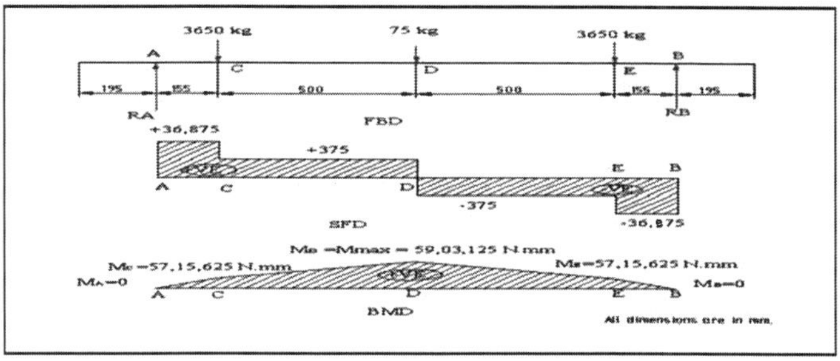

FIGURE 1.8 Shear force and bending moment diagram.

1.5.6 DESIGN OF VARIOUS COMPONENTS

The wheels are rotated by means of the axle. The wheels are fixed to the axle, with bearings or bushings installed at the mounting points where the axle is supported. The axle maintains the position of the wheels. The dimensions of the axle we have used are $40 \times 40 \times 900$ mm.

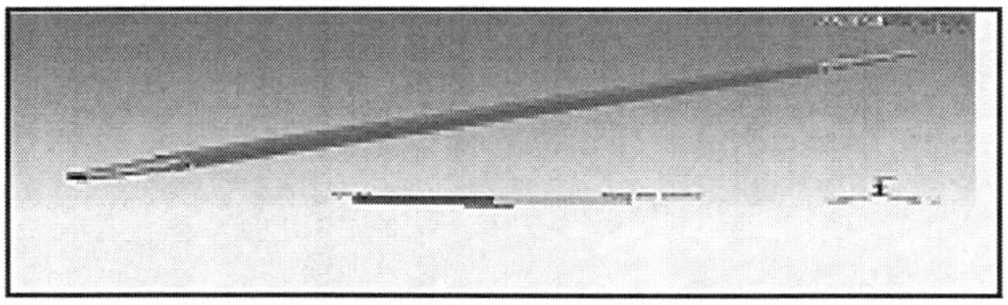

FIGURE 1.9 Axle.

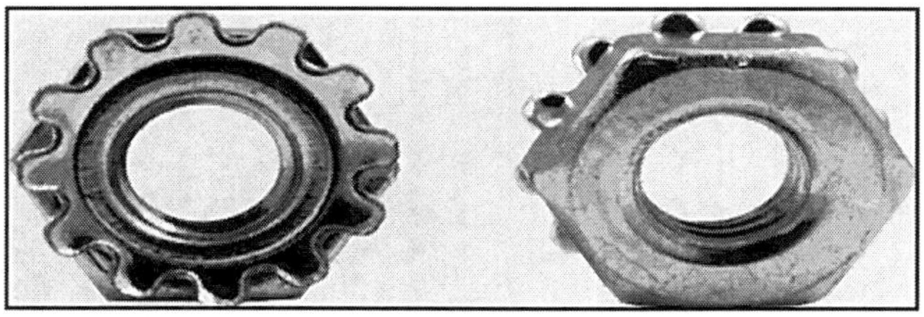

FIGURE 1.10 Spindle nut.

The nut is having internal threads. It is used along with bolt to secure the parts. In applications where vibrations or rotation occur, nut loses so various locking mechanisms such as adhesives, safety pins, lock wires, nylon inserts, and so forth can be employed. Here, we have used the 25 mm spindle nuts to lock the tires with the axle.

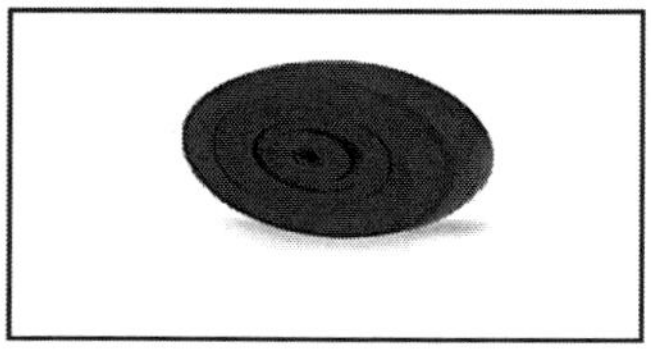

FIGURE 1.11 Rubber tires.

FIGURE 1.12 Steel for structural applications.

A tire is made up of hard rubber that gives cushioning effect. It also acts as a shock absorber. In order to get maximum contact with the road, the treads must be proper on its surface. For this prototype, we are using 200 mm diameter rubber tires. Mild steel contains the lowest percentage of carbon. American Iron and Steel Institute (AISI) grades 1005 through 1025 are suitable for structural applications. We have used mild steel for all the plates required in fabrication. 16 and 14 gauge plates have been used in fabrication.

1.6 RESULTS AND DISCUSSION

In conclusion, the objective of the project has been achieved and by completing the project a lot of knowledge and experience has been gained. It is simple in design, since the other trolley is eliminated and the driving made comparatively easy. It could be adjusted to various lengths depending upon the requirements. The sides can be locked and it is accidental safe trolley. The disadvantage is the capacity of the single collapsible trolley is relatively low compared to two regular trolleys.

1.7 APPLICATIONS

1. Transportation of more goods using the single collapsible trolley.
2. This trolley can be used to transport long length goods such as metal pipes, TMT bars, wooden bars, and so forth which may extend out from a single regular size trolley.
3. Comparatively more quantity of goods can be transported.

1.8 CONCLUSIONS

The maximum bending moment occurs at point D, that is, 5,903,125 N-mm. The maximum allowable bending stress is 430 MPa. The design calculations yield 100.75 MPa. Therefore, it is safe. After designing, it is seen that the weight of the trolley is much reduced and hence the cost. Moreover, the designing method has been made as simpler as possible. The automation could be done by using a double acting hydraulic cylinder, and the trolley size can also be changed by sliding. Even the trolley can be added with the feature of dumping with additional mechanisms of hydraulic means. The additional modification can be used to make the dumping by three ways which could help in dumping in any direction. This trolley can be modified, to be used in factories and enterprises to carry the goods within the factory. Adding brakes to the trolley could again ease the driving situations.

KEYWORDS

- **trolley**
- **hydraulic force**
- **bending moment**
- **shear force**
- **transportation**

REFERENCES

1. Karogal, I.; Ayalew, B. Independent Torque Distribution Strategies for Vehicle Stability Control. SAE International by Clemson University Libraries, Sunday, May 31, 2015, 2009-01-0456, ISSN: 0148–7191.
2. Bhat, K. A.; Katore, H. V. The Failure Analysis of Tractor Trolley Chassis an Approach Using Finite Element Method—A Review. *IOSR J. Mech. Civ. Eng. (IOSR-JMCE)* 24–27. (e-ISSN: 2278–1684, p-ISSN: 2320–334X).
3. Shinde, A. Stress Analysis of Tractor Trolley Chassis with Effect of Various Thickness and Design Optimization for Weight Reduction. *Int. J. Adv. Res. Innovative Ideas Educ.* **2016,** *2*(2), e-ISSN: 2395–4396, DOI: 16.0415/IJARIIE-1894.
4. Bansal, H.; Kumar, S. Weight Reduction and Analysis of Trolley Axle Using Ansys. *Int. J. Eng. Manage. Res.* **2012,** *2*(6), 32–36. (ISSN No.: 2250–0758 www.ijemr.net).
5. Acharya, A.K. Failure Analysis of Rear Axle of a Tractor With Loaded Trolley. *Int. J. Innovative Res. Dev.* **2013,** *2*(10), 241.
6. Manasa, P.; Vijaya, C.; Reddy, B. Static Analysis of Tractor Trolley Axle. *Int. J. Eng. Trends Technol. (IJETT)* **2013,** *4*(9), 4183–4187.

CHAPTER 2

EXPERIMENTAL EVALUATION OF A MAGNETO RHEOLOGICAL BRAKE FOR AUTOMOTIVE APPLICATION

AKSHAY M. DASHWANT[1,*] and SATYAJIT R. PATIL[2]

[1]*Mehanical Engineering Department, Kolhapur Institute of Technology, Kolhapur, Maharashtra 416234, India*

[2]*Automobile Engineering Department, Rajarambapu Institute of Technology, Sakharale, Sangli, Maharashtra 415414, India*

[*]*Corresponding author. E-mail: primeakshay@gmail.com*

ABSTRACT

The upgrading of mechanical systems into electronic systems has made it possible to develop electronically controlled mechanical systems like magneto rheological brakes (MRB). The MRB consist of magneto rheological fluids (MRFs) and operate in the direct shear mode. Due to rapid improvement of such MRB technologies, various configurations of the same have been developed. The paper presents an experimental approach towards the evaluation of the MRB technology. The MRB is designed for a two-wheeler electric bike considering its allocation at the front wheel. The experimentation is carried out by using a designed test rig and planned procedures. The experimental setup design, elements, and assembly have been explained in this article. Experimentation strategy and procedures adopted for experimentation have been illustrated in detail in this paper. Also, the experimental evaluation of MRB is carried out and the data has been presented. The experimental results considering three types of MRFs show lower torque generation than required.

2.1 INTRODUCTION AND LITERATURE REVIEW

A new branch of memory materials known as magneto rheological (MR) fluids has been used in various mechanical systems like dampers, clutches, brakes, and so forth. The Magneto rheological fluid (MRF) properties have the potential to convert mechanical devices into electromechanical devices with greater performance and accuracy. The MRFs consist of magnetic particles suspended in carrier fluids. These fluids behave like non-Newtonian fluids in the presence of an applied magnetic field. When a magnetic field is applied to the MRFs, the suspended particles align together to form a chain, thereby increasing the viscosity of the fluid. As soon as the magnetic field is withdrawn the MRFs return to their normal state.

As there is an electronic interface, it is believed that the MR brake will perform flawlessly and quickly. Also due to the electronic interfaces all the disadvantages of the conventional brake systems have been eliminated. Such a type of brake-by-wire technology has been investigated by some researchers.[1] Jolly[2] discovered the applications of MR brake in velocity control systems. Haung et al.[3] designed cylindrical MRB using the Bingham model. Bydon[4] termed the construction and operation of a rotary MR brake. Huang et al.[3] presented the theoretical design of a cylindrical MRF brake with an annular space for MRF among the rotor and stator. Owing to the cylindrical configuration, brake works in shear mode. They considered a very small amount of MRF along the circumference of the rotor. In their opinion, the torque generated due to the field will be very low in such shape. Li and Du[5] proposed the prototype of new MRB of general nature. They designed, fabricated, and tested the MR brake prototype. They derived the torque equation based on the Bingham plastic model and designed the single disc type brake functioned in the direct shear mode. In design of the MRB, they considered numerous issues such as gaps, electromagnet, fluid behavior, chamber, and seals. They considered the distinctive range of gaps between the rotor and housing (0.25–2 mm) and decided the gap value to be 0.5 mm. Li and Du[5] then studied the effects of supply current and rotary speed on the performances of the MR brake and found that these two factors are directly proportional to the torque output of the MR brake. They concluded that a larger amplifying factor can be gained by either increasing the magnetic field strength or decreasing the rotary speed. From this paper, it can be inferred that the transmitted torque increases gradually with increasing magnetic field strength. It also shows an increasing trend with rotary speed. Bydon and Sapinski[6,7] discovered the applications of MRB in velocity control systems. Bydon termed the construction and operation of

Lord's MRB, which has certain margins. They concluded that Lord brake comprises an extreme 5.65 Nm torque, limits its speed to 1000 rpm, and bounds operating temperature range between −30 and 70°C. Sukhwani et al.[8] developed the experimental setup for the evaluation of MRB. To observe the brake performance, a particular combination of current and speed was selected during the experimentation. Two different torque-measuring techniques were selected. Efforts were made to record the reliable torque value. From this paper it is concluded that, the torque measurement using a torque sensor is a more reliable technique in comparison to the use of a wattmeter. Results indicate a strong shear thinning and thermal thinning behavior of the MRF. This concludes that one needs to include shear-thinning and temperature-thinning effects for analyzing and optimizing any engineering device based on the MRF. Farjoud et al.[9] analytically designed the drum type MRB using the Herschel–Bulkey model. Karakoc et al.[10] developed a disc type MRB for a midsized car and a T-shaped drum type MRB for a midsized motorcycle, respectively. They designed the MRB device with an attention to the magnetic circuit optimization and material selection. In this work, an empirical design approach such as material selection, sealing, working surface area, viscous torque generation, applied current density, and MRF selection was used to select a basic automotive MR brake layout. They examined the MR brake performance using an experimental test rig consisting of a torque sensor, servomotor with control as a part of their experimental setup. From experimental results they found that braking torque increases with the applied current. It was determined that the proposed MRB structure was unable to produce the necessary braking torque to stop a midsized car. The researchers recommended further scope for work to improve the braking torque by increasing the number of disks or totally restructuring the magnetic circuit. Sukhwani and Hirani[11] proposed the design and fabrication of two MR brakes, Brake 1 with MR gap 1 mm and Brake 2 with MR gap 2 mm. From the total theoretical and experimental study, they drew conclusion that high torque MRB require high yield stress MRFs. They concluded that the braking torque increases with increasing magnetic intensity produced by the electromagnet and increasing speed, but they found that there is shear weakening of MRF at higher speeds. They also inferred that the braking torque offered by MRB decreases when MRF gap increases from 1 to 2 mm, so they declared that MRB with 1 mm MRF gap is ideal and the lateral electromagnets are useful to deliver good controllability as compared to a single electromagnet. Sukhwani and Hirani[11] suggested that the design factors like the number of electromagnet turns, MRF gap, saturation limits, and so forth should be seriously considered while configuring the MRB. Hung and

Bok[13] have developed a disc type MRB for a midsized car and a T-shaped drum type MRB for a midsized motorcycle, respectively. Sarkar and Hirani[13] performed theoretical and experimental studies on MRB functioning under compression and shear mode as well as only shear mode. According to their research, it is evident that MR brakes based on compression plus shear mode mechanisms deliver a better torque output.

2.2 EXPERIMENTAL SETUP DESIGN

In this section, the design of the experimental setup for evaluation of an automotive MR brake has been discussed. The various elements required for the experimental setup have been selected by considering specifications of the MR brake designed for an electric bicycle.

2.2.1 MOTOR SUBASSEMBLY

On the basis of analytical evaluation, the maximum torque output of the designed MRB is 50 Nm. The maximum allowable current and operating speed are 4 A and 400 rpm, respectively. One needs a suitable motor to run the MRB efficiently at the maximum torque and speed. Therefore, a Kirloskar make, three phase 3 HP 1440 rpm alternating current (AC) motor is selected for driving the shaft of a MRB in the test rig. The maximum speed of interest is up to 450 rpm, so a gear box having 3:1 ratio has been incorporated in the test rig. An in-line gear box with a reduction of 3:1 is used to obtain a continuous and mandatory torque output at required speeds. A variable frequency control mechanism is selected as the motion control for the selected AC motor.

2.2.2 TORQUE SENSOR AND DATA ACQUISITION SUBASSEMBLY

The torque measurement is an important task while evaluating the performance of MRB. There are various methods for determining the braking torque like torque sensors and load cells. The data acquisition systems have a dominant role to convert the signals coming from the sensors. Also as part of converting the signals from sensors into torque values, the torque converters and indicators have been used by some past researchers. The selection of the best and accurate torque-sensing and data acquisition system is important as it can lead to better evaluations of results. Considering the level of output braking torque of MRB as 50 Nm, ADI-Artech in-line torque sensor (rotary type) is selected.

2.2.3 DATA ACQUISITION

Data acquisition is the process of measuring an electrical or physical phenomenon such as voltage, current, temperature, pressure, or sound. PC-based data acquisition uses a combination of modular hardware and flexible software to transform a standard laptop or desktop computer into a user-defined measurement or control system. While each data acquisition system has a unique functionality to serve application-specific requirements, all systems share common components that include sensors, data acquisition hardware, and a computer. The Dewetron 6.2 system compatible with the ADI-Artech selected torque sensor is preferred as a data acquisition system for the recording and storing of data.

2.2.4 MISCELLANEOUS ELEMENTS

An infrared thermometer (MT-5) is used to measure the heat buildup in the MRB. The motor is provided with a three phase AC power supply while the brakes with direct current (DC) supply.

The infrared thermometer and DC power supply are shown in Figure 2.1.

FIGURE 2.1 Temperature measurement and direct current (DC) supply.

Flange type flexible couplings have been used to connect the motor shaft, sensor shaft, and brake shaft, respectively. Four pillow block bearings are used in total to support the shafts. The total experimental setup has been rested and bolted on a base frame. Figure 2.2 shows the experimental setup incorporating a motor, gearbox, torque sensor, MRB, and control panel.

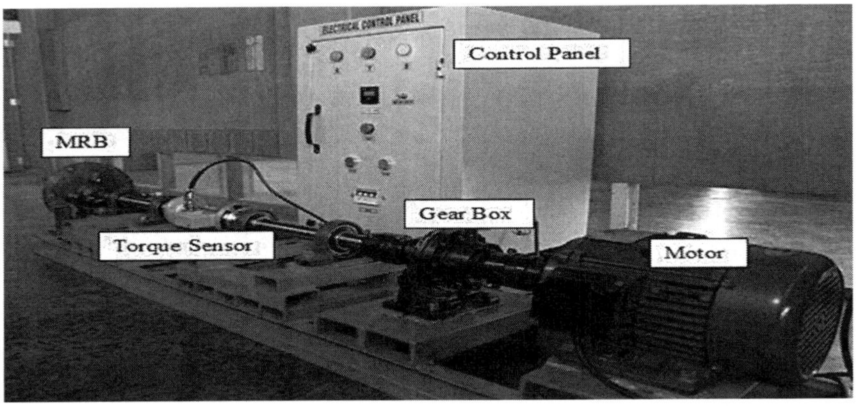

FIGURE 2.2 Experimental setup.

2.3 STRATEGY OF EXPERIMENTATION

Each experimental run is a test. It can define an experiment as a test or series of runs in which purposeful changes are made to the input variables of a process or system so that it may help to observe and identify the reasons for changes that may be observed in the output response. In this experiment, current and speed are the input as well as controllable factors and the temperature is an uncontrollable factor, while torque is the response of the system.

There are many approaches to carry out experimentation of the proposed problem. The one which suits best is to be considered for experimentation. The best guess approach has significant disadvantages, many a times it has no guarantee of success. Also, the major disadvantage of one-factor-at-a-time (OFAT) is that it fails to show any possible interaction between the factors. Furthermore, the problem identified has two factors that is, speed and current. Hence, the correct approach to dealing with this experiment with two factors is to conduct a factorial experiment.[13]

2.3.1 FACTORIAL EXPERIMENT

Many experiments involve the study of effects of two or more factors. In general, factorial designs are the most efficient for this type of experiments, so this technique has been selected to be applied for the proposed problem. By a factorial design, it is meant that in each complete trial or replicate of experiment, all possible combinations of the levels of factors are investigated.[13] If the number of levels in a factorial experiment is more than two for each of the factors in it, then the corresponding design is called full factorial design, where n is the number of factors in the experiment. As the proposed experiment consists of two factors and four levels, it is a $4^2 4^2$ factorial experiment. This design in the factorial series is the one with only two factors; in this case, current (I) and speed (w), each run at four levels that is, low level, average level, moderate level, and high level. Hence there are 4^2 that is, 16 runs in this experimentation as shown in Table 2.1.

TABLE 2.1 Treatment Combination

Factors		Treatment Combination							
Current (I) [A]	Speed (w) [rpm]	I		II		III		IV	
1.0	177	1	177	2	177	3	177	4	177
2.0	235	1	235	2	235	3	235	4	235
3.0	296	1	296	2	296	3	296	4	296
4.0	354	1	354	2	354	3	354	4	354

2.3.2 EXPERIMENTAL PROCEDURE

As the MR brakes have not been commercialized yet, there are no standard procedures for performance testing. Hence, by using the design of experiments (DOE) concepts and previous literature, the testing procedure has been planned. The stepwise procedure adopted for experimentation is as follows.

i. Rotate MRB at a speed of say 100–200 rpm without applying any magnetic field for a few minutes, say 2–3 min, as an initial condition, which would stir the MRF in the brake to distribute it uniformly.

ii. Measure the no field torque generated by the viscosity of MRF for various rotational speeds namely. 177, 235, 296, and 354 rpm.
iii. After calculating the viscous torque, apply the desired current, say 1 A, and wait for 2–3 min, ensuring MRF to form a stable structure at zero rotational speed.
iv. Apply 1 A current to the electromagnet coil of MRB at a constant speed of 177, 235, 296, and 354 rpm respectively, and record the corresponding braking torque by the torque sensor.
v. Repeat the above step, but this time varying the supply current to the electromagnet of MRB from 1 to 4 A with an incremental step of 1 A and record the torque responses.

2.4 EXPERIMENTATION

This section presents the observations during the experimental trials. The experimentation is performed by using the designed setup and planned procedure as mentioned in earlier sections. The observation tables with recorded data have been presented below along with the corresponding torque output plots. The first experimentation trial is carried out using a MRF denoted as CSi 45% which comprises 45% carbonyl iron powder and silicone oil along with additives. Table 2.2 presents the torque responses of MRB using MRF sample CSi 45%.

TABLE 2.2 Torque Response for Magneto Rheological Fluid (MRF) CSi 45%

Speed (rpm)	**177**	**235**	**296**	**354**
Current (A)			**Torque (Nm)**	
1	1.42	1.54	1.63	1.75
2	2.34	2.46	2.57	2.7
3	3.16	3.25	3.4	3.57
4	4.08	4.2	4.33	4.46

Torque versus speed plot for the torque responses of MRB using MRF sample CSi 45% is shown in Figure 2.3.

The next experimentation is carried out using the MRF 132DG from Lord Corporation. Table 2.3 presents the torque responses of MRB using MRF 132DG.

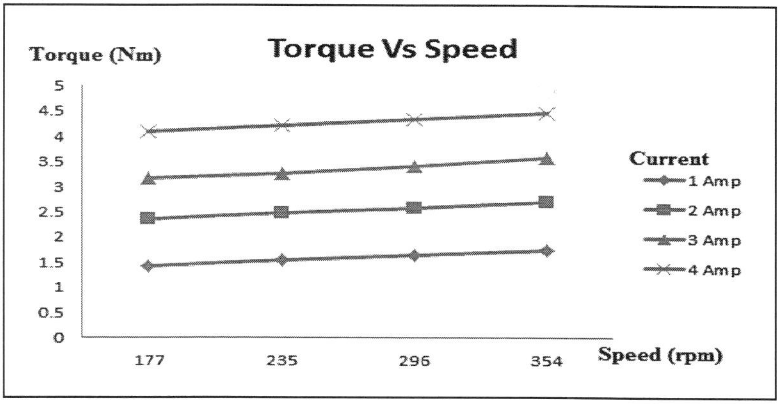

FIGURE 2.3 Torque versus speed plot (CSi 45%).

TABLE 2.3 Torque Responses For MRF 132DG

Speed (rpm)	177	235	296	354
Current (A)			Torque (Nm)	
1	1.27	1.35	1.43	1.57
2	2.05	2.12	2.23	2.35
3	2.69	2.78	2.9	3.01
4	3.77	3.89	4.09	4.21

Torque versus speed plot for the torque responses of MRB using MRF sample 132DG is shown in Figure 2.4.

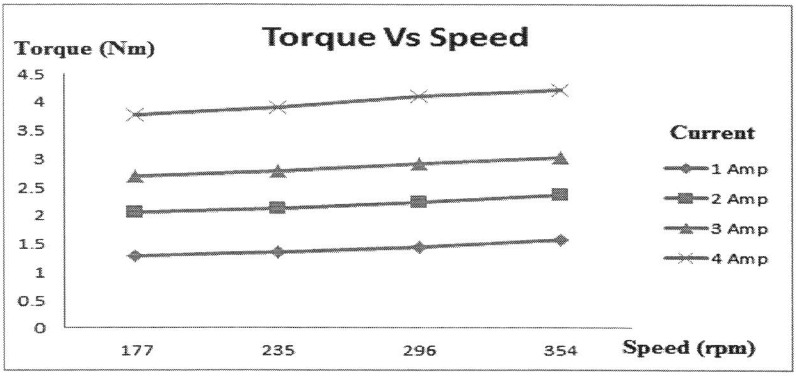

FIGURE 2.4 Torque versus speed plot (MRF132DG).

The succeeding experimental trial is carried out using MRF denoted as ESy 45% which comprises 45% electrolytic iron powder and synthetic oil along with additives. MRF sample ESy 45% suits well for the automotive braking application as it shows the lowest sedimentation ratio. The Table 2.4 presents the torque responses of MRB using MRF sample ESy 45%.

TABLE 2.4 Torque Response for MRF ESy 45%.

Speed (rpm)	177	235	296	354
Current (A)			**Torque (Nm)**	
1	0.617	0.623	0.63	0.69
2	0.89	0.95	1.01	1.08
3	1.25	1.31	1.39	1.46
4	1.67	1.74	1.82	1.89

Torque versus speed plot for the torque responses of MRB using MRF sample ESy 45% is shown in Figure 2.5.

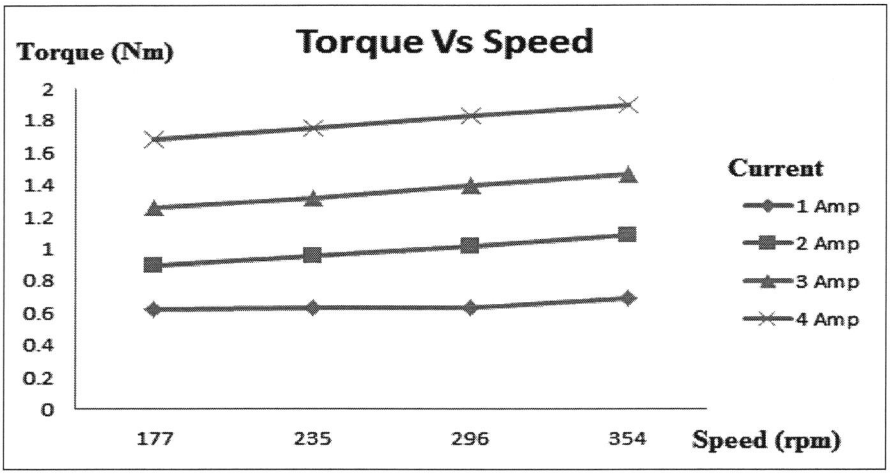

FIGURE 2.5 Torque versus speed plot (ESy 45%).

Figures 2.3, 2.4, and 2.5 show the torque versus speed plots for the proposed MRB using three different MRFs, respectively. It is seen from the plots that as current and speed increase the total torque offered by the

MRB increases and vice versa. The increasing current supply increases the magnetic field intensity, which in turn raises the MRB torque. Increasing rotary speed of the disk makes the MRF exert higher yield stress thereby increasing the braking torque. Also it is clear from the data that the MRF CSi 45% delivers the highest braking torque, while MRF132DG is closer to the same. The ESy 45% MRF sample provides the lower braking torque compared to the other two MRF alternatives. Hence, it is experimentally proved that the total torque offered by the MRB is directly proportional to the supply current, rotary speed of the disk and depends upon the type of MRF used. Also, the result Table 2.3 shows that at maximum current and speed, the total torque developed is equal to 4.21 Nm while the torque requirement for effective braking of an electric bicycle is around 150 Nm. This means that the MRB is not in a position to meet the torque requirement of intended automotive application.

2.5 THERMAL INVESTIGATIONS

Prolonged use of MRB generates thermal energy. This rise in the temperature is due to the friction between rotating disk and the iron particles suspended in the MRF. The thermal investigation of MRB is carried out in this section. Initially, a set of experimental runs have performed and the rise in temperature of MRB has been noted. Table 2.5 gives the temperature after each set of experimental runs.

TABLE 2.5 Thermal Observations I.

Sr. No.	Experimental set	Temperature (°C)
1	1 A; (177, 235, 296 354) rpm	26.9
2	2 A; (177, 235, 296 354) rpm	28.5
3	3 A; (177, 235, 296 354) rpm	30.4
4	4 A; (177, 235, 296 354) rpm	34.8

Hence, it is seen from the above Table 2.5 that the friction in the MRB causes temperature to rise. Also, a typical experiment simulating driving cycle of 10 min and actuation of MRB every single minute is carried out to investigate the temperature rise. The MRB is subjected to run at 235 rpm and it is provided with 2 A current supplies for 5 s. The initial, intermediate, and final temperatures for driving cycle of 10 min are presented in Table 2.6.

TABLE 2.6 Thermal Observations II.

Time (s)	Temperature (°C)	Time (s)	Temperature (°C)
Initial	28.2	390	30.5
65	28.3	455	31.0
130	28.7	520	31.3
195	29.2	585	31.7
260	29.8	Final	32.1
325	30.1		

Figure 2.6 presents the time versus temperature plot based on above data.

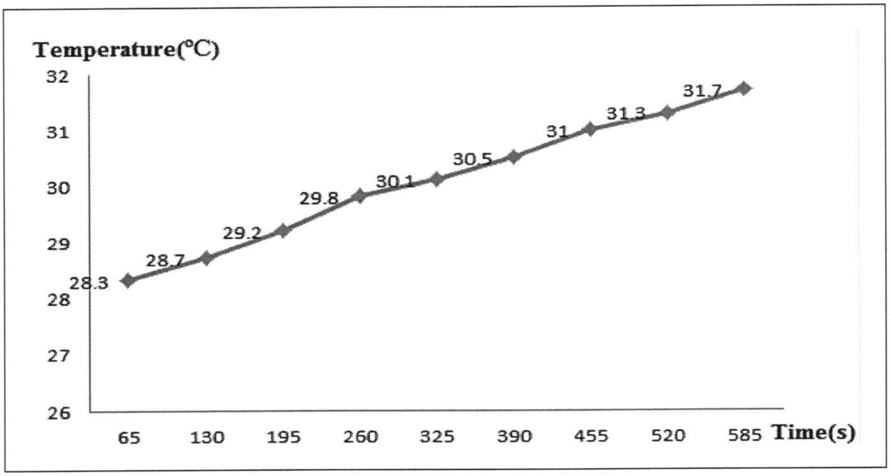

FIGURE 2.6 Time versus temperature plot.

The temperature curve shows a steady temperature rise due to continued use of MRB.

2.6 STATISTICAL ANALYSIS

Regression analysis is made to establish the relationship among parameters of interest that is, braking torque as an output and supply current and speed as inputs for MRB. Minitab 17 is used for obtaining the equations. The regression Equation 2.1 is given below for MRF 132DG along with the main effect plot (Fig. 2.7).

$$\text{Torque} = -0.011 + (0.8412 * I) + (0.001939 * S) \qquad \text{(Eq. 2.1)},$$

where "I" is the current supplied and "S" is the speed in rpm.

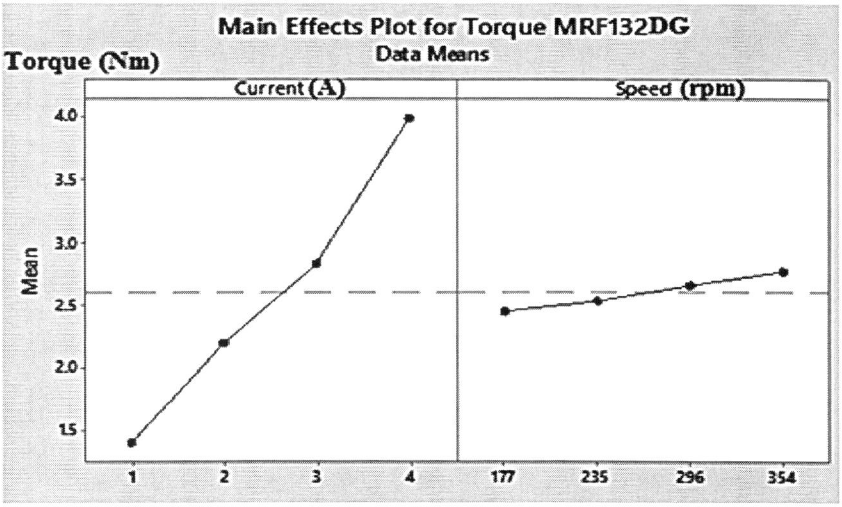

FIGURE 2.7 Main effects plot MRF 132DG.

Similar equations and plots are obtained for other two MRFs that is, CSi 45% and ESy 45%. First term in the regression equation is negative and indicates the presence of a viscous torque in the MRB at zero input current and speed level. It is observed from the plot that the effect of current on braking torque response is significant as compared to the effect of speed. However, the speed is limited to 354 rpm during this experimentation. Higher speed levels may influence the torque response in a higher order

2.7 CONCLUSIONS

MRB is perceived to be a promising substitute for the conventional hydraulic braking system (CHB) in the context of road safety. This work presents the design of experimental setup and experimental evaluation of MRB. The experimental evaluation is carried out for three MRFs. Effects of type of MRF, supply current, and speed on the torque responses of MRB is observed. Furthermore, the temperature rise in the MRB is investigated experimentally.

The conclusions of the study are summed up as follows:

1. The maximum experimental braking torque of MRB is found to be 4.21 Nm. The braking torque requirement of the electric bicycle (150 Nm) could not be met with the present design and MRFs.
2. The synthesized MRF sample CSi 45% delivers 4.46 Nm braking torque, among the three MRFs considered in this study.
3. Temperature rise of 4–5°C is observed during a drive cycle of 10 min. However, the maximum temperature is found to be around 35°C, way below the operating range of MRFs.
4. It is inferred from statistical analysis that the current has significant effect on the torque response while the effect of speed is insignificant.
5. Overall, it is concluded that the present MRB design and MRFs do not meet the braking torque requirements of light weight automotive application like the electric bicycle. Further design modifications in terms of brake geometry, electromagnet circuit design and optimization are essential. There is need to synthesize MRFs with a higher yield stress so as to be able to deliver a higher braking torque. Temperature rise in MRB owing to the application of brake remains within safe operating range of MRFs; thus cooling provisions in the present design is not warrented.

KEYWORDS

- **magneto rheological fluids (MRF)**
- **magneto rheological brake (MRB)**
- **experimental setup**
- **experimental evaluation**
- **statistical analysis**

REFERENCES

1. Carlson, J. D.; Catanzarite, D. N.; St Clair, K. A. In *Commercial Magnetorheological Fluid Devices*, Proceedings of 5th International Conference on ER fluids, MR Suspensions and Associated Technology, Bullough, W., Ed.; 1996, pp. 20–28.

2. Jolly, M. R. Pneumatic Motion Control Using Magnetorheological Technology. *Proceedings of the 27thInternational Symposium on Smart Actuators and Transducers, International Center for Actuators and Transducers (ICAT)*, State College, PA, USA, 1999, pp. 22–23.
3. Huang, J.; Zhang, J. Q.; Yang, Y.; Wei, Y. Q. Analysis and Design of a Cylindrical Magneto-Rheological Fluid Brake. *J. Mater. Process. Technol. Elsevier* **2002,** *129,* 559–562.
4. Bydon, S. *Construction and Operation of Magnetorheological Rotary Brake;* Archiwum Process Control Club, 2002.
5. Li, W. H.; Du, H. Design and Experimental Evaluation of Magneto Rheological Brake. *Int. J. Adv. Manuf. Technol.* **2003,** *21,* 508–515.
6. Bydon, S. *Facility for Induction Motor Velocity Control with Magneto Rheological Brake;* Archiwum Process Control Club, 2003, pp. 12–14.
7. Sapinski, B.; Bydon, S. *Application of Magneto Rheological Brake to Shaft Position Control in Induction Motor;* In AMAS Workshop on Smart Materials and Structures SMART 03, 2003, 169–180.
8. Sukhawani, V. K.; Vijaya, L.; Hirani, H. Performance Evaluation of Magneto Rheological Brake: An Experimental Study. *Indian J. Tribol.* **2006,** *1,* 47–52.
9. Farjoud, A.; Vahdati, N.; Fah, Y. F. Mathematical Model of Drum-Type Mr Brakes Using Herschel–Bulkley Shear Model. *J. Intell. Mater. Syst. Struct.* **2007,** 1–8.
10. Karakoc, K.; Park, E. J.; Suleman, A. Design Considerations for an Automotive Magnetorheological Brake. *Mechatronics* **2008,** *18,* 434–447 (Elsevier).
11. Sukhwani, V. K.; Hirani, H. Design, Development, and Performance Evaluation of High-Speed Magneto Rheological Brakes. *Proc. Inst. Mech. Eng., Part L* **2008,** *222*(1), 73–82.
12. Hung, N. Q.; Bok, C. S. Optimal Design of a T-Shaped Drum-Type Brake for Motorcycle Utilizing Magnetorheological Fluid. *Mech. Based Des. Struct. Mach.* **2012,** *40* 153–162.
13. Sarkar, C.; Hirani, H. Theoretical and Experimental Studies on a Magnetorheological Brake Operating Under Compression Plus Shear Mode. *Smart Mater. Struct* **2013,** *22,* 1–12 (IOP publishing).

CHAPTER 3

SIMULATION FOR ESTIMATION OF A MAGNETO RHEOLOGICAL BRAKE TORQUE-BASED ON FUZZY LOGIC SINGULAR VALUE DECOMPOSITION USING MATLAB

ROMIT M. KAMBLE*, SATYAJIT R. PATIL, and SURESH M. SAWANT

Department of Automobile Engineering, Rajarambapu Institute of Technology, Sakharale, Sangli 415414, India

Corresponding author. E-mail: satyajit.patil@ritindia.edu

ABSTRACT

Magneto rheological fluids (MRF) are intelligent smart materials whose rheological characteristics change rapidly and are under the control of an applied magnetic field. MRF-based devices are successfully commercialized for industrial purpose. Its application for automotive braking has wide scope, and researchers are extending work to bring it to practical use. It has the ability to replace conventional hydraulic brakes (CHB). This controllable yield stress produces shear friction on the rotating disks, generating the braking torque. Magneto rheological brake (MRB) has advantages over CHB in terms of a faster response time. Hence, these brakes are being investigated for vehicular applications. Artificial intelligence is a way of making a computer software think intelligently, in a manner similar to how intelligent humans think. An effort has been made to implement fuzzy logic control strategy for the estimation of MRB torque. In this work, fuzzy logic theory is applied to assign membership function (MF) for inputs and output. Assigned MFs are used to define a rule-based system for fuzzy logic controller proposed for the existing MRB. MFs are assigned for inputs and output. Singular value decomposition method based on Sugeno computation is used for the estimation of brake torque.

3.1 INTRODUCTION

The purpose of this study is to design expert system rules for the control of braking torque of the magneto rheological brake (MRB) developed by Park et al.[1] which depends on the speed of the shaft and the current applied. A new trend called "x-by-wire" is useful for automotive domain which gives able results into the vehicle performance improvement, is highly safe and costs less. In x-by-wire "x" is any system such as braking, steering in which mechanical components will get eliminated and replaced with an electrical system.[2] The conventional hydraulic brake (CHB) system has a delay of 200–300 ms between the brake pedal application and the actual brake response to wheel because of the time required for buildup of pressure in hydraulic lines.[1] An electromechanical brake system has the ability of reducing this time delay effectively, which will ultimately reduce the braking distance of the vehicle. MATLAB's fuzzy logic toolbox is utilized for controlling of MRB torque. Fuzzy logic is easy to implement and logically based on inputs at some points or intervals.

3.1.1 INTRODUCTION TO MAGNETO RHEOLOGICAL FLUID (MRF)

Wang et al.[2] concluded that if a magneto rheological fluid (MRF) is subjected to a magnetic field, it displays a dramatic change in its viscous and elastic properties within the time of 50 ms. In the absence of the magnetic field, MRF behavior is similar to any liquid; but under the effect of a magnetic field, its viscosity increases and becomes two times stronger, forming into a solid-like structure within milliseconds. MRF is a smart fluid in which micron sized magnetic particles are dispersed in a nonmagnetic liquid. The magnetic field induced by the electromagnet generates dipole moment in the suspended particles. MRF has wide range applications such as dampers, brakes, clutches, polishing devices and hydraulic valves that are considered as an alternative to the conventional system.

Kumbhar et al.[3] had done the synthesis of MRF samples which typically meet the requirement of MRB. They characterized various electrolytic irons (EI) and carbonyl iron (CI) powder-based MRF by making grease as a stabilizer, oleic acid as an antifriction additive, and guar gum powder as the surface coating to reduce agglomeration of MRF. They came to the conclusion that the CI powder is better for MRF samples used for the braking application as compared to EI powder. Sarkar and Hirani[4] did analysis of CI-based MRF prepared by mixing oleic acid as an antifriction additive and tetramethyl ammonium hydroxide as the surfactant to reduce agglomeration of the MRF. Experimentation shows that the synthesized MRF is stronger and faster in response compared to the MRF 241ES fluid. Rui et al.[5] prepared a new type of fluid by dispersing nondispersive iron-silica (Fe-SiO_2) composite particles in polyethylene glycol 400. A rheometer was used to study steady state and dynamic rheological properties of the magnetic fluid under different magnetic fields. Experimental results show that Fe-SiO_2 magnetic fluid exhibits a relatively strong magneto viscosity effect for low shear rate.

Research is going on worldwide on MRF to characterize MRF as per the need of application. As seen in the research work[2–5] MRFs principle, characteristics, properties were utilized considering specific application; the analysis of MRF samples has also been done.

3.1.2 INTRODUCTION TO MRB (MAGNETO RHEOLOGICAL BRAKE)

MRB prototype is shown in Figure 3.1. MRB has a disk which rotates with the shaft at the same speed and it is enclosed by a stator. A gap between the disk and stator is filled by MRF and a coil is placed in the stator. When the current is applied to the coil, its magnetic field solidifies the MRF between the gap, the friction between disk and solidified MRF provides the required brake torque which reduces the speed of shaft. Without a magnetic field MRF cannot create restriction on the motion of the rotating disk. MRF particles are freely suspended while the coil is without current. However, when the current is applied to the coil, a magnetic flux path is formed. As a result, the particles of MRF gather to form chain-like structures in the same direction as that of the magnetic flux path. These chain-like structures resists the motion of the MRF and, ultimately, the rotation of the disk.

The brake function can be done by shear force of the MRF. The braking torque values can be adjusted continuously by changing the external magnetic field strength. Karakoc et al.[6] carried out an experiment to measure torque generated by the viscosity of the MRF at different speeds and current variables as well as torque generated by magnetic field, and finally the overall braking torque due to the combined effect with the same configuration. The torque generated due to viscosity of the fluid at various rotational speeds is measured. The relationship between the viscous torque and rotational speed is linear. Then a current was applied to the electromagnet coil and corresponding changes in the braking torque readings were recorded. Results of experimental braking torque[6] against applied current at specific speeds of 100, 200, and 300 rpm are shown in Figure 3.2.

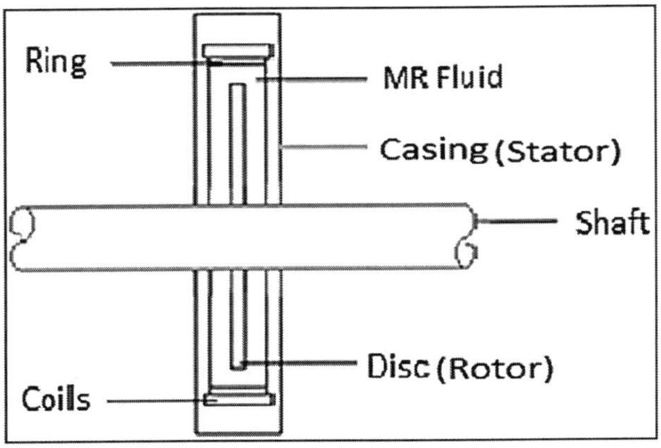

FIGURE 3.1 Basic automotive MRB (magneto rheological brake) design.

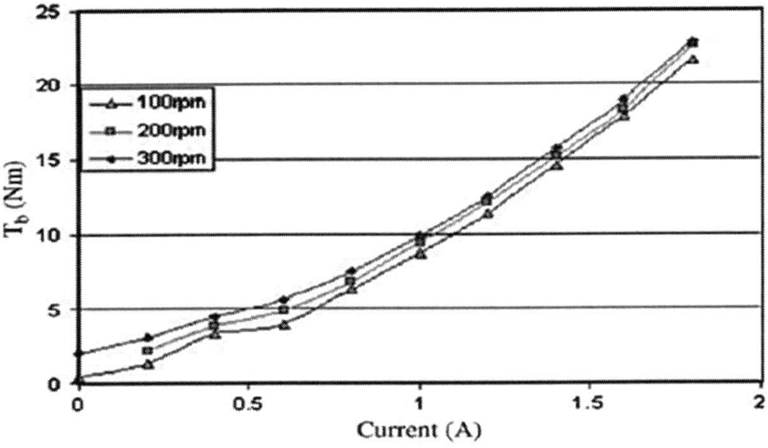

FIGURE 3.2 Torque versus current of MRB. (Reproduced with permission from Karakoc K, Park EJ, Suleman A. Design considerations for an automotive magnetorheological brake. Mechatronics. 2008 Oct. © 2008 Elsevier.)

A mathematical model has been developed by[6] for retarding the brake torque due to friction of MRF and solid surface within the MRB is,

$$T_\beta = T_H + T\mu \tag{3.1}$$

In (3.1) T_β· =· Total braking torque, T_H= Torque due to yield stress induced by applied magnetic field, and T_μ=Torque due to viscosity of MRF. T_H and T_μ are as:

$$T_H = \frac{2\pi}{3} Nk\alpha(r_z^{\ 3} - r_w^{\ 3})i \tag{3.2}$$

$$T\mu = \frac{\pi}{2h} N\mu_p \left(r_z^{\ 4} - r_w^{\ 4}\right)\theta \tag{3.3}$$

In Equations 3.2 and 3.3,

N = number of surfaces of brake disk in contact with MRF, k = constant parameter of MRF,· α· =· proportional gain, r_z and r_w= outer and inner radii of the brake disk, i = applied current, μ_p = viscosity of MRF, θ = rotational speed of the disk, h = thickness of MRF gap.

3.2 FUZZY LOGIC IMPLEMENTATION

Fuzzy logic involves three main stages of fuzzification, rule base, and defuzzification. Many systems show linear behavior of input and output so basic controllers like proportional integral derivative (PID) can give better results for controlling purpose. When it comes to nonlinear relation of system parameters then fuzzy logic is better to choose because of its flexibility to define rules and membership function (MF) of input and output. It is easy to implement the proposed fuzzy logic controller (FLC) of MRB. In this section, the method of applying fuzzy logic on MRB to estimate brake torque is elaborated.

Yam and Yeung[7] give the idea about how to identify input and output as a fuzzy inference system reducible to a singular value decomposition (SVD) system. Also, as an example, they characterize two input and one output MF by the conditions of sum normalization (SN) and non-negativeness (NN). Nonzero values of the output parameter are considered for deciding the number of MF.

Consider a system of two fuzzy variables a and b with input MF $g_i(a)$, i = 1,..., m_a, and h_j (b) = 1,..., m_b and m_a multiplied by m_b rules. SN and NN conditions of input MF need to be satisfied.

SN condition is, (for any value of a and b of fuzzy variable)

$$\sum_{i=1}^{ma} gi(a)=1 \text{ and } \sum_{j=1}^{mb} hj(b)=1$$

NN condition is, (for all a and b within the domain of interest)

$g_i\,(a) \geq 0$ and $h_j\,(b) \geq 0$

Karakoc et al.[6] carried out an experiment on MRB to find out braking torque at different speeds and currents. From Figure 3.2, it is observed that there are two inputs and one output. It is observed in Figure 3.2 that the current varies from 0–2 A and the speed is kept at 100, 200, and 300 rpm. Braking torque at current 0.2, 0.4, 0.6, 0.8, 1, 1.2, 1.4, 1.6, and 1.8 are indicated with symbols for corresponding speed. Hence, current MF is assigned for these values and speed MFs are 100, 200, and 300 rpm. It is a two inputs and one output system. SN and NN conditions are satisfied by MF of current and speed. Hence while applying fuzzy logic, two inputs and one output is selected. Nine MFs of current and three MFs for speed are assigned. Thus, 27 rules are required to be defined for the simulation. MF assignment for inputs and output is necessary so as to define rules for simulation. MATLAB's fuzzy logic toolbox is used for the simulation of braking torque generated by MRB at different speeds and currents. From Figure 3.2, specific speed and the corresponding current give the value of brake torque at every point, but only those highlighted with symbols are taken for simulation to reduce solution time and complexity.

As discussed, Karakoc et al.[6] used current range of 0–2 A and speed is kept at 100, 200, 300 rpm to measure braking torque. Nine MFs assigned for current in the MF editor are shown in Figure 3.3. For efficient results of output, the main concern is about achieving fine approximation and minimum rules to be decided. Baranyi et al.[8] have done work on fuzzy SVD based on the Takagi-Sugeno type fuzzy approximator.

They came to the conclusion that singleton sets give linear interpolated outputs over the grid points which leads to minimized interpolation errors. The Sugeno type fuzzy approximator used in this MRB torque simulation fulfills nonlinear interpolation in between grid points. Speed range is from 0–300 rpm, but results plotted are exactly taken at 100, 200m, and 300 rpm; hence, MF is assigned only at a given speed to minimize the rules as shown in Figure 3.4.

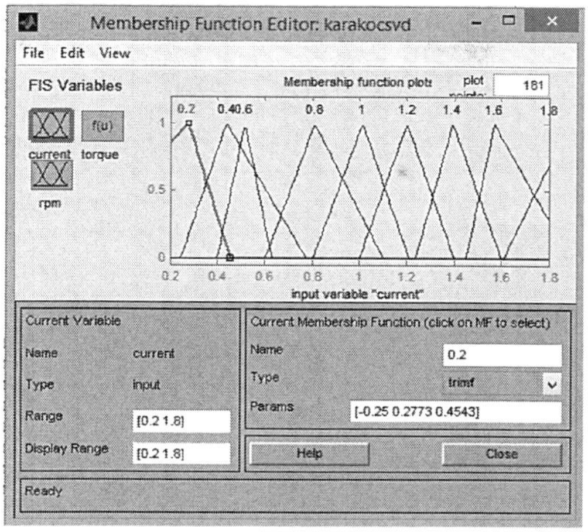

FIGURE 3.3 Current membership functions.

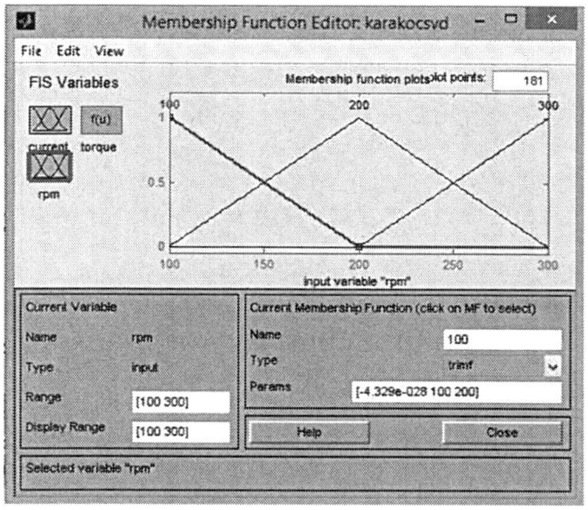

FIGURE 3.4 Speed membership function (MF).

3.3 RULES FOR SIMULATION

Based on experimental results shown in Figure 3.2, a process is carried out further for defining rules. Table 3.1 expresses a rule system defined for subset of input and output. Intervals of points are characterized into fuzzy subset of current C, speed S, and brake torque as BT:

$$C = \{VS, S, L, VL\};$$

$$S = \{S, M, L\};$$

$$BT = \{VS, S, M, L, VL);$$

In the above set, VS=very small, S=small, M=medium, L=large, VL=very large.

Rules are summarized as,

If speed is S and current is VS, then brake torque is VS.

If speed is M and current is S, then brake torque is S.

If speed is L and current is VL, then brake torque is VL.

TABLE 3.1 Rule System for Brake Torque.

Current (A)	**VS**	**S**	**L**	**VL**
Speed (rpm)		Braking torque (BT)		
S	VS	VS	M	L
M	VS	S	M	VL
L	VS	S	L	VL

3.4 SIMULATION RESULTS

The fuzzy system has been implemented using MATLAB's fuzzy logic toolbox.

A Sugeno-type computing algorithm has been selected. The implication type is minimum and the aggregation type is maximum. For the defuzzification, the weighted average method is chosen. In MATLAB fuzzy logic toolbox, rules are simulated for the brake torque. These rules after simulation show a relation between speed, current, and torque in the surface viewer as shown in Figure 3.5.

Twenty-seven rules are defined for simulation and simulation estimates value of brake torque at any value of speed and applied current within a given range. This brake torque estimation is essential to develop FLC because it is important to estimate the required brake torque at a specific speed, so that the necessary amount of current will be given to MRB. Hence, a feedback system is necessary to identify the required amount of current which will be provided by this simulation. As fuzzy logic has advantage of low memory consumption and low computation time, it will be beneficial to implement FLC to apply on MRB to achieve faster response as compared to CHB. After simulation, the value of brake torque vs. the current from Figure 3.2 and from Figure 3.5 are plotted for comparison as shown in Figure 3.6, which shows the behavior of nonlinearity the same as experimentally evaluated plots.

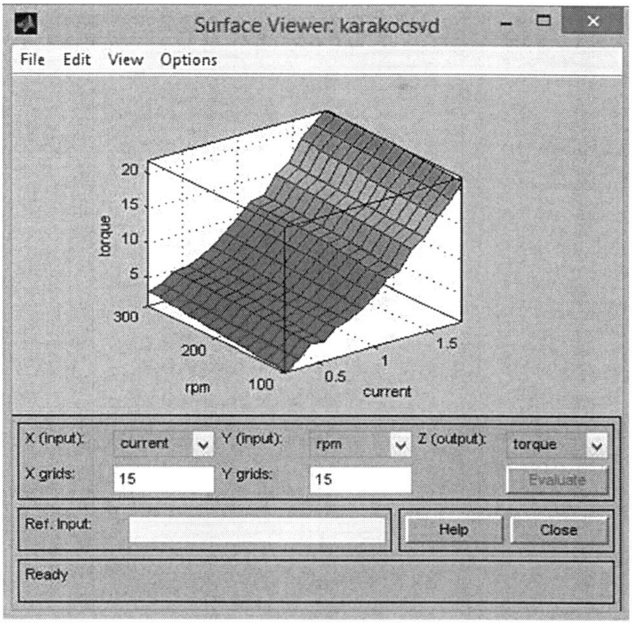

FIGURE 3.5 Surface viewer for brake torque.

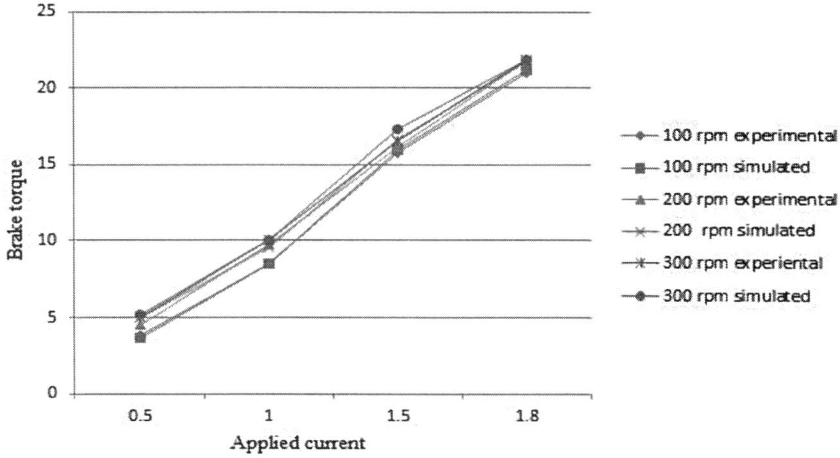

FIGURE 3.6 Comparison of experimental and simulated results.

In simulation, any value of the current and speed entered shows the braking torque which is interpolated by the defuzzification weighted average method. From Figure 3.2 and 3.6, experimental and simulated results of brake torque are shown. In simulation, various manually entered combinations of speed and current applied show brake torque which matches the experimental value.

From experimental and simulated results comparison, it is observed that brake torque is exact and very close at most points, also the brake torque difference between them is maximum at 0.5 N·m at a few points and at other points, it is less than 0.5 N·m. Hence, SVD method is compatible for the given input and output data of MRB for applying fuzzy inference system.

3.5 CONCLUSION

MRB has a quick response time compared to CHB, in addition to that FLC reduces solution time giving more advantage in the sense of effective braking of vehicle to reduce braking time and stopping distance. Hence, MRB with FLC will be beneficial for controlling braking of automotive purpose. Experimentally-evaluated MRB torque is available and this data is compared with the simulated MRB torque using fuzzy logic Sugeno model. The method assumes product-sum-gravity inference and characterizes the input by the conditions of sum normalization and non-negativeness. It is concluded that simulation gives nearly 3% deflections about the experimental value which can be considered by different algorithms while developing proposed FLC. As demonstrated in the paper, it will be possible to develop FLC for MRB. The method is readily applicable to a system with a general number of inputs. It is possible to add more inputs and outputs for effective braking. Fuzzy logic system avoids complexity of computing and its computational cost is relatively low, which may gain its application in time-consuming systems.

KEYWORDS

- **fuzzy logic**
- **MATLAB**
- **membership function**
- **MR brake**
- **MR fluid**

REFERENCES

1. Park, E. J.; Stoikov, D.; daLuz, L. F.; Suleman, A. A Performance Evaluation of an Automotive Magnetorheological Brake Design with a Sliding Mode Controller. Mechatronics. Sep 30, **2006,** *16*(7), 405–416.
2. Wang, J.; Meng, G. Magnetorheological Fluid Devices: Principles, Characteristics and Applications in Mechanical Engineering. *Proc. Inst. Mech. Eng., Part L* Jul 1, **2001,** *215*(3), 165–174.
3. Kumbhar, B. K.; Patil, S. R.; Sawant, S. M. Synthesis and Characterization of Magneto-Rheological (MR) Fluids for MR Brake Application. *Eng. Sci. Tech. Int. J.* Sep 30, **2015,** *18*(3), 432–438.
4. Sarkar, C.; Hirani, H. Synthesis and Characterization of Antifriction Magnetorheological Fluids for Brake. Def. Sci. J. Jul 22, **2013,** *63*(4), 408–412.
5. Gu, R.; Gong, X.; Jiang, W.; Hao, L.; Xuan, S.; Zhang, Z. Synthesis and Rheological Investigation of a Magnetic Fluid Using Olivary Silica-Coated Iron Particles as a Precursor. *J. Magn. Magn. Mater.* Nov 30, **2008,** *320*(21), 2788–2791.
6. Karakoc, K.; Park, E. J.; Suleman, A. Design Considerations for an Automotive Magnetorheological Brake. *Mechatronics* Oct 31, **2008,** *18*(8), 434–447.
7. Yam, Y. Singular Value-Based Identification of Fuzzy System. In *Decision and Control,* Proceedings of the 36th IEEE Conference, Dec 10, 1997, IEEE.(**1997,** *4*, 3341–3346).
8. Baranyi, P.; Yam, Y., Singular Value-Based Approximation with Takagi-Sugeno Type Fuzzy Rule Base. In *Fuzzy Systems,* Proceedings of the Sixth IEEE International Conference, Jul 1, 1997, IEEE. (**1997,** *1*, 265–270).

CHAPTER 4

FREQUENCY RESPONSES OF AUTOMOBILE SUSPENSION SYSTEMS

A. BALA RAJU[1,*] and R. V. CHALAM[2,3]

[1]Madanapalle Institute of Technology and science, Madanapalle, Andhra pradesh, India

[2]National Institute of Technology, Warangal, Telangana, India

[3]chalamrv@yahoo.com

[*]Corresponding author. E-mail: balarajua78@gmail.com

ABSTRACT

The suspension system is one of the most important systems of an automobile. The suspension system of an automobile helps to support the car body, engine, and passengers, and at the same time absorbs shocks received from the ground while the vehicle moves on rough roads. There are three types of suspension systems which are being used in the automobiles such as conventional suspension system, semi-independent, and fully independent suspension system. In this paper, it is attempted to study the behavior of the systems and compare the frequency responses and dynamic responses due to pulse inputs given by the roughness of the road.

4.1 INTRODUCTION

Vehicle suspension system plays the main role in an automobile. The main function of the automobile is to absorb the shocks arising due to the roughness of the road. The usual arrangement consists of supporting the chassis by the axle through springs and dampers, which play an important role in absorbing shocks and keeping chassis affected to a minimum level, and this type of suspension system is called conventional suspension system. In an independent suspension system, the axle which carries the wheel is hinged to the body of the vehicle and is also connected to the body through springs and dampers.

The quarter car model or half car model yield the results very quickly but they are not accurate because they do not represent the system in a realistic way because the roll and/or pitch motions cannot be taken into account by these models. Full car model considers the entire vehicle as it is. The results can be considered to be accurate and realistic. However, the analysis becomes more complex.

The literature reveals that study of the suspension systems has been a very interesting subject to many researchers. For better understanding, we first consider half car model and then full car model. Pater Gaspar[1] considered full car model and proposed a method for identifying suspension parameters taking into account the nonlinear nature of the components. Anil Shirahatt et al.[2] attempted to maximize the comfort level considering a full car model. Genetic algorithms have been employed to perform optimization to arrive at optimum values of suspension parameters. Gao et al.[3] investigated dynamic response of cars due to road roughness treating it as random excitation. Hajkurami et al.[4] studied the frequency response of a full car model as a system of seven degrees of freedom. Ahmed Faheem[5] studied the dynamic behavior using quarter-car model and half car model for different excitations given by the road. Jacquelien et al.[6] used electrical analogy in conjunction with quarter car model and studied the control scheme of the suspension system. Gao et al.[7] also studied the dynamic characteristics considering the mass, damping, and tire stiffness as random variables. Kamalakannan et al.[8] tried adaptive control by varying damping properties according to the road conditions. Sawant et al.[9] developed an experimental procedure for determining the suspension parameters using a quarter car model. Balaraju and Venkatachalam analyzed the dynamic behavior of an automobile using full car model for both fully conventional suspension systems[10] and fully independent suspension systems.[11]

In this paper, an attempt is made to study the frequency response and dynamic response of conventional, semi-independent and fully independent suspension systems.

4.2 FORMULATION

4.2.1 CONVENTIONAL SUSPENSION SYSTEM

Figure 4.1 shows a mass *m* being supported at its four corners. The mass *m* includes mass of the chassis, engine, automobile body, and passengers.

The masses m_1 and m_2 are, respectively, the masses of the front wheel axle and the rear wheel axle. m_1 and m_2 may also be considered to have included the masses of tires. k_1 and c_1 represent the stiffness and damping properties of each of the tires. k_2 and c_2 may represent the stiffness and damping of the supporting springs and dampers near the front wheels. Similarly, k_3 and c_3 represent the springs and dampers near the rear wheels. Seven coordinates are chosen to describe the motion of the entire system. The seven coordinates which describe the vibrating system are x_1, x_2, x, γ_1, γ_2, γ, and λ. The vertical displacements caused by the road roughness may be represented by the variables y_1, y_2, y_3, and y_4.

The vertical linear motion of the center of mass G_1 of the front axle and its roll motion may be represented by x_1 and γ_1, respectively. Similarly, x_2 and γ_2 may be used to describe the linear vertical motion of the center of mass G_2 of the rear axle and its roll motion, respectively. The vertical motion of the center of mass G of the mass m of the main body may be described by the coordinate x. The roll and pitch motions of the mass m may be described by γ and λ, respectively.

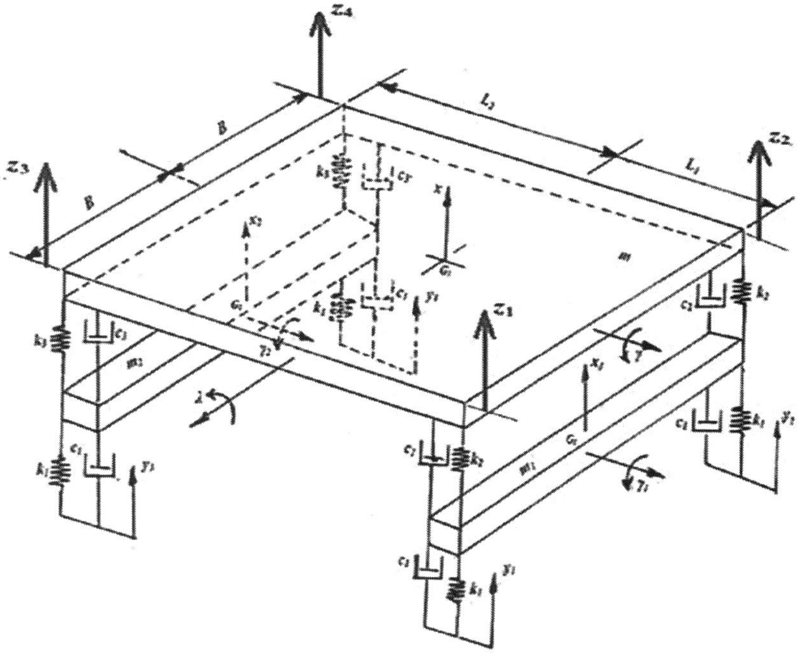

FIGURE 4.1 Schematic arrangement of full car model of the conventional suspension system.

The equations of motion are as follows:

$$m_1 \ddot{x}_1 + 2(k_1 + k_2)x_1 - (2k_2)x - (2k_2L_1)\lambda + 2(c_1 + c_2)\dot{x}_1 - (2c_2)$$
$$\dot{x} - (2c_2L_1)\dot{\lambda} = k_1(y_1 + y_2) + c_1(\dot{y}_1 + \dot{y}_2)$$

$$I_1\ddot{\gamma}_1 + 2(k_1 + k_2)B^2\gamma_1 - (2k_2B^2)\gamma + 2(c_1 + c_2)B^2\dot{\gamma}_1 - (2c_2B^2)$$
$$\dot{\gamma} = -k_1B(y_1 - y_2) - c_1B(\dot{y}_1 - \dot{y}_2)$$

$$m_2\ddot{x}_2 + 2(k_1 + k_3)x_2 - (2k_3)x + (2k_3L_2)\lambda + 2(c_1 + c_3)\dot{x}_2 - (2c_3)$$
$$\dot{x} + (2c_3L_2)\dot{\lambda} = k_1(y_3 + y_4) + c_1(\dot{y}_3 + \dot{y}_4)$$

$$m\ddot{x} - (2k_2)x_1 - (2k_3)x_2 + 2(k_2 + k_3)x + 2(k_2L_1 - k_3L_2)\lambda - (2c_2)\dot{x}_1 - (2c_3)$$
$$\dot{x}_2 + 2(c_2 + c_3)\dot{x} + 2(c_2L_1 - c_3L_2)\dot{\lambda} = 0$$

$$I_r\ddot{\gamma} - (2k_2B^2)\gamma_1 - (2k_3B^2)\gamma_2 + 2(k_2 + k_3)B^2\gamma - (2c_2B^2)$$
$$\dot{\gamma}_1 - (2c_3B^2)\dot{\gamma}_2 + 2(c_2 + c_3)B^2\dot{\gamma} = 0$$

$$I_p\ddot{\lambda} - (2k_2L_1)x_1 + (2k_3L_2)x_2 + 2(k_2L_1 - k_3L_2)x + 2(k_2L_1^2 + k_3L_2^2)$$
$$\lambda - (2c_2L_1)\dot{x}_1 + (2c_3L_2)\dot{x}_2 + 2(c_2L_1 - c_3L_2)\dot{x} + 2(c_2L_1^2 + c_3L_2^2)\dot{\lambda} = 0$$

4.2.2 SEMI-INDEPENDENT SUSPENSION SYSTEM

The semi-independent suspension system is shown in Figure 4.2. Mass of the main body represented by m including with chassis, it is located at all corners of the vehicle. k_2 and c_2 are the spring constant and damping constant, respectively, at the front side. k_3 and c_3 are same as above but at the rear side of the vehicle. The mass m_1 indicates the mass of the tires. The tire properties are represented by the spring constant k_1 and the damping constant c_1. The mass m_2 indicates the mass of the rear wheel axle with tire mass. Pulse inputs are given by the road roughness and described by $y_{i,}$ $i = 1, 2, 3$, and 4.

The displacements of the masses m_1 may be described by using the coordinates, x_1, x_2. The motion of the mass m_2 may be described by using the coordinates, x_3, to describe the up and down motion of its center of mass G_2 and γ_2 to describe its roll motion. The displacement of the center of mass G of the main body of mass m is described by the coordinate x. γ and λ are the roll and pitch rotations of the main body, the entire suspension system is described by seven coordinates, such as, $x_1, x_2, x_3, x, \gamma, \gamma_2$, and λ. The roughness of the road is giving as an input to the vehicle through, y_i and $\dot{y}_i$, $i = 1, 2, 3, 4$ which influences x_1, x_2 and γ_2 only. Moreover, it also turns influence x_3 of the rear axle and, x, γ, and λ motions of the main body.

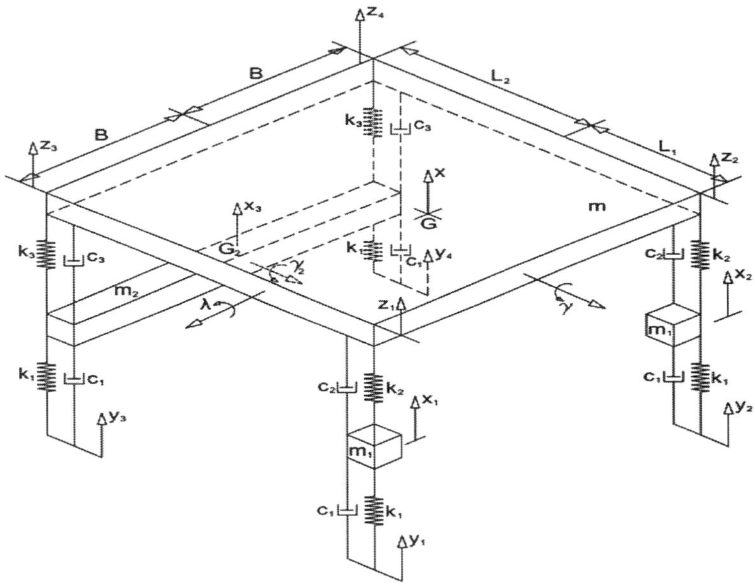

FIGURE 4.2 Schematic arrangement of full car model of semi-independent suspension system.

The equations of motion are as follows:

$$m_1\ddot{x}_1 + (k_1 + k_2)x_1 - k_2x + k_2B\gamma - k_2L_1\lambda + (c_1 + c_2)\dot{x}_1 - c_2\dot{x} + c_2 B\dot{\gamma} - c_2L_1\dot{\lambda} = k_1y_1 + c_1\dot{y}_1$$

$$m_1\ddot{x}_2 + (k_1 + k_2)x_2 - k_2x - k_2B\gamma - k_2L_1\lambda + (c_1 + c_2)\dot{x}_2 - c_2\dot{x} - c_2 B\dot{\gamma} - c_2L_1\dot{\lambda} = k_1y_2 + c_1\dot{y}_2$$

$$m_2\ddot{x}_3 + 2(k_1 + k_3)x_3 - 2k_3x + 2k_3L_2\lambda + 2(c_1 + c_3)\dot{x}_3 - 2c_3\dot{x} + 2c_3L_2\dot{\lambda} = 0$$

$$m\ddot{x} - k_2x_1 - k_2x_2 - 2k_3x_3 + 2(k_2 + k_3)x + 2(k_2L_1 - k_3L_2)\lambda - c_2\dot{x}_1 - c_2\dot{x}_2 - 2c_3\dot{x}_3 + 2(c_2 + c_3)\dot{x} + 2(c_2L_1 - c_3L_2)\dot{\lambda} = 0$$

$$I_r\ddot{\gamma}+k_2Bx_1-k_2Bx_2+2B^2(k_2+k_3)x-2k_3B^2\gamma_2+c_2B\dot{x}_1-c_2B\dot{x}_2+2B^2$$
$$(c_2+c_3)\dot{x}-2c_3B^2\dot{\gamma}_2=0$$

$$I_2\ddot{\gamma}_2-2k_3B^2\gamma+2(k_1+k_3)B^2\gamma_2-2c_3B^2\dot{\gamma}+2(c_1+c_3)B^2\dot{\gamma}_2=k_1$$
$$(y_3-y_4)+c_1(\dot{y}_3-\dot{y}_4)$$

$$I_p\ddot{\lambda}-k_2L_1x_1-k_2L_1x_2+2k_3L_2x_3+2(k_2L_1-k_3L_2)x+2(k_2L_1^2+k_3L_2^2)$$
$$\lambda-c_2L_1\dot{x}_1-c_2L_1\dot{x}_2+2c_3L_2\dot{x}_3+2(c_2L_1-c_3L_2)\dot{x}+2(c_2L_1^2+c_3L_2^2)\dot{\lambda}=0$$

4.2.3 FULLY INDEPENDENT SUSPENSION SYSTEM

The displacements of the masses m_1 may be described by using the coordinates, x_1, x_2, x_3 and x_4 (Fig. 4.3). The displacement of the center of mass G of the main body of mass m is described by the coordinate x. The roll and pitch rotations of the main body are described by the coordinates γ and λ. The total suspension system is described by seven coordinates, such as x_1, x_2, x_3, x_4, x, γ, and λ.

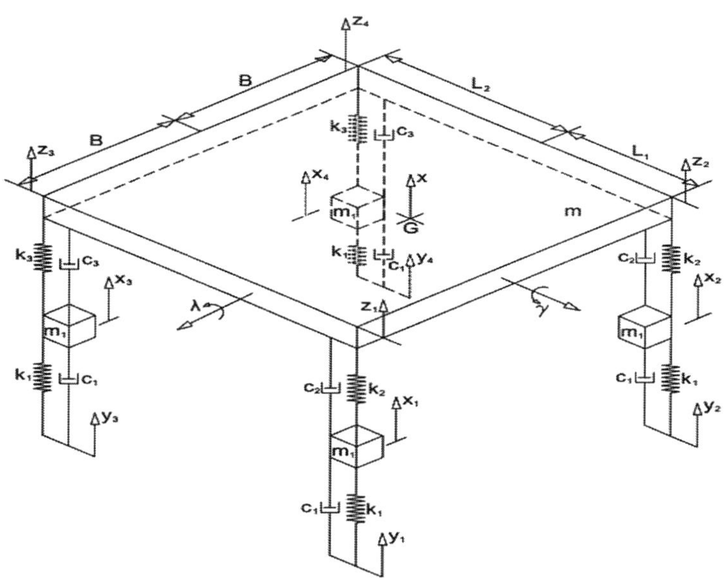

FIGURE 4.3 Schematic arrangement of full car model of the fully independent suspension system.

4.3 ANALYSIS OF THE SUSPENSION SYSTEM

In order to determine the natural frequencies of the system, the damping and the excitation are ignored and the remaining part of equations of motion is considered as,

$$M\ddot{\bar{X}} + K\bar{X} = 0$$

For the sake of numerical computations, a specific set of various parameters, relating to a practical automobile are taken as follows.

$m_1 = 40$ kg, $m_2 = 40$ kg, $m_3 = 100$ kg
$m = 1000$ kg
$I_r = 500$ kg.m^2, $I_p = 1000$ kg.m^2, $I_2 = 20$ kg.m^2
$K_1 = 2 \times 10^5 N/m$, $K_2 = K_3 = 0.5 \times 10^5 N/m$
$C_1 = 1000$ N.s/m $C_2 = C_3 = 1000$ N.s/m
$B = 0.75$ m $L_1 = 1$ m $L_2 = 2.5$ m

The frequency of excitation (*f* in Hz) depends on the road roughness. If the road profile is approximated as a sin wave of wavelength L and the vehicle is moving with a velocity V,

$$m_1\ddot{x}_1 + (k_1 + k_2)x_1 - k_2x + k_2B\gamma - k_2L_1\lambda + (c_1 + c_2)\dot{x}_1 - c_2\dot{x} + c_2 B\dot{\gamma} - c_2L_1\dot{\lambda} = k_1y_1 + c_1\dot{y}_1$$

$$m_1\ddot{x}_2 + (k_1 + k_2)x_2 - k_2x - k_2B\gamma - k_2L_1\lambda + (c_1 + c_2)\dot{x}_2 - c_2\dot{x} - c_2 B\dot{\gamma} - c_2L_1\dot{\lambda} = k_1y_2 + c_1\dot{y}_2$$

$$m_1\ddot{x}_3 + (k_1 + k_3)x_3 - k_3x + k_3B\gamma + k_3L_2\lambda + (c_1 + c_3)\dot{x}_3 - c_3\dot{x} + c_3 B\dot{\gamma} + c_3L_2\dot{\lambda} = k_1y_3 + c_1\dot{y}_3$$

$$m_1\ddot{x}_4 + (k_1 + k_3)x_4 - k_3x - k_3B\gamma + k_3L_2\lambda + (c_1 + c_3)\dot{x}_4 - c_3\dot{x} - c_3 B\dot{\gamma} + c_3L_2\dot{\lambda} = k_1y_4 + c_1\dot{y}_4$$

$$m\ddot{x} - k_2x_1 - k_2x_2 - k_3x_3 - k_3x_4 + 2(k_2 + k_3)x + 2(k_2L_1 - k_3L_2)\lambda - c_2\dot{x}_1 - c_2\dot{x}_2 - c_3\dot{x}_3 - c_3\dot{x}_4 + 2(c_2 + c_3)\dot{x} + 2(c_2L_1 - c_3L_2)\dot{\lambda} = 0$$

$$I_r\ddot{\gamma} + k_2Bx_1 - k_2Bx_2 + k_3Bx_3 - k_3Bx_4 + 2B^2(k_2 + k_3)\gamma + c_2B\dot{x}_1 - c_2B\dot{x}_2 + c_3 B\dot{x}_3 - c_3B\dot{x}_4 + 2(c_2 + c_3)B^2\dot{\gamma} = 0$$

$$I_p\ddot{\lambda} - k_2L_1x_1 - k_2L_1x_2 + k_3L_2x_3 + k_3L_2x_4 + 2(k_2L_1 - k_3L_2)x + 2(k_2L_1^2 + k_3L_2^2)$$
$$\lambda - c_2L_1\dot{x}_1 - c_2L_1\dot{x}_2 + c_3L_2\dot{x}_3 + c_3L_2\dot{x}_4$$
$$+2(c_2L_1 - c_3L_2)\dot{x} + 2(c_2L_1^2 + c_3L_2^2)\dot{\lambda} = 0$$

The frequency of excitation may be expressed as:

$$f = V/L$$

It is to be observed that the frequency of excitation increases with increase in the velocity or decrease in the wavelength.

The natural frequencies, f_n (in Hz) are obtained using MATLAB software as follows

CSS	1.79	2.12	3.88	12.58	12.61	12.62	12.74
SISS	1.79	2.12	3.87	11.43	12.60	12.62	18.88
FISS	1.79	2.12	3.88	12.58	12.61	12.62	12.74

All these seven frequencies are to be taken as natural frequencies of the total system. The resonance may occur whenever the exciting frequency matches with any one of these natural frequencies. It is observed that the first three frequencies are discrete, and are similar to the frequencies obtained for the conventional suspension system. The four higher frequencies are very close to each other.

4.4 FREQUENCY RESPONSE OF THE SYSTEM

The system is excited with a harmonic force at right side front wheel as,

$$y_1(t) = y_{m_1}\, sin(2\pi f)t$$

$$\dot{y}_1(t) = y_{m_1}\,(2\pi f)cos(2\pi f)t$$

Taking $y_{m1} = 0.05m$, the set of seven second order differential equations representing the equations of motion as given in the Equation 4.8 are integrated using fourth order Runge–Kutta method. The response of the system is observed by noting the solutions for $x_1,\ x_2,\ x_3,\ x_4,\ x,\ y,$ and λ. As expected, all these variables are found to vary harmonically with a frequency same as the exciting frequency f. The frequency of excitation is varied, and the maximum values of these variables are noted. For the sake of convenience in discussion, the variables are expressed in nondimensional forms as

$$\tilde{x}_1 = \frac{x_1}{y_{m_1}}, \quad \tilde{x}_2 = \frac{x_2}{y_{m_1}}, \quad \tilde{x}_3 = \frac{x_3}{y_{m_1}}, \quad \tilde{x}_4 = \frac{x_4}{y_{m_1}}$$

$$\tilde{x} = \frac{x}{y_{m_1}}, \qquad \tilde{\gamma} = \frac{B\gamma}{y_{m_1}}, \qquad \tilde{\lambda} = \frac{L_2\lambda}{y_{m_1}}$$

Figure 4.4a shows frequency response, using the nondimensional variables, of x_1, x_2, x, and λ. It may be observed from Figure 4.4a that x_1, x_2, x, and λ are showing the peak responses in the region of the lowest frequency 1.795 Hz. The variables γ_1, γ_2, and γ are showing the peak responses at 2.129 Hz, as shown in Figure 4.4b. In the remaining part, all the variables are showing very low responses. This shows that the first two frequencies are the most important ones in the practical point of view. In order to have a comparative study of various variables, the variations of $\tilde{x}_1$, $\tilde{x}_2$, $\tilde{x}$, and $\tilde{\lambda}$ are shown in Figure 4.4a. This shows that at resonating frequency, $\tilde{x}_1$ and $\tilde{x}$ have almost same values. In the remaining part, $\tilde{x}$ is much less than $\tilde{x}_1$. The figure also shows that the displacement undergone by the rear axle $\tilde{x}_2$ is much smaller than the displacement undergone by the front axle $\tilde{x}_1$. This may be because the excitation is given at one front wheel only. The figure shows that the disturbance given at the front wheel is also producing pitch motion $\tilde{\lambda}$, but this is very small in comparison to the other variables.

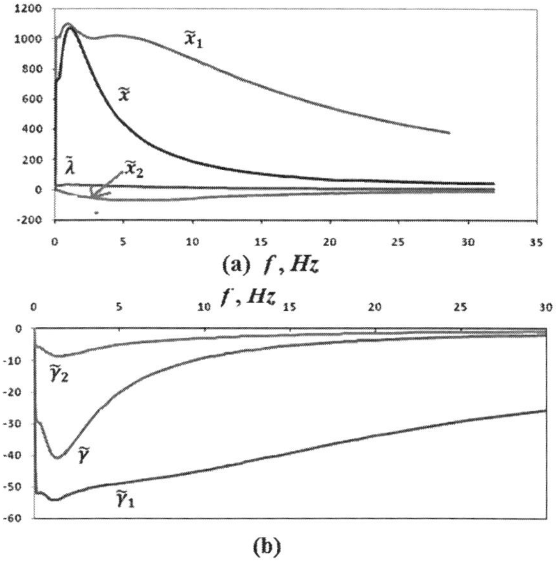

FIGURE 4.4 Frequency responses.

Figures 4.4b shows the variations of $\tilde{\gamma}_1$ $\tilde{\gamma}_2$ and $\tilde{\gamma}s$.

It is observed that the value of the roll motion of the main body γ is in between the values of roll motions of the front axle and rear axle. At high frequencies, γ is almost negligible.

4.5 DYNAMIC RESPONSE OF THE SYSTEM

In order to study the dynamic behavior of the systems, disturbances due to road roughness may be given through y_i and $\dot{y}_i$, $i = 1-4$. and the equations of motion may be integrated. To get a physical feel of the dynamic behavior of system, the absolute displacements of the four corners of the main body may easily be expressed in terms of motions of *m* (Fig. 4.5). Numerical values of various parameters are taken from a practical car. A disturbance is given only to right side front wheel as y_1 (0) = 0.1 and the time responses are obtained for both the suspension systems, that is, z_i, i = 1–4.

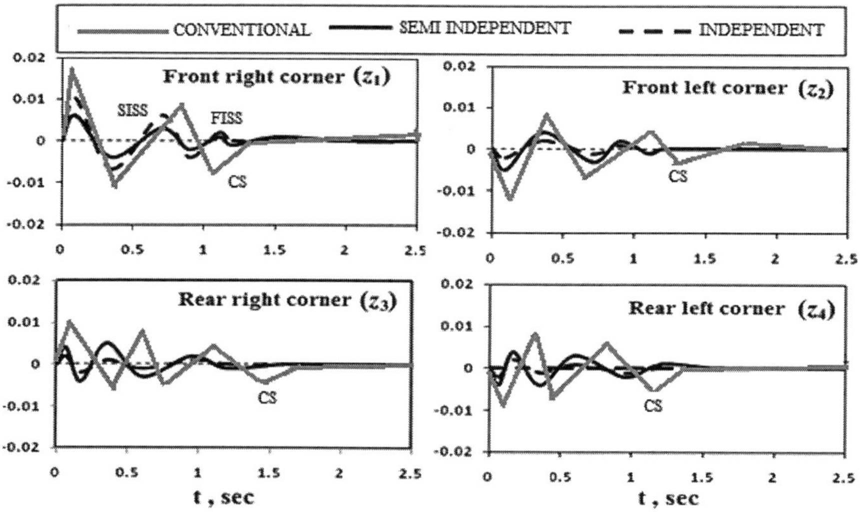

FIGURE 4.5 Absolute displacements of four corners of the main body in *mm*.

4.6 CONCLUSION

The work presented in this paper and significant conclude as follows

i. The equations of motion for all the three types of suspension systems are presented. Each system is modeled as a seven degree freedom system.
ii. Frequency responses were obtained for all the three cases by using MATLAB. The responses were showing maximum values near the first natural frequency. At all other frequencies, the responses were found to be very small. Hence, the study of the frequency responses revealed that the lowest natural frequency is important in the practical point of view.
iii. It is observed that the lower natural frequencies are almost of the same magnitude. A significant difference in values may be noticed among higher frequencies.
iv. A typical disturbance is given to three suspension system and the dynamic behavior is studied. It is observed that in the case of convention suspension system the effect is felt at all the points of the main body. In the case of semi- and independent-suspension system, the effect is confined mostly at the place where the disturbance occurs.

KEYWORDS

- **suspension systems**
- **fully suspension**
- **semi-independent suspension**
- **frequency response**
- **road roughness**
- **dynamic response**

REFERENCES

1. Gaspar, P.; Nadai, L. Estimation of Dynamical Parameters of Road Vehicles on Freeways, 5th International Symposium on Intelligent Systems and Informatics, Serbia, 2007.
2. Shirahatt, A.; Prasad, P. S. S.; Panzade, P.; Kulkarni, M. M. Optimal Design of Passenger Car Suspension for Ride and Road Holding. *J. Braz. Soc. Mech. Sci. Eng.* **2008,** *30*(1), 66–74.

3. Gao, W.; Zhang, N.; Dai, J. A Stochastic Quarter Car Model for Dynamic Analysis of Vehicle with Uncertain Parameters. *Int. J. Veh. Mech. Mobility* **2008,** *46*(12), 1159–1169.
4. Hajkurami, H.; Samandari, H.; Ziaei-Rad, S. Analysis of Chaotic Vibration of Nonlinear Seven Degree of Freedom Full Car Model, 3rd International Conference on Integrity, Reliability and Failure, Portugal, July 2009.
5. Faheem, A.; Alam, F.; Thomas, V. The Suspension Dynamic Analysis for a Quarter Car Model and Half Car Model, 3rd BSME-ASME International Conference on Thermal Engineering, Dhaka, December 2006.
6. Jacquelien, M. A.; Scherpen, D. J.; Maulny, F. Parallel Damping Injection for the Quarter Car Suspension System, *Proceedings of the 17th International Symposium on Mathematical theory of Networks and Systems*, Japan, July 2006.
7. Gao, W.; Zhang, N. Dynamic Analysis of Vehicles with Uncertain Parameter, *14th International Congress on Sound and Vibration*, Australia, July 2007.
8. Kamalakannan, K.; Elaya Perumal, A.; Mangalaramanan, S.; Arunachalam, K. Performance Analysis of Behaviour Characteristics of CVD (Semi Active) in Quarter Car Model. *Jordan J. Mech. Ind. Eng.* **2011,** *5,* 261–265 (June).
9. Sawant, S. H.; Belwakar, V.; Kamble, A.; Pushpa, B.; Patel, D. Vibrational Analysis of Quarter Car Vehicle Dynamic System Subjected to Harmonic Excitation by Road Surface. *Undergrad. Acad. Res. J.* **2012,** *1,* 46–49.
10. Balaraju, A.; Venkatachalam, R. A Study on Conventional Versus Independent Suspension System of an Automobile. *Appl. Mech. Mater.* **2014,** *541–542,* 827–831.
11. Balaraju, A.; Venkatachalam, R. Analysis of Vibrations of Automobile Suspension System using Full Car Model. *Int. J. Sci. Eng. Res.* **2013,** *4*(9), 2305–2311 (September).

CHAPTER 5

INVESTIGATION OF HUMANOID MOVEMENT USING A CLASSICAL APPROACH

PRIYADARSHI BIPLAB KUMAR[1,*], ANIMESH CHHOTRAY[1,2], KRISHNA KANT PANDEY[1,3], and DAYAL R. PARHI[1,4]

[1]*Robotics Laboratory, Mechanical Engineering Department, NIT, Rourkela, Odisha 769008, India*

[2]*chhotrayanimesh@gmail.com*

[3]*kknitrkl@yahoo.in*

[4]*dayalparhi@yahoo.com*

[*]*Corresponding author. E-mail: p.biplabkumar@gmail.com*

ABSTRACT

With the development of science and technology, robotics research has gained immense interest among various scholars. In the current state of technology use, robotics has spread to various industries such as automobiles, production, surgery, industrial automation, and so forth. In the robotics world, robots are used in various shapes and sizes; however, the humanoid form of robot is considered as the most successful form as it is capable of easing human effort in several ways and also mimicking the human behavior. In the study of humanoid robots, path planning and navigation problem is considered as one of the major aspects of design. During path planning of a humanoid robot to reach a certain goal position, problems like obstacle avoidance, goal-following behavior are encountered. To reduce the path length of travel and time taken to reach the goal position, some optimization techniques can be used. Those techniques can be based on both the classical approach and computational intelligence. In the current investigation, focus has been given to the use of classical approach for the navigation problem. Different classical approaches may include regression analysis, mathematical techniques, statistical approaches, and so forth. A humanoid NAO has been used for the current analysis. A regression controller has been designed for the NAO robot considering the challenges involved in the navigation problem. The regression controller is designed considering the principles of regression technique. By implementing the regression controller in the robot, simulation of several environments has been carried out. A real experimental setup has also been designed with the similar positioning of obstacles as was in the case of simulation. Parameters like path length and time taken are noted down, and finally, a comparison has been done between the simulated and experimental results. This work can also be extended toward the use of modern techniques for the same navigation problem. Dynamic environment problems such as multiple NAO navigation can also be achieved by this approach.

5.1 INTRODUCTION

With the advancement in modern technology, robots are becoming an integral part of our life. A robot irrespective of its size and features is primarily designed to ease human life and assist human beings in all the possible ways. Recent advancements in the robotic designs have placed robotics research as one of the most promising fields. Robots are categorized into several categories such as mobile robots, biped robots, and so forth based on their performance measures. Out of all the forms of robots, humanoid robots are accepted as the best forms as they can mimic the human behavior when trained accordingly. As the size, configuration, and features of the humanoid robot match with the humans, they are also able to replace humans in some fields. They can also be advantageous in some areas where human intervention is not possible due to the complex environment and the critical obstacles present in the path. Humanoid robots can work in these complex terrains provided they are equipped with some technology that will help them to avoid the obstacles present in the path and successfully reach the goal position.

In the current work, focus has been given toward navigational approach of humanoid robots. Here, humanoid NAO is considered for the analysis. Several techniques are available for path planning and navigation of humanoids. They may be based on classical approaches or computational intelligence. Regression analysis is considered as the technique to be used in the humanoid NAO. A regression controller is designed for the humanoid considering the basic rules of regression technique. After implementing the regression controller for the NAO, it is tested both in simulated and real environments. A comparison is done between the results obtained to draw the concluding remarks.

5.2 BACKGROUND AND REVIEW

Robot path planning and navigation is regarded as the most challenging problem in robotics research area. Several attempts have been made in the past to demonstrate this problem by the use of different navigational methods. Both classical techniques and computational intelligence techniques have been developed to navigate the robot. Some of the classical approaches can be cited here. Li et al.[1] gave an idea about finding out a global optimal or a suboptimal path if information is available about the environmental conditions. This approach was based on an artificial potential field. An algorithm has been developed by Lazaro et al.[2] to carry out the navigation task starting with the information given by a sensor. Alternative routes can be searched by this algorithm if an obstacle is detected and optimal path can be chosen by smoothly tracking the trajectory, without rapid deviations in the robot movement and orientation. The navigation control of a robotic agent in static as well as dynamic environments is studied by Kashmiri et al.[3] by a kernel regression approach. Asano etal.[4] presented a method for construction of a specific path by using disjoint polygons. Benamati et al.[5] developed a new technique called flat potential field, for the navigation problem using a static environment. Jollyb et al.[6] presented a path-planning approach based on the Bezier curve technique. Sohei et al.[7] studied about recharging techniques for a group of robots working in an environment. Keshmiri et al.[8] enhanced the same previous problem without sticking to a particular recharging station. A nonholonomic path-planning method is studied by Liang et al.[9], designed to take into consideration both constraints and minimization of path. Masehian et al.[10] reviewed almost 35 years of development regarding motion planning of different types of robots based on moment point concept. Minguez et al.[11] described the design of a divide and conquer strategy applicable to path planning approach. A control methodology for nonholonomic mobile manipulators is developed by Papadopoulos et al.[12] in presence of obstacles. Takahashi and Schilling[13] proposed an algorithm to plan a collision-free path in a rectangular workspace containing polygon obstacles. Zhang et al.[14] proposed a model based on artificial potential field for dynamic path planning. Qi et al.[15] modified the previously available potential field approach and designed it as an artificial potential field approach. Shi and Zhao[16] proposed an improved potential field approach to overcome the problem of trapping at the local minima of artificial potential field. Tingbin and Qisong[17] proposed a mature and stable artificial potential field (APF) method for robot path planning. He et al.[18]proposed a method to overcome the limitations of trapping a local minimum. Sheng et al.[19] proposed an improved APF algorithm to solve the goal of no reachability and trap at local minimum. Hwang and Ahuja[20] proposed a method that plans the path at two levels.

5.3 OUTCOME OF LITERATURE REVIEW

Robot path planning has been attempted by several researchers previously. Some of the navigational techniques are based upon computational intelligence and some are based on the classical approach. However, very few works have been reported in dealing with a humanoid robot in a complex environment where the robot has to reach the goal avoiding all the obstacles that are present in the path. The behavior of the humanoid robot while avoiding the obstacles also demands certain attention toward the different criteria they follow during the navigation. Based upon the outcome of the existing literature, the objectives of the current work have been decided.

5.4 OBJECTIVES

The current work is primarily aimed at dealing with the navigational problem of a humanoid robot. Here, humanoid NAO is considered for the analysis. Regression analysis is considered as one of the classical approaches used in the current work. To design a regression controller for a humanoid robot, it is necessary to show some light toward the basic principles of the regression technique. After the design of the regression controller, the humanoid robot needs to be tested for navigation, both in simulated and actual environments. Finally, a comparison is aimed to be performed between both the results to show a coherence between the results.

5.5 HUMANOID NAO

NAO is a medium-sized, programmable, and integrated humanoid robot designed and developed by Aldebaran Robotics. The used NAO in the current work is of version V3.3. It is of 58 cm height, 5 kg weight equipped with x86 AMD Geode processor working at 500 MHz. NAO has 256 MB synchronous dynamic (SD)RAM and a 2 GB flash memory. It gets its power from a 6-cell lithium ion battery.[21] NAO is capable of almost 30 min of uninterrupted operation using its battery power. A total of 21 degrees of freedom (DOF) are present in the NAO. Head possess 2, each arm carries 4, each leg has 5 and pelvis has 1 DOF. Different types of sensors are associated with NAO. Two cameras are set on the head which work on alternate basis. These two cameras enable near and distant view respectively. Two receiving and two emitting sonars are present on the chest that can sense the obstacles present within a certain range. One 3-axis accelerometer and one 2-axis gyroscope are equipped within the NAO that can provide information about instantaneous movements. Force sensitive resistors are also associated with the foot of the NAO that can give the amount of forces that are applied on the ground by the foot at any time. Encoders present provide the value of each joint at any time of operation.

5.6 OUTLINE OF THE REGRESSION ANALYSIS

Regression analysis is a statistical method to establish a relation among various variables. This technique is used to model and analyze the correlation among the independent and dependent variables. It helps one to know the change in dependent variables with respect to the independent variables, if the independent variables are kept fixed. It is used to predict and forecast some output data by giving some input data. In linear regression, the dependent variables are represented as linear combination of the parameters but not linear to independent variables. Suppose, we want to formulate an equation of n data, y_i is a dependent variable and x_i is an independent variable with some parameters α_1 and α_2.

Then the equation can be written as

$$y_i = \alpha_1 + \alpha_2 x_i + e_i \qquad i = 1,2,\ldots\ldots n \tag{5.1}$$

Where, e is an error term.

For an example, how a basic equation can be represented in a regression analysis is shown in the following Figure 5.1.

Here, in x-axis, some ranges have been taken from −20 to 60 and in y-axis from 0–15. The straight-line equation with some variables can be represented as shown in the above figure. The accumulation of scattered data to a straight line is the beauty of regression equation. Smoothing and regression can be applied to different areas like science, engineering, economics, medical research, and so forth.

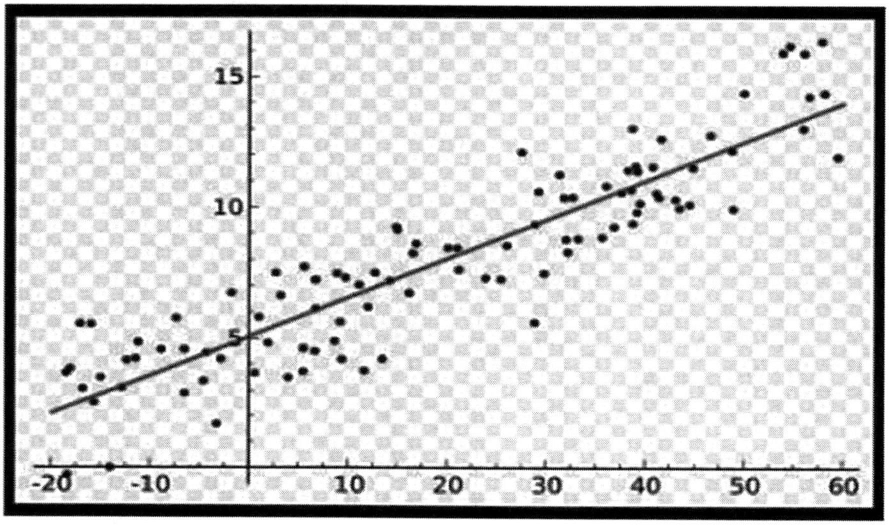

FIGURE 5.1 Description of linear regression.

5.7 CONTROL ARCHITECTURE OF REGRESSION ANALYSIS FOR HUMANOID NAVIGATION

To predict the turning angle in order to avoid the obstacles present during path planning is the primary objective of regression controller. Here in this analysis, three input parameters are considered left obstacle distance (LOD), right obstacle distance (ROD), and front obstacle distance (FOD). With the help of ultrasonic sensors mounted on the humanoid robot these parameters can be observed and noted down by the microprocessor. There is one output parameter called as the turning angle (TA). Based on the data obtained from the sensors regarding input parameters, the controller is supposed to generate output data. Figure 5.2 demonstrates the initial position of the NAO robot in the specified environment. Different parameters for the humanoid such as LOD, ROD, FOD and TA are represented for the initial position of the robot. The humanoid robot follows a particular sign convention which is demonstrated in the following Figure 5.3.

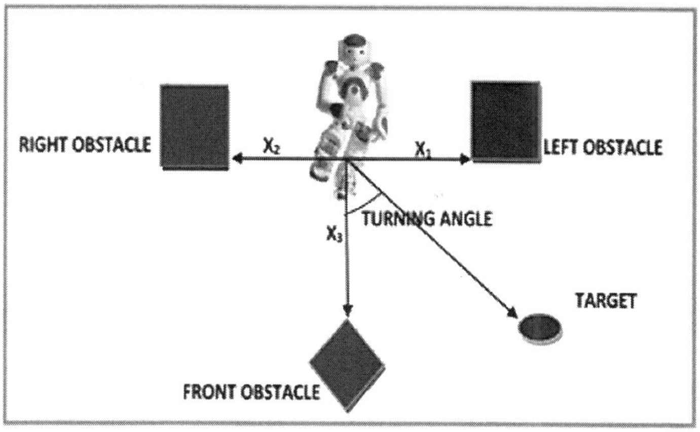

FIGURE 5.2 Initial position of the humanoid robot in the environment.

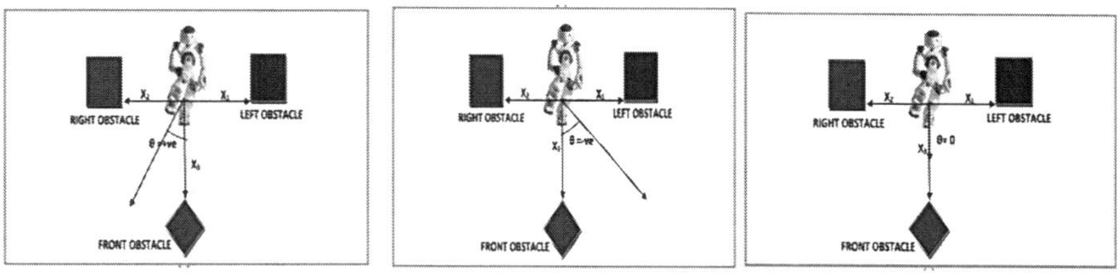

FIGURE 5.3 Sign conventions used in terms of TA.

The robot is always headed toward the target as the robot has the prior information about the source and target positions. When any obstacle is sensed by the sensors of the robots, then the proposed algorithm works as per the training pattern. By taking data from the training pattern provided in the following table, it can be illustrated in a better way. By taking the eighth data from Table 5.1 for the values of FOD=35, LOD=55, ROD=40, the TA is assumed as −15. When the humanoid senses any obstacle, it will take a decision as per the algorithm. Illustrating the example from the eighth data, the front side distance from the obstacle is 35 cm whereas the humanoid is 55 cm away from the left side and 40 cm away from right side. As the threshold distance here to take a turn is defined as 30 cm, the humanoid should take a right or left turn. As the left obstacle distance is greater than right obstacle distance; hence, the humanoid goes for a left turn.

TABLE 5.1 Examples of Some Training Data for Regression Controller.

Sr. no.	Front obstacle distance (FOD)	Left obstacle distance (LOD)	Right obstacle distance (ROD)	Turning angle (TA)
1	60	30	50	0
2	30	40	60	10
3	35	70	50	20
4	30	55	30	−15
5	70	30	40	0
6	40	40	30	−25
7	45	35	50	20
8	35	55	40	−15
9	60	30	40	16
10	50	35	70	12
11	30	30	55	15
12	40	70	30	−17
13	30	40	40	−25
14	50	45	35	−18
15	36	39	48	12
16	65	46	58	0
17	46	66	52	−19
18	51	65	33	0
19	41	53	62	16
20	48	36	39	15
21	58	65	46	−11
22	80	45	55	0
23	38	59	42	10
24	55	80	45	−22
25	42	38	59	18
26	33	51	65	−12
27	62	41	53	18
28	50	60	30	−14

After implementing the data in the regression toolbox of the Minitab software, we get the equation of regression as follows.

$$C_4 = 23.0228 - 0.006183C_1 - 0.28508C_2 + 0.786367C_3 \qquad (5.2)$$

Where C_1= front obstacle distance (FOD), C_2= left obstacle distance (LOD), C_3= right obstacle distance (ROD), C_4=turning angle.

This governing equation is used to program a regression navigational controller and implement it for the optimization of the path travelled by the humanoid robot. As we are considering this as a global optimization problem, humanoid has the prior information of source and target positions. When it starts the journey toward its target, it just follows the target-seeking behavior and heads toward the target. When its sensors sense any obstacles in its path, it will calculate the FOD, LOD, and ROD distances. Then the proposed regression algorithm will start working. According to the algorithm, after checking that any of the distances is coming near to the threshold value, the program will calculate and give a C_4 value. After getting the turning angle value, the humanoid robot starts moving accordingly until any more obstacles come in its path or it reaches the target.

There are three types of reactive behaviors developed and implemented on the humanoid robot. Those are target-seeking behavior, obstacle-avoidance behavior, and barrier-following behavior, as depicted in the following Table 5.2. The first two behaviors are mandatory for the path planning, whereas the third one is a complimentary one. By applying the third one, the efficacy of the algorithm can be increased by reduction of power consumption. It can be introduced in some situations, for example, one long obstacle or barrier is present, and opposite to that is the target. In this case, the humanoid has to follow a straight path parallel to the barrier and only after crossing that it can go to the target. So here the barrier-following behavior works good and by doing this less power is consumed. The reason for less power consumption is that the microprocessor need not work until it crosses that barrier. When there is no obstacle found in the way of the humanoid, it can move easily toward the target following the target-seeking behavior. And when it is obstructed by any obstacle, it can follow the obstacle-avoidance behavior.

TABLE 5.2 Description of Various Reactive Behaviors Implemented by the Humanoid Robot.

Types of reactive behaviors	Explanation of the reactive behaviors	Activity of the robot
Obstacle Avoidance (OA)	To avoid the obstacles present in the path when sensed by the sensors of a robot	Turning angle is set accordingly to avoid the obstacles
Target Seeking (TS)	To seek for the target when there are no obstacles present in the path	Turning angle is adjusted accordingly to reach the goal position
Barrier Following (BF)	To follow a barrier when searching for a target if a series of obstacles are present near the robot	Robot moves in parallel to the barrier keeping a fixed turning angle

5.8 DEMONSTRATIONS OF THE REGRESSION NAVIGATIONAL CONTROLLER

The proposed regression navigational technique has been implemented in both the softwares for simulation and experiment. For simulation, a program was developed in Lua the language with the same logic, and for the experiments, it has been written in Python for the NAO robot. For simulation and experiments, environments of the same area have been considered with the same size of obstacles at the same positions. Lastly, both the simulation and experimental results have been analyzed, discussed, and compared to optimize the path of the humanoid robots.

5.8.1 SIMULATION AND EXPERIMENTS WITH NAO ROBOT

Virtual robot experimentation platform (VREP) is a simulation software best suited for the NAO robot navigation. It follows the programming language Lua, which is based on the ANSI C language. VREP is chosen here for the simulation of the humanoid due to its properties like collision detection, minimum distance calculation, and better path or motion planning. A program has been designed and written in the VREP software. The navigational controller for the biped robot is based on three main strategies which are target-seeking behavior, obstacle-avoidance behavior and barrier-following behavior. By considering the three main behaviors and regression algorithm, a program has been developed in the Lua language and deployed by the software. An environment (200 × 250 unit) with five static obstacles is designed and developed in the software for the simulation. Then simulations have been executed by taking a single NAO robot, and the target-seeking and obstacle-avoidance behavior has been observed. The following Figure 5.4 demonstrates the results obtained from the simulation platform.

To validate the effectiveness of the regression navigational controller in an environment with static obstacles, a comparison is required among the simulation and experimental results. For the experiments, NAO humanoid robot is taken as the platform for the programs, written with the logic of regression analysis for path planning. Numbers of real-time experiments have been carried out in an environment of size 200 × 250 cm for the NAO robot, taken similar to the simulation environment. An initial and final point is fixed in the environment with five static obstacles same as the simulation environment. The journey starts from the start point with a target-seeking behavior while the robot has sensory data about any obstacles. Based upon the training pattern, it tries to avoid the obstacles. After receiving the data from the ultrasonic sensors, various input parameters are recorded for the obstacle position and based upon the reactive behaviors, the robot follows a path which leads it to the target. The following Figure 5.5 demonstrates the results obtained from the experimental platform.

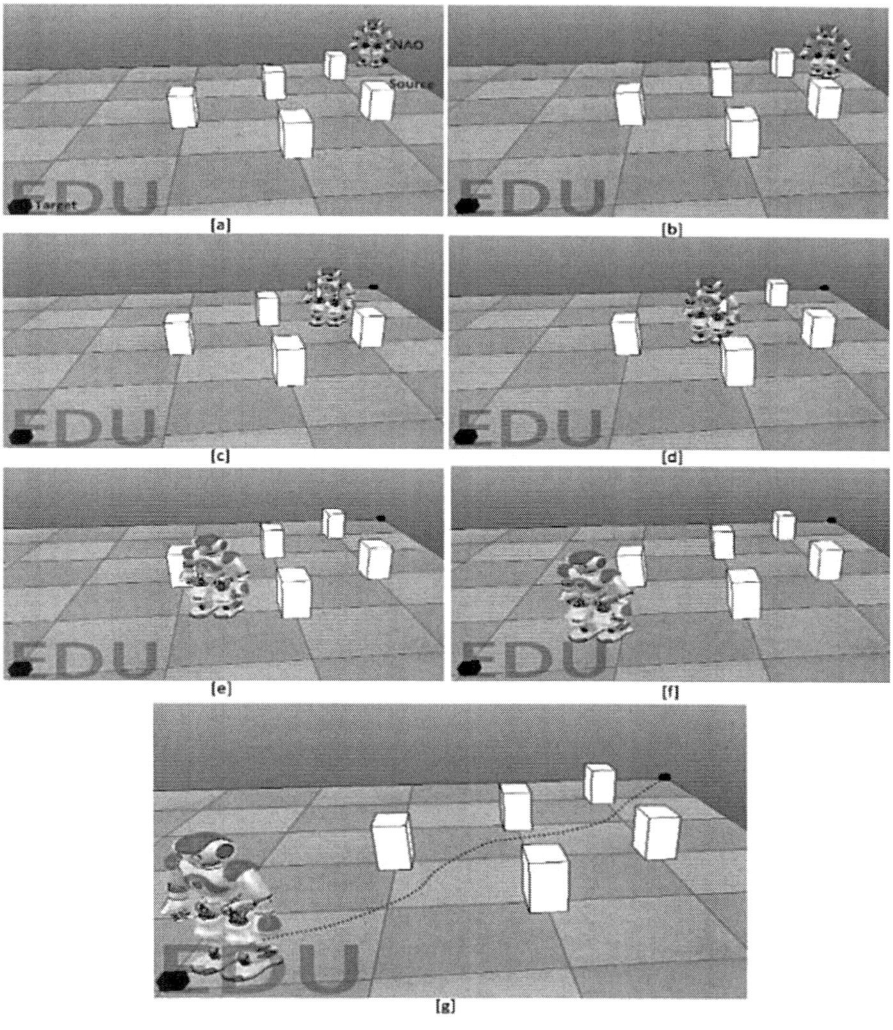

FIGURE 5.4 Simulation results for a humanoid NAO using a regression controller.

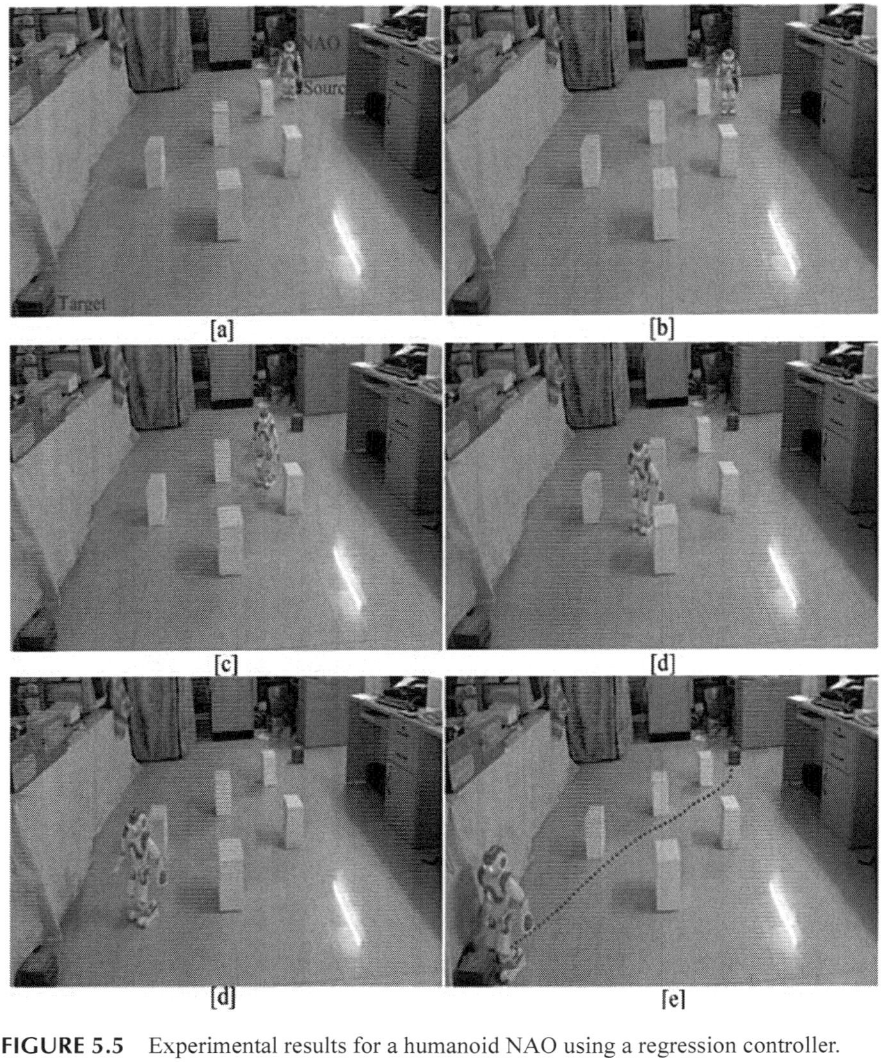

FIGURE 5.5 Experimental results for a humanoid NAO using a regression controller.

To show a comparison between the simulation results and the experimental results, the data is illustrated in the following Tables 5.3 and 5.4. It is observed that the path length and time required to reach the goal position show a relatively higher value in experiments as compared to simulated results. It happens due to some errors like the slipping effect between the floor surface and the feet of the robot, error in signal transmission that maybe in the data cable or the Wi-Fi connection, presence of friction at the surface of floor, and so forth. By a measuring tape, the path length is measured and by using a stopwatch, time is noted down and depicted in the following tables.

TABLE 5.3 Path Length Comparison.

No. of scenario	Path length covered during simulation (in cm)	Path length covered during the experiment (in cm)	% of error
1	269.88	286.15	5.69
2	380.53	404.90	6.02
3	337.39	360.26	6.35

TABLE 5.4 Time Required Comparison.

No. of scenario	Time taken during simulation (s)	Time taken during experiment (s)	% of error
1	33.90	36.11	6.12
2	47.79	51.08	6.44
3	42.45	45.58	6.86

5.9 RESULTS AND DISCUSSIONS

The regression controller implemented on the NAO robot follows the basic principles of the basic regression. It can successfully avoid the obstacles present in the path and reach the goal position. By the simulation and experiment of NAO in complex environments, parameters such as path length and time taken to cover the path have been recorded. The errors between the simulated data and experimental data have been calculated by taking the experimental data as the reference one. The calculated errors are well within the permissible limits. This implies that the regression technique is capable of giving satisfactory results for working in complex environments. The smooth navigational pattern followed by the NAO also proves that a significant amount of stability is present in the NAO and it can face complex terrains also.

5.10 CONCLUSIONS

Robots are becoming very essential for human assistance and path planning, with the navigational problem of the robots being the most discussed area of research. Humanoid robots are considered dominant over other forms of robots due to their configurative features and the ability to mimic human behavior. In the current investigation, path planning approach for a humanoid NAO robot has been considered. By using regression analysis as a classical technique, a regression controller was designed to be implemented into a NAO robot. By the use of the regression controller, the NAO was able to successfully avoid the obstacles present in the path and reach the goal position. The experiments were tested both in simulated and real environments. A comparison was also performed between both the environments and the errors calculated were well within the permissible limits. This technique can also be applied to perform multiple NAO experiments, where along with the static obstacles, dynamic obstacles are also present.

KEYWORDS

- **humanoid**
- **regression**
- **path planning**
- **navigation**

REFERENCES

1. Li,G.; Yamashita,A.; Asama,H.; Tamura,Y. An Efficient Improved Artificial Potential Field Based Regression Search Method for Robot Path Planning. In *IEEE International Conference on Mechatronics and Automation,* Aug 2012; pp 1227–1232.
2. Lazaro, J. L.; Gardel, A.; Mataix, C.; Rodriguez, F. J.; Martin, E. Adaptive Workspace Modeling, Using Regression Methods, and Path Planning to the Alternative Guide of Mobile Robots in Environments with Obstacles. In *7th IEEE International Conference on Emerging Technologies Factory Automation,* 1999; Vol. 1, pp 529–534.
3. Kashmiri, S.; Payandeh, S. Robot Navigation Controller: A Non-Parametric Regression Approach. *IFAC Proc.* **2010,** *43*(22), 22–27.
4. Asano, T.; Asano, T.; Guibas, L.; Hershberger, J.; Imai, H. Visibility-Polygon Search and Euclidean Shortest Paths. In *IEEE26th Annual Symposium on Foundations of Computer Science,* Oct1985; pp 155–164.
5. Benamati, L.; Cosma, C.; Fiorini, P. Path Planning Using Flat Potential Field Approach. In *IEEE 12th International Conference on Advanced Robotics,* July 2005; pp 103–108.
6. Jolly, K. G.; Kumar, R. S.; Vijayakumar, R. A Bezier Curve Based Path Planning in a Multi-Agent Robot Soccer System Without Violating the Acceleration Limits. *Rob. Auton. Syst.* **2009,** *57*(1), 23–33.

7. Keshmiri, S.; Payandeh, S. Regression Analysis of Multi-Rendezvous Recharging Route in Multi-Robot Environment. *Int. J. Soc. Rob.* **2012,** *4*(1), 15–27.
8. Keshmiri, S.; Payandeh, S. Multi-Robots, Multi-Locations Recharging Paradigm: A Regression Route Technique. *Proceedings of the 14th Lasted International Conference, Robotics And Applications*, Cambridge, MA, USA, 2009, pp 160–165.
9. Liang, T. C.; Liu, J. S.; Hung, G. T.; Chang, Y. Z. Practical and Flexible Path Planning for Car-Like Mobile Robot Using Maximal-Curvature Cubic Spiral. *Rob. Auton Syst.* **2005,** *52*(4), 312–335.
10. Masehian, E.; Sedighizadeh, D. Classic and Heuristic Approaches in Robot Motion Planning-A Chronological Review World Academy of Science. *Eng. Technol.* **2007,** *29*(1), 101–106.
11. Minguez, J.; Montano, L. Nearness Diagram (Nd) Navigation: Collision Avoidance in Troublesome Scenarios. *IEEE Trans. Rob. Autom.* **2004,** *20*(1), 45–59.
12. Papadopoulos, E.; Papadimitriou, I.; Poulakakis, I. Polynomial-Based Obstacle Avoidance Techniques for Nonholonomic Mobile Manipulator Systems. *Rob. Auton. Syst.* **2005,** *51*(4), 229–247.
13. Takahashi, O.; Schilling, R. J. Motion Planning in a Plane Using Generalized Voronoi Diagrams. *IEEE Trans. Rob. Autom.* **1989,** *5*(2), 143–150.
14. Hong, Z.; Liu, Y.; Zhongguo, G.; Yi, C. The Dynamic Path Planning Research for Mobile Robot Based on Artificial Potential Field. In *IEEE International Conference on Consumer Electronics, Communications and Networks (CECNet),* June 2011; pp 2736–2739.
15. Qi, N.; Ma, B.; Liu, X. E.; Zhang, Z.; Ren, D. A Modified Artificial Potential Field Algorithm for Mobile Robot Path Planning. In *IEEE 7th World Congress on Intelligent Control and Automation, WCICA 2008,* June 2008; pp 2603–2607.
16. Shi, P.; Zhao, Y. Global Path Planning for Mobile Robot Based on Improved Artificial Potential Function. In *IEEE International Conference on Automation and Logistics,* Aug 2009; pp 1900–1904.
17. Tingbin, C.; Qisong, Z. Robot Motion Planning Based on Improved Artificial Potential Field. In*3rd International Conference on Computer Science and Network Technology (ICCSNT),* Oct 2013; pp 1208–1211.
18. Bing, H.; Gang, L.; Jiang, G.; Hong, W.; Nan, N.; Yan, L. A Route Planning Method Based on Improved Artificial Potential Field Algorithm. In *International Conference on Communication Software and Networks (ICCSN),* May 2011; pp 550–554.
19. Sheng, J.; He, G.; Guo, W.; Li, J. An Improved Artificial Potential Field Algorithm for Virtual Human Path Planning. In *International Conference on Technologies for E-Learning and Digital Entertainment,* Aug 2010; pp 592–601, Springer: Berlin Heidelberg, 2010.
20. Hwang, Y. K.; Ahuja, N. A Potential Field Approach to Path Planning. *IEEE Trans. Rob. Autom.* **1992,** *8*(1), 23–32.
21. Kofinas, N.; Orfanoudakis, E.; Lagoudakis, M. G. Complete analytical inverse kinematics for NAO. In *Autonomous Robot Systems (Robotica), 2013 13th International Conference,* April 2013; pp 1–6, IEEE.

CHAPTER 6

TOPOLOGY OPTIMIZATION OF A SPUR GEAR

ARUN J. KULANGARA*, P. SUBHASH CHANDRA BOSE, and C. S. P. RAO

Department of Mechanical Engineering, National Institute of Technology, Warangal 506004, India

**Corresponding author. E-mail: arunjkulangara@gmail.com*

ABSTRACT

Topology optimization (TO) finds the best use of material for a body subjected to different loads with respect to some objectives. Additive manufacturing processes enable manufacturing of complex geometries with multimaterials. Useful distribution of material in a geometric domain to match target mass, displacement, and stiffness is made possible with TO. TO was performed by CAESS ProTOp software and Creo 2.0. In this work, TO of a spur gear is carried out. It is optimized for tangential force acting on the gear tooth. Stresses and displacements of topology-optimized spur gear are compared with a fully solid part. The displacements on the two parts are comparable. It is observed that after TO, stresses are reduced and a volume reduction of 25% is obtained.

6.1 INTRODUCTION

Topology optimization (TO), a mathematical tool used in the conceptual design stage, has been found to fulfill the part weight reduction problem by optimal distributing of the material throughout the components body.[1] Material distribution is done without any preconceived shape using different computational techniques for innovative and high-performance structures. Most of the TO problems are aimed at maximizing stiffness for a loading condition and within prescribed material usage. TO has wide application in aerospace and automotive sectors for lightweight structures. Reducing weight leads to less part manufacturing cost. The application of the TO in various fields of engineering will improve design cost and quality.

In additive manufacturing, a three-dimensional model is used to make objects by the process of joining materials layer upon layer. Parts of significantly greater complexity can be produced compared with traditional processes and this increased complexity generally does not have a significant effect on the cost of the process. This provides the designer with significantly greater design freedom. This chapter discusses the application of TO by considering a spur gear as computer-aided design (CAD) model for optimization.

6.2 LITERATURE REVIEW

TO is a relatively new field of structural optimization and has a great impact on the performance of structures. Bendsoe and Sigmund explain TO of solid structures as the determination of features such as the number, location, and shape of holes and the connectivity of the domain.[2] Investigation of the issues and opportunities for the application of TO methods for additive manufacturing has been carried out.[1] Moreover, conversion of topology-optimized output files to usable averaged Mises (AM) input data for production of mesoscale structures for realizing intermediated density regions is studied. Ground structure approach, solid isotropic material with penalization, homogenization, level set method, evolutionary structural optimization, and genetic algorithms are few popular methods proposed for implementing TO. Most of these methods, first developed for structural engineering problems, are now used for fluid flow optimization, vibration analysis, and heat transfer areas.[3]

TO of Airbus A320 nacelle hinge bracket was carried out.[4] It shows that significant proportion of weight could be saved in the part while reducing maximum stress and maintaining stiffness. Altair OptiStruct software was used for TO of a brake pedal and weight of the part was reduced by 22% compared to the original design.[5] In-depth research about TO has been carried out on the design of an automobile engine bracket[6] in which 40% mass got reduced. It is observed that the stiffness has been greatly improved with better load-carrying capacity.

6.3 ASSUMPTIONS

For developing the design and TO model, the following assumptions are taken:

- Full load is applied on the surface of each single tooth in a static condition.
- The load is distributed uniformly across the full-face width.

6.4 DESIGN CONSIDERATIONS

In this chapter, two spur gears of similar material are taken. They are required to transmit the power to the parallel shaft while mating. Moreover, the gears have similar number of teeth and the material taken is steel. The upper gear is subjected to a torque of 1,694,772.5 N-mm at shaft. As per Rathore and Tiwari,[7] the data given are as follows:

- Pitch circle = 63.5 mm
- Pressure angle = 20°
- Number of teeth N = 20
- Radius of addendum = 69.85 mm
- Radius of dedendum = 55.88 mm
- Shaft radius = 31.75 mm
- Root fillet radius = 2.54 mm
- Face width = 25.4 mm
- Torque applied at shaft = 1,694,772.5 N-mm
- Young's modulus = 200 GPa
- Poisson's ratio = 0.3
- Face width b = 25.4 mm
- Pitch diameter = 127 mm

$$\text{Module } m = \frac{\text{Pitch circle diameter}}{\text{No. of teeth}} = \frac{25.4}{127} = 6.35\,\text{mm}$$

$$\text{Tangential Force } F_t = \frac{2\times M_t}{P_d} = \frac{2\times 1694772.5}{127} = 26689.33\text{ N}$$

6.5 MATERIAL PROPERTIES

The detailed material data is given in Table 6.1.

TABLE 6.1 Material Properties of Spur Gear.

Sr. no.	Parameters	Value	Unit
1	Young's modulus	200	GPa
2	Poisson's ratio	0.3	
3	Tensile strength ultimate	1962	MPa
4	Tensile strength yield	1500	MPa
5	Density	7850	kg/m^3
6	Thermal expansion	11×10^{-6}	$°C^{-1}$

6.6 PROCESS OF TOPOLOGY OPTIMIZATION (TO)

A CAD model of spur gear was modeled initially on which TO was carried out (Fig. 6.1).

FIGURE 6.1 Computer-aided design model of spur gear.

6.6.1 DESIGN

- During the design stage, a CAD model with volume regions was created in a solid modeling software.
- Loads, constraints, and materials were defined as per the design considerations.

- Meshing was generated with maximum 0.01-mm and minimum 0.01-mm mesh size resulting in nearly 7,900,000 elements.
- The mesh model was exported for finite element method analysis.

6.6.2 *TOPOLOGY OPTIMIZATION*

- The main objective of the study is volume reduction.
- The primary target of 30% volume reduction was considered and the optimization cycles were carried out.
- The part was optimized for reducing the displacement to a value near to that of the completely solid part after primary volume reduction.
- Stresses and displacements were taken as secondary objectives while optimization.
- Finally, optimization cycles were repeated for limiting displacement to a value of 0.8 mm.

6.7 TO RESULTS

TO of spur gear has resulted in a volume reduction of 25.6%. Displacements and stresses on the solid and optimized part during optimization cycle are tabulated in Table 6.2.

TABLE 6.2 Stresses of Solid and Optimized Models.

Sr. no.	Model	Displacement (mm)	Averaged Mises stress (MPa)	Signed average effective stress (MPa)
1	Solid model	0.7348191	2722	±2722
2	Optimized model	0.7992178	2722	±2722

For the same loading condition, the displacement was constrained to 0.8 mm and it is observed that a volume reduction of 25.6% is obtained. The stresses on the optimized spur gear are compared with the fully solid model. For optimization, each tooth is loaded simultaneously. For optimization process, the part is loaded more than 10 times of practical loading condition. Therefore, the stresses on the solid and optimized parts in finite element analysis during optimization are above the yield strength. Here, the work compares solid part with optimized part of similar material and under same boundary conditions, so the material properties need not be considered. It is observed that the stresses on the optimized spur gear are less than fully solid spur gear.

Figure 6.2 shows averaged Mises stress for the optimized and solid part. Figures 6.3–6.5 are displacement, averaged Mises stress, and signed average effective stress, respectively, for the optimized spur gear, where:

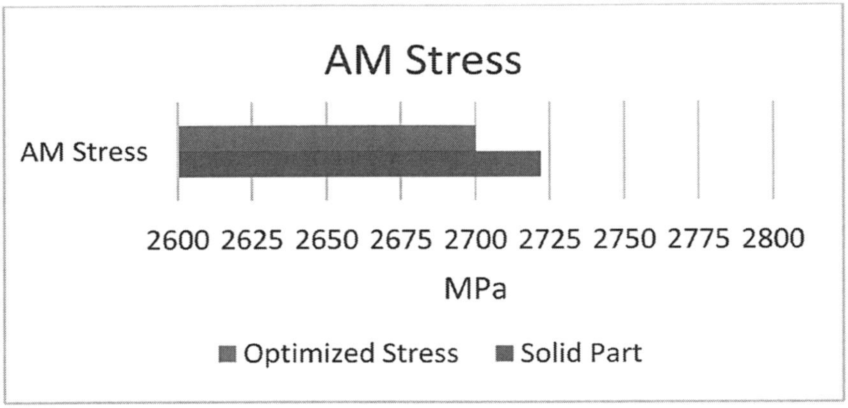

FIGURE 6.2 Averaged Mises (AM) stress for the solid and optimized model.

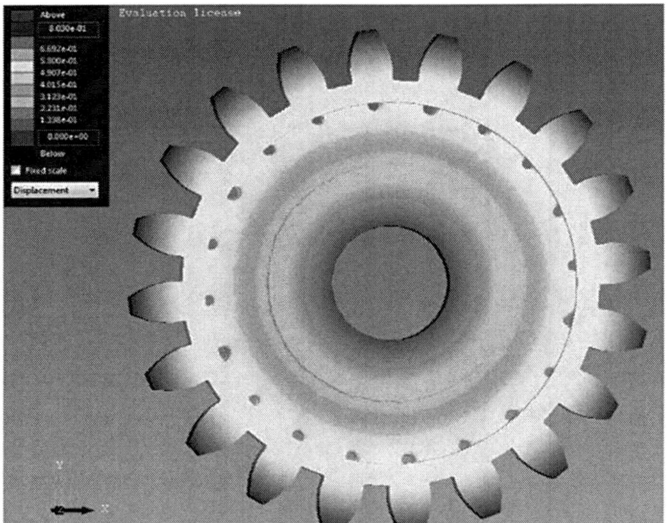

FIGURE 6.3 Displacement

FIGURE 6.4 AM stress

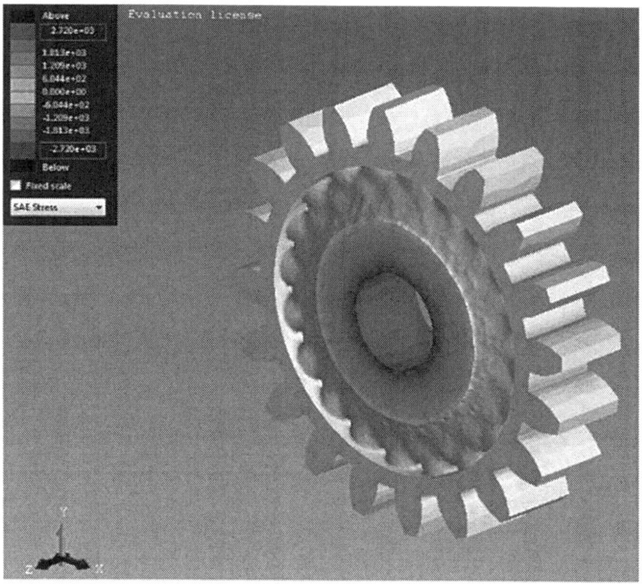

FIGURE 6.5 Signed average effective stress

AM stress being "averaged Mises" stress. (Stress quantity computed at finite element nodes by assembling contributions of averaged Mises stress computed in individual finite elements).

SAE stands for "signed average effective" stress. Here the effective stress is simply the Mises stress; this quantity is actually AM stress multiplied by the sign of the volumetric deformation. (AM stress with a positive sign if the material is mainly in tension [red color] or a negative sign if the material is mainly in compression [blue color]).

6.8 CONCLUSION

TO of spur gear for volume reduction as primary objective was carried out. A volume reduction of 25.6% is obtained with a constraint of 0.8-mm displacement, which is a mass reduction of 1.3 kg, from 5 to 3.7 kg. Stresses after TO are less than that of fully solid spur gear even after 25.6% volume reduction. Optimized part puts forward a lightweight structure. Significant reduction in weight, material, and manufacturing cost can be achieved through TO. Mechanical simulation and practical analysis on optimized spur gear should be carried out for in-depth research on TO.

KEYWORDS

- **topology optimization**
- **additive manufacturing**
- **weight reduction**

REFERENCES

1. Rezaie, R.; Badrossamay, M.; Ghaie, A.; Moosavi, H. Topology Optimization for Fused Deposition Modeling Process, the Seventeenth Cirp Conference on Electro Physical and Chemical Machining (ISEM). *Proc. CIRP* **2013,** *6,* 521–526.
2. Bendsoe, M.; Sigmund, O. *Topology Optimization Theory, Methods and Applications,* 2nd ed.; Springer: Germany, 2003.
3. Sundararajan, V. G. Topology Optimization for Additive Manufacturing of Customized Meso-Structures Using Homogenization and Parametric Smoothing Functions (Unpublished Doctoral Dissertation), The University of Texas at Austin, 2010.

4. Tomlin, M.; Meyer, J. Topology Optimization of an Additive Layer Manufactured (ALM) Aerospace Part, *The 7th Altair CAE Technology Conference,* 2011.
5. Sudin, M. N.; Tahir, M. M.; Ramli, F. R.; Shamsuddin, S. A. Topology Optimization in Automotive Brake Pedal Redesign. *Int. J. Eng. Technol. (IJET)* **2014,** *6*(1), 398–402.
6. Wu, P.; Ma, Q.; Luo, Y.; Tao, C. Topology Optimization Design of Automotive Engine Bracket. *Energy Power Eng.* **2016,** *8,* 230–235.
7. Rathore, R. K.; Tiwari, A. Bending Stress Analysis and Optimization of Spur Gear. *IJERT Int. J. Eng. Res. Technol.*, **2014,** *3*(5), ISSN: 2278–0181.

CHAPTER 7

DYNAMICS AND STRUCTURAL ANALYSIS OF A REENTRY MODULE FOR A SAMPLE RETURN MISSION TO 2010TK7

ATHOTA RATHAN BABU[1,*], NEVILLE ANTON GEORGE[2], V. HARSHA VARDHAN REDDY[2], and D. GOVARDHAN[1]

[1]*Department of Aeronautical Engineering, Institute of Aeronautical Engineering, Hyderabad, Telangana, India*

[2]*Department of Aerospace Engineering Indian Institute of Space Science and Technology, Trivandrum, Kerala, India*

[*]*Corresponding author. E-mail: athota.aero@gmail.com*

ABSTRACT

The mission detailed in this chapter intends to perform a soft landing on the object, collecting some sample material, and safely return it back to Earth for further scientific studies. This chapter highlights the dynamics of reentry module using 3-degree-of-freedom code and structure analysis using Abaqus 6.13, which is a commercial finite element method-based software. The major loads acting in the reentry module are the inertia loads due to acceleration, drag during reentry, and internal pressure in propellant tanks. The maximum longitudinal (axial) deceleration expected is 13 g and the maximum drag expected is 40 kN. Therefore, we have constraints on maximum "g" level, which is governed by initial flight path angle and entry velocity of the module. An optimal flight path angle has to be chosen for obtaining the "g" level within the limits. Loads obtained from the trajectory of reentry module give input to perform structural analysis. The major constraint in this aspect is maximum allowable stress, thickness of the shell, and thickness of thin sheet used for partition of propellant tanks. Therefore, an optimal solution has to be carried out for meeting all constraints.

7.1 INTRODUCTION

The Earth-based Trojan asteroid, later named as 2010TK7, was discovered by the Wide-field Infrared Survey Explorer space telescope in the year 2010. The detailed orbit analysis carried out later identified it as an object sharing the Earth's orbit and revolving around the Sun at the L4 Lagrangian point of the Earth. Though many Trojans were identified for the outer planets, 2010TK7 became the first Earth-based Trojan to be identified and studied. Most Trojans are expected to be in the L4 or L5 points of the Earth. Hence, they are most likely to appear in the daytime sky and are extremely difficult to be identified and observed from Earth's surface. The study of this object can not only throw light onto the origin of the Earth and the Solar System but also can be of huge socioeconomical significance. Though the object is comparatively nearer to the Earth, it is hard to carry out any study using Earth-based or space telescopes due to its small size and rather odd orbital parameters. Detailed analysis of the orbital parameters of the target object revealed that the mission would be characterized by the very large delta-V requirement. The mission detailed in this chapter intends to perform a soft landing on the object, collecting some sample material, and safely return it back to the Earth for further scientific studies.

This chapter concerns with the dynamics and structural analysis of a body which is entering into the Earth. A 3-degree-of-freedom code is been developed in MATLAB for evaluation of dynamics. In this mission, reentry is ballistic reentry at an altitude 140 km with an entry velocity of 8 km/s. The reentry module is attached to the propulsion module and carries a payload to capture material from particle cloud created by the impact of propulsion module with debris cloud. After material capture, the reentry module provides thrust to return back to the Earth and reenters and lands on the Earth's surface. The propulsion module and reentry module are kept inside the heat shield, and hence, it is protected from all aerodynamic loads in atmospheric flight regime of the launch vehicle. The major load coming on the reentry module is the inertia loads due to acceleration and drag load during reentry. The reentry module consists of the outer isogrid shell, thermal protection, propellant tanks (inner volume partitioned by Al alloy sheets), feed lines, engine, navigation and guidance, and electronic systems and payload (material capture system).

7.2 METHODOLOGY

The equations of motion[1,2] for the atmospheric phase are standard and have been developed in a non-inertial frame of the planet. The kinematic equations of motion of point mass are given by:

$$\dot{\delta} = \frac{V\cos\gamma\cos\chi}{R\cos\lambda} \tag{7.1}$$

$$\dot{\lambda} = \frac{V\cos\gamma\sin\chi}{R} \tag{7.2}$$

$$\dot{R} = V\sin\gamma \tag{7.3}$$

The kinetic equations of motion of point mass are given by[3]

$$\dot{v} = -g\sin\gamma + \frac{F_Y}{m} + \Omega^2 R\cos\lambda(\sin\gamma\cos\lambda - \cos\gamma\sin\chi\sin\lambda) \tag{7.4}$$

$$\dot{\gamma} = \left(\frac{V}{R} - \frac{g}{V}\right)\cos\gamma + \frac{F_X}{V\,m} + 2\Omega\cos\chi\cos$$

$$\lambda + \frac{\Omega^2 R\cos\lambda(\cos\gamma\cos\lambda + \sin\gamma\sin\chi\sin\lambda)}{V} \tag{7.5}$$

$$\dot{\chi} = \frac{-V\cos\lambda\cos\chi\tan\lambda}{R} + \frac{F_Z}{m\,V\cos\gamma} + 2\Omega$$

$$(\tan\gamma\sin\chi\cos\lambda - \sin\lambda) - \frac{\Omega^2\cos\chi\cos\lambda\sin\lambda}{V\cos\gamma} \tag{7.6}$$

Figure 7.1 shows ballistic reentry of the module from an altitude of 140 km (circular orbit) to the surface of the Earth. The major load acting on the module during reentry is drag and heat flux. Since drag is proportional to the square of module velocity, the deceleration also occurs proportionally to it. Since heat flux is proportional to the cube of module velocity, the material used for the module is very important. The reason for using the blunt body for reentry is to have detached shock, such that heat loads do not come on the body. During reentry, the heat load is ejected into the atmosphere with the use of blunt body. In structural aspects, the module's stress distribution plays a major role. Figure 7.2 shows the computer-aided design model of reentry module; during reentry, the base area is taken as maximum projected area.

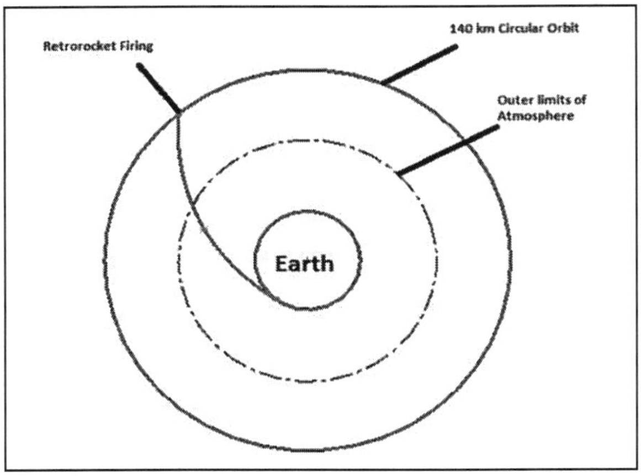

FIGURE 7.1 Ballistic reentry.

FIGURE 7.2 Computer-aided design model of the reentry module.

7.3 RESULTS AND DISCUSSION

Input data for trajectory analysis are mass of module, 250 kg, surface area, 1.7671 m^2, temperature of air, 298 K, coefficient of drag, 1.1 (assumed constant), nose radius, 0.32 m, entry velocity, 8 km/s, and initial flight path angle, $-0.6°$.

Figure 7.3 shows a 4-node doubly curved shell element which is used for meshing the element. Figure 7.4 explains about deceleration level varying with altitude. Since reentry has started from 140 km, the deceleration will remain constant till 100 km, because in that regime, the density of atmosphere is very low. As the density of atmosphere starts increasing from 100 km, the deceleration level starts increasing till maximum value, because of high velocity and increase in density of atmosphere. It increases till maximum value and decreases because in this regime, the decrement of velocity is dominant.

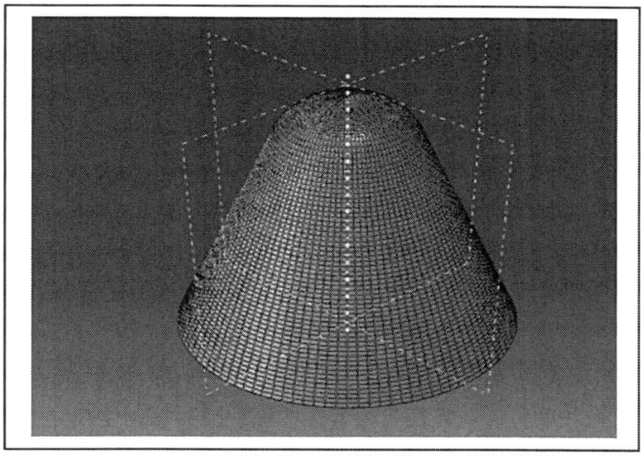

FIGURE 7.3 Finite element model of the reentry module.

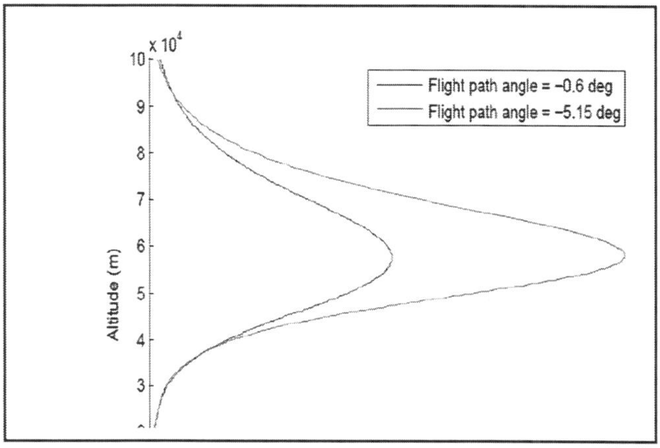

FIGURE 7.4 Deceleration versus altitude.

Similarly from Figures 7.5, 7.7, and 7.8, the dominance of increase in density in one regime and dominance of velocity in other regime are observed. Initial flight path angle is an interesting parameter which affects the maximum deceleration level and heat load.[1] From Figures 7.4–7.8, it is seen that steepest entry has maximum deceleration level, maximum drag, maximum heat flux, and minimum time for landing. The loads which are obtained during trajectory analysis will be input loads for structural analysis of module.

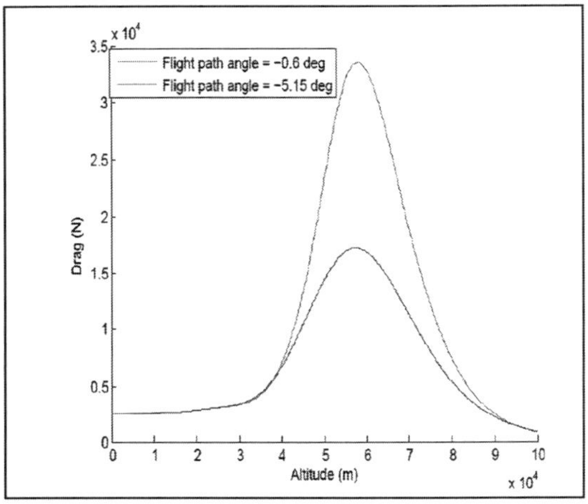

FIGURE 7.5 Altitude versus drag.

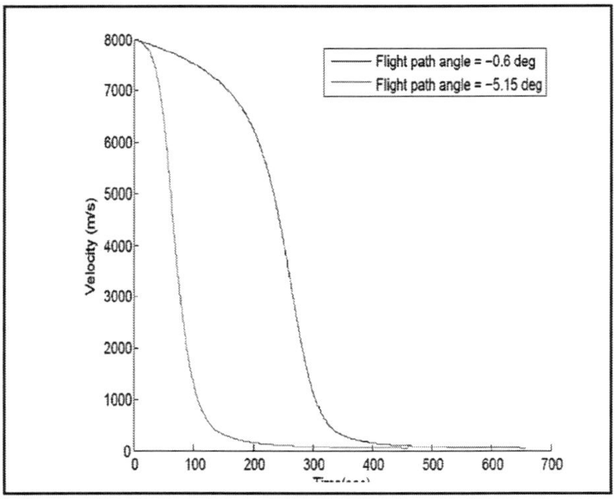

FIGURE 7.6 Time versus velocity.

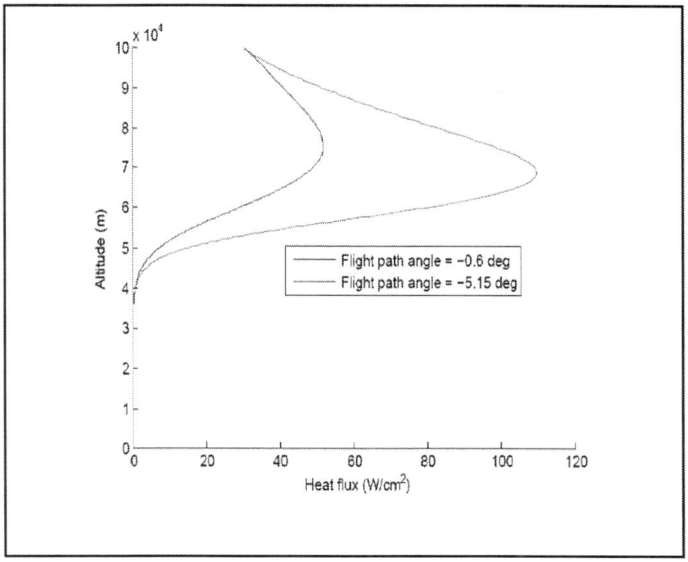

FIGURE 7.7 Heat flux versus altitude.

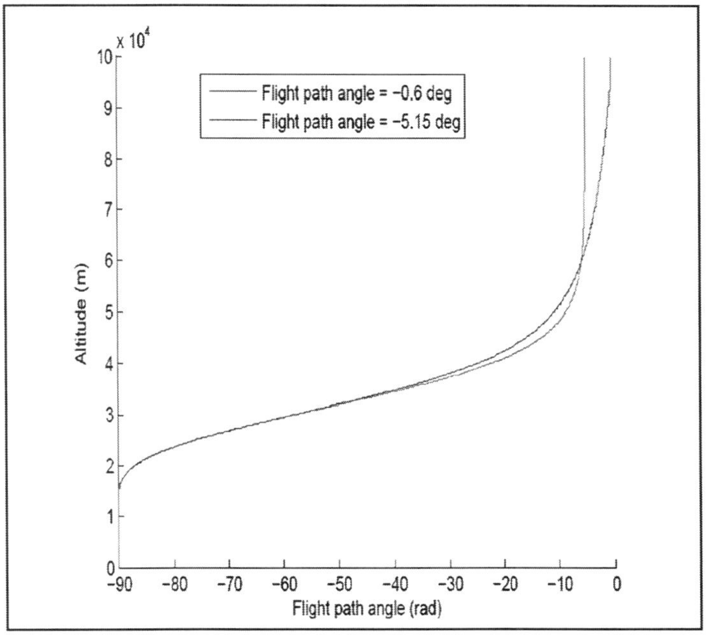

FIGURE 7.8 Flight path angle versus altitude.

Figure 7.3 shows the outer shell which is made up of isogrid shell with 15-mm equivalent thickness. The base is isogrid shell with 30-mm equivalent thickness. Partitioning for propellant tanks is a thin sheet of 3-mm thickness. The material of both isogrid and thin sheet is aluminum alloy AA 2219[4] in T87 condition. Maximum allowable stress is 200 N/mm^2. Finite element analyses are carried out for complete model. Abaqus 6.13 commercial finite element method-based software is used for analyzing the structure. Figures 7.9 and 7.10 show the stress and displacement distribution in the atmospheric regime.

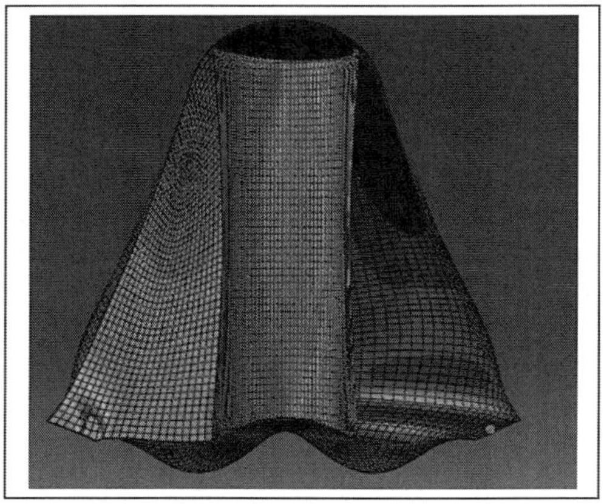

FIGURE 7.9 Stress distribution in reentry module.

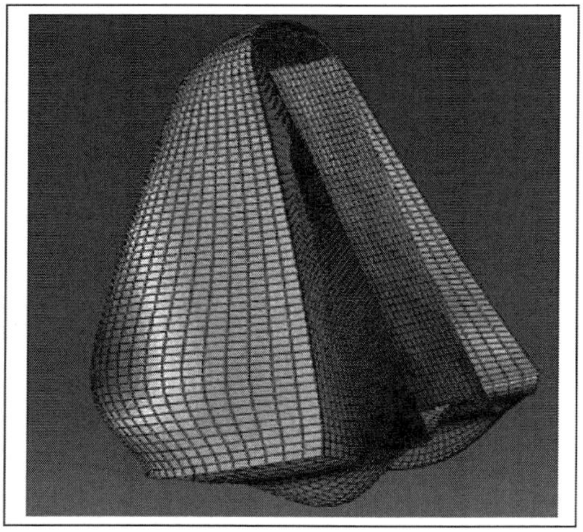

FIGURE 7.10 Displacement distribution in reentry module.

The structure is fixed at its linkage to the propulsion module and 15-g axial load and 5-bar tank pressure were applied to the model. Stress and displacement distributions were obtained. A 40-kN drag (22-kPa downward pressure on top surface) and 2.5-bar tank pressure were applied to the model. Stress and displacement distributions were obtained. Maximum stress is 73 MPa. Maximum displacement is 9.8 mm.

7.4 CONCLUSION

From the above results, it is seen that flight path angle plays a major role in determining the maximum deceleration and heat flux. Therefore, a steepest dip of reentry module is not preferred. This led to a reduction of g loads and drag force acting on the module, which, in turn, resulted in stress and displacement distribution within limits. In future, more optimization studies, interface design, and assembly details are to be focused in the detailed design phase.

KEYWORDS

- **non-inertial frame**
- **reentry**
- **stress**
- **displacement**

REFERENCES

1. Regan, F. J.; Anandakrishnan, S. M. Flow Field Description; Equations of Motion of a Point Mass. In *Dynamics of Atmospheric Re-entry,* 1st ed.; Przemieniecki, J. S., Eds.; AIAA Education Series; AIAA: Washington, DC, 1993; pp 318–321.
2. Kumar, M.; Tewari, A. Trajectory and Attitude Simulation for Aerocapture and Aerobraking. *J. Spacecr. Rockets* **2005,** *42*(4), 684–693.
3. Vinh, N. X. Equations of Motion, In *Flight Mechanics of High Performance Aircraft,* Cambridge Aerospace Series 4; Cambridge University Press: New York, **1993**.
4. Bruhn, E. F. *Analysis and Design of Flight Vehicle Structures;* Gran Corporation: Whitney, Texas, Second Edition, 2008.

CHAPTER 8

DESIGN AND ANALYSIS OF VARIOUS HOMOGENEOUS INTERCONNECTED SCAFFOLD STRUCTURES FOR TRABECULAR BONE

V. PHANINDRA BOGU[1,*], M. N. MADHU[1,3], Y. RAVI KUMAR[1,4], and K. ASIT KUMAR[2,5]

[1]*Department of Mechanical Engineering, NIT Warangal, Warangal, Telangana, India*

[2]*Department of Metallurgical and Material Engineering, NIT Warangal, Warangal, Telangana, India*

[3]*Madhumn10@gmail.com*

[4]*raviykumar@yahoo.com*

[5]*asit@nitw.ac.in*

[*]*Corresponding author. E-mail: phanibogu@gmail.com*

ABSTRACT

Tissue engineering aims at the development of biological substitutes that restore, maintain, and improve tissue function. Hard tissue (bone) damage occurs due to trauma and diseases and has to be regenerated or replaced with another suitable tissue or implants. Conventional treatment techniques such as autograft and allograft procedures are expensive, complicated, and infectious. Hence, 3D printed scaffolds are used to overcome these complications and problems. Scaffolds provide temporary structural support for cell attachment and the subsequent tissue growth. In this work, homogeneous interconnected (pores) scaffold structures were modeled by varying pore size, porosity, strut cross section, and by keeping the same unit cell size. Computational fluid dynamics (CFD) analysis was done for different scaffold structures by considering Darcy's equation to know the permeability of the scaffold with blood properties. Studies were carried out to find out the impact of the design parameters such as porosity, strut cross section, and size on the scaffold functional factor like permeability. Through the ANSYS software, analysis is done to calculate the nature and properties of fluid flow through the scaffold structure. Permeability is calculated using Darcy's permeability equation for incompressible laminar flow equation and was found to be in the range of 4.8×10^{-9} to 2.3×10^{-8} m^2 for strut cross section of 300–500 μm. Results showed that strut cross section and porosity were the influencing parameters and play an important role in the scaffold design.

8.1 INTRODUCTION

Cancellous bone, also known as spongy or trabecular bone, is one of the two types of bone tissue found in the human body. The trabecular structure is found at the ends of long bones as well as in the pelvic bones, ribs, skull, and the vertebrae in the spinal column. It is very porous and contains the red bone marrow, where blood cells are made. It is weaker and easier to fracture than the cortical bone, which makes up the shafts of long bones. Conventional treatment techniques such as autograft (a patient's own tissue) and allograft (tissue from another person) that are used to repair or restore damaged tissues are expensive, complicated, and infectious.[1] To overcome these problems 3D printed scaffolds are developed.[2,3] Advances in fabrication technologies now enable the strategic designing of scaffolds with complex, biomimetic structures and properties. Tissue engineering scaffold assists as a medium to facilitate cell adhesion and proliferation providing temporary mechanical support to newly grown tissue.

Tissue engineering aims to produce patient-specific biological substitutes to overcome the limitations of conventional clinical therapies. Fabrication of scaffolds using additive manufacturing techniques has more popular advantages like control over the pore size, establishment of homogeneous structures, and interconnected pores. Custom-made scaffolds could be developed by appropriate incorporation of a computational model and rapid prototyping (RP) technology. The microstructure of the scaffold determines the cell adhesion and tissue vascularization. Properties such as compressive strength (to ensure the mechanical stability of the scaffold structure) and permeability[4,5] (to ensure the required transportation of nutrients, gases, and allow cell proliferation) of scaffolds are considered for a successful design of scaffolds.

8.2 MATERIALS AND METHODS

The important or influencing parameters which need to be considered while designing scaffolds for any biomedical application, which in turn controls important factors like biocompatibility, biodegradability, mechanical properties, manufacturing technology, and so forth are discussed below (Table 8.1).

TABLE 8.1 Permeability Studies from Several Authors, using Different Biomaterials, Pore Sizes and Porosities.

Authors	Materials	Pore size (avg. μm)	Porosity (%)	Permeability (m^2)
Sanz-herrera et al. (2009)[6]	Bio glass-based foams	510–720	90–95	1.96×10^{-9}
Li et al. (2003)[7]	Z-Biphasic calcium phosphate (BCP); D-BCP I-BCP Hydroxyapatite (HA)-60 HA-50	565,300 528 450 250	75,78 60 60 50	[0.018×10^{-9}; 0.10×10^{-9}] 0.35×10^{-9} 0.37×10^{-9} 0.075×10^{-9}
Lee and Wang (2006)[8]	Poly(propylene fumarate) (PPF)	300 600 900	7–20 38–50 50–65	[0.02×10^{-9}–0.09×10^{-9}] 0.55–1.50×10^{-9} 1.50–2.3×10^{-9}
Jeong et al. (2011)[9]	Poly(1,8-octanediol–co-citrate) (POC)	900 900	50 60	0.351×10^{-9} 4.74×10^{-9}
Truscello et al.[10]	Ti alloy (Ti6Al4V)	750 800 900	56.7 70.5 79.6	$0.052 \times 10^{-9} \pm 3.9 \times 10^{-12}$ $1.69 \times 10^{-9} \pm 1.78 \times 10^{-10}$ $3.61 \times 10^{-9} \pm 8.5 \times 10^{-10}$

8.2.1 PORE CROSS SECTION

To study the effect of cross section on permeability, different strut sections are chosen (circular, hexagonal, and square).

8.2.2 PORE SIZE

Scaffolds with a smaller pore size or a higher pore density degraded slower than the scaffolds with a larger pore size or lower pore density (controls the degradation rate).[11] Temenoff et al. (2000) used pore size in the range 300–500 μm.

8.2.3 POROSITY

The complex structure of the pores assists in guiding and promoting new tissue formation. Materials with high porosity[12] enable the effective release of biofactors such as proteins, genes, or cells and provide good substrates for nutrient exchange.[6]

$$\text{Porosity} = 1 - \frac{V_{scaffold}}{V_{unitcell}} \tag{8.1}$$

8.2.4 UNIT CELL SIZE

Unit cell size of all models was kept uniform to have homogeneity in the dimension.

8.2.5 SURFACE-AREA-TO-VOLUME RATIO

Ratio defines the effectiveness of cell attachment and proliferation. Higher ratio is better because area available for the cell culture is more.

8.2.6 MODELING OF SCAFFOLDS

Scaffolds are modeled by following a unit cell-based approach or lattice technique. The geometrical properties of the scaffold required for the design are tabulated in Table 8.2. 3D computer-aided design (CAD) modeling software is used to model scaffolds. Unit cell with uniform cross sections (circular, hexagonal and square) and architecture is maintained to have homogeneous and interconnected pores as shown in Figure 8.1.

The 3D scaffold structures are developed through repeating the unit cells in x, y, z directions. These repetitive unit cells are unioned to get the required 3D scaffold structures with variable dimensions and cross-sections, which is shown in Figures 8.2 and 8.3.

TABLE 8.2 Geometrical Features of Scaffolds Chosen for Modeling.

Strut cross section	Strut size (mm)	Unit cell volume(mm^3)	Scaffold volume (mm^3)	Porosity (× 100%)	Surface area (m^2)
Circular (cir)	0.3	1	0.1738	0.826	2.4878
	0.4	1	0.2865	0.714	3.1662
	0.5	1	0.4122	0.587	3.7692
Hexagonal (hex)	0.3	1	0.1466	0.853	2.3869
	0.4	1	0.2436	0.756	3.0433
	0.5	1	0.3541	0.646	3.6302
Square (squ)	0.3	1	0.2160	0.784	3.0600
	0.4	1	0.3520	0.648	3.8400
	0.5	1	0.5014	0.498	4.5054

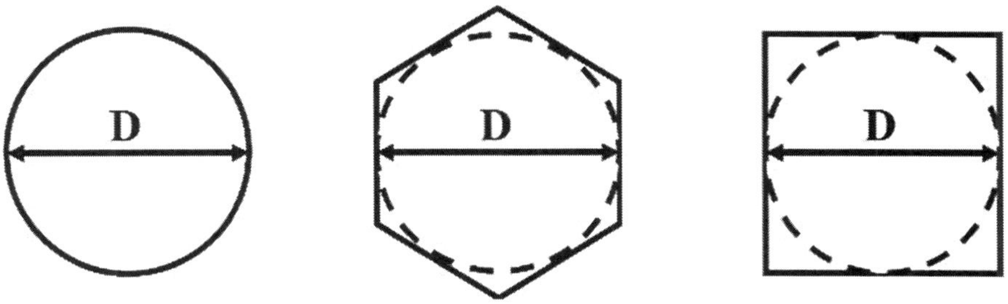

FIGURE 8.1 Strut cross sections/dimensions.

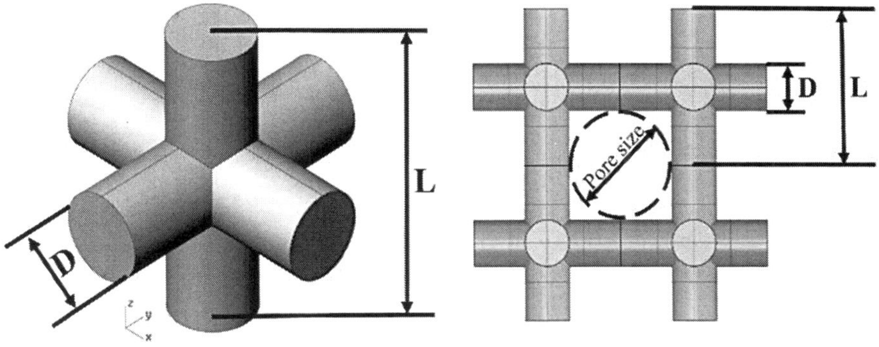

FIGURE 8.2 Unit cell dimensions.

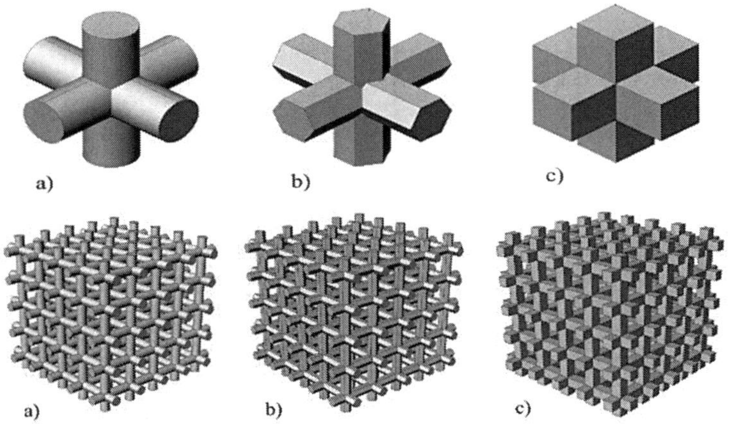

FIGURE 8.3 Scaffolds (a) circular, (b) hexagonal, (c) square, respectively, with dimensions $10 \times 10 \times 10$ mm^3.

8.3 ANALYSIS

A computational model to simulate porous and mechanical characteristics of 3D tissue engineering scaffolds was developed. The simulation is performed by manipulating the design inputs and analyzing the influences of change of the parameters on porous and mechanical characteristics of the scaffolds.

Figure 8.4 shows the setup modeled for the analysis of scaffold structure for fluent studies to find out the pressure gradient, velocity profile, and other relevant data. The setup includes an inlet, an outlet, a wall, and a fluid section. The fluid domain is of dimensions $40 \times 10 \times 10$ mm^3 and is 3D modeled in the CAD modeling software. Scaffold structure has been subtracted from the domain to get the setup for fluent analysis by maintaining the upstream and downstream accordingly.

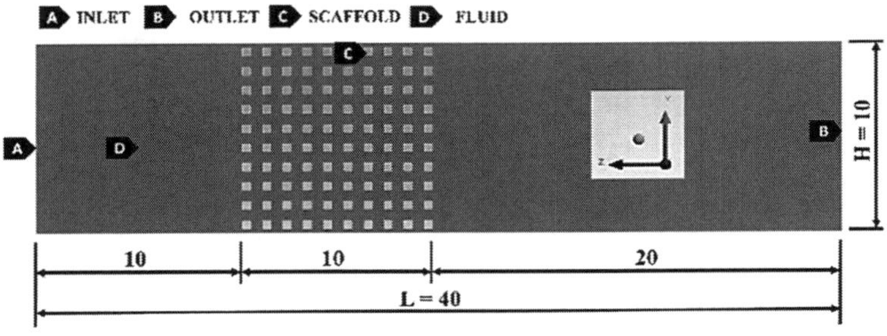

FIGURE 8.4 Complete setup for fluid flow analysis with boundaries.

8.3.1 COMPUTATIONAL FLUID DYNAMICS (CFD) ANALYSIS

CFD approach to find permeability: The Navier–Stokes equations, mass and momentum conservation were solved for the incompressible laminar flow by means of a finite volume discretization approach used by the ANSYS Fluent. The governing equations are solved simultaneously across all domain nodes. A cascade of successively coarser grids allows the solution information to propagate rapidly across the computational domain. The linear set of equations that arises by applying the finite volume method to all elements in the domain is a set of discrete conservation equations. The convergence criteria used in the simulations performed in this study was 1×10^{-9}. This factor was used to reduce the initial residual mass flow during the simulation progress.[14] The properties of the fluid considered were similar to that of blood, that is, viscosity of blood is 0.0035 Pas and density of blood is 1080 kg/m^3.

The fluid flows in at 0.1 m/s through the model outlet which is at zero gauge pressure, assuming stationary conditions as well as rigid walls. No-slip conditions were imposed at the walls of the scaffold as well as at the channel walls. Steady-state Navier–Stokes equations were used to describe the flow problem. The pressure drop across the scaffold was computed by means of the commercial software package for all models and used to calculate the permeability coefficient.

Darcy equation: Permeability is a measure of the ability of a fluid medium to flow through it.

$$k = \frac{Q\mu l}{A\Delta P} \tag{8.2}$$

Q is the volumetric flow rate, ΔP is the pressure drop, A and L are the cross-sectional area and the length of the porous material, respectively, and μ is the dynamic viscosity of the fluid.[15]

8.4 RESULTS

The streamline velocity profiles of fluid particles flowing through the scaffold along with the inlet, outlet, and fluid wall is shown in Figure 8.5. Here, fluid is assumed to undergo a laminar steady flow and is incompressible. The variation of pressure or pressure gradient generated due to the abstraction in flow medium can be observed from Figure 8.6. This is due to the introduction of scaffold in the fluid medium to find out its permeability by measuring the pressure gradient across it.

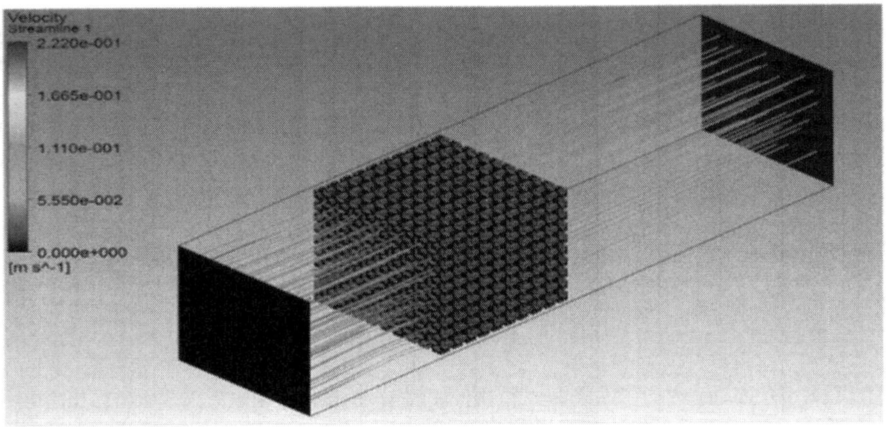

FIGURE 8.5 Variation of streamline velocity along the scaffold.

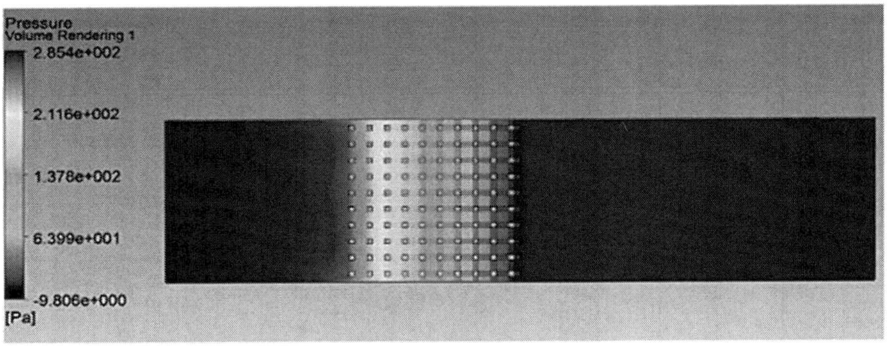

FIGURE 8.6 Variation of pressure of fluid flowing through the scaffold.

Darcy's equation is used to calculate the permeability of the scaffold. The process of measuring the pressure gradient repeated for various specimens and results is plotted for the pressure gradient versus the strut size which is reported in Figure 8.7.

Permeability of the scaffold depends on the size, shape, architecture and porosity of the medium in which the fluid is flowing. In our analysis, size and shape of the strut has been varied to study the intrinsic permeability of the scaffold structure. Results showed that as we increase the strut size of the scaffolds, the intrinsic permeability of the scaffolds has decreased which is shown in Figure 8.8. Since size of the scaffold structure is kept constant, increase in the strut size leads to decrease of the pore size. This shows that as we increase the pore size, porosity of the structure increases. This leads to the increase in intrinsic permeability of the scaffold which is shown in Figures 8.8 and 8.9.

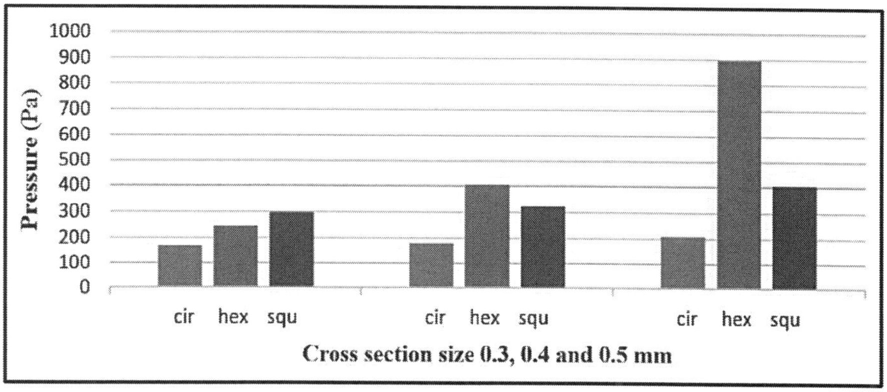

FIGURE 8.7 Graph showing pressure difference for scaffold structure with different cross sections.

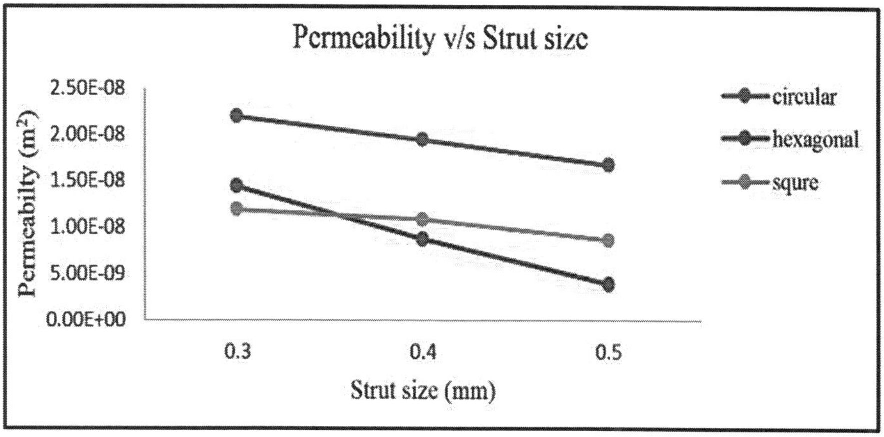

FIGURE 8.8 Graph showing the variation of permeability versus the strut size.

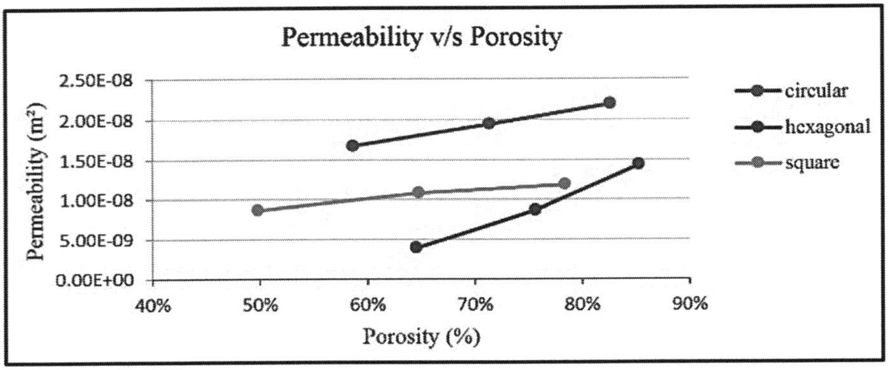

FIGURE 8.9 Graph showing the effect of porosity on permeability.

8.5 DISCUSSION

The design of the scaffold for tissue engineering applications, essentially for the trabecular bone is a complex problem that depends on many parameters and constraints. In this paper, scaffold structures were designed by considering parameters such as porosity, strut cross section, strut size which controls the pore size and by maintaining the same domain for all the designs. The main objective of the research work was to have homogeneous interconnected pores. This was achieved using modeling of scaffold structure by a unit cell-based approach. Here, parameters such as strut size and strut cross section were varied keeping the unit cell size same.

A total of nine variant specimens were modeled by changing the cross sections (circular, hexagonal, and cubic), strut size (0.3, 0.4, and 0.5 mm), and keeping the unit cell dimensions constant for all models. The computational studies were carried out using a software tool to find out the mechanical stability and permeability of the specimens designed.

Permeability of the designed scaffolds was found to be in the range of 4.8×10^{-9} to 2.3×10^{-8} m^2 for circular, cubic, and hexagonal cross sections, where the hexagonal cross section shows comparatively less permeability and the circular one shows high permeability, and the cubical cross section has intermediate permeability. The results obtained matched with the available resources.[4]

8.6 CONCLUSION

In this paper, the design of a homogeneous interconnected scaffold structure for cancellous bone is carried out. The geometry of the scaffold structure is simplified using regular and homogeneous distribution of pores. The effect of the strut cross-section, pore size on the pressure variation, and the magnitude of velocity is studied. From the literature, the trabecular (cancellous) bone pore size ranges from 300 to 500 μm, so the strut size is maintained to be in this range which directly controls the pore size.

Computational fluid dynamical analysis shows the effect of porosity, pore size, cross section, and other geometrical parameters which are not considered in this study on the pressure in the scaffold, velocity, and shear stress. Permeability is the important design parameter to consider because it determines many functions of scaffolds like the transportation of gas, nutrients, and fluids. Designing a scaffold, keeping permeability as a design factor, will also encompass the effect of pore size and distribution. Cross section of the pore also affects the permeability of the scaffold structure. Results show that scaffolds with circular cross sections have higher permeability of 4.8×10^{-9} to 2.3×10^{-8} m^2 for the pore size of 300–500 μm, respectively. Whereas, scaffolds with hexagonal cross sections have comparatively less permeability and cubical cross-sections have the permeability intermediate between circular and cubical cross sections. The results obtained are acceptable, since they match with the permeability obtained through computational and experimental analysis in the literature.[16] This approach is based on Darcy's law and this study gives importance to geometric parameters on the permeability and other parameters where it eliminates the use of time-consuming experiments. This justifies that the method followed is valid and the designs can be used for further stages like in vitro and in vivo studies, implantation, and so forth.

KEYWORDS

- **additive manufacturing**
- **tissue engineering**
- **scaffold**
- **permeability**
- **trabecular bone**
- **computational fluid dynamics**

REFERENCES

1. Arifin, A.; Sulong, A. B.; et al. Material Processing of Hydroxyapatite and Titanium Alloy (Ha/Ti) Composite as Implant Materials Using Powder Metallurgy: A Review. *Mater. Des.* **2014,** *53,* 165–175.
2. Chua, C. K.; Leong, K. F.; et al. Development of a Tissue Engineering Scaffold Structure Library for Rapid Prototyping. Part 1: Investigation and Classification. *Int. J. Adv. Manuf. Technol.* **2003,** *21,* 291–301.
3. Chua, C. K.; Leong, K. F.; et al. Development of a Tissue Engineering Scaffold Structure Library for Rapid Prototyping. Part 2: Parametric Library and Assembly Program. *Int. J. Adv. Manuf. Technol.* **2003,** *21,* 302–312.
4. Pennella, F.; Cerino, G.; Massai, D.; Gallo, D.; FalvoD'UrsoLabate, G.; Schiavi, A.; Deriu, M. A.; Audenino, A.; Morbiducci, U. A Survey of Methods for the Evaluation of Tissue Engineering Scaffold Permeability. *Ann. Biomed. Eng.* **2013,** *41*(10), 2027–2041.
5. Dias, M. R.; Fernandes, P. R.; Hollister, S. J.; et al. Permeability Analysis of Scaffolds for Bone Tissue Engineering. *J. Biomech.* **2012,** *45,* 938–944.
6. Sanz-herrera, A.; Garcı, M.; Doblare, M.; Ochoa, I.; Yunos, D. M.; Boccaccini, A. R. Permeability Evaluation of 45S5 Bioglass-Based Scaffolds for Bone Tissue Engineering. *J. Biomech.* **2009,** *42,* 257–260.
7. Li, S.; De Wijn, J. R.; Li, J.; Layrolle, P.; De Groot, K. Macroporous Biphasic Calcium Phosphate Scaffold with High Permeability/Porosity Ratio. *Tissue Eng.* **2003,** *9*(3), 535–548.
8. Lee, K.; Wang, S. Fabrication and Characterization of Poly(Propylene Fumarate) Scaffolds with Controlled Pore Structures Using 3-Dimensional Printing and Injection Molding. *Tissue Eng.*, **2006,** *12*(10), 2801–2811.
9. Jeong, C. G.; Zhang, H.; Hollister, S. J. Permeability Effects on Sub-Cutaneous in Vivo Chondrogenesis using Primary Chondrocytes. *Acta Biomater.* **2011,** *7*(2), 505–514.
10. Truscello, S.; Kerchkhofs, G.; et al. Prediction of Permeability of Regular Scaffolds for Skeletal Tissue Engineering: A Combined Computational and Experimental Study. *Acta Biomater.* **2012,** *8,* 1648–1658.
11. Luo, Z.; Zhang, Q.; Shi, M.; Zhang, Y.; Tao, W.; Li, M. Effect of Pore Size on the Biodegradation Rate of Silk Fibroin Scaffolds. *Adv. Mater. Sci. Eng.* **2015,** *2015,* AID 315397, 7.
12. Bogu, V. P.; Kumar, Y. R.; Khanra, A. K. Homogenous Scaffold-Based Cranial/Skull Implant Modelling and Structural Analysis—Unit Cell Algorithm-Meshless Approach. *Med. Biol. Eng. Comput.* **2017,** *55*(11), 2053–2065.
13. Loh, Q. L.; Choong, C. Three-Dimensional Scaffolds for Tissue Engineering Applications: Role of Porosity and Pore Size. *Tissue Eng., Part B* **2013,** *9*(6), 485–502.
14. Santamari, V. A. A.; Malve, M.; Duizabo, A.; Mena Tobar, A.; GallegoFerrer, G.; Garcia Aznar, J. M.; Doblare, M.; Ochoa, I. Computational Methodology to Determine Fluid Related Parameters of Non Regular Three-Dimensional Scaffolds. *Ann. Biomed. Eng.* **2013,** *41,* 2367–2380.
15. Sadir, S.; Kadir, M. R. A.; Öchsner, A.; Harun, M. N. Modeling of Bio Scaffolds: Structural and Fluid Transport Characterization. *World Acad. Sci. Eng. Technol. Int. J. Mech. Aerosp. Ind. Mech. Manuf. Eng.* **2011,***5*(2).
16. Viana, T.; Biscaia, S.; et al. Permeability Evaluation of Lay-Down Patterns and Pore Size of PCL Scaffolds, 3rd International Conference on Tissue Engineering. *ICTE* **2013,** *59,* 255–262.

PART II
Materials and Manufacturing (MM)

CHAPTER 9

THERMOGRAVIMETRIC/DIFFERENTIAL THERMAL ANALYSIS (TG–DTA) OF BINARY METAL CATALYST FOR MULTIWALL CARBON NANOTUBE (MWCNT) SYNTHESIS

V. J. PILLEWAN[1,*], D. N. RAUT[1,3], K. N. PATIL[2], and D. K. SHINDE[1,4]

[1]*Department of Production Engineering, Veermata Jijabai Technological Institute, Matunga, Mumbai, Maharashtra 400019, India*

[2]*Department of Mechanical Engineering, K. J. Somaiya College of Engineering, Vidyavihar, Mumbai, Maharashtra 400077, India, E-mail: kashinath@somaiya.edu*

[3]*dnraut@vjti.org.in*

[4]*dkshinde@vjti.org.in*

[*]*Corresponding author. E-mail: v_pillewan@rediffmail.com*

ABSTRACT

The formation of carbon nanotubes (CNTs) occurs typically between 600 and 1000°C. Due to the low-temperature process, chemical vapor deposition (CVD) method is generally better for the production of the multiwall carbon nanotubes (MWCNTs) as compared to other synthesis methods. The transition metals used for CNT synthesis are Fe, Co, and Ni. The wet impregnation methods are used for binary catalyst preparation and the starting materials are ferric nitrate, cobalt nitrate, and nickel nitrate. The samples are prepared with 1, 3, and 5% catalyst weight percentage with a $CaCO_3$ substrate. Thermogravimetric and differential temperature analysis (TG–DTA) is carried out to study the effect of weight loss as well as oxide transition in the presence of inert gas. It is observed that at 417°C, the crystallization of γ-Fe_2O_3 occurs; also, the reduction of the Fe_3O_4 to FeO occurs at 712°C. Similarly, nickel nitrate starts its water separation at 268.25°C, partial oxidation starts at 388.25°C, and finally nickel oxide formation takes place at 503.25°C. The $CaCO_3$ reduction to CaO occurs at 700°C with 40% weight reduction. TG–DTA curve shows catalytic decomposition temperature. Scanning electron microscopy (SEM) and transmission electron microscopy (TEM) images substantiate the result reflecting uniform diameter of CNTs produced by CVD method.

9.1 INTRODUCTION

The catalyst is an essential element for the formation of nanotubes in chemical vapor deposition (CVD) process.[1] Generally, catalysts are used for the synthesis are transition metals. Fe, Co, and Ni are the most effective and widely used metals. They have certain common properties such as they act as catalysts for hydrocarbon cracking, form stable carbides, and carbon diffuses readily through them.

Lee et al.[2] investigated the effect of the catalyst on carbon nanotubes (CNTs) prepared by CVD using acetylene with Fe, Co, and Ni catalysts. They found that the CNT growth rate is a function of catalyst particle size and diffusion coefficient. As catalyst particle size increases, the growth rate decreases. The higher diffusion rate of carbon through the catalyst increases the growth rate of CNTs. The average diameter of the nanotubes is found to be below the catalyst particle size. The crystallinity of the nanotubes is found to decrease in the following order: Fe>Ni>Co. CNTs grown on Fe show the broadest distribution and largest diameters. Yu et al.[3] also studied the effect of Fe catalyst particle size on the growth of CNTs.

They found that the optimum particle size of the catalyst is 13–15 nm, whereas, Chen et al.[4] concluded that the optimum size of Ni particle is around 34 nm. The amount of carbon deposit obtained per 100 g of Ni catalyst is 0.25 g. The highest yield obtained is at 50 min of precursor (i.e., methane) flow. Ermakova et al.[5] show that for methane precursor, the amount of carbon deposit obtained per gram of Ni catalyst is 7.68 g/h on SiO_2 support. Ren et al.[6] discussed the effect of different types of catalysts on the growth of CNTs. It is seen that growth for CNTs synthesis is in the order Ni>Fe–Co. Kathyayini et al.[7] investigated Fe, Co, and Fe–Co catalysts and found that the highest weight percentage of carbon is obtained when the Fe–Co catalysts are used as a catalyst, the individual Fe and Co give less yield of carbon deposit. The different research groups[8–11] have reported that there exists an optimum particle size of the catalyst ranging from 10 to 40 nm. The very-large-size catalyst particles are not suitable for CNTs synthesis. Singh et al.[12] use Co as a catalyst and MgO as a substrate. They showed that the highest CNT yield obtained is with 50 wt.% of catalyst loading in MgO. The highest carbon yield obtained per gram of catalyst is 7.88 g/h.

The different substrates used for the CNT formation are $CaCO_3$, MgO, Al_2O_3, and SiO_2. Magnesium oxide substrate is widely used because it is easy to dissolve after the deposition of carbon in the purification process with HCl.[13] $CaCO_3$ decomposes into CaO at 712°C. This is easily removed with the help of acids during the purification process. For example, use of ferric nitrate as the transition metal source leads to an acidic solution when dissolved in either water or ethanol. This solution reacts with the basic $CaCO_3$ powder and forms complexes of Fe_3^+ as well as with Co_2^+.[14] The interaction of these metal catalyst elements with support is not as strong as compared to Si and MgO. Therefore, bigger multiwall CNTs (MWCNTs) are preferred to grow with abundantly $CaCO_3$. Thus, for high yield and uniform growth of CNTs, SiO_2 substrate is used. In the case of Fe catalyst with SiO_2 form a silicate, it is assumed that the carbon production depends on the amount of silicate that remains after the reduction. The transition of Fe_2O_3 to Fe_3O_4 takes place during the reduction reaction which starts at 312°C and continues up to 1223°C.[15] There is always a strong interaction between the support and metal catalyst, which gives crystalline growth of carbon nanostructures (nanotubes and/or nanofibers) irrespective of the type of carbon precursor used for the synthesis.[16,17] However, for purification, carbon deposit obtained with SiO_2 substrate needs the hazardous chemical reduction and it is difficult to clean the CNTs.

This chapter studies the effect of the various transition metal catalysts for MWCNT synthesis. The $CaCO_3$ substrate is used for the catalyst preparation. The transition temperature of various oxide formations, the weight loss analysis, and the TG–DTA are carried out in inert gas (argon) atmosphere. Further, prepared catalyst is used for CNT synthesis, and the scanning electron microscopy/transmission electron microscopy (SEM/TEM) imaging is carried out to evaluate the diameter of CNT.

9.2 EXPERIMENTAL PROCEDURE

The starting materials used for the catalyst preparation are listed in Table 9.1. A mixture of ferric nitrate ($Fe(NO_3)_3 \cdot 9H_2O$), cobalt nitrate ($Co(NO_3)_2 \cdot 6H_2O$), and nickel nitrate ($Ni(NO_3)_2 \cdot 6H_2O$) is used as catalyst precursors, while $CaCO_3$, in fine powder form, is used as a substrate.

TABLE 9.1 Different Reagents and Chemicals Used to Prepare the Catalyst for Deposition.

Material	Purity (%)	Source
Ferric nitrate ($Fe(NO_3)_3 \cdot 9H_2O$)	99.8	S. D Fine
Cobalt nitrate ($Co(NO_3)_2 \cdot 6H_2O$)	99.8	S. D Fine
Nickel nitrate ($Ni(NO_3)_2 \cdot 6H_2O$)	99.8	S. D Fine
Calcium carbonate ($CaCO_3$)	99.9	S. D Fine

The sample is prepared by mixing a measured quantity of $Fe(NO_3)_3 \cdot 9H_2O$, $Co(NO_3)_2 \cdot 6H_2O$, $Ni(NO_3)_2 \cdot 6H_2O$ separately with $CaCO_3$ in deionized water. Thereafter, water is evaporated at 100°C followed by subsequent cooling in moisture-proof atmosphere. The amounts of catalyst precursor and substrate required for different weight percentage catalyst loadings are calculated for Fe, Co, and Ni.

9.3 RESULTS AND DISCUSSION

TG–DTA of the different catalysts is carried out to understand the decomposition states of the active catalyst oxides at a particular temperature.

Figure 9.1 shows the TG–DTA of the Co catalyst. It shows that up to 710°C, the Co catalyst is stable and after that, a sudden weight loss occurs. This may be due to the reduction of $CaCO_3$ substrate to the CaO and release of CO_2.

Figure 9.2 shows the TG–DTA of the Fe catalyst. It is observed that at 417°C, crystallization of γ-Fe_2O_3 occurs. The second peak occurs at 710°C, which corresponds to the reduction of $CaCO_3$ to CaO. The initial weight loss occurs up to 100°C due to evaporation of water (less than 2%).

Figure 9.3 shows the TG–DTA of the Ni catalyst. It is observed that nickel nitrate starts its water separation at 268.25°C. Further partial oxidation starts at 388.25°C (peak I). The nickel oxide (Ni_2O_3) formed at 503.25°C (peak II). This information is consistent with the literature.[18] This catalyst is used for CNT synthesis.

Figure 9.4 shows the weight loss of the various catalysts prepared from 600 to 750°C for the CNT synthesis. It is observed that as the temperature increases, the weight of the samples is reduced, which is consistent with the TG–DTA. Highest weight loss (39.20%) is observed in 1% cobalt catalyst, whereas, least weight reduction (24.85%) is observed in 5% cobalt catalyst.

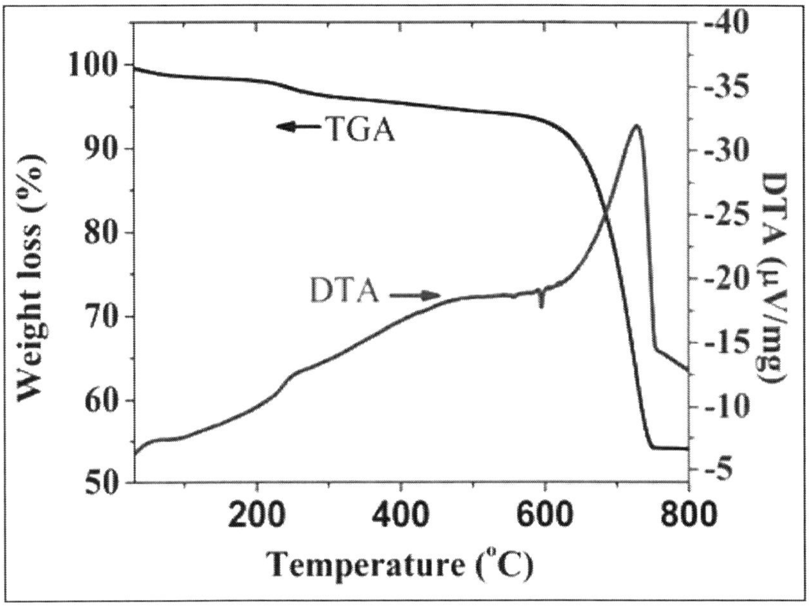

FIGURE 9.1 Thermogravimetric–differential thermal analysis (TG–DTA) curves of Co catalyst heated from 30°C at a heating rate of 10°C/min up to 820°C.

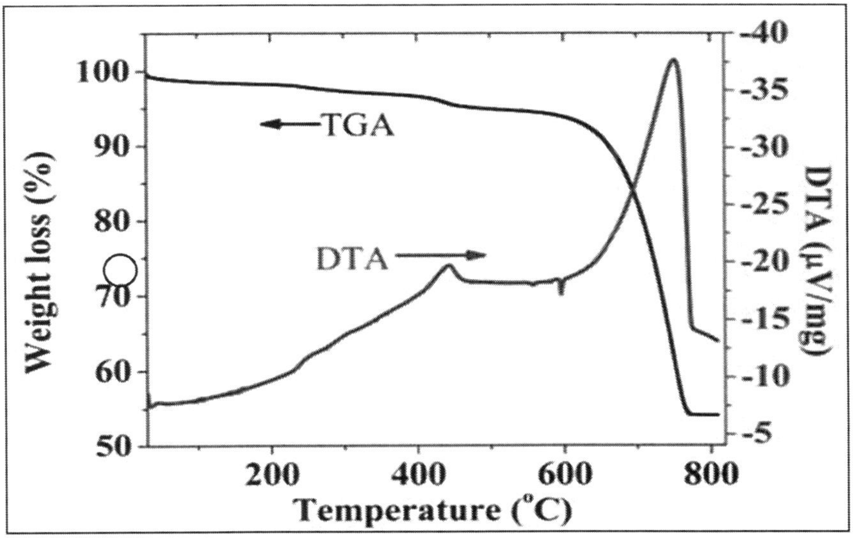

FIGURE 9.2 TG–DTA curves of Fe catalyst heated from 30°C at a heating rate of 10°C/min up to 820°C.

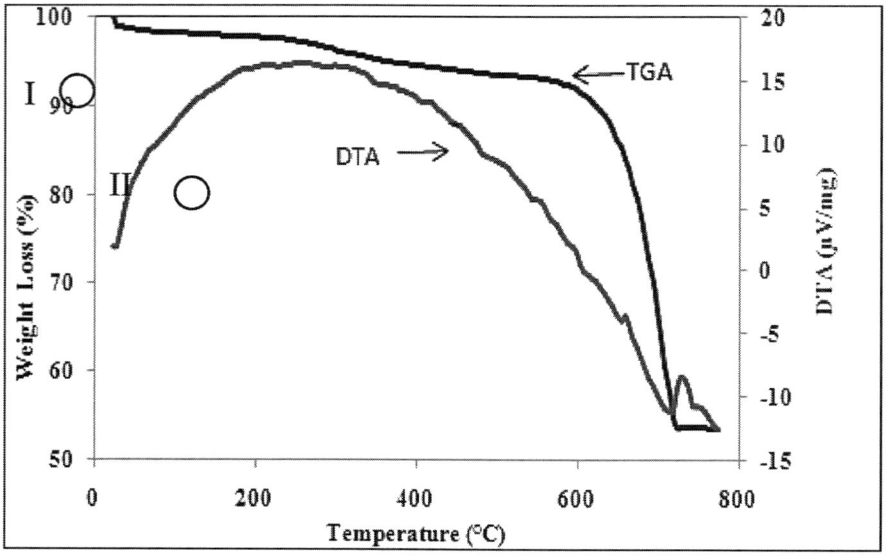

FIGURE 9.3 TG–DTA curves of Ni catalyst heated from 30°C at a heating rate of 10°C/min up to 820°C.

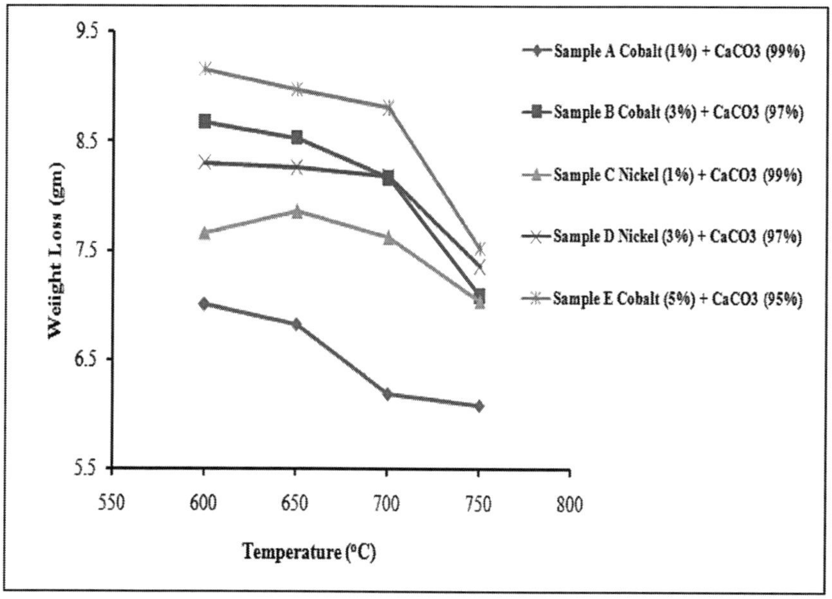

FIGURE 9.4 Weight losses of various catalysts at various temperatures.

Figure 9.5a shows SEM images of unpurified CNTs with some catalyst and substrate impurities and it is observed that the diameter of the CNTs varies in the range of 25–55 nm, whereas, the length of the nanotubes is in few microns.

Figure 9.5a shows large bundles of twisted purified CNTs without catalyst and substrate impurities. Table 9.2 observed that the diameter of the CNTs varies in the range of 19–45 nm. The average diameter of the purified CNTs is 31.20 nm, evaluated by ImageJ software, which is within the range of nanoparticle size of Fe, Co, and Ni.

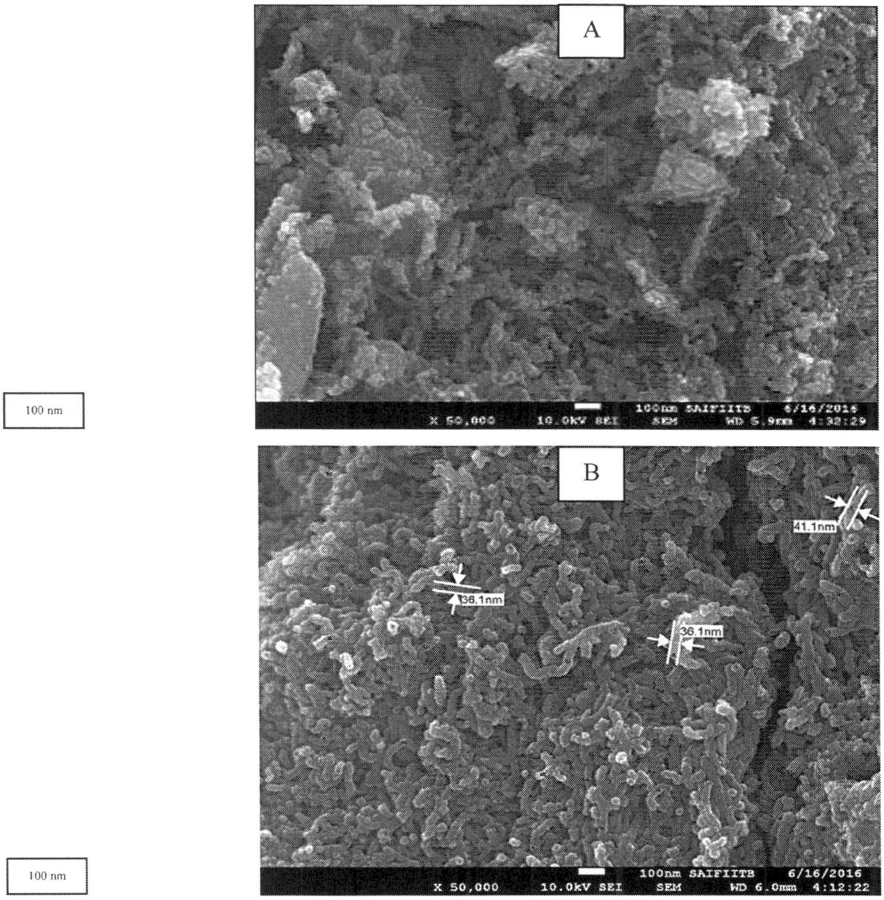

FIGURE 9.5 Scanning electron microscopy (SEM) images of (a) unpurified carbon nanotube (CNT) and (b) purified CNT.

Source: Sophisticated Analytical Instrument Facility (SAIF), IIT Bombay.

TABLE 9.2 Diameter of Purified Carbon Nanotubes (CNTs) (Source: Randomly Selected from Figure 9.5b).

CNT no.	Diameter (nm)
1	34.74
2	44.33
3	42.45
4	26.33
5	43.59
6	34.39
7	38.84
8	30.89
9	19.99
10	28.56
11	28.70
12	29.12
13	32.56
14	28.12
15	30.76
16	34.27
17	28.12
18	32.56
19	26.33
20	34.74
21	22.84
22	24.23
23	28.56
24	30.35
25	26.33
26	37.12
27	25.86
28	29.12

In Figure **9**.6, TEM images indicated that CNTs are not straight and small defects on these tubes are observed. An amorphous carbon is also observed on the surface of the CNT. This could be reduced if the CNTs are annealed in air ambient.

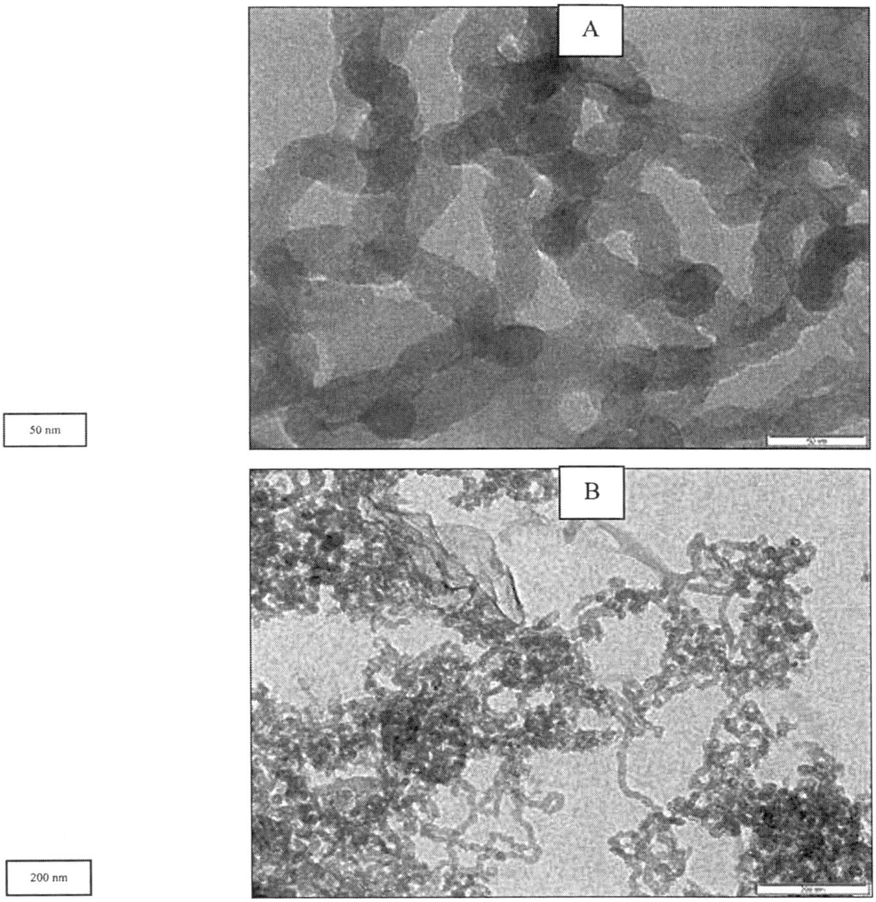

FIGURE 9.6 Transmission electron microscopy (TEM) images of CNT produced on $CaCO_3$-supported catalysts at (a) 700°C and (b) 750°C.
Source: Sophisticated Analytical Instrument Facility (SAIF), IIT Bombay.

It is reported that the reduction of the Fe_3O_4 to FeO occurs at 712°C and it is stable above 750°C without showing any catalytic activity.[19] At 700°C, approximately 52% weight is reduced and has shown the complete reduction of the $CaCO_3$ to CaO. Thus, no further weight reduction takes place when temperature increases up to 900°C. However, the weight percentage of the metal catalyst into the support is too less. Thus, it is difficult to see the DTA peaks for the transition of these metal oxides.

9.4 CONCLUSIONS

It is observed that at 710°C, $CaCO_3$ reduces to CaO, which leads to the sudden reduction in the weight loss, which is consistent with all the samples. At 417°C, it is observed that the crystallization of γ-Fe_2O_3 occurs; also, the reduction of the Fe_3O_4 to FeO occurs at 712°C. The FeO is stable above 570°C and does not have any catalytic activity. Similarly, it is observed that nickel nitrate starts its water separation at 268.25°C. Further partial oxidation starts at 388.25°C and nickel oxide forms at 503.25°C. Around 20–40% weight loss is observed during catalytic decomposition. The average diameter of the purified CNTs is measured by ImageJ software and found to be 31.20 nm, which is within the range of average nanoparticle size of Fe, Co, and Ni. SEM and TEM images of synthesized CNTs reflecting uniform diameter and reasonable purity are produced by CVD method and show the relationship of TG–DTA.

ACKNOWLEDGMENT

Authors acknowledge the financial support by A. T. E. Industries Pvt. Ltd., Mumbai, for the given project.

KEYWORDS

- **multiwall carbon nanotubes**
- **metal catalyst**
- **chemical vapor deposition**
- **scanning electron microscope**
- **transmission electron microscope**

REFERENCES

1. Sen, R.; Govindraju, A.; Rao, C. N. R. Carbon Nanotubes by the Metallocene Route. *Chem. Phys. Lett.* **1997,** *267,* 276.
2. Lee, C. J.; Park, J.; Yu, J. A. Catalyst Effect on Carbon Nanotubes Synthesized by Thermal Chemical Vapor Deposition. *Chem. Phys. Lett.* **2002,** *360,* 250.

3. Yu, Z.; Chen, D.; Totdal, B.; Holeman, A. Effect of Catalyst Preparation on Carbon Nanotube Growth Rate. *Catal. Today* **2005,** *100,* 26.
4. De, C.; Kjersti, O. C.; Ester, O.; Zhixin, Y.; Bård, T.; Nieves, L.; Antonio, M.; Anders, H. Synthesis of Carbon Nanofibers: Effects of Ni Crystal Size During Methane Decomposition. *J. Catal.* **2005,** *229,* 82.
5. Ermakova, M. A.; Ermakov; D. Y.; Kuvshinov; G. G. Effective Catalysts for Direct Cracking of Methane to Produce Hydrogen and Filamentous Carbon Part I Nickel Catalysts. *Appl. Catal. A* **2000,** *201,* 61.
6. Huang, Z. P.; Wang, D. Z.; Wen, J. G.; Sennett, M.; Gibson, H.; Ren, Z. F. Effect of Nickel, Iron and Cobalt on the Growth of Aligned Carbon Nanotubes. *Appl. Phys. A* **2002,** *74,* 387.
7. Kathyayini, H.; Nagaraju, N.; Fonseca, A.; Nagy, J. B. Catalytic Activity of Fe, Co, and Fe/Co Supported on Ca and Mg Oxides, Hydroxides, and Carbonates in the Synthesis of Carbon Nanotubes. *J. Mol. Catal. A* **2004,** *223*(1–2), 129.
8. Baker, R. T. K.; Harris, P. S.; Thomas, R. B.; Waite, R. J. Formation of Filamentous Carbon from Iron, Cobalt, and Chromium Catalyzed the Decomposition of Acetylene. *J. Catal.* **1973,** *30,* 86.
9. Chitrapu, P.; Lund, C. R. F.; Tsamopoulos, J. A. A Model for the Catalytic Growth of Carbon Filaments. *Carbon* **1992,** *30,* 285.
10. Ajayan, P. M. Nanotechnology: How Does Nanofiber Grow? *Nature* **2004,** *427,* 402.
11. Nerushev, O. A.; Dittmar, S.; Morjan, R.-E.; Rohmund, F.; Campbell, E. E. B. Particle Size Dependence and Model for Iron-Catalyzed Growth of Carbon Nanotubes by Thermal Chemical Vapor Deposition. *J. Appl. Phys.* **2003,** *93,* 4185.
12. Singh, B. K.; Hojin, R.; Rajeev, C. C.; Nguyen, D.; Soo, J. P.; Seok, K.; Jae, R. L. Growth of Multiwalled Carbon Nanotubes from Acetylene Over In Situ Formed Co Nanoparticles on MgO Support. *Solid State Commun.* **2006,** *139,* 102.
13. Md, S.; Mo, Y. H.; FazleKibria, A. K. M; Kim, M. J., Nahm, K. S. High Growth of SWNTs and MWNTs from C_2H_2 Decomposition Over Co–Mo/MgO Catalysts. *Carbon* **2004,** *42,* 2245.
14. Chee, H. S.; Andrew, T. H., Particle Technology and Fluidization $CaCo_3$ Supported Co-Fe Catalysts for Carbon Nanotube Synthesis in Fluidized Bed Reactors. *AlChE J.* **2008,** *54*(3), 657.
15. Crisan, M.; Jitianu, A.; Crisan, D.; Balasoiu, M.; Dragan, N.; Zaharescu, M. Sog-Gel Monocomponent Nano-Sized Oxide Powders. *J. Optoelectron. Adv. Mater.* **2000,** *2*(4), 339.
16. Ermakova, M. A.; Ermakov, D. Y. Ni/SiO_2, and Fe/SiO_2 Catalysts for the Production of Hydrogen and Filamentous Carbon Via Methane Decomposition. *Catal Today* **2002,** *77*(3), 225.
17. de los Arcos, T.; Vonau, F.; Garnier, M. G.; Thommen, V.; Boyen, H. G.; Oelhafen, P.; Düggelin, M.; Mathis, D.; Guggenheim, R. Influence of Iron–Silicon Interaction on the Growth of Carbon Nanotubes Produced by Chemical Vapor Deposition. *Appl. Phys. Lett.* **2002,** *80*(13), 2383.
18. Wolfgang, B.; Claus, E.; Mimoza, G. Thermal Decomposition of Nickel Nitrate Hexahydrate, $Ni(NO_3)_2$ $6H_2O$ in Comparison to $Co(NO_3)_2$ $6H_2O$ and $Ca(NO_3)_2$ $4H_2O$. *Thermochim. Acta* **2007,** *456,* 64–68.
19. Patil, K. N.; Solanki, C. S. Precursor to High Purity Carbon Nanotubes: A Step by Step Evaluation of Carbon Yield. *J. Nano Res.* **2009,** *6,* 75–87.

CHAPTER 10

CHARACTERIZATION OF FE–CO–NI-TERNARY METAL CATALYST FOR HIGH-YIELD MULTIWALL CARBON NANOTUBES

VILAS G DHORE[1,*], W. S. RATHOD[1,2], and K. N. PATIL[3]

[1]*Department of Mechanical Engineering, Veermata Jijabai Technological Institute, Mumbai, Maharashtra 400019, India*

[2]*wsrathod@vjti.org.in*

[3]*Department of Mechanical Engineering, K. J. Somaiya College of Engineering, Mumbai, Maharashtra 400077, India, E-mail: kashinath@somaiya.edu*

[*]*Corresponding author. E-mail: vilasdhore@gmail.com*

ABSTRACT

A carbon nanotube (CNT) is a promising material for various applications in the field of medical science, energy engineering, electronics, sports and automobiles, and so forth due to their excellent properties. These applications demand large-quantity and high-quality CNTs. The yield and quality of CNTs depend on different parameters such as the type of catalyst, percentage of the catalyst and type of substrate, the synthesis temperature, type of precursor gases and the method of purification, and so forth. The literature survey suggested that the binary catalyst increases CNTs yield. However, very few studies discussed the effect of ternary catalyst for high-yield CNTs synthesis. This chapter highlights the behavior of ternary catalyst (Fe, Co, Ni) at various temperatures in the range of 600–750°C in the presence of argon atmosphere. The material characterization was carried out with the help of X-ray diffraction (XRD) to identify the active catalyst species for the

CNT synthesis. It is observed that Fe_2O_3 and Fe_3O_4 were present at $2\theta = 24.33°$ (012), 33.44° (104), 35.88° (110), 41.17° (113), 43.82° (202), 49.86° (024), 54.55° (116), 58.17° (018), 62.94° (214), 64.48° (300), and 72.72° (101). Similarly, Co_3O_4 was present at $2\theta = 31.26°$ (220), 36.84° (311), 38.54° (222), 44.80° (400), 55.64° (422), 59.64° (511), 65.22° (400), and 75.32° (533). Further, Ni_2O_3 was present at $2\theta = 31.2$, 32.3, 43.4, and 63.2°. The X-ray diffraction (XRD) analysis carried out at 600, 650, and 700°C shows a similar pattern, whereas, it was observed that Fe catalyst starts losing its activity as the temperature rises toward 750°C and it is converted to FeO.

10.1 INTRODUCTION

Carbon nanotubes (CNTs) are getting more attention after their discovery by Iijima in 1991 due to their exceptional mechanical and electrical properties.[1] Volder et al. discussed various applications of CNTs with diversified products ranging from automotive, rechargeable batteries, sports material, electronics, large-area coating to water filter, and so forth.[2] A high aspect ratio of CNTs enables it as a prominent reinforcement material in metal matrix and polymer matrix composites than glass fiber.[3] However, the yields of CNTs are significantly low to cater the demand for the various applications.

Couteau et al.[4] proposed large-scale synthesis of multiwall nanotubes (MWNTs) on $CaCO_3$ substrate with Fe–Co catalyst. The rotary chemical vapor deposition (CVD) furnace is used for the constant production of high-quality MWNTs at a rate of 0.0694 g/min. Choi et al.[5] proposed a cost-effective synthesis method for CNTs for large-scale production using dispersal of fine metal particles and magnetic fluid as catalysts. The metal catalyst is deposited on the plane substrate which has limited area. Thus, this method is not useful for high-yield CNTs synthesis. To overcome this restriction, spin coating of catalyst nanoparticles of Si particles is done.

Teknaka et al.[6] investigated the effect of catalyst loading. There is an increase in the CNTs yield when the metal catalyst loading increased without major increase in the particle size due to the existence of more surface area for the reactions. Nevertheless, if the metal support is too high, it leads to bigger particle sizes and; thus, long diffusion length. In this case, the growth rate is lower. Yu et al.[7] discussed the effect of metal loading on the growth of CNTs. The growth rate of the CNTs significantly depends on the amount of metal. The growth rates of CNTs at 20, 40, and 60% metal loadings are evaluated and it is observed that the growth rate significantly increases at 60% loading. They concluded that an increase in metal loading without increasing the catalyst size increases the CNT yield. The reason for the high activity at 60% metal loading is diverse particle size dispersal available for the synthesis of CNTs. Yu et al. also concluded that the increase in catalyst particle size adversely affects the growth rate. The higher diffusion rate of carbon through the catalyst also surges the growth rate of CNTs. The maximum yield they attained was with methane as a precursor.

The catalyst is an essential element for the formation of nanotubes in CVD process. Patil et al.[8] discussed the properties of the catalyst particle that affect the CNT formation, which are the diameter and diffusivity of carbon through it. Generally, the catalysts used for synthesis are transition metals. The most effective and extensively used metals are Fe, Co, and Ni. These metals have certain common properties as follows:

1. They act as catalysts for hydrocarbon cracking.
2. They all form stable carbides.
3. Carbon diffuses readily through them.

Ermakova et al.[9] have reported that the method of catalyst preparation plays an important role. The maximum yield is obtained on the $NiO–SiO_2$ catalyst prepared by impregnation method. Oncel and Yurum[10] discussed the growth rate of nanotube for corresponding catalyst in the order of Ni>Co>Fe. However, Fe catalyst provides the best crystallinity of nanotubes. It is essential to have a small-sized catalyst particle for the synthesis of better CNTs. Nickel, iron, cobalt, molybdenum, and copper are used successfully in CVD for CNT synthesis. The study reveals that the growth rate of nanotube for corresponding catalyst is in the order of Ni>Co>Fe; however, Fe catalyst provides the finest crystallinity of nanotubes.[11] The catalysts are used as individual, binary, or ternary mixture. Patil et al.[12] investigated that the highest yield of nanotube occurs using binary metal catalyst. In the present study, the ternary combination of Ni, Fe, and Co catalysts and $CaCO_3$ is used as a substrate material. The temperature in the range of 600–750°C is considered suitable for the growth of nanotubes. Ni et al.,[13] in their study, observed that the thick CNT was formed at a lower temperature range of 550–650°C with Co/MgO catalyst using methane as carbon precursor, while thin CNTs of the order of 3–10 nm diameter size were obtained using Mo/Co/MgO catalyst at a temperature above 650°C. The rate of carbon precursor is responsible for the growth of nanotubes. Although the growth rate depends on hydrocarbon supply, much higher supply adversely affects the growth rate.[14]

This chapter discusses the effect of the ternary metal catalyst, that is, Fe, Co, and Ni with a variable weight percentage of the metal catalyst loading ranging from 3 to 10% on the $CaCO_3$ substrate. The characterization was evaluated by X-ray diffraction (XRD) analysis to investigate the active metal oxide species for CNT synthesis. The surface morphology and the elemental oxide contribution are evaluated with the help of scanning electron microscopy-energy dispersive spectroscopy (SEM-EDS). Further, the $CaCO_3$ reduction was studied.

10.2 EXPERIMENTAL PROCEDURE

A ternary mixture of ferric nitrate hexahydrate, cobalt nitrate, and nickel nitrate hexahydrate was used as a catalyst precursor on a $CaCO_3$ substrate. The contribution of each catalyst precursor is presented in Table 10.1.

TABLE 10.1 The Catalyst Loading.

	3%	5%	7%	10%
Cobalt nitrate	10.39	17.32	24.25	34.64
Nickel nitrate hexahydrate	9.44	15.73	22.03	31.47
Ferric nitrate hexahydrate	13.75	22.91	32.07	45.82
$CaCO_3$	184.83	181.02	177.21	171.49

These nitrates are dissolved in deionized water. The $CaCO_3$ paste is prepared and the nitrate solution is added to it. Thus, the ternary (Co–Ni–Fe) precipitate is formed. The precipitate is dried in oven at 1200°C till it is converted to dry powder. While drying the mixture, it is continuously stirred so as to obtain the uniform dispersion of the catalyst in the substrate. The time required for drying of various samples varied between 5 and 7 h. This catalyst is used as a seed for CNT synthesis.

The weight reduction study was carried out in the presence of inert atmosphere using argon with flow rate of 30 ml/min as a carrier gas. To understand the effect of temperature on the catalyst, the catalyst was heated at 600–750°C as the CNT synthesis at this temperature range varies. The catalyst is kept inside the furnace for 20 min. Then, it is furnace-cooled. It is observed that after the heating, the weights of the samples reduced from 16% to almost 38%, depending on the percentage of the catalyst and the furnace temperature. The result shows that maximum weight reduction occurred with 7% catalyst and at 7500°C.

10.3 RESULTS AND DISCUSSION

10.3.1 X-RAY DIFFRACTION (XRD) ANALYSIS

XRD test was conducted using Cu–Kα (λ=1.54060 Å) radiation in an Empyrean X-ray Diffractometer operating at 45 kV and 40 mA. Figure 10.1 shows the XRD results of catalyst prior to the furnace-heating. The graph shows that the sample of 5% catalyst has more crystalline peaks compared with other two samples, whereas the nature of all the three graphs is similar. Figure 10.2 shows the XRD plots for ternary catalyst after furnace-heating at temperatures ranging from 600 to 750°C. From the XRD results, it is observed that 700°C it shows better peaks.

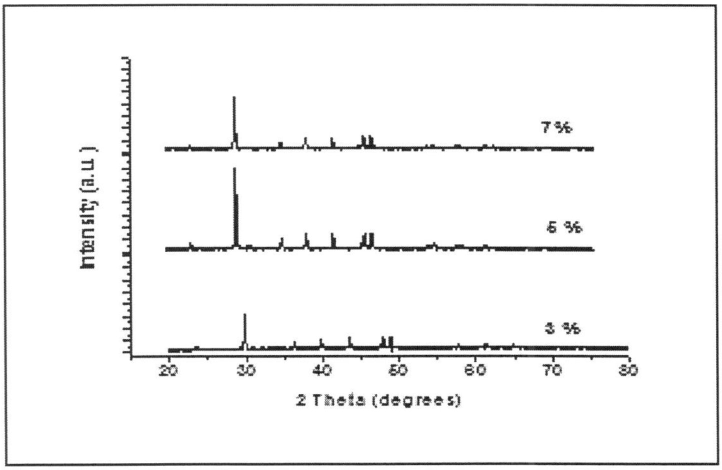

FIGURE 10.1 X-ray diffraction (XRD) pattern of precipitate of Ni/Co/Fe catalyst at various percentages without furnace-heating.

The material characterization was carried out with the help of XRD to identify the active catalyst species for the CNT synthesis. It is observed that Fe_2O_3 and Fe_3O_4 were present at 2θ=24.33° (012), 33.44° (104), 35.88° (110), 41.17° (113), 43.82° (202), 49.86° (024), 54.55° (116), 58.17° (018), 62.94° (214), 64.48° (300), and 72.72° (1010). Similarly, Co_3O_4 were present at 2θ=31.26° (220), 36.84° (311), 38.54° (222), 44.80° (400), 55.64° (422), 59.64° (511), 65.22° (400), and 75.32° (533). Further, Ni_2O_3 was present at 2θ=31.2, 32.3, 43.4, and 63.2°. The XRD analysis carried out at 600, 650, and 700°C showed a similar pattern, whereas it was observed that Fe catalyst lost its activity as temperature rose toward 750°C and it was converted to FeO. It is observed that at room temperature, the crystallinity of the catalyst samples is inferior and which further improves with the increasing temperature. At 700 and 750°C, the catalyst peaks are sharper, which leads to small diameter of the catalyst elements, which in turn help smaller diameter CNTs. The XRD analysis of the catalyst loading ranging from 3 to 10 wt.% suggests that the sharper peaks were observed at 5 wt.% catalyst loading. This will be a favorable condition for the smaller diameter CNTs. Further, yield evaluation of the CNTs on these weight percent of catalyst is yet to be studied.

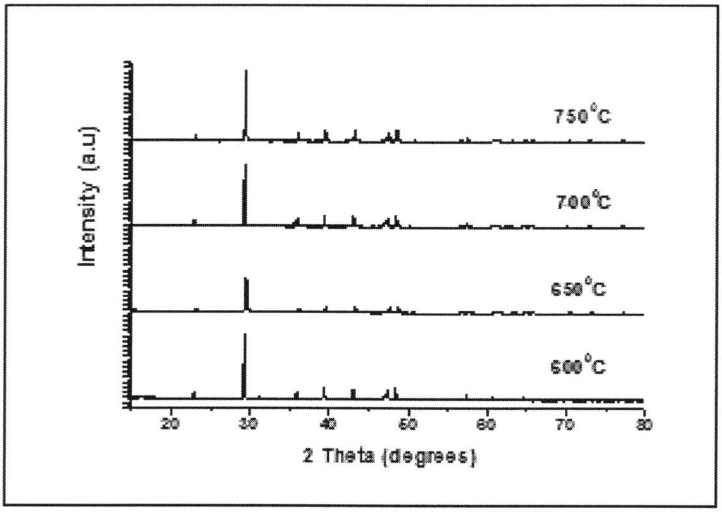

FIGURE 10.2 XRD pattern of 7% Ni/Co/Fe catalyst at various furnace temperatures in argon atmosphere.

10.3.2 SCANNING ELECTRON MICROSCOPY-ENERGY DISPERSIVE SPECTROSCOPY (SEM-EDS) ANALYSIS

Figure 10.3 shows the SEM-EDS analysis of various elements present in the catalyst. The traces of S, Mg, and C were observed in it. It may be because some trace of these elements is used as the ingredient for catalyst preparation. The other elements such as Fe, Co, Ni, and O are also present. The surface morphology is porous in nature, which is agglomerated. Then, this catalyst is used for the CNT synthesis (refer Fig. 10.4).

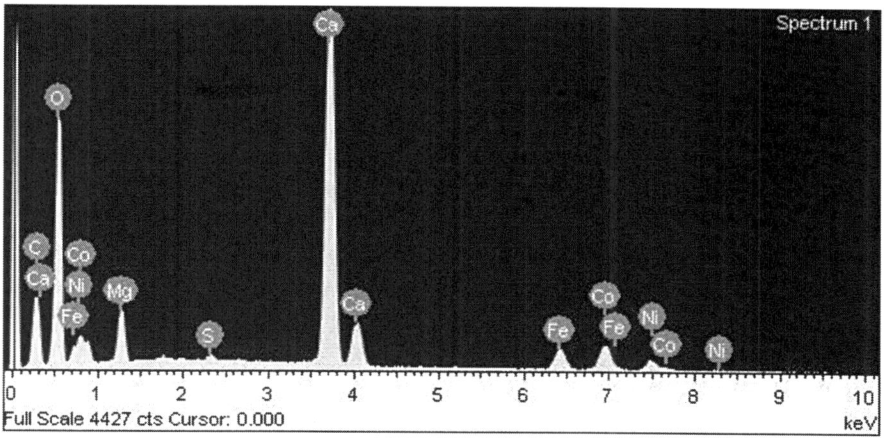

FIGURE 10.3 Scanning electron microscopy-energy dispersive spectroscopy (SEM-EDS) analysis of catalyst.

FIGURE 10.4 SEM image of sample Ni/Co/Fe-$CaCO_3$ catalyst.

10.4 CONCLUSIONS

In this study, the ternary (Co–Ni–Fe) catalyst was prepared and its characterization is presented. The XRD suggests that 5 wt.% catalysts loading will be better in terms of small seed dimensions. This helps in CNT synthesis of better quality. Further, the catalyst is heated in the argon atmosphere up to 750°C. The active range of CNTs synthesis is 600–750°C in CVD method. The XRD characterization is carried out for the same range of temperature and it is observed that Fe_2O_3 and Fe_3O_4 are present in the catalyst samples, along with Ni_2O_3 and Co_3O_4 catalyst elements. The SEM-EDS analysis suggests that traces of S, Mg, and C were present in the catalyst samples. Further, the agglomerated morphology observed, which is porous in nature.

ACKNOWLEDGMENT

The authors acknowledge the financial support by A.T.E. Industries Pvt. Ltd., Mumbai, for the given project.

KEYWORDS

- **multiwall carbon nanotubes**
- **metal catalyst**
- **chemical vapor deposition**

REFERENCES

1. Gulino, G.; Vieira, R.; Amadou, J.; Patrick, N.; Ledoux, M. J.; Galvagno, S.; Centi, G.; Pham-Huu, C. C_2H_6 as an Active Carbon Source for a Large Scale Synthesis of Carbon Nanotubes by Chemical Vapour Deposition. *Appl. Catal. A* **2005,** *279,* 89–97.
2. De Volder, M. F.; Tawfick, S. H.; Baughman, R. H.; Hart, A J. Carbon Nanotubes: Present and Future Commercial Applications. *Science* **2013,** *339,* 535–539.
3. Koziol, K.; Boskovic, B. O.; Yahya, N. *Carbon* **2010,** *77,* 23–49. DOI: 10.1007/8611_2010_12.
4. Couteau, E.; Hernadi, K.; Seo, J. W.; Thien-Nga, L.; Miko, C.; Gaal, R.; Farro, F. CVD Synthesis of High-Purity Multiwalled Carbon Nanotubes Using $CaCo_3$ Catalyst Support for Large-Scale Production. *Chem. Phys. Lett.* **2003,** *378,* 9.
5. Choi, G. S.; Cho, Y. S.; Son, K. H.; Kim, D. J. Mass Production of CNTs Using Spin-Coating of Nanoparticles. *Microelectron. Eng.* **2003,** *66,* 77.
6. Takenaka, S.; Kobayashi, S.; Ogihara, H.; Otsuka, K. Ni/SiO_2 Catalyst Effective for Methane Decomposition Into Hydrogen and Carbon Nanofiber. *J. Catal.* **2003,** *217,* 79.

7. Yu, Z.; Chen, D.; Totdal, B.; Holeman, A. Effect of Catalyst Preparation on Carbon Nanotube Growth Rate. *Catal. Today* **2005,** *100,* 26.
8. Patil K. N. Development and Modeling of Carbon Nanotubes (CNTs) Synthesis Process. Ph.D. Thesis, IIT Bombay, Mumbai, 2011.
9. Ermakova, M. A.; Ermakov, D. Y.; Kuvshinov, G. G. Effective Catalysts for Direct Cracking of Methane to Produce Hydrogen and Filamentous Carbon Part I. Nickel Catalysts. *Appl. Catal. A* **2000,** *201,* 61.
10. Oncel, C.; Yurum, Y. Carbon Nanotube Synthesis Via the Catalytic CVD Method: A Review on the Effect of Reaction Parameters, Fullerenes. *Nanotubes Carbon Nanostruct.* **2006,** *14,* 17–37.
11. Magrez, A.; Seo, J. W.; Smajda, R.; Mionić, M.; László, F. Catalytic CVD Synthesis of Carbon Nanotubes: Towards High Yield and Low Temperature Growth. *Materials* **2010,** *3,* 4871–4891.
12. Patil, K. N.; Mahajani, S.; Solanki, C. Optimization of Catalyst Loading for Effective Utilization of Feed Stock for Carbon Nanotubes Synthesis Process. *Int. J. Nanosci.* **2011,** *10,* 1–5.
13. Lei, N.; Kuroda, K.; Zhou, L.-P.; Kizuka, T.; Ohta, K.; Matsuishi, K.; Nakamura, J. Kinetic Study of Carbon Nanotube Synthesis Over Mo/Co/MgO Catalysts. *Carbon* **2006,** *44,* 2265–2272.
14. Meyyappan, M.; Delzeit, L.; Cassell, A.; Hash, D. Carbon Nanotube Growth by PECVD: A Review. *Plasma Sources Sci. Technol.* **2003,** *12,* 205–216.

CHAPTER 11

EFFECT OF MOLD VIBRATIONS ON MECHANICAL AND METALLURGICAL PROPERTIES OF ALUMINUM 356 CASTING

NAGARAJU TENALI*, B. KARUNA KUMAR, and K. CH. KISHOR KUMAR

Mechanical Engineering Department, Gudlavalleru Engineering College, Sheshadri Rao Knowledge Village, Gudlavalleru, Andhra Pradesh 521356, India

**Corresponding author. E-mail: tenali.n1830@gmail.com*

ABSTRACT

Casting is one of the versatile and most commonly used production methods in industry. The traditional casting has certain disadvantages like poor strength due to hot tears and shrinkage. The property of casting process mostly depends on the microstructure after solidification. Providing mold vibration during casting is one of the latest techniques employed in order to get a better structure of casting. Mold vibration during casting gives less chance of hot tear, better morphology, surface finish, and reduced amount of shrinkage. In this research work, the effect of mold vibration during solidification of A356.0 ANSI (UNS-A13560) alloys for different wavelengths of mold vibration and at a fixed pouring temperature was investigated to understand the changes in microstructure and mechanical properties of casting. The A356.0 casting has been prepared in a graphite mold with and without vibrations. The frequencies are varied from 0 to 20 Hz during mold vibrated casting process. A casting has been made without vibration as well to compare the results of casting with vibration. The experimental result showed significant grain refinement and remarkable improvement in compression strength and hardness of mold vibrated casting.

11.1 INTRODUCTION

Out of all the manufacturing processes, casting process is best and cheapest manufacturing process due to its simplified procedure. At present, more than 85% of the products are produced by casting processes. Casting is carried out basically by three important stages: heating, poring, and solidification. The quality of casting depends on the flow behavior of molten metal and other process parameters. The quality of metals and alloys is determined by a very important operation, that is, of solidification. Solidification is the operation that gives shape and structure. Currently, the solidification technique has experienced a rapid development. Due to the progresses made, the castings are used in high-security parts in the aerospatial industry, automotive, chemical, and metallurgical equipment.[1] Recent techniques suggest that mold vibration during pouring and till solidification is one of the important methods to produce casting for better morphology, surface finish, and reduced amount of shrinkage.[2] Mold conditions, pouring temperature, frequency of vibration, and other process variables are factors that would have a definite effect on the microstructure and properties of the cast.[3] It must be noted that above-mentioned factors are chosen considering the requirement that the material solidifies in a manner that would maximize the properties desired while simultaneously preventing potential defects such as shrinkage, porosity, voids, and trapped inclusions. The most effective method of early fault detection in a metal cast is high-frequency vibration analysis as its parameter changes quickly in the early stages of defect development. There are mainly three types of vibration, such as ultrasonic vibration, electromagnetic vibration, and mechanical vibration.[4] Out of above three methods, mechanical vibration is simple one due to its easier control over its parameters. A number of researchers have employed ultrasonic and electromagnetic vibrations and studied their effect on casting product.[5–8] Experimentation with mold vibration in order to alter the Al-cast microstructure of cast components date back to 1868. Naoki Omura[9] observed that the average density of AC4C aluminum alloy increased by imposition the mechanical vibration. The casting defects involved in the specimen reduced and became smaller with the increase of vibration frequency. Olufemi and Ademola[10] vibration of moulds during melt solidification gives refining on the grain structures of the alloy and improve the mechanical properties of material. Cambell[11] concluded that the mechanical vibration causes improvement in mechanical and corrosion properties of alloys. Dommaschk[12] studied the effect of vibration on pure aluminum (Al; wt.%)–Si–Mg alloys along with other nonferrous alloys. His research focused on the grain refinement and reported that the dependence of the casting wall thickness on casting characteristics could be minimized with the use of mechanical vibration. S. S. Mishra et al.[4] observed that the mechanical mold vibration on liquid metallic materials and their crystallization has a significant effect on the grain structure casting. They observed that the grain size reduced by vibration and grain became more compact and spheroidal in shape. Pillai[8] used very-low-frequency vibration to the study the effect on A-356 and Al12Si alloy. He concluded that mechanical vibrations improve the density and elongation of cast component. Abu-Dheir[13] used an electromagnetic shaker to induce mechanical vibration in a permanent mold and concluded that vibration homogenizes the temperature distribution in the mold and promotes more uniform dendrite structure and less porosity in the castings. The study of Ciućka[14] is related to the determination of the location of shrinkage cavities; it clarifies the influence of conditions of heat dissipation by the casting on the shape and location of shrinkage cavities.

11.2 MATERIAL SELECTION

ANSI-A356.0 (UNS A13560) is selected for the examination. The alloy is used in many industrial applications such as airframe castings, machine parts, truck chassis parts, aircraft and missile components, and structural parts requiring high strength. In this alloy, aluminum (Al) is the major metal and the distinctive alloying elements are copper, magnesium, manganese, silicon, and zinc. The basic chemical composition and mechanical properties are shown in Tables 11.1 and 11.2, respectively.

TABLE 11.1 Basic Chemical Composition of A356.0.

Component	Si	Cu	Mg	Ti	Fe	Mn	Zn
%.wt	6.5–7.5	0.2	0.25–0.45	0.2	0.2	0.2	0.1

TABLE 11.2 Typical Mechanical Properties of A356.0.

Tensile strength (N/mm^2)	Hardness (RBH)	Modulus of elasticity (N/mm^2)
230	75	71

11.3 METHODOLOGY

In this work, methodology consists of two steps. Step one is specimen preparation by casting and step two is testing the specimen as per the standard.

11.3.1 SPECIMEN PREPARATION

To prepare casting specimen, EN8 metallic die was prepared with mold diameter of 20 mm and length 160 mm. The die is robust to withstand the vibration. The specimen preparation work was performed in two phases. In the first phase, the castings were prepared without vibration. The A356.0 alloy bar was cut into small pieces. These small pieces were placed in the graphite crucible and this alloy was heated until it reached suitable pouring temperature (720°C) in the muffle furnaces. To this molten A356.0, small amount of degassing agent C_2CL_6 was added, and then the molten metal was stirred with a motor-operated stirrer. The molten metal poured into the metallic mold was coated with zirconium paint and prepared baseline preparation for microstructure and mechanical properties. This baseline was used to compare corresponding information obtained in second phase. In the second phase, the castings were prepared by adding mechanical mold vibrations during solidification. The mechanical mold vibrations were created using predetermined vibrating table (Fig. 11.1).

FIGURE 11.1 Vibrating table setup.

The vibrations were applied till the solidification of molten metal. Five test samples (two for tensile test and three for hardness test and microstructure) were casted at each frequency of vibration. In addition, other five test samples were casted without vibration—named as casting with 0 Hz. Castings were made at different vibration frequencies as 0, 10, and 20 Hz.

After casting, all the samples were processed by heat treatment. Some of the (two for each frequency) heat-treated samples were converted into required tensile test specimen as per E8 standard. The samples are shown in Figure 11.2. Remaining heat-treated samples were processed for hardness test and microstructure evaluation. Hardness and microstructure observed at four different (10 mm distance from top, bottom, and two intermediate) locations. Hardness samples are just ground at the surface with belt grinder, followed by polishing operation. The microstructure samples were processed under different metallography stages such as polishing on different silicon graded paper, fine polishing on double-disc polishing machine, and finally applying suitable etchant.

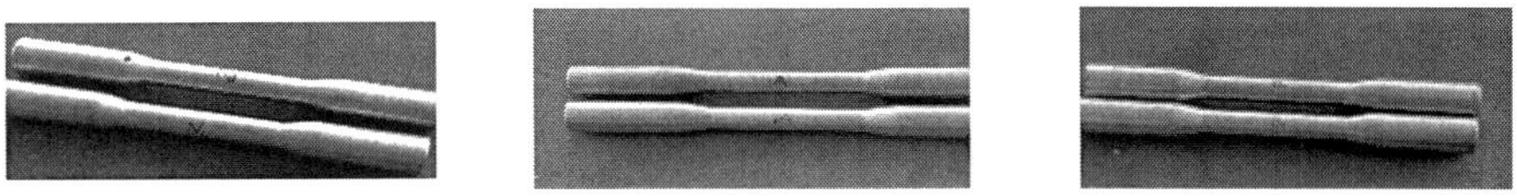

FIGURE 11.2 Tensile test specimen. (a) 0 Hz mold vibrations, (b) 10 Hz mold vibrations, (c) 20 Hz mold vibrations.

11.3.2 TESTING

The second step of the methodology is specimen testing. Three different tests were conducted in this research work: tensile test, hardness test, and microstructure test. The ultimate tensile strength of the samples was examined with computerized universal testing machine. The test was conducted on the samples of all the frequencies (0, 10, and 20 Hz). The results are tabulated in Table 11.3. Hardness of the samples was tested with rock well hardness testing machine. The hardness test was conducted on all hardness samples. The test results are tabulated in Table 11.4. Microstructure of each sample was tested with metallurgical microscope. The microstructure images are shown in Figure 11.3.

TABLE 11.3 Tensile Test Results without and with Vibration.

Mold vibrations (Hz)	Ultimate tensile strength (N/mm^2)		
	Trail 1	Trail 2	Average
0	225	258	241.5
10	221	229	225
20	222	202	212

TABLE 11.4 Hardness Values of the Samples at Different Levels of Each Frequency of Mold Vibrations.

Mold vibrations (Hz)	Number of specimens	Hardness (RHB) Number of trials			Hardness (RHB)
		1	2	3	Average
0	1	62	60	58	64.5
	2	35	58	74	
	3	62	60	74	
	4	64	67	66	
10	1	84	97	86	88.5
	2	90	87	89	
	3	81	84	83	
	4	90	97	93	
20	1	91	93	93	91.5
	2	86	95	92	
	3	88	92	88	
	4	97	88	91	

11.4 RESULTS AND DISCUSSIONS

11.4.1 TENSILE TEST

Graph 11.1 was plotted for the average results obtained from the tensile test. It is observed that the tensile strength decreased as the mold vibration frequency increased. It is a known fact that tensile strength and compressive strength are reciprocal to each other—as tensile strength is decreased, the compressive strength is increased. Therefore, based on this, we can conclude that if mold vibration increases, the compressive strength of the component also increases.

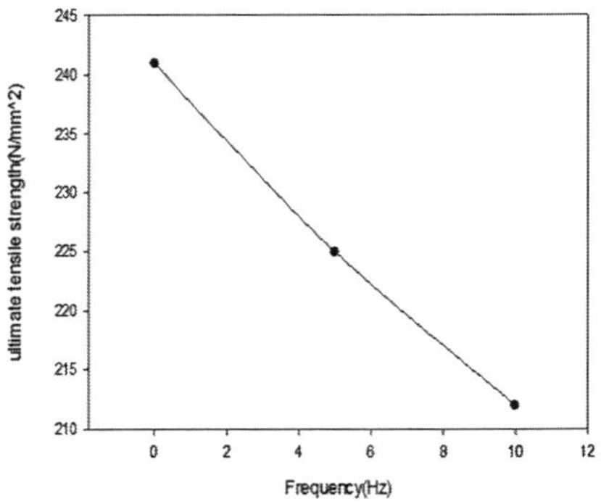

GRAPH 11.1 Frequency (Hz) versus tensile strength.

11.4.2 HARDNESS TEST

Hardness results of the cast samples are shown in Table 11.4. The graph (Graph 11.2) plotted for average hardness values at each frequency shows that hardness of casting increases with the increase in the mold vibration frequency because vibration causes refinement of grains. The highest hardness is obtained at the bottom of each and every sample because highest cooling rate and highest intensity of mold vibration occur at the bottom of the sample, which is evident from the microstructure. More cooling action promotes creation of fine grain, and at the same time, the vibration produces refined grains. It means that more the grain boundary, more is the hardness. Therefore, due to the combined effect of cooling rate and intensity of vibration, we can observe the remarkable increase in the hardness. The variation of hardness with mold vibrations is shown in Graph 11.2 and Table 11.4.

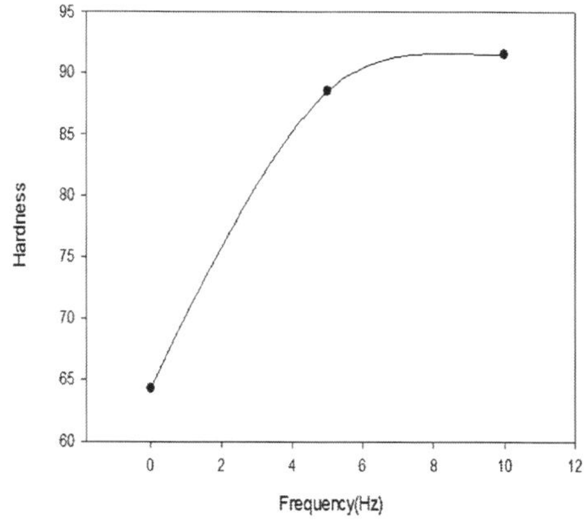

GRAPH 11.2 Frequency (Hz) versus hardness (RHB).

11.4.3 METALLOGRAPHY RESULTS

The microstructure images of different samples are shown in Figures 11.3a–c. Figure 11.3a is a microstructure of the sample without vibration; in this, we observe little bit coarse grains. In Figure 11.3b (microstructure of the sample 10 Hz vibration), refined grains are observed which are closer to each other. This effect is observed more in Figure 11.3c (microstructure of the sample 20 Hz vibration). Therefore, the refinement of grains can be observed with the increase of mold vibration frequency. Moreover, the compactness of the grains increases by increasing the mold vibration frequency.

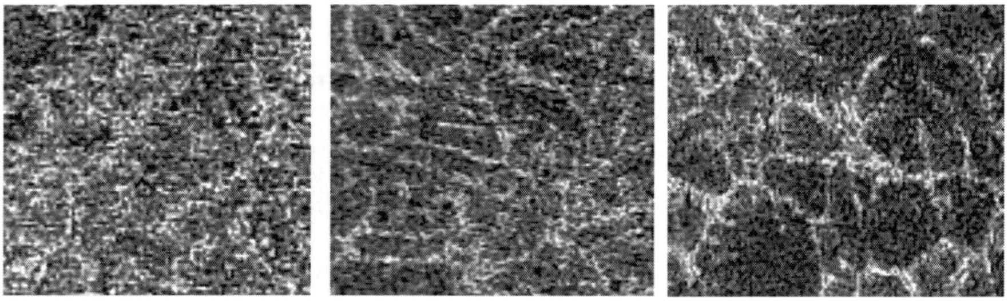

(a) (b) (c)

FIGURE 11.3 Microstructure structure at each frequency. (a) Without vibration, (b) at 10 Hz vibration, (c) at 20 Hz vibration.

11.5 CONCLUSION

The effect of mechanical mold vibration on mechanical properties and casting characteristics of A356.0 alloy was evaluated. Based on the experimental results, there are three main conclusions; first conclusion is by increasing the mold vibration frequency of the casting, the compressive strength increases while tensile strength is decreased. The increasing of compressive strength is best sign of the A356.0 casted product. Second conclusion is by increasing the mold vibration frequency of the casting, the hardness increases. This is also considered as a very good sign. Third and last conclusion is that grain refinement was achieved with the mold vibration, which leads to good strength of the casting. Further study can be focused on the effect of pouring temperature and effect of frequency of more than 20 Hz. Apart from the above, the effect of size variation of the casting and use of different materials can be studied as well.

KEYWORDS

- **mold vibrations**
- **solidification**
- **microstructure**
- **frequency**
- **grain refinement**

REFERENCES

1. Pîrvulescu, C. M.; Bratu, C. Mechanic Vibrations Generation System and Effect on the Casting Alloys Solidification Process. *U.P.B. Sci. Bull., Series B* **2010,** *72*(3), ISSN 1454–2331.
2. Deshpande, J. The Effect of Mechanical Mold Vibration on the Characteristics of Al-Alloy. Ph.D. Thesis, Worcester Polytechnic Institute. 2006.
3. Sujith Kumar, P.; Abhilash, E.; Joseph, M. A. Solidification under Mechanical Vibration: Variation in Metallurgical Structure of Gravity Die Cast A356 Aluminium Alloy", *International Conference on Frontiers in Mechanical Engineering (FIME)*, NIT Karnataka, 2010, 140-146.
4. Mishra, S. S.; Sahu, S. S.; Ray, V. Effect of Mold Vibration on Mechanical and Metallurgical Properties of Al-Cu Alloys. *Int. J. Technol. Res. Eng.* **2015,** *3*(1), ISSN (Online): 2347–4718.
5. Jian, X. Refinement of Eutectic Silicon Phase of Al A356 Alloy using High Ultrasonic Vibration. *Scr. Mater.* **2006,** *54,* 893–896.
6. Jian, X.; XU, H.; Meek, T. T.; Han, Q. Effect of Power Ultrasound on Solidification of Al A356 Alloy. *Mater. Lett.* **2005,** *59,* 190–193.
7. Zhi-qiang, Z.; Qi-chi, L.; Jian-zhong, C. *Trans. Nonferrous Met. Soc. China* **2008,***18* s113.
8. Pillai, R. M. *J. Mater. Process. Technol.* **2004,** *146,* 338–348.
9. Omura, N.; Murakami, Y.; Mingjun, L.; Tamura, T.; Miwa, K. Effects of Mechanical Vibration on Macrostructure and Mechanical Properties of AC4C Aluminum Alloy Castings. *Mater. Trans.* **2009,** *50*(11) 2578–2583.
10. Olufemi, A. F.; Ademola, I. S. Effects of Melt Vibration During Solidification on the Mechanical Property of Mg-Al-Zn Alloy. *Int. J. Metall. Eng.* **2012,** *1*(3), 40–43.
11. Cambell, J. *Int. Met. Rev.* **1981,** *26*(2), 71–108.
12. Dommaschk, C. Beitrag zur Gefugebeeinflussung Erstarrender Metallschmelzen Durch Vibration (Contribution to the Influence of Vibration on the Microstructure of Solidifying Molten Metals), Ph.D. Thesis, University of Freiberg, Germany, 2003.
13. Abu-Dheir, N. Solidification of Al-alloy, TMS, 2004, pp 361–368.
14. Ciućka, T. Influence of Vibrations during Crystallization on Mechanical Properties and Porosity of EN AC-AlSi17 Alloy. *Archives of Foundry Eng.* **2013,** *13*(1), 5–8.

CHAPTER 12

DEFORMATION AND FRACTURE BEHAVIOR AND THEIR IMPLICATIONS ON THE DESIGN OF A FORGED ALUMINUM ALLOY AA2219 IN T652 CONDITION

D. V. MADHURI, KRANTHI KUMAR GUDURU*,
SANTHOSHA KUMARI, and P. VARALAXMI

Mechanical Engineering, CJITS, Warangal, India

**Corresponding author. E-mail: Kranthicjits1@gmail.com*

ABSTRACT

T652 is the thermo mechanical treated (TMT) condition used to improve the mechanical properties and microstructure of metals. T652 is obtained by the stepwise procedure of heat treatment, quenching, and aging. Samples of AA2219 are taken for conducting this experiment. AA2219 is the alloying element of Cu with aluminum where AA stands for American associate. After the process is done, these samples are cut according to the American society for testing and materials (ASTM) specimen for the tensile test. Electrical discharge machining (EDM) wire cut is the machine used to cut the tensile and fracture samples. Observations show that the yield strength, ultimate tensile strength, and fracture toughness increases (under certain heat treatment conditions and total deformation takes place). Microstructures are viewed under the optical microscope after the metallographic procedure is done and also the Vickers micro hardness values are examined.

12.1 INTRODUCTION

High performance aluminum alloy is widely used in the fields of aviation and aerospace to manufacture some main structures and key components of an airplane or aircraft for its characteristics of low density and high strength. Generally, the high performance mechanical properties of this kind of aluminum alloy can be achieved by thermo mechanical treated (TMT) process, which includes three main procedures of the solid solution—quenching, plastic deformation, and aging. TMT process can give excellent performance of material by taking advantage of its deformation strengthening during plastic deformation and the transformation strengthening during heat treatment.[1] T652 is the temper designation for "artificial aging" condition, mentioned in Table 12.1. For T652 condition, heat treatment is conducted below the melting point and solution treatment is carried out with a temperature ranging from 55°C to 60°C. Then aging is done at a temperature where ample time is given for the alloy to retain properties according to the conditions it is treated in.[3] Metallography is done to know the difference in grain size and grain boundaries after the T652 condition. A piece of the main sample after the heat treatment is polished and etched with an appropriate reagent to reveal its grain boundaries in a microstructural view. These microstructure images are then captured. The Al alloy is aged at different conditions of limited under aging (LUA), under aging (UA), peak aging (PA), and over aging (OA). This is to make a difference in its hardness values. Microhardness test is performed on the sample under peak aging condition. The type of hardness measurement done is the Vickers hardness test, which is a simple and easy test to get the microhardness values. The amount of weight applied on it depends on its hardness and ductility. As aluminum has high ductility, an amount of 500 g of indent is applied for the test. Tensile deformation and fracture behavior of the AA2219 alloy is to be determined after artificial aging treatment.[5,6] For the tensile test, samples are cut into tensile sheet samples as per the ASTM standard.[9] The tensile sheet is fixed into the universal testing machine (UTM) for testing with holders at its ends. Load is applied according to the strength it can bear, that is, yield strength. After this, the sample starts necking at the plastic deformation and then breaking off. The ultimate tensile strength (UTS), yield strength, and % elongation readings are observed.

Metallography is done to know the difference in grain size and grain boundaries after the T652 condition. A piece of the main sample after the heat treatment is polished and etched with an appropriate reagent to reveal its grain boundaries in microstructural view. These microstructure images are then captured. The Al alloy is aged at different conditions of LUA, UA,

PA, and OA. This is to make a difference in its hardness values. Microhardness test is performed on the sample under peak aging condition. The type of hardness measurement done is the Vickers hardness test, which is a simple and easy test to get the microhardness values. The amount of weight applied on it depends on its hardness and ductility. As aluminum has high ductility, an amount of 500 g of indent is applied for the test. Tensile deformation and fracture behavior of the AA2219 alloy is to be determined after the artificial aging treatment.[5] For the tensile test, samples are cut into tensile sheet samples as per the ASTM standards.[9] The tensile sheet is fixed into the UTM for testing with holders at its ends. Load is applied according to the strength it can bear, that is, yield strength. After this, the sample starts necking at the plastic deformation and then breaking off. The UTS, yield strength, and % elongation readings are observed.

TABLE 12.1 Implications of T652 Condition.

Temper designation	Process	Temper alloys (T)	Product form
T652	Solution heat treated and artificially aged	2014, 2219, 6061, 6151, 7075	Forging and forging stock

12.2 EXPERIMENTAL STUDIES

12.2.1 MATERIAL

The material used in this experiment is 2219 aluminum alloy[2] sheet under annealed condition with 6.5 mm thickness, cut using electrical discharge machining (EDM).[7] The main chemical constituent is listed in Table 12.2 and parameters of TMT in Table 12.3.

TABLE 12.2 Chemical Composition of AA2219.

Elements	Al	Cu	Mn	Ti	V	Zr
Weight (%)	93	6.3	0.3	0.06	0.10	0.18

TABLE 12.3 Parameters of TMT Process of 2219 Aluminum Alloy.

Process	Temperature	Time
Heat treatment	535°C	90 min
Quenching	50–60°C, Water quenching	30 min
Aging	191°C	26 h

12.2.2 DEFORMATION METHOD

Uniaxial tensile test is known as a basic and universal engineering test to achieve material parameters such as ultimate strength, yield strength, % elongation, % area of reduction and Young's modulus.[9] These important parameters obtained from the standard tensile testing are useful for the selection of engineering materials for any applications required. The specimen is subjected to an external tensile loading, the metal will undergo elastic and plastic deformation, as shown Figure 12.1. Initially, the metal will elastically deform giving a linear relationship of load and extension.[4] These two parameters are then used for the calculation of the engineering stress and engineering strain to give a relationship as illustrated in the figure below using equations as follows

$$\sigma = P / A_o \tag{12.1}$$

$$\epsilon = (L_f - L_o) / L_o = \Delta L / L_o \tag{12.2}$$

FIGURE 12.1 Deformed tensile samples.

12.2.3 FRACTURE METHOD

Fracture toughness is defined as the amount of stress required to propagate a preexisting flaw. Flaws may appear as cracks, voids, metallurgical inclusion, weld defects, and design discontinuities. The orientation of crack and the direction of the crack, which is relative to loading, determine the load of fracture. Fracture occurs when the tensile stress at some crack tip exceeds the theoretical cohesive strength of the material.[3] A parameter called stress intensity factor (K) determines the fracture toughness of the material. The stress intensity factor is a function of loading, crack size, and geometry of the sample, as shown in Figures 12.2 and 12.3 using ANSYS, and in Figures 12.4 and 12.5 after the facture. The stress intensity factor is calculated as:

$$K_1 = \sigma\sqrt{\pi a \beta}$$

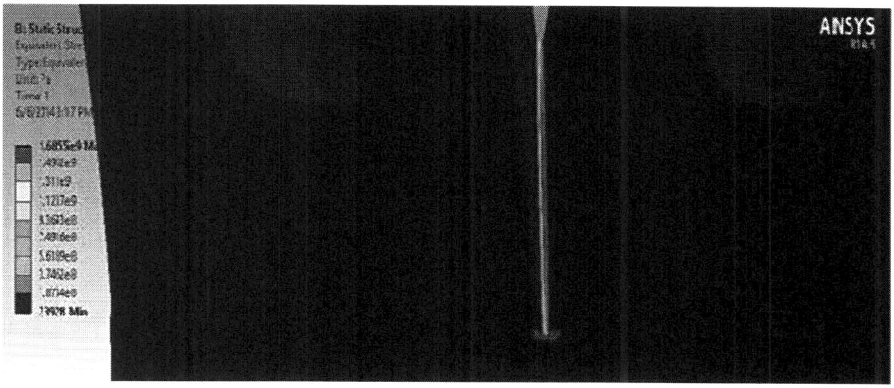

FIGURE 12.2 von Mises stress in ANSYS.

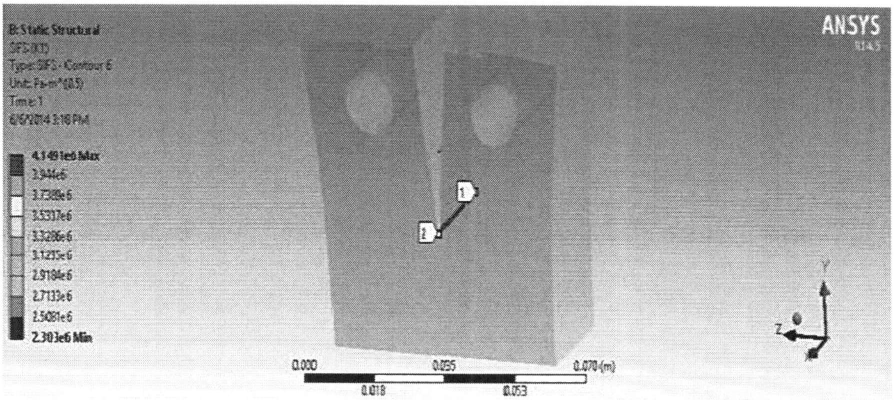

FIGURE 12.3 Stress intensity factor (K1) in ANSYS.

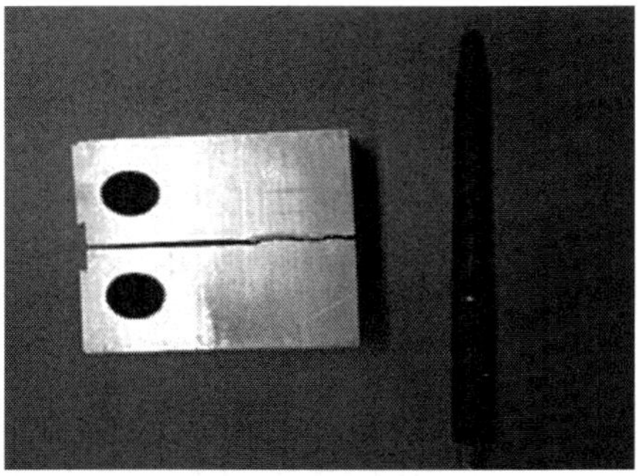

FIGURE 12.4 Tested fracture sample 1.

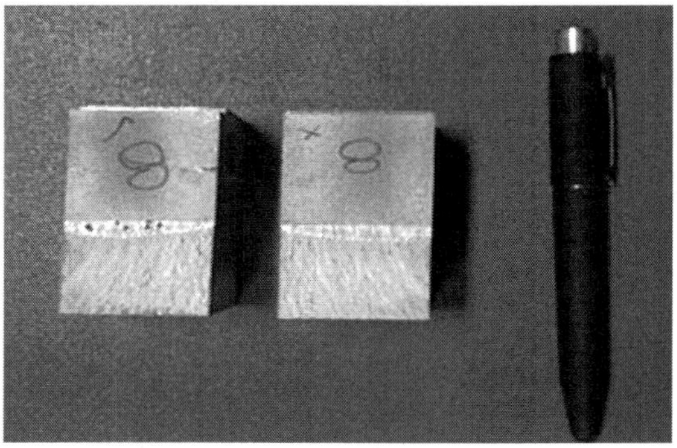

FIGURE 12.5 Tested fracture sample 2.

12.3 RESULTS

12.3.1 MECHANICAL PROPERTIES

The results of the mechanical properties[8] by a uniaxial tensile test are given in Tables 12.4 and 12.5.

TABLE 12.4 Mechanical Properties of Tensile Tested Samples before Aging.

Samples	Yield strength	Ultimate tensile strength (UTS)	Elongation
	(MPa)	(MPa)	(%)
1.	222.5	363.62	5
2.	219.17	352.48	4

TABLE 12.5 Mechanical Properties of Tensile Tested Samples after Aging.

Samples	Yield strength	UTS	% elongation
	(MPa)	(MPa)	
1.	288	377	6
2.	287	371	5.3

The results of mechanical properties by Fracture test for a load of 15 kN are given.

12.3.2 NONAGED RESULTS IN ANSYS

Displacement = 0.16 mm
where K1 = K1c

$$K1 = K1c = 4.1MPa$$

K1 (stress intensity factor) can be considered approximately equal to K1C (plain strain fracture toughness)[10] when the thickness of the component is the same. In the analysis of nonaged conditions, the thickness taken is equal to the practical design, mechanical properties of fracture-tested samples after aging mentioned in Table 12.6.

TABLE 12.6 Mechanical Properties of Fracture-Tested Samples After Aging.

S. No.	Displacement	Critical stress intensity factor (KIC) (MPa)
1	0.52	27Ms
2	0.49	23
Average	0.505	25

12.3.3 UNITS AND NUMBERS

The international system of units (SI units) is used in the Institute of Electrical and Electronics Engineers (IEEE) publications. Unit symbols should be used with measured quantities, for example, 1 mm, but not when unit names are used in the text without quantities, for example, “a few millimeters.” Use a zero before decimal points: “0.25,” not “0.25.” Include a space between the number and the unit label when used as a noun. Replace the space with a hyphen when used as an adjective. For example, “The 10-GHz antennas now operate at 9.8 GHz." The hyphen makes it clear that we are specifying frequency and not the number of antennas.

12.3.4 MICROSTRUCTURES

These microstructures of AA2219 are taken after the aging is done. The enlarged grain boundaries are revealed and precipitate formation is observed. Microstructures of AA2219 before aging are shown in Figures 12.6 and 12.7, and after aging are shown in Figures 12.8 and 12.9.

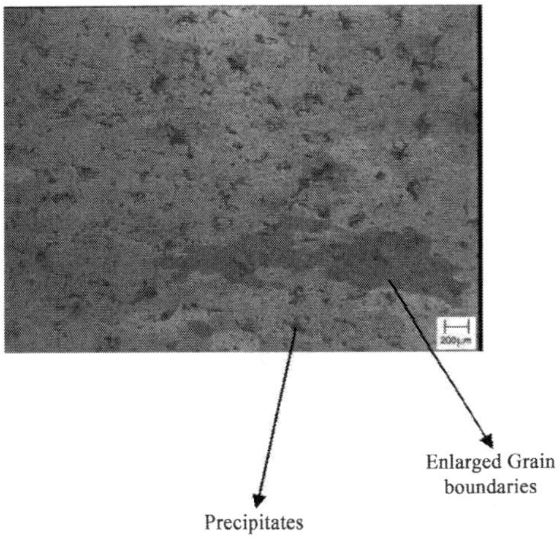

FIGURE 12.6 Sample 1 at 50X enlarged grain boundaries precipitates.

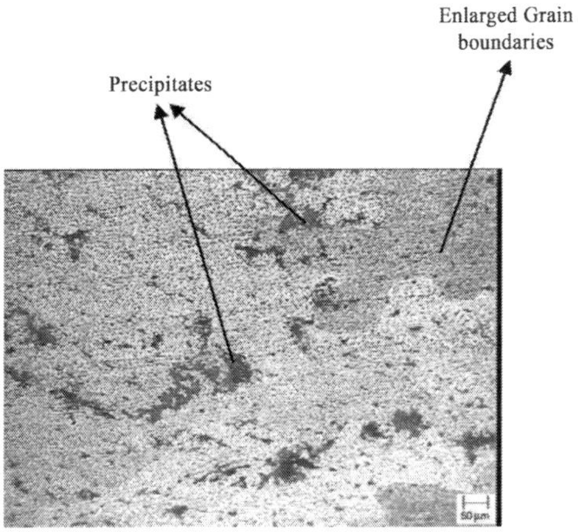

FIGURE 12.7 Sample 1 at 200X.

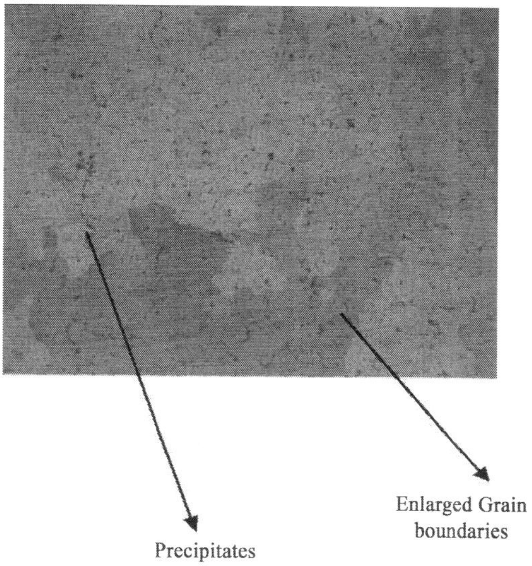

FIGURE 12.8 Sample 2 at 50X enlarged grain boundaries precipitates.

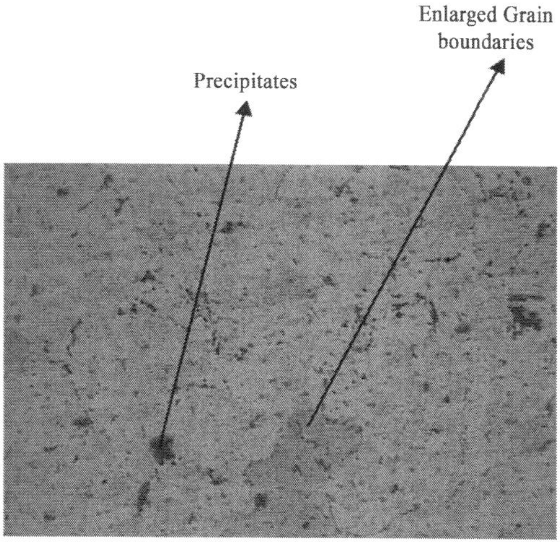

FIGURE 12.9 Sample 2 at 200X.

From the seismic electron microscope the above microstructures are found in precipitates and grain sizes which enlarged with their grain boundaries in 50 and 200X with each two samples.

12.4 CONCLUSION

The effects on the microstructures and tensile properties of an Al-Cu AA2014-based alloy were investigated in the heat treatment and aging condition. From the analysis of the results obtained, the following conclusions may be drawn. There have been a substantial number of experiments conducted on the Al-Cu AA2219 to obtain the optimum results. This review shows vital importance to the peak aging condition where experiments have been conducted on samples to determine the yield strength, UTS, % elongation, displacement and K1C values of the alloy by varying temperature conditions during heat treatment and aging. In addition to this, static analysis is conducted in ANSYS for measuring the fracture toughness in the nonaged condition where the standard properties of AA2219 were considered. There is a clear improvement in all the values of mechanical properties for the peak aged condition after both these procedures.

ACKNOWLEDGMENT

I sincerely acknowledge our guide, technicians, and friends for their valuable contribution towards this research work.

KEYWORDS

- **T652 condition**
- **AA2219**
- **tensile deformation**
- **fracture toughness**
- **metallography**
- **aging**

REFERENCES

1. An, L.-h.; Cai, Y.; Liu, W.; Yuan, S.-J.; Zhu, S.-Q.; Meng, F.-C. Effect of Pre-Deformation on Microstructure and Mechanical Properties of 2219 Aluminum Alloy Sheet by Thermo Mechanical Treatment. *Trans. Nonferrous Met. Soc. China* **2012,** *22,* 370–375.

2. Chen, Y.; Liu, H.; Feng, J. Friction Stir Welding Characteristics of Different Heat-Treated-State 2219 Aluminum Alloy Plates. *Mater. Sci. Eng.* **2006,** *420,* 21–25.
3. Christopher, T.; Sankaranarayanasamy, K.; Rao, N. B. Failure Assessment on Tensile Cracke Specimens of Aluminum Alloys. *ASME* **2004,** *19,* 123–126.
4. Floreen, S.; Ragone, D. V. The Fluidity of Some Aluminium Alloys. *AFS Trans.* **1958,** *70,* 391–393.
5. Patel, M.; Pandey, P. M.; Rao, V. Study on Machinabilty of Al_2O_3 Ceramic Composite in EDM Using Response Surface Methodology. *J. Met. Mater.* **2014,** *45,* 345–350.
6. Ravi, K. R.; Pillai, R. M.; Amaranathan, K. R.; Pai, B. C.; Chakraborty, M. Fluidity of Aluminum Alloys and Composites: A Review. *J. Alloys Comp.* **2008,** *456,* 201–210.
7. Rzychon, T.; Kielbus, A. The Influence of Pouring Temperature on the Microstructure and Fluidity of AE42 Alloy. *Arch. Mat. Sci. Eng.* **2007,** *28,* 345–348.
8. Sabatino, M. D.; Arlberg, L.; Brusethaug, S.; Apelian, D. Fluidity Evaluation Methods for Al-Mg-Si Alloys. *Int. J. Cast Met. Res.* **2006,** *19,* 94–97.
9. Sheshadri, M. R.; Ramachandran, A. Casting Fluidity and Fluidity of Aluminium and Its Alloys. *AFS Trans.* **1965,** *73,* 292–304.
10. Wu, L.-M.; Wang, W.-H.; Hsu, Y.-F.; Trong, S. Effects of Microstructure on the Mechanical Properties and Stress Corrosion Cracking of an Al-Zn-Mg-Sc-Zr Alloy by Various Temper Treatments. *Jpn. Inst. Met. Mater. Trans.* **2007,** *48*(3), 600–609.

CHAPTER 13

SAMPLING STRATEGY OF FREE-FORM SURFACE EVALUATION USING A COORDINATE MEASURING MACHINE

GOITOM TESFAYE* and REGA RAJENDRA

Department of Mechanical Engineering, College of Engineering, Osmania University, Hyderabad, Telangana, India

Corresponding author. E-mail: goitom.tesfaye@gmail.com

ABSTRACT

Free-form features find wider applications in the dies and molds, patterns and models, plastic products, automotive, aerospace, biomedical, entertainment, and geographical data processing. If free-form surfaces have such wide application, evaluation of free-form using coordinate measuring machine (CMM) is the crucial topic. One method of free-form surface evaluation is taking discrete sample points from the surface by CMM, but this sampling strategy found trade-off relationship between a number of samples and surface deviation of the actual from nominal. Moreover, the good strategy is needed that optimize this trade-off relation. In this chapter, sampling strategy based on the contour of the free-form surface is proposed to take a discrete sample by CMM form surface, since the contour curves are denser at the area with high curvature and less dense with less surface curvature which is the driving force of this method. The virtual surface is compared with the actual one. The sample points are distributed on the actual free-form surface by data point parallel projection method and iteratively sampled from the surface. A comparison is performed between the proposed method and the well-known sampling methods, which is equi-parametric method combined with mean patch Gaussian curvature method; experiment results show the effectiveness and robustness of this method.

13.1 INTRODUCTION

Free-form surfaces have wide application in industries such as aerospace, automotive, consumer goods, and die and mold making. Recently, manufacturing industries are involved in the batch production, high-variety production, and tight tolerance high-quality products. To make the production industry competitive, reliable, and efficient, product inspection of the free-form surface is important to meet the strict quality requirements and to keep up with the frequent variations in product design. In surface manufacturing, coordinate measuring machines (CMM) is one method to achieve this quality. The main problem with using CMM is existing trade-off relationship between the numbers of samples and measuring time. In addition, the strategy should be easily adapted at the shop floor level and good space filling properties to reduce the number of sample points. Machining of the free-form surface is carried out by computer numerical control (CNC) machine and the data for machining is extracted from the computer-aided design (CAD) model, involving various sources of error, which cause undesirable deviations in the machined surface. This gave rise to coordinate metrology as a crucial means to test the conformity of the physical object to its CAD model.[1] To test the conformity of the surface, the first step is to obtain the measurement data. Measured data can be acquired either by discrete data point using mechanical contact probes such as CMM touch probes or 3D topography measuring systems using scanning method which are widely used in practical applications.

CMMs with touch-triggered probes can provide high measurement accuracy at the submicron level. However, the measurement speed is much lower than a 3D vision system. A scanning system can acquire thousands of data points over a large spatial range at a time.[2] However, the resolution of the measured data point is relatively low.[3] Therefore, in practical applications, using one of the techniques means that the user has to suffer its limitations. The CMM with touch-triggered technique involves determining the coordinate values of the measurement points located on the surface of the analyzed object. As a result, a set of discrete data in the form of coordinate measurement points is obtained. When using CAD/CAM techniques, an essential aspect of coordinate measurements is to provide relevant digital data concerning the geometry of the workpiece.[4] During sampling of the surface using coordinate metrology, the number of measuring points that accurately represent the surface and the distribution of those points are the important factors, which help us to minimize sampling effort and (non-value-added) measurement time.[5]

Measurement strategy in coordinate measuring technique includes a lot of aspects. The best-known ones are the issues concerning probing strategy; it means the problems referring to the number and distribution of measurement points.[6] Many researches have been done to optimize the trade-off relation between the numbers of samples and measuring times as well as inspection cost; some of these researches are summarized as below. Diaa and Stephen[7] introduced sampling plane of the surface using CMM probe head based on scanning isoparametric lines on the sculptured surface. Suleiman et al.[8] proposed sampling strategy on small-scale surface based on the complexity and the size of the patches of the surface. Mingrang et al.[9] introduced new sampling strategy on basis of uniform surface area and dominant points principle. Rajamohan et al[10] recommended an adaptive sampling approach based on advanced path-detecting method. Poniatowska[11] advised the sampling of the free-form surface based on the deviation model of the free-form surface. Generally, sampling strategies are divided into three major categories: blind sampling strategy, adaptive sampling strategy, and manufacturing signature-based sampling strategy.[12]

In this chapter, a third-degree free-form surface is modeled using Rhinoceros 5 (student version) software and same is machined on CNC mill using CNC code. Discrete samples are collected by using touch-triggered probe CMM. A comparison is made with the nominal free-form surface (design) and the machined free-form surface. In this research, the sample number and their location are determined from the contour intersection. The contribution of the chapter is easily determining the number, locations of samples data that approximately describe the surface to the desired accuracy level without searching for complex methods, ease of sampling from the surface, and easy to adapt to a production environment.

13.1.1 NONUNIFORM RATIONAL BASIS SPLINE FREE-FORM SURFACE FOR MACHINING

In this research, the free-form surfaces are nonuniform rational basis spline (NURBS) surfaces which the control points have generated from the design mentor and then fed into SolidWork 2015 (student version) for further surface modeling as shown in Figure 13.1. As the degree of the model increases, it is difficult to control, but in design mentor, we can easily control the knots and control points of the surface.

The equation of NURBS surface is

$$S(u,v)=\sum_{i=0}^{n}\sum_{j=0}^{m}R_{ij}(u,v)P_{ij} \tag{13.1}$$

$$R_{ij}(u,v) = \frac{N_{i,p}(u)N_{j,q}(v)w_{i,j}}{\sum_{k=0}^{n}\sum_{i=0}^{m} N_{k,p}(u)N_{i,q}(v)w_{k,i}} \tag{13.2}$$

where $N_{i,p}(u)$ and $N_{i,q}(v)$ are the base functions along u and v direction, $P_{i,j}$ is control points, and $W_{i,j}$ is the weight of each control points.

FIGURE 13.1 Free-form surface model of third degree.

13.1.2 OVERVIEW CONTOUR OF FREE-FORM SURFACE

The overall process of surface contouring, shown in Figure 13.2, is described through analytical analysis, first attempt to extract sectional contours curves from the 3D CAD model. A vector normal to the slicing plane direction (n) is defined by the user using two points (p_1 and p_2). The 3D model length along this directional normal is defined as overall model length (L).[13]

For a given N number of parallel planes (10, 20, 30… which is the distance between the parallel planes from the reference plane; Fig. 13.2a)

$$L = n\,(p_1 - p_2) \tag{13.3}$$

where L: overall model length of the surface,
n: directional normal,
p_1 and p_2: the two extreme points on the surface.

The gap between the parallel plane that contouring the solid model is constant and can be determined by using this formula

$$t = \frac{L}{N-1} \tag{13.4}$$

where t is the gap between the parallel planes.

After contouring of free-form surface, the contoured sectional curves are intersected by perpendicular parallel planes as shown in Figure 13.2c to find the sample points of the free-form surface in Figure 13.2d. Using these intersection points, the discrete data point is measured by CMM.

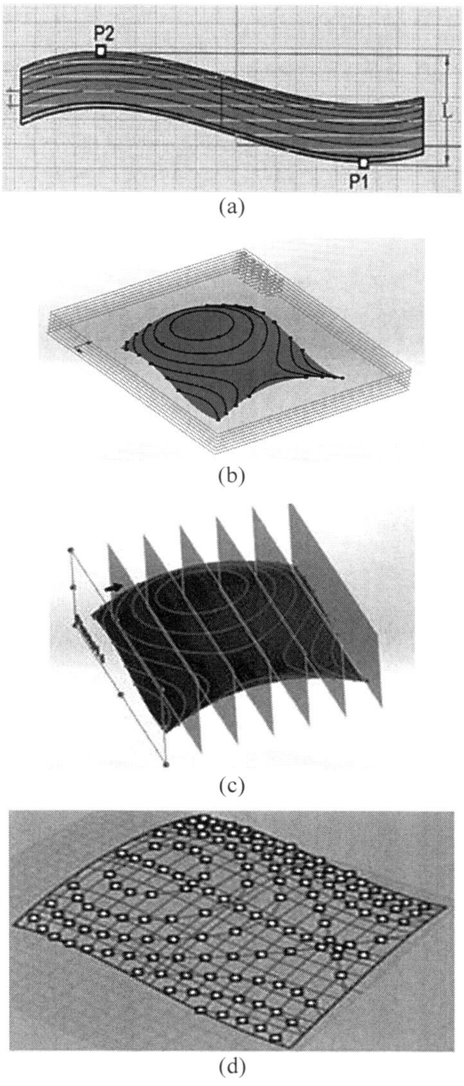

FIGURE 13.2 (a–d) Procedure of contour making of the free-form surface by parallel plane slicing and creating intersection points for sampling in solid work and Rhinoceros 5 (student version).

13.1.3 METHODOLOGY OF THE EXPERIMENTAL

In our experiment, the free-form surface contour is generated by parallel plane slicing principle. The maximum height distance of the surface is considered during slicing the surface. For detail clarification, we express the procedure that is followed through the flow chart in Figure 13.3 an the below steps.

a) Machining of the free-form surface is carried out by three-axis vertical milling CNC machine model of VML800 at a spindle speed of 2000 rpm with accuracy ± 0.005 mm.
b) We create the free-form surface contour of length "*L*" from the 3D solid model as shown in Figure 13.2a.
c) To generate the sample point location and number, we intersect the surface contour by using "*N*" number of parallel planes (Fig. 13.2c).
d) The intersection between the "*N*" parallel plane and the surface contour is considering as samples points of the surface. Moreover, we transfer these sample points from the CAD model to the actual machined surface by using point parallel projection method (Fig. 13.2d).
e) Then the CMM is used to take sample points by using these intersection points.
f) We measure the free-form surface deviation (actual one from the nominal surface) and check if the deviation is within the tolerance or not.
g) If the deviation is not within the tolerance limit, then we increase the contour curve until the deviation is within the tolerance.
h) Finally, we save the number of sample points as output.

The procedure of the methodology is explained in a summarized way by using flow diagram.

13.1.4 DEVIATION CALCULATION

Prior to the determination of the surface deviation, it is necessary to fit the measurement data to the nominal design surface. For each measurement point, the value of the local geometric deviation, that is, the distance of this measurement point from the CAD model of the nominal surface in the normal direction is established. In this chapter, the maximum surface deviation of 0.1 mm is considered and uncertainty of the CMM is evaluated as per ISO 15530-2 for multiple strategies in the calibration of artifacts.[14] The average

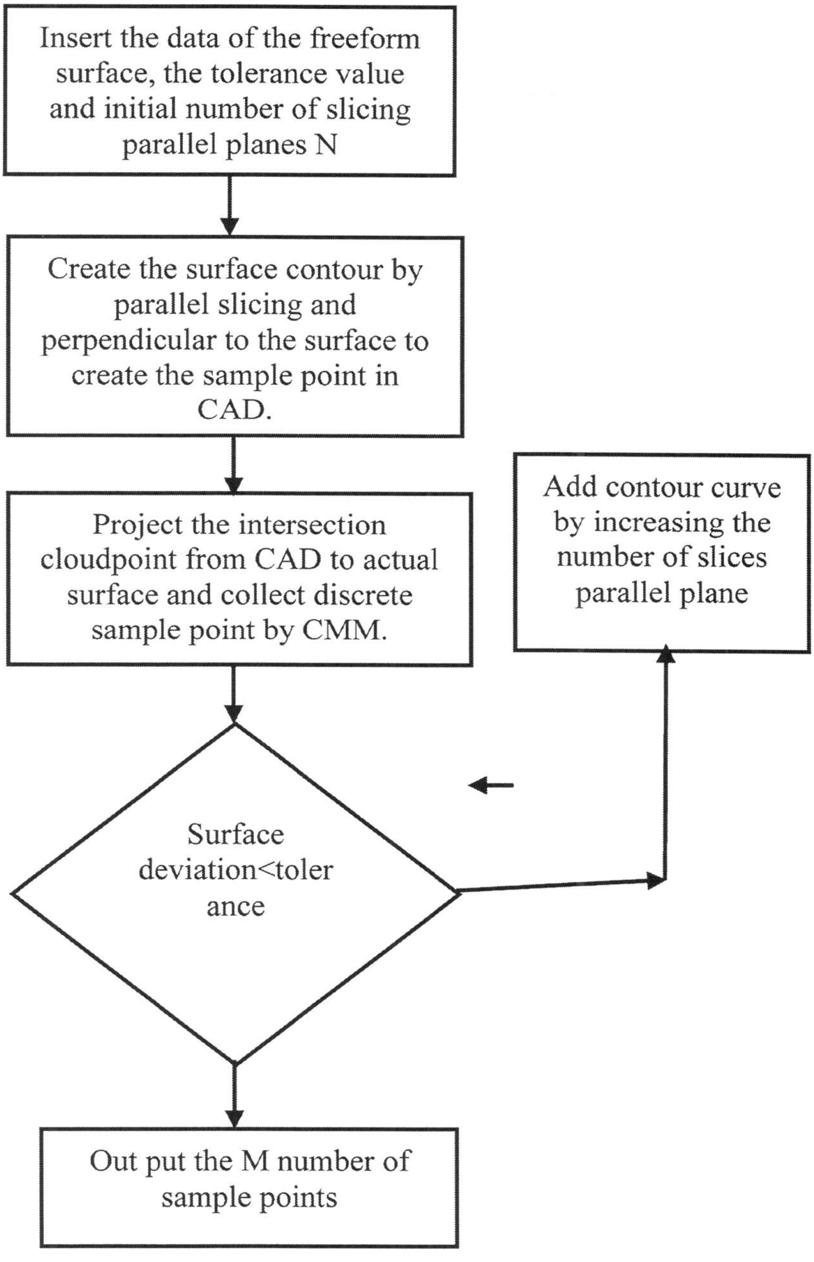

FIGURE 13.3 Flow diagram of the methodology of our work.

deviation of the surface is considered for the actual deviation between the surfaces. The machined surface is described as follows: an ideal (nominal) shape of the surface element can be described by the shape function *N(p)*, where *p* denotes feature variables describing the surface.

$$A\,(p) = N\,(p) + d\,(p) \tag{13.5}$$

$$d(p) = \hat{n}\left[A(p) - N(p)\right] \tag{13.6}$$

where: *A(p)*: the actual geometric form of the surface,
N(p): nominal geometric form of the surface,
$\hat{n}$: normal vector to the surface.

In the coordinates of measurement, points are sampled on the surface. Before the surface deviation is calculated, the sample points should fit to be able to construct the surface as shown in Figure 13.4. In this research, reconstruction and fitting are done by the mesh\Delaunay 2.5D (best-fit plane) method. The principle of this method is the 3D cloud point and is first projected on the best-fitting plane (least squares). Then the corresponding 2D points are triangulated and the mesh structure is applied to the 3D points.[15] Finally, surface smoothing is applied to approximate the surface (Fig. 13.4b).

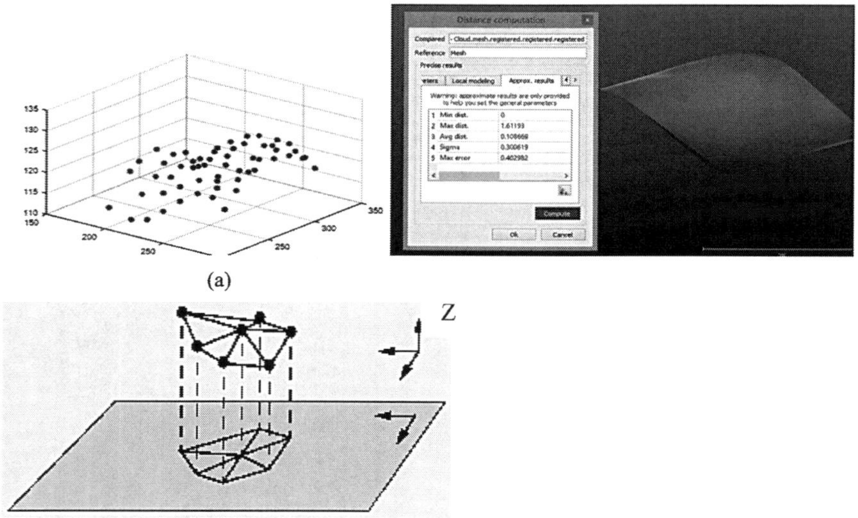

FIGURE 13.4 (a) Coordinate measuring machine (CMM) cloud data points before surface fitting in MATLAB (student version) and (b) concept mesh\Delaunay 2.5D CMM cloud data points fitting.

13.2 EXPERIMENTAL RESULTS

The free-form surfaces are measured by MITUTOYO: a model of Crysta-Plus M544 CMM with TP200 probe and stylus diameter of 2 mm calibrated by a master ball of diameter 19.99620 mm. Measurement is taken at an environmental temperature of 20°C ± 0.2, depending on the ISO 1036-2 standard accuracy. The maximum permissible error is E = (3.5 + 4.5L/1000) μm. The samples are distributed based on the abovementioned methods as shown in Figure 13.5. Models are made up of aluminum alloy (Al6061) with a surface dimension of 100 mm × 100 mm. The sample distribution of the equi-parametric combined with Gaussian curvature as shown in Figure13.6 and the contour based sampling explained in detail in Figure 13.7.

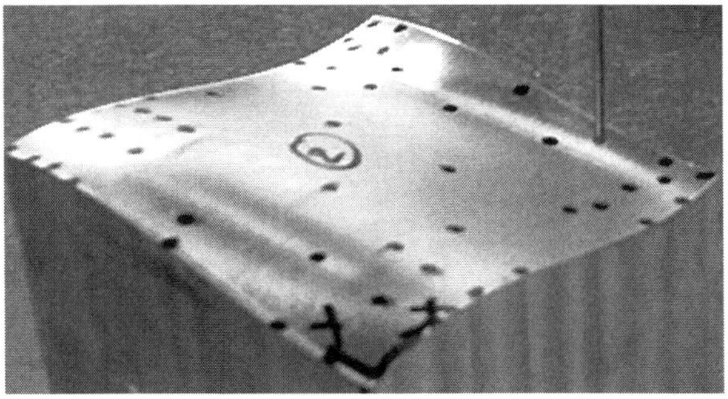

FIGURE 13.5 Actual free-form surface with discrete sample distribution.

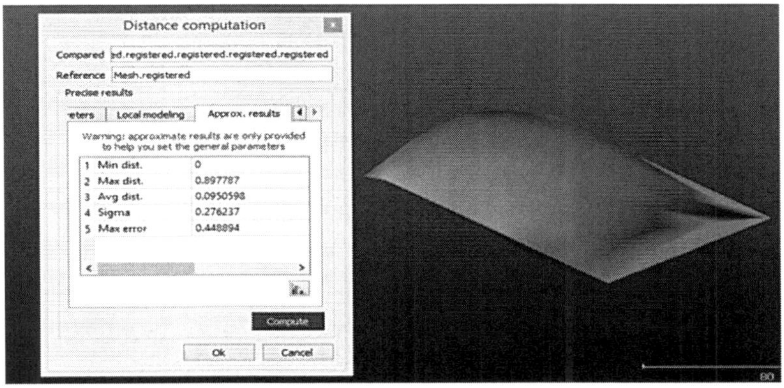

FIGURE 13.6 Equi-parameter and Gaussian curvature patch sampling surface deviation (a) Model 2 and (b) Model 1 in cloud compare software.

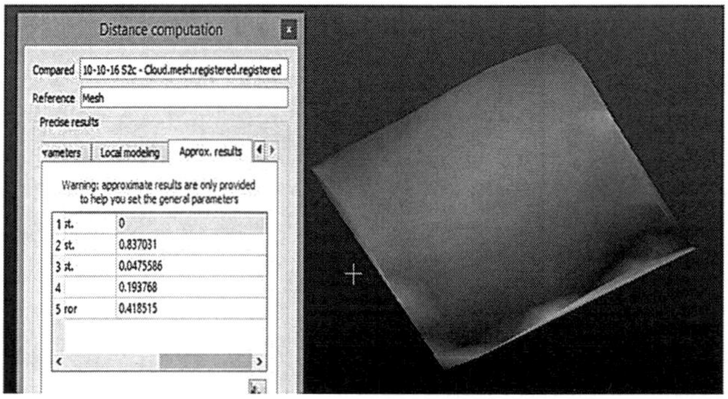

(a)

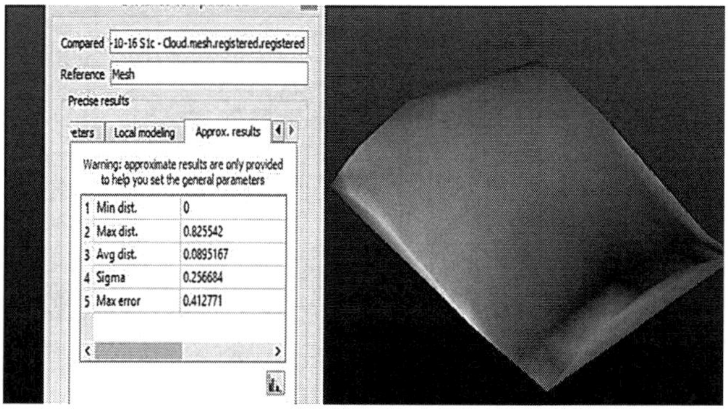

(b)

FIGURE 13.7 Contour sampling surface deviation in cloud compares (a) model 2 and (b) model 1 in cloud compare software.

13.3 DISCUSSIONS AND CONCLUSIONS

From Figures 13.8 and 13.9, Gaussian distribution, histogram surface deviation of each model has a deviation with almost zero means. This indicates us that the distribution of our cloud data point as compared to the actual one is a normal distribution and the form tolerance of the free-form surface is single-handed or upper specification limit (ISO 14253-1). Approximately 95% of the samples are within the tolerance value. From Table 13.1 comparison of the surface deviation of Model 1 using equi-parametric combined mean Gaussian curvature patch method and contour method, the number of samples and surface deviation are almost equal. However, in Model 2, the contour method reduces the deviation by 50%. This is shown that the surface curvature increases, the equi-parametric combined with Gaussian curvature patch and the contour method has equal effectiveness.

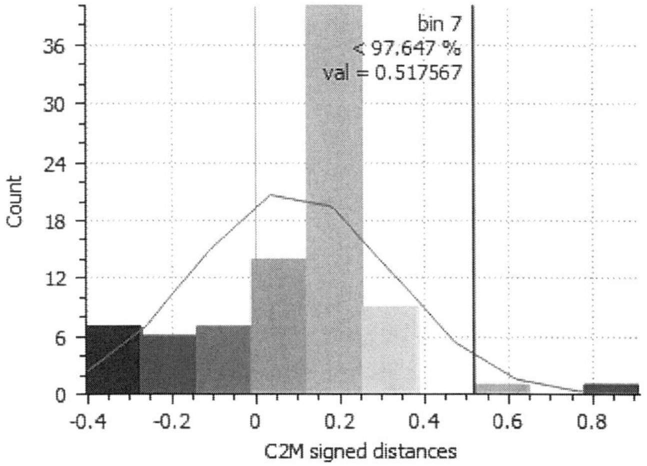

(a)

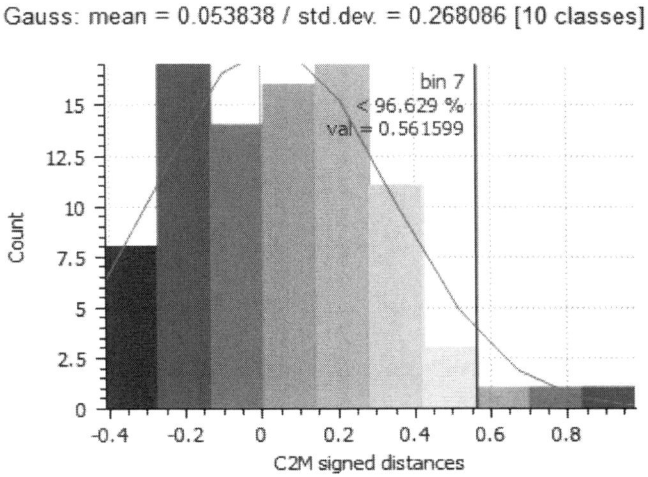

(b)

FIGURE 13.8 Gaussian distribution of samples using equi-parametric and Gaussian curvature patch method (a) model 1 and (b) model 2.

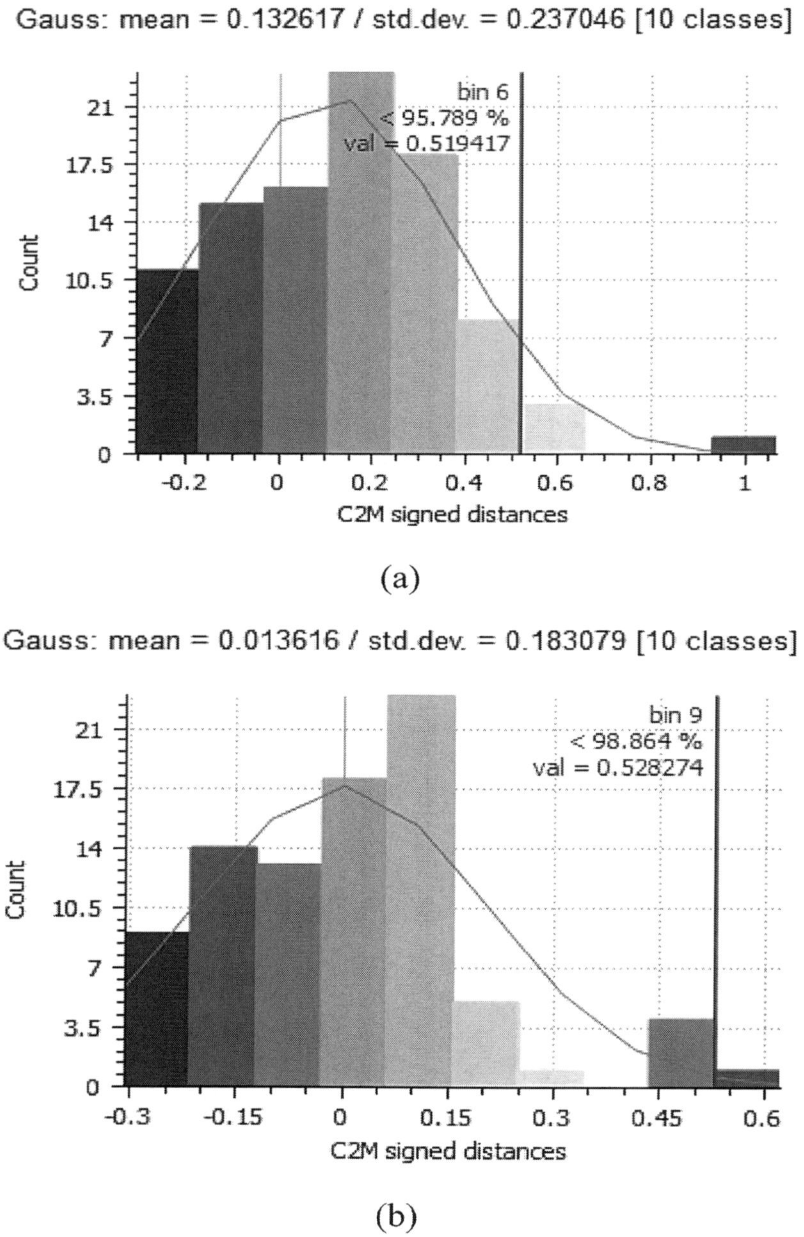

FIGURE 13.9 Gaussian distribution of sample using contour method (a) model 1 and (b) model 2.

TABLE 13.1 Actual Free-form Surface Deviation from the Computer-Aided Design (CAD) Model and Number of Samples.

	Model 1		Model 2	
	Contour method	**Equi-parametric and Gaussian curvature patch**	**Contour method**	**Equi-parametric and Gaussian curvature patch**
Number of samples	83	84	88	89
Average distance (mm)	0.0895	0.0951	0.0476	0.101

The proposed method is tested on two models with two different free-form surfaces for the distributed scatter points and the same is evaluated for comparison with equi-parametric and Gaussian curvature patch sampling method for performance prediction of complex surface deviation. As our method uses surface contour sampling strategy, the samples are nearly equal to equi-parametric and Gaussian curvature patch sampling method, but on both models, the surface deviation is reduced by nearly 25%. It is easy to adapt to a production environment for measuring complex surfaces and avoid the complex calculation and algorithm for allocation and determination number of sample point of the surfaces.

KEYWORDS

- **coordinate measuring machine**
- **sampling**
- **free-form surface**
- **form**
- **deviation**

REFERENCES

1. ElKott, D. F.; Veldhuis, S. C. CAD-Based Sampling for CMM Inspection of Models with Sculptured Features. *Eng. Comput.* **2007,** *23,* 187–206.
2. Yi, X.; Zexiang, L. Method for Determining the Probing Points for Efficient Measurement of Freeform Surface. *Int. J. Mech. Aerosp. Ind. Mech. Manuf. Eng.* **2008,** *2*(2), 230–235.
3. Li, Y.; Nomula, P. R. Surface-Opening Feature Measuring Using Coordinate Measuring Machines. *Int. J. Adv. Manuf. Technol.* **2015,** *79,* 1915–1929.

4. Poniatowska, M. Characteristics of Geometric Errors Determined Using Discrete Measurement Data. *Archiwum Technologii Maszyn i Automatyzacji* **2008,** *28*(2), 51–58.
5. Suleiman, M.; Shivakumar, R. An Intelligent Sampling Method for Inspecting Free-Form Surfaces. *Int. J. Adv. Manuf. Technol.* **2009,** *40,* 1125–1136.
6. Marcin, S.; Wladyslaw, J. Optimisation of Measuring Strategies in Coordinate Measuring Technique. *Meas. Sci. Rev.* **2001,** *1*(1), 191–194.
7. Diaa, F.; Stephen, C. Isoparametric Line Sampling for the Inspection Planning of Sculptured Surfaces. *Comput. Aided Des.* **2005,** *37,* 189–200.
8. Suleiman, M.; Rami, H.; Nabeel, M. Verification of New Sampling Methods on Small Scale Freeform Surfaces. *Jordan J. Mech. Ind. Eng.* **2012,** *6*(1), 1–9.
9. Mingrang, Y.; Yingjie, Z.; Yunlong, L.; Zhang, D. An Adaptive Sampling Approach for Digitizing Unknown Free-Form Surfaces Based on Advanced Path Detecting. *12th CIRP Conference on Computer Aided Tolerancing*, Xiasha Higher Education Zone, Hangzhou 310018, P.R. China, 2013.
10. Rajamohan, G.; Shunmugam, M.; Samuel, G. Practical Measurement Strategies for Verification of Freeform Surfaces Using Coordinate Measuring Machines. *Metrol. Meas. Syst.* **2011,** *XVIII*(2), 209–222.
11. Poniatowska, M. Deviation Model Based Method of Planning Accuracy Inspection of Free-Form Surfaces Using CMMs. *Measurement* **2012,** *45,* 927–937.
12. Bianca Maria, C.; Giovanni, M.; Stefano, P. A Tolerance Interval Based Criterion for Optimizing Discrete Point Sampling Strategies. *Precis. Eng.* **2010,** *34,* 745–754.
13. Sareen, K. K.; Knopf, G. K.; Canas, R. Contour-Based 3D Point Cloud Simplification for Modeling Freeform Surfaces. *IEEE International Conference on Science and Technology for Humanity (TIC-STH),* Toronto, Canada, 2009.
14. Igor, V.; Rudolf, P.; Miodrag, H.; Branko, S. Different Approaches in Uncertainty Evaluation for Measurement of Complex Surfaces Using Coordinate Measuring Machine. *Meas. Sci. Rev.* **2015,** *15*(3), 111–118.
15. Schall, O.; Samozino, M. Surface from Scattered Points: A Brief Survey of Recent Developments. Unpublished.

CHAPTER 14

INFLUENCE OF SiC AND B4C PARTICLES ON MICROSTRUCTURE AND MECHANICAL PROPERTIES OF COPPER SURFACE COMPOSITES FABRICATED BY FRICTION STIR PROCESSING

N. RAMA KRISHNA[1,*], L. SUVARNA RAJU[2], G. MALLAIAH[3], and A. KUMAR[4]

[1]*Department of Mechanical Engineering, KITS, Singapur, Huzurabad, Telangana, India*

[2]*Department of Mechanical Engineering, VFSTR University, Vadlamudi, A.P, India, E-mail: rajumst@gmail.com*

[3]*Department of Mechanical Engineering, KITS Singapur, Huzurabad, E-mail: gmallaih_kits@yahoo.co.in*

[4]*Department of Mechanical Engineering, NITW, Warangal, Telangana, India, E-mail: adepu_kumar7@yahoo.co.in*

[*]*Corresponding author. E-mail: rk.me1611@gmail.com*

ABSTRACT

Friction stir processing (FSP) is a novel technique used for the production of surface composites, refinement of microstructure, and enhancing the mechanical properties. In this work, surface matrix composite was fabricated on the surface of copper with ceramic reinforcement using FSP technique. SiC and B4C were used as reinforcement. The chosen FSP parameters were traverse speed of 40 mm/min, tool rotational speed of 1120 rpm, and 10 kN of axial load. The FSP tool made of high carbon high chromium (HcHCr) with cylindrical tapper threaded profile pin having shoulder diameter of 24 mm, pin length of 3.8 mm, and pin diameter of 8 mm was used. The FSP was carried out using friction stir welding machine. Six combinations of surface composites (Cu/2Vol.%SiC, Cu/4Vol.%SiC, Cu/6Vol.%SiC, Cu/2Vol.%B4C, Cu/4Vol.%B4C, and Cu/6Vol.%B4C) were fabricated. The fabricated surface composites were examined by optical microscope for dispersion of reinforcement particles. The tensile properties of the surface composites are better at 4 vol.% of SiC and B4C compared to other vol.% of reinforcements. This is due to the grain refinement of copper in the surface composite which can be related to the interaction between powder particles and dislocations within the matrix. It is also observed that with an increase in the addition of B4C reinforcement particles recrystallization temperature increases by pinning grain boundaries of the copper matrix and blocking the movement of dislocations and; thus, improving the strength. The observed mechanical properties are correlated with microstructure. The microhardness of friction stir processed plates was analyzed using a Vickers hardness tester. B4C reinforced Cu surface composite resulted in higher microhardness.

14.1 INTRODUCTION

Friction stir processing (FSP) is a solid state technique based on the principle of friction stir welding (FSW) and is used for material processing in order to modify the microstructure, mechanical properties, and to fabricate surface composites.[1] In many engineering applications, the surface properties decide the life of the components rather than their bulk properties.[2] The surface layer reinforced with ceramic particles is normally called surface composite.[3] In recent years, copper-based surface composites are gaining widespread importance in several applications due to their good mechanical, thermal, and tribological properties.[4] An extensive study on copper-based surface composites is; therefore, needed without much loss in bulk properties of the matrix material.[5] Though several techniques are available to fabricate surface composites, FSP is a simple, green, and low energy consumption route based on the principles of FSW to fabricate surface composites with superior results.[6] SiCp (sillicon carbide particles) are of great technological importance because of their application as reinforcement for metal matrix composites and structural ceramics with exceptional thermal shock resistance qualities. The heat generation during FSP is due to friction between tool and workpiece, which softens the metal matrix and the intense stirring action of the tool and aids in the distribution of the reinforcement particles within the plasticized metal matrix zone. But the main difficulty in fabrication of particulate composites is the agglomeration of fine reinforcement particles.[7] The tendency of particle agglomeration can be notably reduced by appropriate designing of particulate deposition technique and tool design. Recently, Asadi et al.[8] and Barmouz et al.[9] have investigated the effectiveness of designing net holes as particulate deposition technique instead of conventional groove method in the copper substrate and obtained agglomeration free composites with enhanced mechanical properties using microsized SiC and B4C particles, respectively. Heat generation during FSP plays the key role in producing a defect-free surface composite. So, the heat generation, process parameters, particulate deposition technique, and tool pin profiles are very important while fabricating surface composites by FSP route. The successful research on the influence of SiC and B4C reinforced particulate deposition on microstructure and mechanical properties of copper surface composites via FSP are rarely reported so far. Hence, the objective of the present investigation is to achieve the surface composite via FSP and also to study the influence of SiC and B4C reinforced particles on microstructure and mechanical properties of Cu surface composites fabricated by single pass FSP.

14.2 EXPERIMENTAL PROCEDURE

In this study, pure copper plates with dimensions of $200 \times 120 \times 4$ mm were used as the base metal. The chemical composition and mechanical properties of the base metal are given in Tables 14.1 and 14.2, respectively. SiC and B4C particles (20 μm) were used as reinforcement. The surface of plates was cleaned with acetone before FSP. The process experiments were performed using vertical milling machine with a tapper cylindrical threaded pin shape made of high carbon high chromium (HCHCr) tool steel with given dimensions according to Figure 14.1. The tool was tilted 2° from the plate normal direction. According to Figure 14.2, a net of holes was designed on the surface of the plate, and the SiC and B4C were located into them before processing. The constant rotational and traverse speeds were 1120 rpm and 40 mm/min, respectively. The SiCp/B4Cp were compressed into the holes and the upper surface of the holes was closed with an FSP tool without the pin to prevent SiCp to escape from the holes. In the next stage, the tool is plunged with the pin into the plate to stir the material along with the reinforcement to produce the surface composites. The specimens were clamped on to the backing plate and fixed by the bolts. Single pass FSP was used to fabricate the Cu/SiC and Cu/B4C surface composites.

TABLE 14.1 Chemical Composition of Copper Plate (wt.%)

Elements	Amount (wt.%)
Zn	0.03
C	0.02
Co	0.003
Fe	0.0027
Al	0.002
Ni	0.001
Cu	Balance

TABLE 14.2 Mechanical Properties of Base Material.

Material	UTS (MPa)	YS (MPa)	%EL	Impact toughness (J)	Hardness (HV)
Pure copper	260	231	31	18	110

UTS: ultimate tensile strength; YS: yield strength.

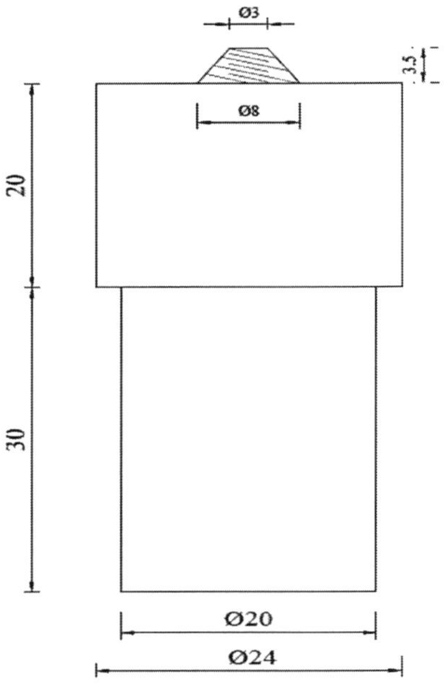

FIGURE 14.1 Schematic of shoulder and pin with specific dimensional.

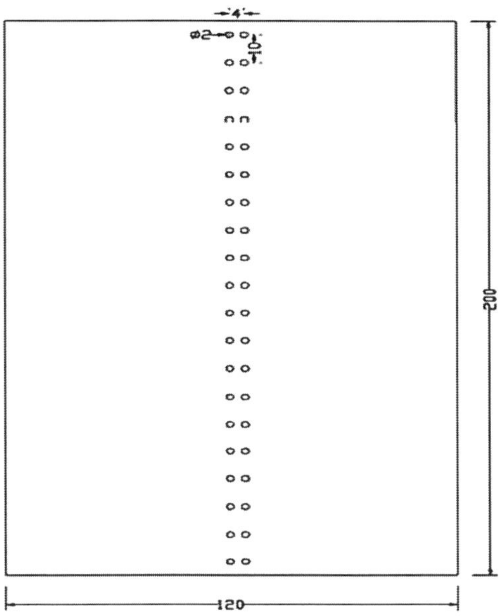

(All dimension are in mm)

FIGURE 14.2 Schematic representation of the design of blind holes on Cu plate (depth 3 mm).

After FSP, microstructural observations were carried out at the cross section of stir zone (SZ) of the surface composites normal to the FSP direction, mechanically polished and etched with 100 ml distilled water, 15 ml HCl, and 2.5 g ferric chloride. Microstructure changes were observed by an optical microscope in the SZ. The tensile specimens were prepared by Wire cut Electrical Discharge Machine to the required dimensions which are normal to the FSP direction. The tensile test was conducted with the help of a computer-controlled universal testing machine at a cross head speed of 0.5 mm/min. Similarly, the impact specimens were taken in transverse to the processing direction. The charpy "V" notch impact test was carried out using pendulum type impact testing machine at room temperature. The schematic sketch of both tensile and impact specimens is shown in Figures 14.3 and 14.4, respectively. Microhardness test was carried out by using Vickers digital microhardness tester (Model: Autograph, Make: Shimadzu) with a 15 g load for 15 s duration at the cross section (SZ) of surface composites normal to the FSP direction.

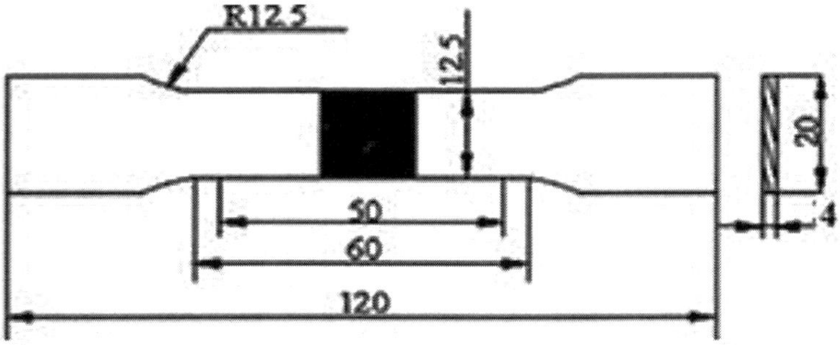

FIGURE 14.3 Schematic sketch of both tensile specimens.

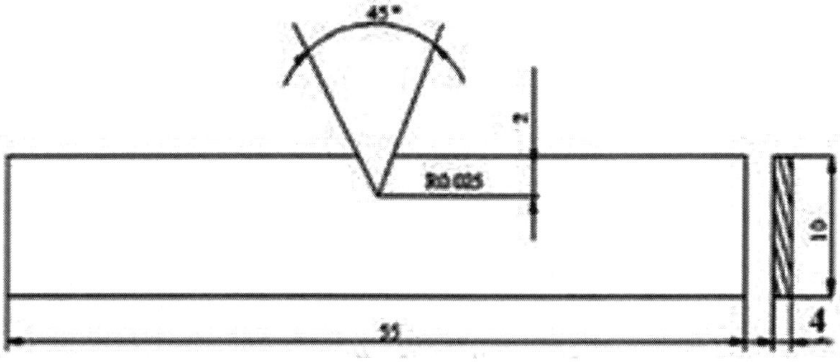

FIGURE 14.4 Schematic sketch of both impact specimens.

14.3 RESULTS AND DISCUSSION

14.3.1 MICROSTRUCTURAL STUDIES

The surface morphology of Cu/SiC and Cu/B4C surface composites is shown in Figure 14.5. The specimens for metallographic examination were sectioned to the required size from the SZ which is transverse to the processing zone. The optical micrographs of the defect-free FSPed specimens produced by one pass FSP are shown in Figures 14.6 and 14.7. The reinforcement of ceramic particles and the severe plastic deformation refined the grains of the copper matrix. It is observed that at Cu-4%SiC and Cu-4% B4C surface composites, the reinforced particles are dispersed uniformly in the processed zone due to the severe dynamic recrystallization and fairly homogeneous distribution was observed. No cluster particles were seen. The uniform distribution of B4C particles was seen in the micrographs as shown in Figure 14.7. This can be attributed to intense stirring and sufficient material flow which reduces the possibility of formation of agglomerations. At higher volume percentage, the B4C particles were not mixed with the plasticized copper properly. Hence, agglomerations were formed.

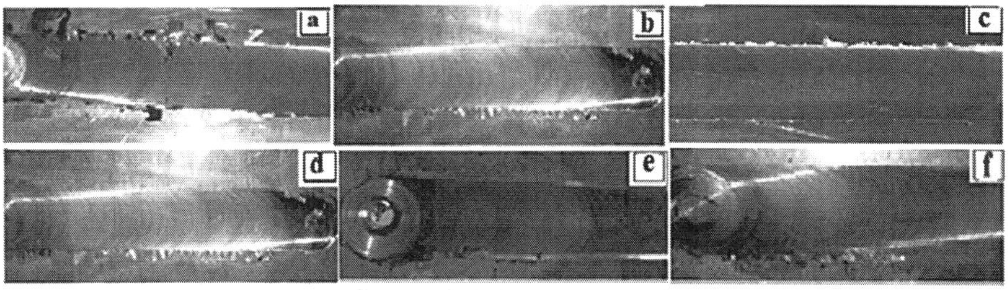

FIGURE 14.5 Surface morphology of copper composites (a) Cu-2%SiC, (b) Cu-4%SiC, (c) Cu-6%SiC, (d) Cu-2%B4C, (e) Cu-4%B4C, (f) Cu-6%B4C.

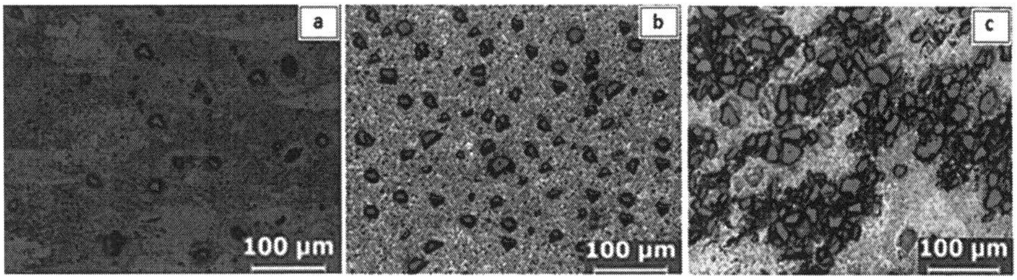

FIGURE 14.6 Microstructures of processed samples (a) Cu-2%SiC, (b) Cu-4%SiC, (c) Cu-6%SiC.

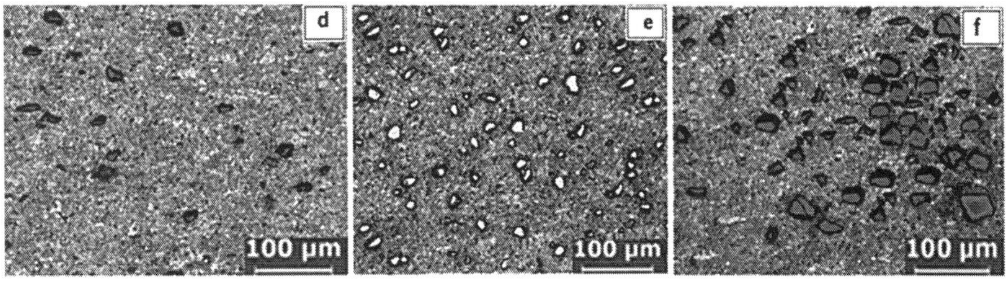

FIGURE 14.7 Microstructures of processed samples (d) Cu-2%B4C, (e) Cu-4%B4C, (f) Cu-6%B4C.

14.3.2 MECHANICAL PROPERTIES

Microhardness. The microhardness of samples was measured by Vickers microhardness tester. The base metal showed an average hardness of 110HV. It was observed that the average microhardness of a surface composite by Cu-6% SiC was higher than that of the pure copper matrix. The grain refinement in the surface composites produced by FSP is related to the presence of the SiC and B4C particles which were uniformly distributed among the grain boundaries of the matrix and restricted the grain growth during the processing. In all the processed composites, the increase in hardness in SZ region of surface composites was due to (i) hard phase dispersion of B4C (Orowan strengthening mechanism) and (ii) grain refinement in SZ (Hall–Petch relationship).

The tensile properties such as ultimate tensile strength (UTS), yield strength (YS), and % elongation (EL) are presented in Table 14.3. From the results, the better mechanical properties such as UTS, YS, and % EL were observed at Cu-4%SiC and Cu-4% B4C surface composites; this may be due to homogeneous mixture of SiC and B4C particles in a copper matrix and also grain refinement of copper in the surface composite compared to other conditions of surface composites. The YS is corporately low as compared to pure copper due to the inclusion of SiC and B4C content. As the volume percentage increases mechanical properties are deteriorated. This may be caused due to improper mixing of plasticized copper with SiC and B4C particles and which aids the formation of agglomerations.

TABLE 14.3 Mechanical Properties of the Copper Surface Composites.

S. no	Vol %	Ceramic particles	UTS (MPa)	YS (MPa)	% elongation EL (MPa)	IT (J)	Microhardness (HV)
1	2	SiC	229	151	7.04	12	124
2	4	SiC	248	163	9.07	14	132
3	6	SiC	208	138	7.82	8	153
4	2	B4C	210	119	8.40	8	131
5	4	B4C	262	159	11.96	16	139
6	6	B4C	229	132	10.10	12	147
7	Base metal		260	231	31	18	110

14.4 CONCLUSIONS

From these investigations, the following conclusions can be derived:

- FSP is a successful low energy consumption route for preparing surface composites.
- A set of blind holes of 2 mm in diameter and 3 mm in depth was used as a successful particulate deposition technique to prepare surface composites by FSP route with a uniform distribution of microsized SiC and B4Cp in the copper matrix.
- The microstructure at the processed zone of Cu/SiC and Cu/B4C surface composites with 4 vol.% of SiC and B4C particles exhibits fine grain size and better mechanical properties compared to the other conditions of surface composites.

ACKNOWLEDGMENT

The authors would like to thank the authorities of Defense Metallurgical Research Laboratory (DMRL), Hyderabad, India and, also, one of the authors N. Rama Krishna is thankful to the principal and the management of KITS, Huzurabad for their constant support during this work.

KEYWORDS

- **friction stir processing**
- **Cu/SiC composite**
- **microstructure**
- **mechanical properties**

REFERENCES

1. Chang, C. I.; Du, X. H.; Huang, J. C. Achieving Ultrafine Grain Size in MgeAleZn Alloy by Friction Stir Processing. *Scr. Mater.* **2007,** *57,* 209–212.
2. Jiang, J. T.; Zhen, L.; Xu, C. Y.; Wu, X. L. Microstructure and Magnetic Properties of SiC/Co Composite Particles Prepared by Electroless Plating. *Surf. Coat. Technol.* **2006,** *201*(6), 3139–3146.

3. Attia, A. N. Surface Metal Matrix Composites. *Mater. Des.* **2001,** *22*(6), 451–457.
4. Romankov, S.; Hayasaka, Y.; Shchetinin, I. V.; Yoon, J.-M.; Komarov, S. V. Fabrication of Cu-SiC Surface Composite Under Ball Collisions. *Appl. Surf. Sci.* **2011,** *257*(11), 5032–5036.
5. Mishra, R. S.; Ma, Z. Y. Friction Stir Welding and Processing. *Mater. Sci. Eng. R: Rep.* **2005,** *50*(1–2), 1–78.
6. Mishra, R. S.; Mahoney, M. W. *Friction Stir Welding and Processing*. ASM International Materials, January 2007, ISBN: 978-0-87170-840-3.
7. Cartigueyen, S.; Mahadevan, K. Role of Friction Stir Processing on Copper and Copper Based Particle Reinforced Composites—A Review. *J. Mater. Sci. Surf. Eng.* **2015,** *2*(2), 133–145.
8. Asadi, P.; Givi, M. K. B.; Abrinia, K.; Taherishargh, M.; Salekrostam, R. Effects of Sic Particle Size and Process Parameters on the Microstructure and Hardness of AZ91/SiC Composite Layer Fabricated by FSP. *J. Mater. Eng. Perform.* **2011,** *20*(9), 1554–1562.
9. Barmouz, M.; Asadi, P.; Givi, M. K. B.; Taherishargh, M. Investigation of Mechanical Properties of Cu/SiC Composite Fabricated by FSP: Effect of SiC Particles' Size and Volume Fraction. *Mater. Sci. Eng. A* **2011,** *528*(3), 1740–1749.

CHAPTER 15

STUDIES ON THE PROPERTIES OF CLAY-REINFORCED EPOXY/ POLYESTER COMPOSITE

K. V. P. CHAKRADHAR

Department of Mechanical Engineering, Madanapalle Institute of Technology and Science, Madanapalle, Andhra Pradesh, India

Corresponding author. E-mail: chakra.komanduri@gmail.com, drchakradharkvp@mits.ac.in

ABSTRACT

Unsaturated polyester (UP)-toughened epoxy blends were developed. Montmorillonite (MMT) clay was dispersed into the above system to prepare blended epoxy/UP/clay nanocomposites in different weight ratios namely, 0, 1, 2, 3, and 5%. The specimen was characterized for mechanical properties and damping property analysis. Blended nanocomposites were fabricated by high shear mechanical mixing followed by ultrasonication process to get homogeneous mixing under the aid of in situ polymerization. The above studies were conducted as per the American Society for Testing and Materials (ASTM) standards. Data obtained from the above studies indicated that the introduction of UP into the epoxy resin and further reinforcement with clay improved the mechanical and damping properties to an appreciable extent. The homogeneous morphologies of the UP-toughened epoxy and epoxy/UP/clay nanocomposite systems were ascertained using scanning electron microscope (SEM) and transmission electron microscopy (TEM) studies. The objective of this study was to identify a suitable nanocomposite which offers a low-cost, high strength material that can be applied for engineering and structural applications to provide better performance.

15.1 INTRODUCTION

Polymer-polymer blends have gained significant commercial growth, as they save nearly 36% weight of the total polymer consumption without compromising weight. Epoxy is a versatile and widely-accepted matrix material used for the fabrication of advanced composites, hardware components, electrical circuit board materials and missile equipments because of its excellent bonding, physicochemical, thermal, mechanical, dielectric, and aging characteristics.[1,2] To improve its performance characteristics in advanced engineering applications, toughening of epoxy material is essential to improve its impact strength. The toughness of the epoxy resin has been increased by blending it with flexible polymers and elastomers. However, modification of the epoxy resin with elastomers improves its impact properties with decrease in some of the physical properties of the cured epoxy at higher temperatures.[3–5] Hence, a suitable polymeric material is needed to improve the impact resistance, stress–strain properties by retaining stiffness, glass transition temperature, and thermal stability of the epoxy matrix. To achieve this, development of an intercross-linked polymer network of thermoset—thermoset blends have been extensively studied due to their enhanced mechanical properties.[6–10] Polyester resins have a wide range of industrial relevance, such as in industrial finishes and maintenance, architectural uses, paints, and surface coatings. These resins have a number of advantages, including versatility in the structure and properties, overall low cost, and the ease of application. However, they suffer from some drawbacks such as low alkali resistance and low hardness. So, to overcome these drawbacks of polyester resins, blending with other suitable resins, such as epoxy resin, amino resin, silicone resin, and ketonic resin, can be performed as polyester resins have good compatibility with a wide variety of other resins. The better compatibility comes from the relatively low viscosity of the resin and from the structure of the resin, which contains a relatively polar, aromatic backbone and aliphatic side chains with low polarity. The blending technique maybe used effectively to overcome the inferior properties of both the components. Miscible polymer blends produce a new improved material from the less superior individual components. Again, because of the enhancement of the properties such as the mechanical, thermal, structural and barrier properties, even at low concentrations, the nanocomposites of such blends have drawn much attention. It is also well known that polymer-clay nanocomposites could offer better mechanical, thermal and damping properties compared to the conventional composites. The advantages of nanoparticles over conventional macro- or microparticles are their higher surface area and aspect ratio which could improve adhesion between nanoparticles and polymers and lower the amount of loading to achieve equivalent properties. Nowadays, the most widely used clay mineral for polymer-clay nanocomposites is montmorillonite. Various studies were conducted on epoxy-clay nanocomposites (ECN) under different curing conditions. The exfoliated clay structure possesses superior properties and gives a few advantages over other nanofillers like in terms of cost and environment-friendly matters.[2,8,9] This paper presents studies on the mechanical and damping properties and the scanning electron microscope (SEM) and transmission electron microscopy (TEM) analysis of the clay-filled nanocomposite. The objective of this study was to identify a suitable nanocomposite which offers low-cost, high strength material that can be applied for engineering and structural applications to provide better performance.

15.2 EXPERIMENTAL PROCEDURE

15.2.1 MATERIALS AND METHODS

The resins used in this study were (i) Epoxy (Ciba-Geigy, Araldite-LY 556 and Amine hardener HY-951) with the resin-hardener ratio as 100:10 and (ii) Unsaturated polyester (Ecmalon 9911, Ecmas Hyderabad) with 2% cobalt naphthenate as the accelerator, 2% methyl ethyl ketone peroxide (MEKP) as the catalyst in 10% dimethylaniline (DMA) solution as promoter, in the ratio of the resin/accelerator/catalyst/promoter: 100/2/2/2. In addition, exfoliated montmorillonite clay (product No.:682,659; brand: Aldrich, USA; product name: nanoclay, hydrophilic bentonite; formula: $H_2Al_2O_6Si$; molecular weight: 180.1 g/mol; appearance (color): light tan to brown; appearance (form): powder; loss on drying: 18.0%; density: 600–1100 kg/m^3; size: 25 μm), surface modified with 25–30% trimethyl stearyl ammonium was used as filler material.

Mechanical tests like tensile and flexural tests were conducted. Tensile testing samples were prepared in dumbbell shapes with dimensions $100 \times 20 \times 3$ mm as per the ASTM D 638 standards. In each case, five samples were tested and the average value was tabulated. The samples were loaded in tension at a cross-head speed of 1 mm/min to determine their tensile behavior. A three point bend test was carried out to determine the Interlaminar shear strength (ILSS) of the samples with dimensions $100 \times 10 \times 3$ mm as per ASTM D 2344. The testing was performed on short beams of $24 \times 63.5 \times 3$ mm at the crosshead speed of 1 mm/min maintaining the sample size of five for each experiment and the average value has been reported. Tensile strength (TS) test, ILSS test were carried out using Instron universal testing machine (IUTM series-3369). In each case, five identical samples were tested and their average load at first deformation was noted and tabulated.

The damping analysis was conducted through impact testing of nanocomposite specimens as per the ASTM E-756 standards. The purpose of the test was to measure the damping ratio of epoxy/UP/clay blend nanocomposite materials. The above property was tested on epoxy/UP blend system as a function of clay with different weight variations of clay (i.e. 0, 1, 2, 3, 4 and 5%). The nanocomposite specimens were fixed like a cantilever and an impact test was carried out to determine the damping ratio. One end of the plate was fixed in a bench vice with the help of a C-clamp and other end was free. Accelerometer was put on the fixed-end side of the plate. It was excited by an impact hammer, with a force of 80 N and the response values at different frequencies were tabulated for different compositions of the nanocomposite blend. These values were used to determine the damping ratios. The specifications of the test setup and the conditions under which the test was carried out are given below.

Specimen size [ASTM E-756]	300 × 30 × 3 mm
Accelerometer	Brüel & Kjær (B&K) Type 4393 + 2647 (charge converter)
Impact hammer	PCB Type 86D05
Software	B&K PULSE Type7536
Temperature	25°C

A JEOL JSM-6400 JAPAN SEM at 15 kV accelerating voltage equipped with energy dispersive spectroscopy (EDS) was used to study the dispersion of clay particles in the blended nanocomposites. The fractured surfaces were coated with a thin film of gold to increase the electrical conductance for SEM analysis. Finally, the phase structures of blended nanocomposite were studied using TEM-2000EX, Japan.

15.2.2 FABRICATION OF BLENDED NANOCOMPOSITES

Firstly, clay was dried in an oven at a temperature of 80°C for 24 h. Then, a pre-calculated amount of clay and epoxy/polyester (i.e., 85/15 wt/wt ratio) were mixed together in a suitable beaker. Clay was mixed in a stipulated quantity to epoxy/polyester and was mixed thoroughly with a mechanical shear stirrer for about 1 h at ambient temperature conditions. Then the mixture was placed in a high intensity ultrasonicator for one and half hour with pulse mode (15 s on/15 s off). External cooling system was employed by submerging the beaker containing the mixture in an ice bath to avoid temperature rise during the sonication process. Once the process was completed, hardener/accelerator/catalyst/promoter (100:10/2/2/2) parts by weight were added to the modified epoxy/polyester mixture. A glass mold with the required dimensions was used for making samples on par with the ASTM standards. The glass mold was coated with mold-releasing agent (wax) to enable easy removal of the sample. The nanocomposite mixture was poured over the glass mold. A brush and roller were used to impregnate the nanocomposite. The closed mold was kept under pressure for 24 h at room temperature. To ensure complete curing, the blended nanocomposite samples were post cured at 70°C for 1 h and the test specimens of required sizes were cut out from the sample sheet.

15.2.3 MECHANICAL PROPERTIES

The primary goal of the blended nanocomposite material, as a product, was to withstand the applied mechanical forces. It was achieved by load transfer between the matrix and the nanoparticle induced by the shear deformation of the matrix around the fillers. This shear deformation was produced because of the high Young's modulus of the nanoparticle and the larger differences between the mechanical properties of the composite constituents. Table 15.1 shows effects of the epoxy/polyester (85/15, W/W) nanocomposites as a function of the clay content. It was observed that, mechanical properties were optimized at 3 wt.% clay content. When compared with neat blend samples, 3 wt.% clay-filled blend samples gave the following results.

TS showed a 30% increase and ILSS showed a 52% increase in strengths. Mechanical properties improved as the clay content was increased up to 3 wt.%. This was evidently attributed to the elastomeric nature of the added clay content. In the absence of clay content, miscible epoxy/polyester blends exhibited somewhat inferior mechanical properties as compared with the blends containing clay, indicative of poor interfacial adhesion between two phases. Compared to the neat blends of epoxy/polyester, the binary blends on the addition of clay exhibited remarkable improvement in the mechanical properties.

TABLE 15.1 Mechanical Properties of Blended Epoxy/Polyester Nanocomposite as a Function of Clay.

Clay content (wt.%)	Interlaminar shear strength (ILSS) (MPa)	Tensile strength (TS) (MPa)
1	5.12±3.2	20.71±2.5
2	6.69±2.5	24.31±4.6
3	7.34±1.8	26.29±5.7
4	10.67±2.2	29.71±3.91
5	8.63±4.2	25.28±2.0

Note: Value shown after "±" is standard deviation.

It was observed that properties were optimized at 3 wt.% clay and it was due to maximum elastomeric clay content. In literature, it has been indicated that adding small amounts of nanoclays into polymer-based materials could potentially enhance their strengths such as, the impact strength, toughness, hardness, and flexure of the current samples with the clay content less than 5 wt.%.[6–8] This substantiates the existence of an optimal limit because the physical properties between these nanostructural materials and matrix are different. The gradual increase in TS and ILSS is shown in Figure 15.1 as a function of the clay content.

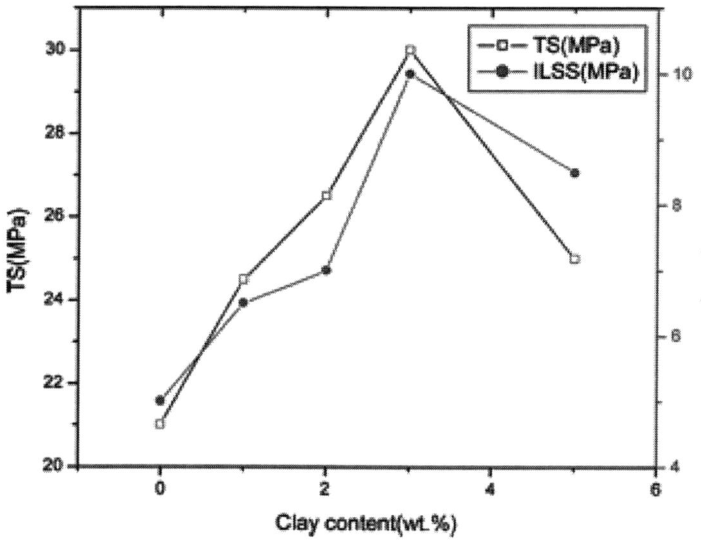

FIGURE 15.1 Interlaminar shear strength (ILSS) and tensile strength (TS) of the blended epoxy/polyester as a function of clay.

15.2.4 DAMPING PROPERTIES

One of the goals of the developed nanocomposites was to withstand vibrations when used in structural applications. The property of vibrations is the damping ratio. The knowledge of the above property is also essential to an engineer for the proper selection of materials for specific applications. The study of the above property of polymers enables one to decide their field of application. Especially, the study of damping properties may decide the application of nanocomposites in areas like automobiles, machines, equipment that are predominantly affected by vibrations. Moreover, the study is also useful to decide the application of these nanocomposites in buildings, structures, and so forth, that fall because of vibrations. Due to faulty design and poor fabrication, there is an unbalance in the components and parts of the above-discussed applications that lead to heavy stress and failure of components and parts.[7] Excessive vibrations are dangerous to the life of human beings. Thus keeping in view the above devastating effects, the study of vibration and damping properties of materials is essential for an engineer to minimize the vibration effects over machinery, structures and so forth by designing them suitably. The objective of this study was to investigate the natural frequency and damping ratio of epoxy/UP/MMT clay nanocomposites.

It was observed in Table 15.2 that the damping properties for epoxy/UP system as a function of clay were increased appreciably at 4 wt.% clay content. Among all clay variations, 4 wt.% clay samples show better damping properties. When compared to pure blend the 4 wt.% nanocomposite showed 23% (Mode 2) and 57% (Mode 3) increase in damping ratios. This was due to the better dispersion of clay particles in the blend at that ratio. Moreover, clay is said to have a reasonably good vibration damping property.[9–10]

TABLE 15.2 Damping Property of Blended Epoxy/Polyester Nanocomposite as a Function of Clay.

S. No.	Material compositions (Epoxy/UP with varying clay content)	Modal Frequencies		Damping Ratio (%)
		Modes	Frequencies (Hz)	
1	0% (pure blend)	2	147	2.13
		3	397	1.37
2	1%	2	156	2.47
		3	421	1.90
3	2%	2	154	1.92
		3	407	1.37

TABLE 15.2 *(Continued)*

S. No.	Material compositions (Epoxy/ UP with varying clay content)	Modal Frequencies		Damping Ratio (%)
		Modes	Frequencies (Hz)	
4	3%	2	195	1.97
		3	520	1.50
5	4%	2	144	2.62
		3	383	2.16
6	5%	2	272	1.69
		3	810	1.66

Note: As per the theory, first mode is not considered for damping.

15.2.5 MORPHOLOGY STUDIES

15.2.5.1 SCANNING ELECTRON MICROSCOPY (SEM) STUDIES

Examination of impact-fractured surfaces can provide information related to the interfacial property and the mode of involved dissipation of materials. SEM micrographs of various epoxy/polyester blends containing varying clay concentration are shown in Figures 15.2a–e. The neat blend sample shows brittle fracture surface, indicative of miscible characteristics between the epoxy and polyester as in Figure 15.2a Varada Rajulu et al.[11] have successfully proved that epoxy and polyester have good miscibility. In Figure 15.2b, it was observed that brittle fracture turned to ductile fracture due to the addition of clay particles. In Figure 15.2c, brittle fracture can no longer be seen as the clay content is gradually increased from 1 to 2 wt.% as a result of which strong ductile nature of the composite was observed. In Figure 15.2d, another strong ductile-fractured surface can be observed. This is an indication of good dispersion of nanoparticles that brought out maximum improvement of mechanical properties in 3 wt.% clay loading. On the other hand agglomeration of several clay particles, poor adhesion, and bonding were observed in Figure 15.2e. Increase in clay particles correspondingly increases viscosity of the modified polymer that might be a reason for the failure at 5 wt.% of clay.[12] With the increase in clay loadings, the well-dispersed layers or lines also increased in the matrix as were observed in the earlier studies. The crack initiation is caused by the stress concentrations caused by the agglomerated clay. The high stress concentrations caused by the agglomerated particles might affect the mechanical properties, which is low deformation property and reduced strength by initiating early failure in the epoxy/UP blends. In addition to blends of polymers, so far, no report has been published on the utilization of synthetic-based resins as the sole blend components. Because of the presence of long chain fatty acids in synthetics, the used blend components have advantages such as ease of processability and compatibility.

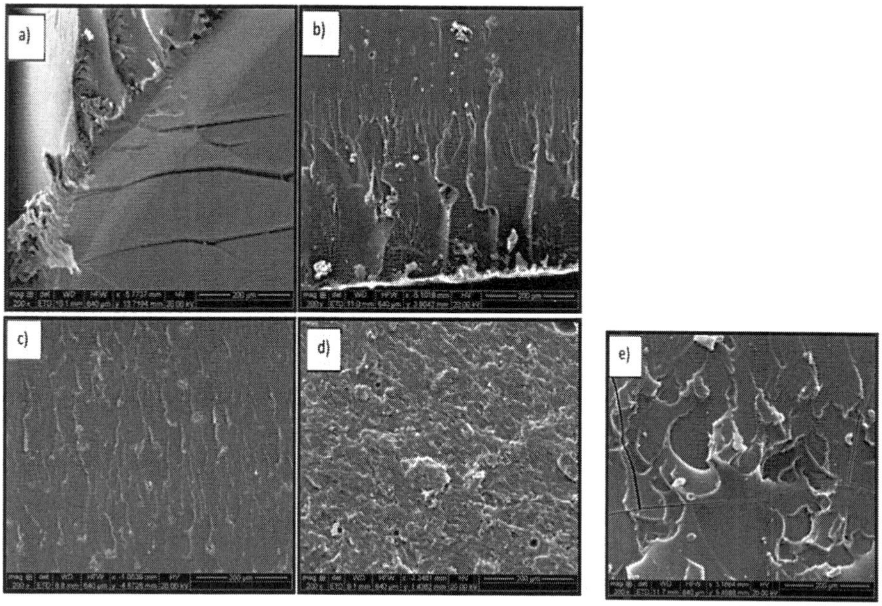

FIGURE 15.2 SEM analysis of epoxy/polyester blend as a function of nanoclay at (a) 0 wt.% clay, (b) 1 wt.% clay, (c) 2 wt.% clay, (d) 3 wt.% clay, (e) 5 wt.% clay.

15.2.5.2 TRANSMISSION ELECTRON MICROSCOPY (TEM) STUDIES

The morphology and the actual structure or pattern of the clay layer dispersion in the polyester/epoxy blends were further analyzed by TEM. It has been reported in the literature that increased clay loading enhances the ordering of clay platelets and gradually degrades the exfoliation potential of the polymer. Figures 15.3a and b show a homogeneous dispersion of clay platelets for 3 wt.% clay loading. Upon nanocomposite formation, the individual clay layers were found to be disintegrated or partially exfoliated and well dispersed in the polymer matrix. Also, the individual clay layers and zones with more than one clay layer were noted in the TEM images. The reason for this excellent distribution was the strong interactions between the polar carboxylic ester groups of the polymer and the –OH group of clay.

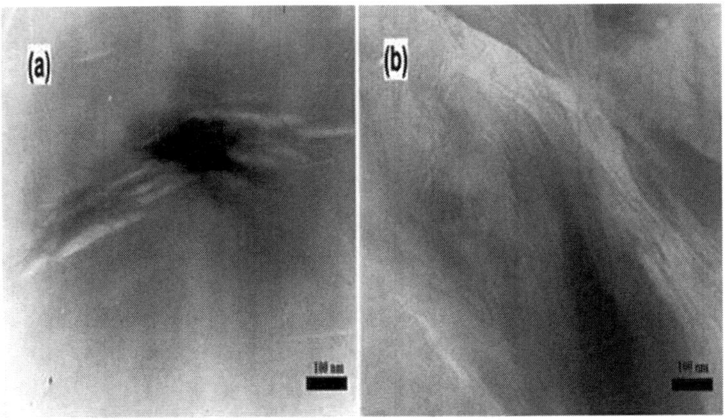

FIGURE 15.3 TEM analysis.

15.3 CONCLUSIONS

In this study, the mechanical, morphological, and damping properties of epoxy/polyester blend reinforced with MMT clay were studied and the following conclusions drawn:

1. UP-toughened epoxy reinforced with clay as nanocomposites were prepared for studying the mechanical and damping properties.
2. ILSS and TS increased on increasing the clay content from 0–3 wt.% and decreased on further increase in clay.
3. Damping ratio was found to vary between 0 and 3 wt.% clay and optimized at 4 wt.% clay. It decreased on further increase in the clay content.
4. SEM and TEM analysis revealed that excellent adhesion and interfacing between the matrices and clay was the main reason for optimum improvement of the mechanical and damping properties of clay-filled nanocomposites.
5. It was observed from the SEM analysis that agglomeration of clay and microvoids are the indication of poor performance at 5 wt.% clay-filled nanocomposite. The crack initiation was due to the stress concentrations caused by the agglomeration of clay.

6. This nanocomposite can be applied in making lightweight components for automobile parts, transportation systems, structural parts, and consumer products.

KEYWORDS

- **clay**
- **epoxy**
- **unsaturated polyester**
- **mechanical properties**
- **vibration analysis**

REFERENCES

1. Li, J.; Jian Guo, Z. The Influence of Polyethylene-Polyamine Surface Treatment of Carbon Nanotube on the TPB and Friction and Wear Behavior of Thermoplastic Polyimide Composite. *Polym–Plast. Technol. Eng.* **2011,** *50,* 996–999.
2. Chinnakkannu, K. C.; Muthukaruppan, A.; Rajkumar, J. S.; Periyannan, G. Thermo Mechanical Behaviour of Unsaturated Polyester Toughened Epoxy–Clay Hybrid Nanocomposites. *J. Polym. Res.* **2007,** *14,* 319–328.
3. Uday, K.; Gautam, D.; Niranjan, K. *Mesua ferrea* L Seed Oil Based Highly Branched Polyester/Epoxy Blends and Their Nanocomposites. *J. Appl. Polym. Sci.* **2011,** *121,* 1076–1085.
4. Guo, J.; Han-Xiong, H.; Zhao-Ke, C. Rheological Responses and Morphology of Polylactide/Linear Low Density Polyethylene Blends Produced by Different Mixing Type. *Polym. Plast. Technol. Eng.* **2011,** *50,* 1035–1039.
5. Song, Z.; Li, Z.; Yan-Ying, W.; Yi, Z.; Shi-Bo, G.; Yu-Bao, L. Fabrication of Hydroxyapatite/Ethylene-Vinyl-Acetate/ Polyamide 66 Composite Scaffolds by the Injection-Molding Method. *Polym. Plast. Technol. Eng.* **2011,** *50,* 1047–1054.
6. Jha, A.; Bhowmick, A. K. Mechanical and Dynamic Mechanical Thermal Properties of Heat and Oil Resistant Thermoplastic Elastomeric Blends of Poly (Butylene Terephthalate) and Acrylate Rubber. *J. Appl. Polym. Sci.* **2000,** *78,* 1001–1008.
7. Colakoglu, M. Damping and Vibration Analysis of Polyethylene Fiber Composite Under Varied Temperature. *Turk. J. Eng. Environ. Sci.* **2006,** *30,* 351–357.
8. Vijaya Kumar, K. R.; Sundareswaran, V. Mechanical and Damping Properties of Epoxy Cyanate Matrix Composite Under Varied Temperatures. *J. Eng. Appl. Sci.* **2010,** *5,* 106–111.
9. Alam, N.; Asnani, N. T. Vibration and Damping Analysis of Fibre Reinforced Composite Material Cylindrical Shell. *J. Compos. Mater.* **1987,** *21,* 348–361.

10. Chandra, R.; Singh, S.; Gupta, K. Experimental Evaluation of Damping of Fiber-Reinforced Composites. *J. Compos. Technol. Res.* **2003,** *25,* 1–12.
11. Varada Rajulu, A.; Ganga Devi, L.; Babu Rao, G. Miscibility Studies of Epoxy/Unsaturated Polyester Resin Blend in Chloroform by Viscosity, Ultrasonic Velocity, and Refractive Index Methods. *J. Appl. Polym. Sci.* **2003,** *89*(11), 2970–2972.
12. Benny Cherian, A.; Varghese, L. A.; Thachil, E. T. Epoxy-Modified Unsaturated Polyester Hybrid Networks. *Eur. Polym. J.* **2007,** *43,* 1460–1469.

CHAPTER 16

EVALUATION OF MECHANICAL BEHAVIOR OF GLASS PARTICULATE-CONTAINING AL6061 ALLOY COMPOSITES

Y. C. MADHUKUMAR* and UMASHANKAR

Siddaganga Institute of Technology, Tumkur, Karnataka, India

**Corresponding author. E-mail: madhuyc3@gmail.com*

ABSTRACT

In this work, the mechanical behavior of Al6061 metal matrix composites has been investigated. Al6061 is used as matrix. Glass particulates are used as reinforcements of 88 μm with weight percentages of 3, 6, 9, and 12 wt.%. The stir casting method was used to produce the composites. The composite was then studied with respect to its microstructure, X-ray diffraction (XRD), microhardness, and tensile strength. The distribution of glass particulates in the matrix alloy was examined using the optical microscope and XRD analysis. Microhardness and tensile strength of composites increase up to 9 wt.% and suddenly decrease up to 12 wt.% because the particles are agglomerated in the composites.

16.1 INTRODUCTION

Composites are materials that are the combination of two or more different metals. Composites are of two types: metal matrix composites (MMCs) or polymer matrix composites. High hardness, Young's modulus, yield strength, and ultimate strength of aluminum alloys are achieved for aluminum matrix composites by adding hard ceramic particles. MMCs can be fabricated by various fabrication routes such as liquid-state processing (stir casting and squeeze casting), solid-state processing (powder metallurgy). Among these, liquid-state processing, in particular, stir casting process, gives better bonding between matrix and reinforcement because of stirring action. The challenge in the fabrication of MMCs is difficulty in achieving uniform distribution of particles. Much work has been done by fabricating Al-based MMCs, being SiC, Al_2O_3, and B_4C a reinforcement material, and very limited research work has been carried using glass particulate. Glass particulate has good hardness, wear resistance, and is less dense.[1–4]

The main objective of the work is to study the effect of reinforcement in Al matrix composites on hardness, tensile strength, and microstructure.

16.2 EXPERIMENTAL DETAILS

16.2.1 MATERIALS

Aluminum alloy Al6061 is used as matrix material. The chemical composition of matrix material is shown in Table 16.1. The glass particulate with a size of 88 μm and with varying weight percentages of 3, 6, 9, and 12 wt.% are being used as reinforcing material for the fabrication of composites.

TABLE 16.1 The Chemical Composition of the Al6061 Alloy.

Elements	Cu	Mg	Si	Fe	Mn	Cr	Zn	Al
Wt.%	0.4	1.2	0.8	0.7	0.15	0.45	0.25	Bal

16.2.2 PROCESSING OF COMPOSITES

In the present work, melt stir casting technique is used to fabricate. The required amount of Al 6061 alloy was fed into a graphite crucible and melted in a furnace to a temperature of 750°C. Before adding the reinforcement to the molten slurry, it is preheated at 200°C in the oven to remove the absorbed gases in reinforcement. The slurry is continuously stirred for 5–10 min at a certain speed to get a vortex, then the preheated reinforcement is added in a created vortex and stirred with the help of stirrer, and then is transferred to preheated permanent mold.[5–7]

The cast samples were subjected to microstructural and X-ray diffraction (XRD) analysis to identify the presence of glass particulates in Al6061 matrix. The tensile strength of the prepared composites was measured using a computerized uniaxial tensile testing. Microhardness test measurement was performed on composites using Zwick micro-Vickers hardness tester under a load of 50 g for 10 s.

16.3 RESULTS AND DISCUSSION

16.3.1 MICROSTRUCTURE ANALYSIS

The optical microstructures of Al–glass particles of different weight percentages are prepared using optical microscopy. Figure 16.1a–d shows the optical microstructure of the fabricated Al6061–glass particulate. Due to preheating of the reinforcement, it is observed that uniform distribution of reinforcement in the aluminum matrix is achieved.

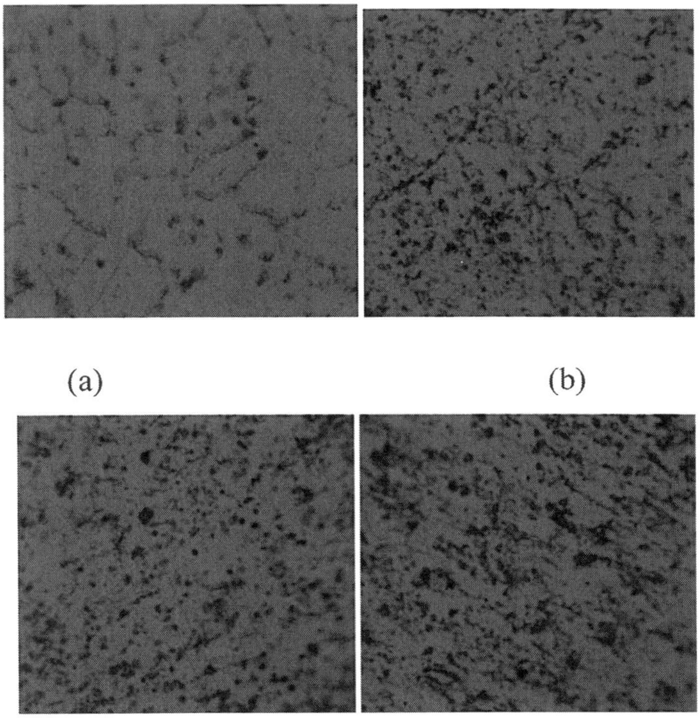

(a) (b)

(c) (d)

FIGURE 16.1 Microstructure of Al6061—glass particulate metal matrix composites (MMCs) (a) 3, (b) 6, (c) 9, and (d) 12 wt.% at 100X.

16.3.2 HARDNESS

Hardness is defined as the resistance of materials to indentation. Figure 16.2 shows that adding the reinforcement into Al6061 matrix composites increases the hardness of MMCs. Hardness value increases with addition of glass particulates; it has been also noted that when % of glass particulate increases up to 9% and decrease in hardness value because of agglomeration which leads to porosity.[9–10]

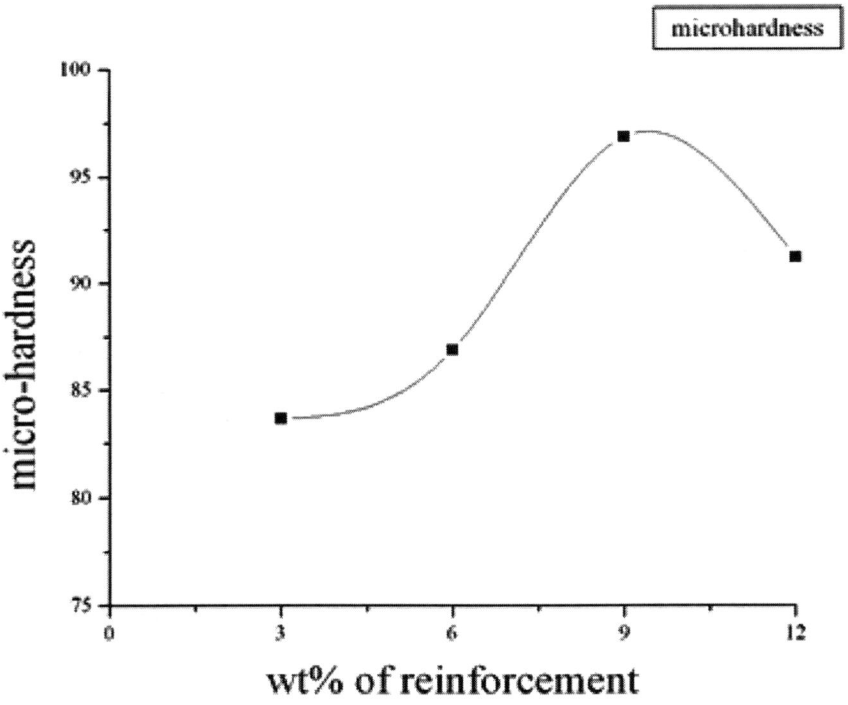

FIGURE 16.2 Microhardness of glass particles-reinforced MMCs.

16.3.3 TENSILE STRENGTH

Figure 16.3 shows that adding the reinforcement into Al6061 matrix increases the tensile strength of the composites. The tensile strength of composites is increased from 119 to 160 MPa by adding the reinforcement. The dislocation movements are stopped in the matrix by the addition of glass particulates. The coefficient of thermal expansion difference between the matrix and reinforcement results in high dislocation density in the MMCs. Since the difference in the coefficient of thermal expansion between matrix and reinforcement increases, number of dislocations in the MMCs also increases. The increased number of dislocations stops dislocation movement. Hence, the tensile strength of the MMCs increases. Increasing the applied stress increases a number of grain boundaries and thus prevents the dislocation movement and ends up with dislocation pile-up at the grain boundary region. This increases the tensile strength of the MMC by these two obstacles.[11–12]

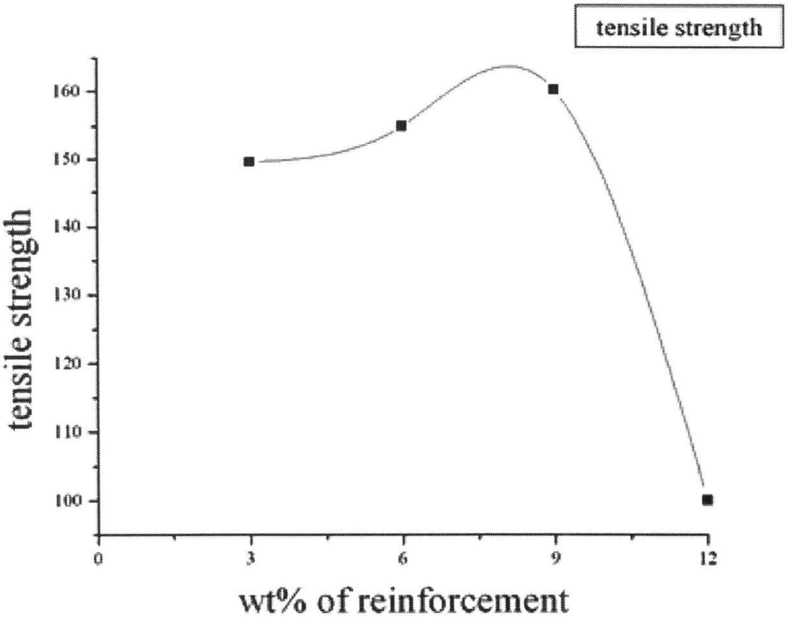

FIGURE 16.3 Tensile strength of glass particles-reinforced MMCs.

16.3.4 X-RAY DIFFRACTION ANALYSIS

In this work, the prepared composites are analyzed using X-ray powder diffraction analysis method used to identify the reinforcement. From Figure 16.4, X-Ray powder diffraction analysis confirms the presence of reinforcement.

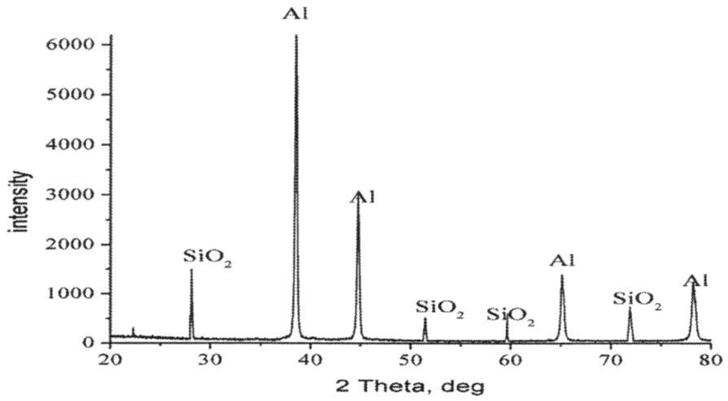

FIGURE 16.4 X-ray diffraction pattern of glass particles-reinforced MMCs.

16.4 CONCLUSIONS

In this work, the Al6061–glass particulates MMCs were prepared using melt stir casting method with different weight percentages (3, 6, 9, and 12 wt.%). The uniform distribution and agglomeration of reinforcement are observed by the microstructure. The presence of reinforcement in the composites is conducted by XRD analysis. Adding the reinforcement into the Al6061 matrix increases the microhardness and tensile strength of the composites up to 9 wt.% of reinforcement.

KEYWORDS

- **Al 6061**
- **stir casting**
- **microstructure**
- **tensile strength**
- **microhardness**

REFERENCES

1. Sekar, K.; Allesu, K.; Joseph, M. A Mechanical and Wear Properties of Al-Al_2O_3 Metal Matrix Composites Fabricated by the Combined Effect of Stir and Squeeze Casting Method. *Trans. Indian Inst. Met.* **2015,** *68*, S115–121.

2. Erturk, A. T.; Sahin, M.; Aras, M. Tribological Behavior of SiC Particulate Reinforced AA5754 Matrix Composite under Dry and Lubricated Conditions. *Trans. Indian Inst. Met.* **2016.** DOI: 10.1007/s12666–016–0915–7.
3. Kala, H.; Mer, K. K. S.; Kumar, S. A Review on Mechanical and Tribological of Stir Cast Aluminum Matrix Composites. *Proc. Mater. Sci.* **2014,** *6*, 1951–1960.
4. Rahman, M. H.; Mamun Al Rashed, H. M. Characterization of Silicon Carbide Reinforced Aluminum Matrix Composites. *Proc. Eng.* **2014,** *90*, 103–109.
5. Harrigan, W. C. Commercial Processing of Metal Matrix Composites. *Mater. Sci. Eng. A* **1998,** *244*, 75–79.
6. Shorowordi, K. M.; Laoui, T.; Haseeb, A. S. M. A.; Celis, J. P.; Froyen, L. Microstructure and Interface Characteristics of B_4C, SiC and Al_2O_3 Reinforced Al Matrix Composites a Comparative Study. *J. Mater. Process. Technol.* **2003,** *142*, 738–743.
7. Ramesh, C. S.; Keshavamurthy, R. Microstructure and Mechanical Properties of Ni-P Coated Si_3N_4 Reinforced Al6061 Composites. *Mater. Sci. Eng. A* **2009,** 99–106.
8. Jebeen Moses, J.; Dinaharan, I.; Joseph Sekhar, S. Characterization of Silicon Carbide Particulate Reinforced AA6061 Aluminium Alloy Composites Produced via Stir Casting. *Proc. Mater. Sci.* **2014,** *5*, 106–112.
9. Rajesh, G. L.; Auradi, V.; Kori, S. A. Processing of B_4C Particulate Reinforced 6061 Aluminium Matrix Composites by Melt Stirring Involving Two-Step Addition. *Proc. Mater. Sci.* **2014,** *6*, 1068–1076.
10. Kalaiselvan, K.; Murugan, N. Production and Characterization of AA6061-B_4C Stir Cast Composite. *Mater. Des.* **2011,** *32*, 4004–4009.
11. Gopalakrishnan, S.; Murugan, N. Prediction of Tensile Strength of Friction Stir Welded Aluminium Matrix TiC_p Particulate Reinforced Composite. *Mater. Des.* **2011,** *32*, 462–467.
12. Hashim, J.; Looney, L. Metal Matrix Composites Production by the Stir Casting Method. *J. Mater. Process. Technol.* **1999,** *92–93*, 1–7.

CHAPTER 17

PREPARATION AND ELECTRIC DISCHARGE MACHINING OF CERAMIC PARTICULATE-REINFORCED ALUMINUM MATRIX COMPOSITE

ANIL KUMAR BODUKURI*, K. ESWARAIAH, KATLA RAJENDAR, and V. SAMPATH

Department of Mechanical Engineering, Kakatiya Institute of Technology & Science, Warangal, Telangana, India

*Corresponding author. E-mail: anil.kuphd@gmail.com

ABSTRACT

Metal-matrix composites (MMCs) have been increasingly used in automotive and aerospace industries because of their superior properties compared to unreinforced alloys. These materials are difficult to manufacture by the conventional machining methods because of the hard and abrasive reinforced particles. The present work focuses on the preparation of MMCs and investigates the effect of current, pulse on time and pulse off on the metal removal rate of drilling by electrical discharge machining (EDM) of the prepared Al-Al_2O_3p metal matrix composites. Experiments have been conducted on the electric discharge machine, according to the principles of Taguchi's design of experiments method.

17.1 INTRODUCTION

Electrical discharge machining (EDM) is one of the most basic nontraditional, potential, major manufacturing processes of budding intricate surface geometry and integral angles. The process is applicable to any conductive material regardless of its hardness, toughness, and strength. The electrode is moved toward the workpiece until the gap is little enough in order that the voltage is great enough to ionize the dielectric fluid in the gap between the tool and workpiece. The material is removed with the erosive effect of the electrical discharges of the tool and workpiece.

Aluminum and aluminum-based metal matrix composites (MMCs) have turned out to be striking engineering materials and used in aerospace, military and automobile industries, and on the addition of ceramic particles to aluminum improves its high toughness, strength, wear resistance, and corrosion resistance. Due to possession of superior hardness and reinforcement strength, MMCs are difficult to be machined by traditional techniques. Whereas, the machinability concerned with the applications of existing MMCs are limited due to their poor machinability, poor surface finish, and excessive tool wear. Hence electrical discharge machining (EDM) process becomes a viable method to these kinds of MMC's, even materials with poor machinability such as cemented tungsten carbide and composites can also be processed without much difficulty by the EDM process.[1,2]

The presence of reinforced particles adds additional complexity during machining process when compared to that of monolithic matrix material. But there are still deficiencies in understanding the mechanisms of the machining processes. Tool breaking, tool wear and, recast layer are the main problems in electric discharge machining.[3] Rajkumar et al.[4] suggested that current and pulse on time are the most influential parameters in both conventional and microwave heat treatment. Talla et al.[5] observed in their article that powder-mixed EDM process using Al-suspended kerosene dielectric for the machining of resulted in better metal removal rate (MRR) when compared to conventional EDM process. Mohith Tiwari et al.[6] investigated in their work that peak current contributes the most significantly towards MRR and tool wear rate (TWR).

Sameh S. Habib[7] developed a model between the parameters, and MRR and TWR. These models have been quite good, unique, and powerful and establish good relations between the different process parameters which are a major influence on the MRR and TWR. These models are justified through the various tests conducted by experiments.

Assarzadeh and Ghoreishi[8] investigated that TWR can be lowered by applying small current levels with long pulse durations. The adjustment results in high erosion of the workpiece than the tool, which results in less TWR. The surface roughness response is mostly affected by pulse on time, discharge current, pulse off time, duty cycle, and the interaction effect between the first two parameters. Less rough surfaces can be obtained by setting short pulse durations along with relatively high enough discharge currents.

The most commonly applied techniques in optimization are Taguchi technique, artificial neural network (ANN), response surface methodology (RSM), genetic algorithm, gray relational approach, and the fuzzy logic approach.[9] Gray relational analysis is used primarily for multi-response optimization to obtain a corresponding level of input parameters for better performance characteristics. In general, these are various techniques used for optimization in different applications such as drilling, turning, milling, EDM.[10,11]

Shabgarda et al.[12] presented the comparison and validation of fuzzy results with the experiment findings verifying the high accuracy of models. Fuzzy modeling technique could be an economical and successful method for the prediction of EDM and output parameters according to input variables.

17.2 EXPERIMENTAL WORK

17.2.1 MATERIAL SELECTION

Aluminum 7075 is known for exceptional performance in extreme environments. 7075 is highly resistant to attack by both seawater and industrial chemical environments. The chemical composition of aluminum alloy 7075 is given in Table 17.1. Alloy 7075 also retains exceptional strength after welding. Physical and mechanical properties of aluminum alloy 7075 are shown in Table 17.2. It has the highest strength of the non-heat treatable alloys but is not recommended for use in temperatures in excess of 65°C.

TABLE 17.1 Chemical Composition of Aluminum Alloy 7075.

Element (%)	Present
Magnesium (Mg)	2.1–2.9
Manganese (Mn)	0.2–0.3
Iron (Fe)	0.50
Typical Silicon (Si)	0.0–0.40
Titanium (Ti)	0.05–0.25
Chromium (Cr)	0.18–0.28
Copper (Cu)	1.2–2.0
Typical Others (Total)	0.0–0.15
Zinc (Zn)	5.1–6.1
Other (Each)	0.0–0.05
Aluminum (Al)	Balance

TABLE 17.2 Physical and Mechanical Properties of Aluminum Alloy 7075.

Property	Value
Density	2.81 g/cm^3
Melting point	530–630°C
Thermal expansion	25 × 10–6/K
Modulus of elasticity	72 GPa
Thermal conductivity	121 W/m K
Electrical resistivity	0.058 × 10–6 Ω m
Proof stress	125 MPa
Tensile strength	272 MPa
Brinell hardness	150 HB
Other (each)	0.0–0.05
Aluminum (Al)	Balance

17.2.2 REINFORCED MATERIAL

Alumina (Al_2O_3) is the only oxide produced by the metal aluminum and occurs in nature as the natural mineral corundum (Al_2O_3). Physical and mechanical properties of alumina (Al_2O_3) are shown in Table 17.3. Fused alumina is identical in chemical and physical properties with the natural corundum. It is very hard and alumina lends itself for use as an abrasive material. Another useful property of the material is its high melting point, that is, above 2000°C, which makes it useful as a refractory and as linings of special furnaces. The mechanical, chemical, and electrical properties are represented in following tables.

TABLE 17.3 Physical and Mechanical Properties of Alumina (Al_2O_3).

Property	Value
Density	3.96
Melting Point	2000°C
Thermal Expansion	8.1 10–6/°C
Modulus of Elasticity	375 GPa
Thermal Conductivity	28 W/m°K
TENSILE STRENGTH	220 MPa
BRINELL HARDNESS	14 kg/mm^2

17.2.3 FABRICATION OF THE COMPOSITE

Stir-casting apparatus consists of a cylindrical-shaped graphite crucible as it can withstand high temperature which is much higher than the required temperature [680°C] and also the aluminum does not react with graphite at this temperature. This crucible is mounted in a muffle which is made up of high ceramic alumina. This type of furnace is known as a resistance heating furnace. Aluminum, in liquid stage is very reactive with atmospheric oxygen and there is formation of oxide when it comes in contact with the open air. So it is necessary to carry out the process of stirring in a closed chamber with nitrogen gas as an inert gas in order to avoid the oxidation. Closed chamber is made with the help of steel sheets. This reduces heat loss and gas transfer as compared to an open chamber. Due to the corrosion resistance to atmosphere, steel is selected as the stirrer shaft material. One end of the shaft is connected to a 5 hp motor with flange coupling and at the other end blades are welded. Aluminum alloy matrix is formed by heating the aluminum alloy ingots in the furnace. A stirring action is started and increased slowly from 30 to 350 rpm with a speed controller. The reinforcement (Al_2O_3) is incorporated in the metal matrix at the semisolid stage near 640°C. Schematic diagram of stir casting process is shown in Figure 17.1.The dispersion time taken is about 5 min. After that the slurry is reheated to a temperature above melting point to make sure the slurry is fully liquid and then it is poured into a mold.

The stir casting process starts with the introduction of an empty cylindrical crucible in the furnace. At first, the furnace temperature is set to 500°C and then it is steadily increased up to 900°C. The high temperature of the muffle furnace melts the aluminum alloy Al7075 quickly, reduces the oxidation level, and also enhances the wettability of the reinforcing Al_2O_3 particles in the matrix metal. Aluminum alloy Al7075 is cleaned to remove the dust particles, weighed, and then placed in the cylindrical crucible for melting. During melting, nitrogen gas is passed into the furnace to create an inert atmosphere around the molten matrix. Powder of alumina (Al_2O_3) is used as reinforcement. At a time a total of 700 gm of molten composite was processed in the crucible. A required quantity of reinforcement powder is weighed on the weighing machine and heated for half an hour at the temperature of 500°C. Then it is thoroughly mixed with each other with the help of a stirrer.

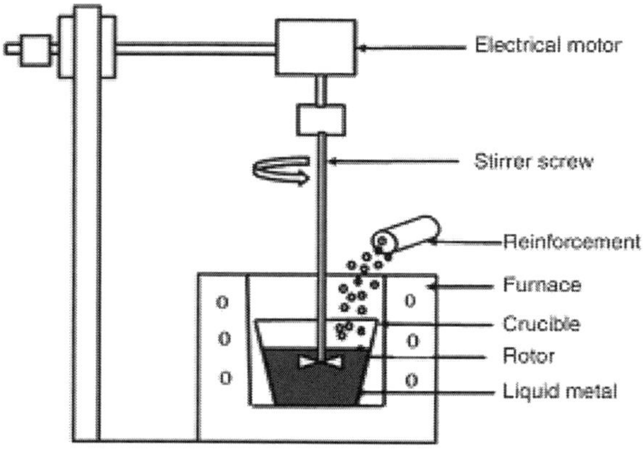

FIGURE 17.1 schematic diagram of stir casting process.

17.2.4 ELECTRICAL DISCHARGE MACHINING (EDM) PROCESS

EDM is one of the most extensively used nonconventional manufacturing processes used for hard materials which are very difficult to machine with in conventional techniques. EDM is sometimes referred to as spark machining, spark eroding, burning, die sinking or wire erosion. Electrical discharge machine is shown in Figure 17.2.This is a manufacturing process whereby a desired shape is obtained using electrical discharges (sparks).

FIGURE 17.2 Electrical discharge machine.

17.2.5 DESIGN OF EXPERIMENTS

Design of experiments (DOE) is a statistical technique used to study the effects of multiple variables on performance measures simultaneously. It provides an efficient experimental schedule and statistical analysis of the experimental results. Taguchi method is a statistical method developed by Genichi Taguchi. It is used to address the DOE for the analysis of experiments.

Product robustness, pioneered by Taguchi, uses experimental design to study the response surfaces associated with both the product means and variances to choose appropriate factor settings so that variance and bias are both small simultaneously. Designing a robust product means learning how to make the response variable insensitive to uncontrollable manufacturing process variability or to the usage conditions of the product by the customer.

Taguchi defines quality characteristics in terms of signal to noise (S/N) ratio which can be formulated for different categories which are as follows:

17.2.6 SMALL ARE BEST CHARACTERISTICS

Data sequence for TWR, which is smaller-the-better performance characteristic, is preprocessed as per Equation 17.1.

$$S/N = -\ 10 \log ((1/n)\ (\Sigma y2)) \qquad (17.1)$$

17.2.7 LARGER IS BEST CHARACTERISTIC

Data sequence for MRR, which is higher-the-better performance characteristics, is preprocessed as per Equation 17.2.

$$S/N = -\ 10 \log ((1/n)\ (\Sigma(1/y2)) \qquad \text{(Eq. 17.2)}$$

Where, y is the value of response variables and n is the number of observations in the experiments

17.2.8 DESIGN VARIABLE

Design parameter, process parameter, and constant parameter are the following:

17.2.1.1 DESIGN PARAMETER

1. MRR
2. TWR

17.2.1.2 MACHINING PARAMETER

The machining parameters are shown in Table 17.4 and the basic L9 orthogonal array is shown in Table 17.5.

1. Discharge current (Ip)
2. Pulse on time (T on)
3. Pulse off time (T off)

TABLE 17.4 Machining Parameters.

Current (Amps)	T on	T off	Material removal rate (MRR)	Tool wear rate (TWR)
6	100	20	13.40	2.400
6	200	50	17.00	2.300
6	500	100	22.40	2.390
9	100	50	45.80	2.291
9	200	100	77.80	2.220
9	500	20	52.00	2.690
12	100	100	110.20	2.210
12	200	20	77.66	2.527
12	500	50	91.28	2.676

TABLE 17.5 Basic Taguchi L9 (3^3) Orthogonal Array.

Process parameters	Level 1	Level 2	Level 3
Current (Amp)	6	9	12
Pulse on time (μs)	100	200	500
Pulse off time (μs)	20	50	100

17.2.1.3 CONSTANT PARAMETER

1. Voltage
2. Flushing pressure
3. Polarity
4. Electrode diameter

17.3 RESULTS AND DISCUSSIONS

The analysis of experimental data was done by using the Minitab 17 software. In general it is used for DOE applications. The experimental observation was transformed to S/N ratios for measuring the quality characteristics. The means of S/N ratio for workpiece is shown in Figure 17.3 and the mean values of S/N ratio for tool is shown in Figure 17.4.

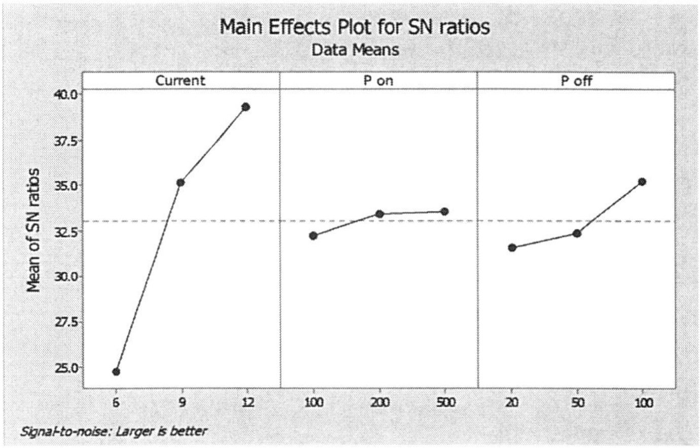

FIGURE 17.3 Means of S/N ratio for workpiece.

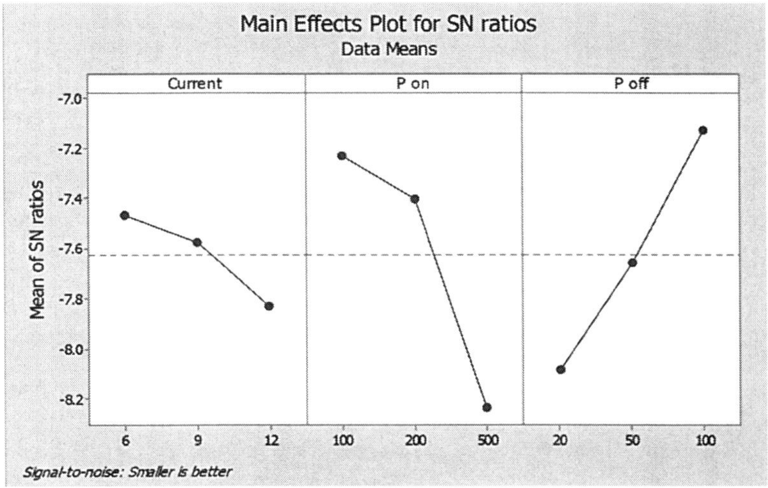

FIGURE 17.4 Mean values of S/N ratio for tool.

17.4 CONCLUSIONS

In this paper, a study of the influence of three erosion parameters (current intensity, pulse on time, and pulse off time) on MRR and electrode wear rate in EDM of MMC AL-Al_2O_3 has been performed.

Regarding electrode wear, current intensity and pulse on time are the factors which have more influence on the response. For minimal electrode wear, high pulse time and high current intensity should be applied.

For high current intensity values, the best results are achieved with high pulse time and for low current intensity values, best results are obtained with low pulse time.

KEYWORDS

- **clectrical discharge machining**
- **metal-matrix composite**
- **Taguchi method**
- **stir casting**
- **metal removal rate**
- **tool wear rate**

REFERENCES

1. Ho, K. H.; Newman, S. T. State of the Art Electrical Discharge Machining (EDM). *Int. J. Mach. Tool. Manuf.* **2003,** *43,* 1287–1300.
2. Gopalakannan, S.; Senthilvelan, T.; Kalaichelvan, K. Modeling and Optimization of Al 7075/10 wt % Al_2O_3 Metal Matrix Composites by Response Surface Method. *Adv. Mater. Res.* **2012,** *488–489,* 856–860.
3. Pramanik, A. Developments in the Non-Traditional Machining of Particle Reinforced Metal Matrix Composites. *Int. J. Mach. Tools Manuf.* **2014,** *86,* 44–61.
4. Rajkumar, K.; Santosh, S.; Javed Syed Ibrahim, S. Gnanavelbabu, A. Effect of Electrical Discharge Machining Parameters on Microwave Heat Treated Aluminum-Boron Carbide-Graphite Composites: 12th Global Congress on Manufacturing and Management, GCMM 2014. *Proc. Eng.* **2014,** *97,* 1543–1550.
5. Talla, G.; Sahoo, D. K.; Gangopadhyay, S.; Biswas, C. K. Modeling and Multi-Objective Optimization of Powder Mixed Electric Discharge Machining Process of Aluminum/Alumina Metal Matrix Composite: Full Length Article. *Eng. Sci. Technol., Int. J.* **2015,** *18,* 369–373.

6. Tiwari, M.; Mausam, K.; Sharma, K.; Singh, R. P. Investigate the Optimal Combination of Process Parameters for EDM by using a Gray Relation Analysis. (International Conference on Advances in Manufacturing and Materials Engineering, ICAMME 2014) *Proc. Mater. Sci.* **2014,** *5,* 1736–1744.
7. Habib, S. S. Study of the Parameters in Electrical Discharge Machining through Response Surface Methodology Approach. *Appl. Math. Modell.* **2009,** *33,* 4397–4407.
8. Assarzadeh, S.; Ghoreishi, M. Statistical Modeling and Optimization of Process Parameters in Electro-Discharge Machining of Cobalt-Bonded Tungsten Carbide Composite (WC/6%Co) (The Seventeenth CIRP Conference on Electro Physical and Chemical Machining (ISEM)) *Proc. CIRP* **2013,** *6,* 463–468.
9. Aggarwal, A.; Singh, H. Optimization of Machining Techniques—A Retrospective and Literature Review. *Sadhana* **2005,** *30,* 699–711.
10. Lin, C. L.; Lin, J. L.; Ko, T. C. Optimisation of the EDM Process Based on the Orthogonal Array with Fuzzy Logic and Grey Relational Analysis Method. *Int. J. Adv. Manuf. Technol.* **2002,** *19*(4), 271–277.
11. Lin, Y.-C.; Lee H.-S. Optimization of Machining Parameters using Magnetic Force-Assisted EDM Based on Gray Relational Analysis. *Int. J. Adv. Manuf. Technol.* **2009,** *42,* 1052–1064.
12. Shabgard, M. R.; Badamchizadeh, M. A.; Ranjbary, G.; Amini, K. Fuzzy Approach to Select Machining Parameters in Electrical Discharge Machining (EDM) and Ultrasonic-Assisted EDM Processes. *J. Manuf. Syst.* **2013,** *32,* 32–33.

CHAPTER 18

EROSIVE BEHAVIOR OF ALUMINA-FILLED ZINC ALUMINUM ALLOY METAL MATRIX

MAMATHA T. G.[1,*], HARSHA B. P.[1,2], and AMAR PATNAIK[3]

[1]*Department of Mechanical Engineering, JSSATE, Noida, Uttar Pradesh 201301, India*

[2]*harshabp@jssaten.ac.in*

[3]*Department of Mechanical Engineering, MNIT, Jaipur Rajasthan, India, E-mail: patnaik.amar@gmail.com*

[*]*Corresponding author. E-mail: mamathatg@jssaten.ac.in*

ABSTRACT

The erosion behavior of alumina-reinforced ZA-27 alloy composites fabricated by stir casting technique is studied by means of erosive wear-test rig. In this investigation, the influence of independent parameters such as impact velocity, filler content, erodent temperature, impingement angle, erodent size on erosive wear behavior of composites was studied. A finite element (FE) model (ANSYS/AUTODYN) of erosive wear was established for damage assessment and validated by a well-designed set of design of experiment (DoE) technique. It is observed that there is a good agreement between the computational and experimental results, and that the proposed simulation model is very useful for the evaluation of damage mechanisms. The results obtained in this work enable alumina-reinforced ZA-27 alloy composites to exhibit better wear resistance as compared with unfilled ZA-27 alloy composite.

18.1 INTRODUCTION

Metal-matrix composites (MMCs) are in constant growing stage over the years because they have better physical, mechanical and tribological properties comparing to matrix materials. Tribological properties of composite comparing to the matrix material are of importance in all machine components where the parts are in contact and relatively sliding motions. The fact is that tribological properties define the possible application of material far more than their mechanical properties. Particle-filled MMCs offer significant improvements relative to monolithic alloys in room temperature, that is, mechanical properties such as strength, modulus, fatigue resistance, and wear resistance. The majority of investigations into the creep behavior of particle-reinforced aluminum alloys have been performed at temperatures greater than 250°C.[1–4] Xiandong et al.[5] have studied both ZL-I09 alloy and ZA-27 alloy as the matrices of the composites, and the reinforcements such as SiC, Si_3N_4, B_4C, $Ab0_3$, and the graphite flakes without coating. The particulate-reinforced MMCs showed excellent tribological properties, with the wear becoming more uniform, the wear rate and friction coefficient decreasing appreciably, and the relative seizure resistance increasing. During past decades, there were a number of computational models proposed to simulate wear processes at macro/meso- and atomic/nanoscales. The most widely used macro/meso-scopic model is finite element method (FEM),[6–9] which can be used to deal with elastic/plastic contact problems and failure processes in the contact region. Adler[10] established the two-dimensional (2D) waterdrop impact model with DYNA3D code in 1995. The calculation did not involve any solid erodent.

ZA-27 is a high-strength performer of the zinc alloys whether for gravity, pressure die casting or sand casting, and so forth, and it is also the lightest alloy offer excellent bearing and wear resistance properties. ZA-27, however, requires care during melting and casting to assure sound internal structure, particularly for heavy wall sections.

With the addition of alumina into ZA-27 alloy composite, the aluminum percentage increases as aluminum is chemically reactive and more adhesive.[11]

Recent finite element modeling has suggested that these differences in behavior may be dependent on the bonding strength between the particles and matrix material.[12]

In the present work, alumina is taken as a reinforcing material and ZA-27 is used as a base material for tribological study. Such study can provide vital relevant information on the effect of control factors on erosive wear environment. A Taguchi design for the experiments has been used to assess data in a controlled way and the individual effect of these parameters on wear behavior have been established systematically and compared with the proposed nonlinear FE model (ANSYS/AUTODYN) results. The utility of the new formulation and development of composites filled with alumina particulates is proven through their various tribological applications. This model is simple and flexible, which helps to gain insight into various wear processes.

18.2 EXPERIMENTAL PROCEDURE

18.2.1 PREPARATION OF THE COMPOSITES

In the present work, ZA-27 alloy having the chemical composition as per the ASTM B669–82 ingot specification used as the base matrix alloy is given in Table 18.1. Alumina particles of size 80 μm are used as the reinforcement. The percentage of alumina is varied from 0 to 12 wt.% in a step of 4% by weight. In this process, the matrix alloy ZA-27 was first superheated above its melting temperature and stirring was initiated to homogenize the temperature. The addition of alumina into the molten ZA-27 alloy above its liquidus temperature at 500°C was carried out by using a mechanical stainless steel stirrer coated with aluminite. The stirrer was rotated at a speed of 450 rpm in order to get uniform mixing of alumina particulate in the matrix material for 2–3 min. The molten metal was then poured into permanent molds for casting and the temperature was then lowered gradually.

TABLE 18.1 Chemical Composition of Za-27 Alloy in Weight Percent (ASTM B669–82).

Element	Aluminum	Magnesium	Copper	Zinc
Percentage composition (wt.%)	25–28	0.01–0.02	2.0–2.5	Balance

18.2.2 EROSION TEST APPARATUS

The solid particle erosion experiments are carried out on the erosion test rig (ASTM G76) as shown in Figure 18.1. The setup is capable of creating reproducible erosive situations for assessing erosion wear resistance of the composite samples. The erodent particles impact the specimen which can be held at different angles with respect to the direction of erodent flow using a swivel and an adjustable sample clip. The apparatus is equipped with a heater which can maintain the erodent temperature (40, 60, 80, and 100°C) during an erosion trial. In the present study, dry silica sand of different particle sizes (250, 350, 450, and 600 μm) are used as erodent. The samples are cleaned in acetone, dried, and weighed to an accuracy of ± 0.001 mg before and after the erosion trials using an electronic balance. The weight loss is recorded for subsequent calculation of erosion rate.

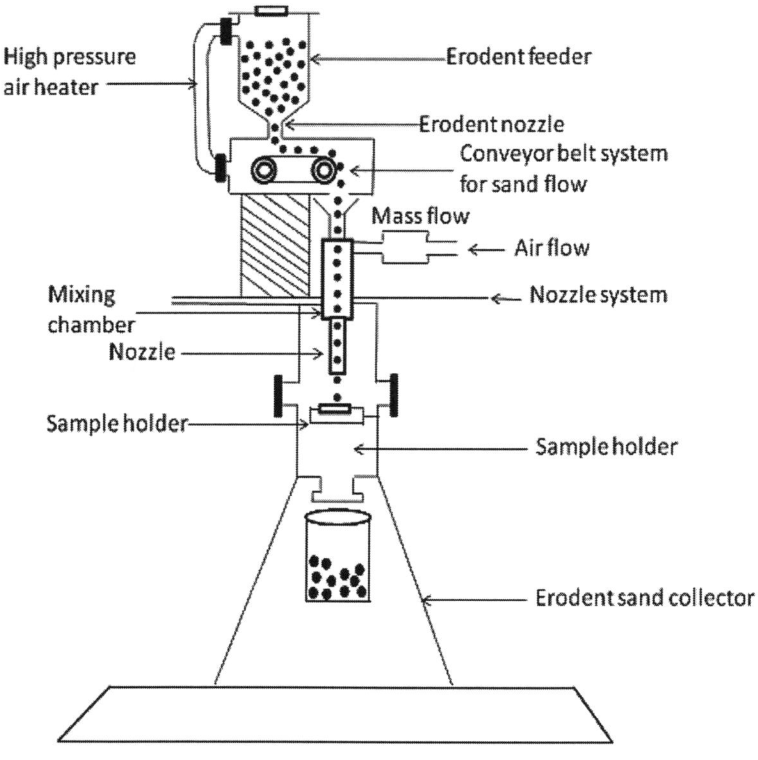

FIGURE 18.1 Schematic diagram of an erosion test rig.

18.2.3 EXPERIMENTAL DESIGN

DoE is an evaluation tool for determining robustness and the signal-to-noise (S/N) ratio is the most important component of a parameter design. In the present study, the experimental layout of the L_{16} (4^5) orthogonal array is taken consisting of five control factors and each factor at four levels without considering the outer array for noise factors (Table 18.2). There are several S/N ratios available depending on the type of performance characteristics. The S/N ratio for minimum erosion rate can be expressed as "lower is better" characteristic, which is calculated as logarithmic transformation of loss function as shown below.

Smaller is the better characteristic:

$$\frac{S}{N} = -10 \log \frac{1}{n}\left(\Sigma y^2\right) \tag{18.1}$$

where 'n' is the number of experiments in the L_{16} orthogonal array design and y_i is the measured values. The plan of the experiments is as follows: the first column of this orthogonal array is assigned to impact velocity (A), the second column to alumina content (B), the third column to erodent temperature (C), the fourth column to impingement angle (D), and the fifth column to erodent size (F) at constant standoff distance (75 mm) for all the test runs respectively to estimate the erosion rate.

TABLE 18.2 Levels of the Variables Used in the Experiment.

Control factor	Level				Units
	I	II	III	IV	
A. Velocity of impact	25	43	66	77	m/s
B. Alumina content	0	4	8	12	%
C. Erodent temperature	40	60	80	100	°C
D. Impingement angle	30	45	60	90	Degree
E. Erodent size	250	350	450	550	μm

18.2.4 FINITE ELEMENT (FE) MODEL

In the present work, the FE model has been developed to calculate erosive wear rate during solid particle impact on target material through explicit dynamic code ANSYS/AUTODYN. The geometrical modeling was done by Pro/Engineer software and was imported in IGES format. This file was opened by ANSYS Design modeler and meshed in meshing module by choosing explicit environment at ANSYS workbench platform. The explicit hex meshing was done which consist of 80,000 nodes and 75,000 finite elements. The solid particles were modeled using smooth particle hydrodynamics (SPH) with properties of silica sand. The material model applied SPH sand particles were modeled using compaction equation of state, MO Granular strength, and Hydro (Pmin) failure material model. ElTobgy et al.[13] indicated that simulating a single particle was not sufficient and three or more particles were needed to simulate the erosion process. In this study, we use 50 spherical sand SPH particles so as to ensure the accuracy of the model. All of the particles strike the target area at random locations. All the bottom nodes of the target material were fixed. The rotation degree of freedom for all the particles has been constrained. According to the research of Woytowitz and Richman,[14] the distance between any two alternative particles' centers in the same group is no less than 0.6r to avoid the damage interaction (r is the radius of the erodent particles). The erosion rate is defined as the ratio of cumulative mass loss of target materials and amount of erodent used during each experimental test.

18.3 RESULTS AND DISCUSSION

18.3.1 WEAR PROPERTIES

18.3.1.1 INFLUENCE OF IMPINGEMENT ANGLE ON WEAR RATE

The steady-state erosion rates are measured as a function of impingement angle of alumina-filled ZA-27 alloy composites as shown in Figure 18.2. It is observed that the unfilled composite shows maximum erosion rate as compared to particulate-filled ZA-27 alloy composites under constant impact velocity. At 60° impingement angle, the erosion rate variation is marginal among the particulate-filled composites, whereas 8 wt.% alumina-filled MMC shows least erosion rate as compared with other particulate-filled composites (Fig. 18.2). The difference between the experimental results and finite element simulated results following same trends but only in case of 12 wt.% alumina-filled composites shows slight variation between experimental results with the simulated results. Figure 18.3 shows Von Mises equivalent stress distribution of all the composites as a function of the impingement angle. The maximum erosion rate appeared at about 60° impingement angle and causes savior plastic deformation within that region. The large plastically deformed elements are removed from target material due to high strain energy. The removal of elements is based on shearing action on elements by solid particle impacting.

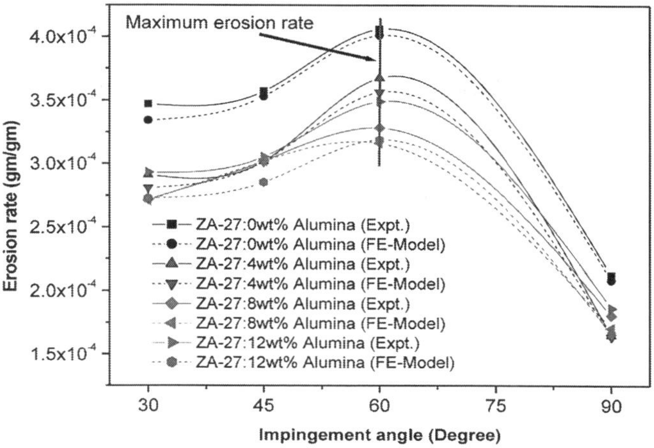

FIGURE 18.2 Variation of erosion rate with impingement angle for the alumina-filled ZA-27 composites.

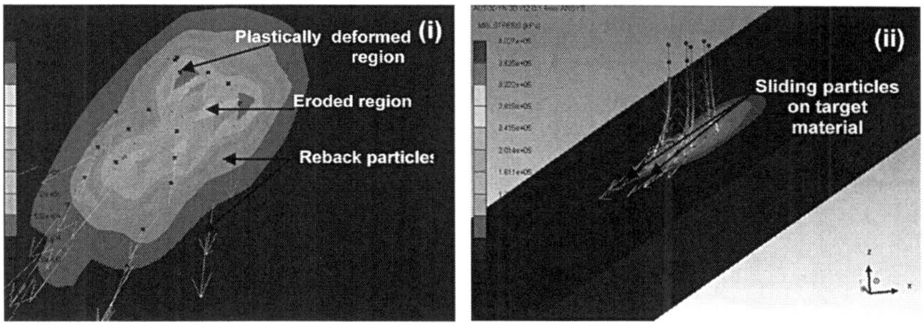

FIGURE 18.3a Von Mises equivalent stress distribution on ZA-27 target material: (i) plastic deformation and element failure region, (ii) sliding particles after impact.

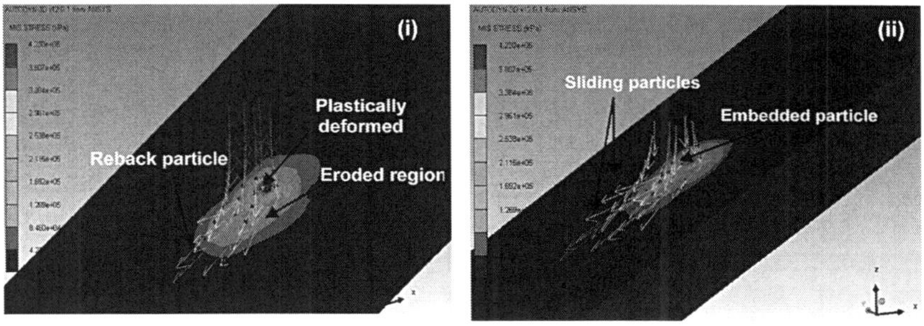

FIGURE 18.3b Von Mises equivalent stress distribution on ZA-27 with 4 wt.% alumina target material: (i) plastic deformation and element failure region, (ii) sliding and embedded particles after impact.

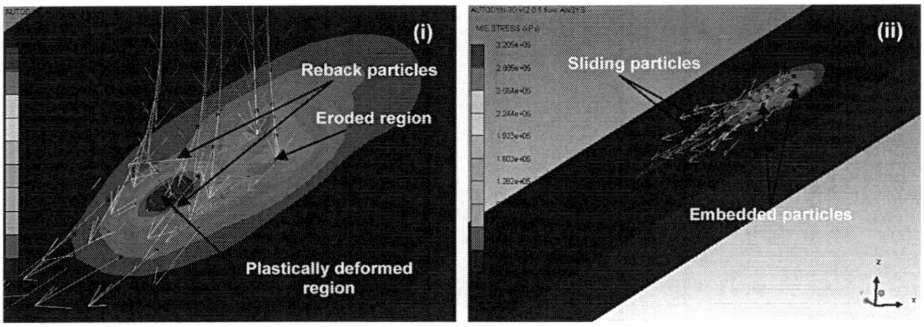

FIGURE 18.3c Von Mises equivalent stress distribution on ZA-27 with 8% alumina target material: (i) plastic deformation and element failure region, (ii) Sliding and embedded particles after impact.

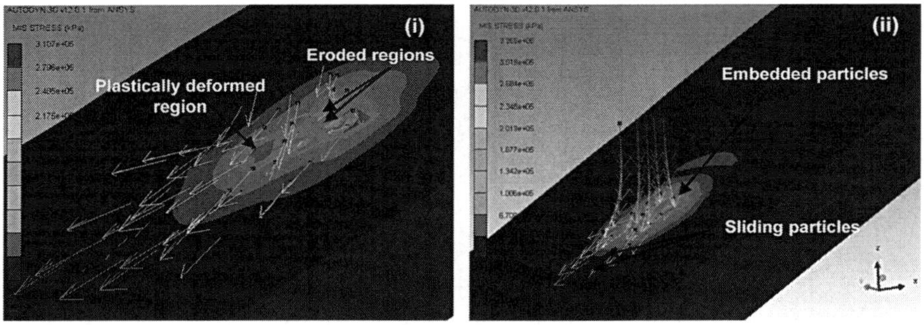

FIGURE 18.3d Von Mises equivalent stress distribution on ZA-27 with 12% alumina target material: (i) plastic deformation and element failure region, (ii) sliding and embedded particles after impact.

18.3.1.2 INFLUENCE OF IMPACT VELOCITY ON WEAR RATE

The steady-state erosion rates are measured at four different velocities (i.e., 25–79 m/s), which enables the determination of one of the major control factors, that is, impact velocity dependence of the erosion rate. The experimental results of the erosion rates for the materials related to the impact velocity are shown in Figure 18.4.

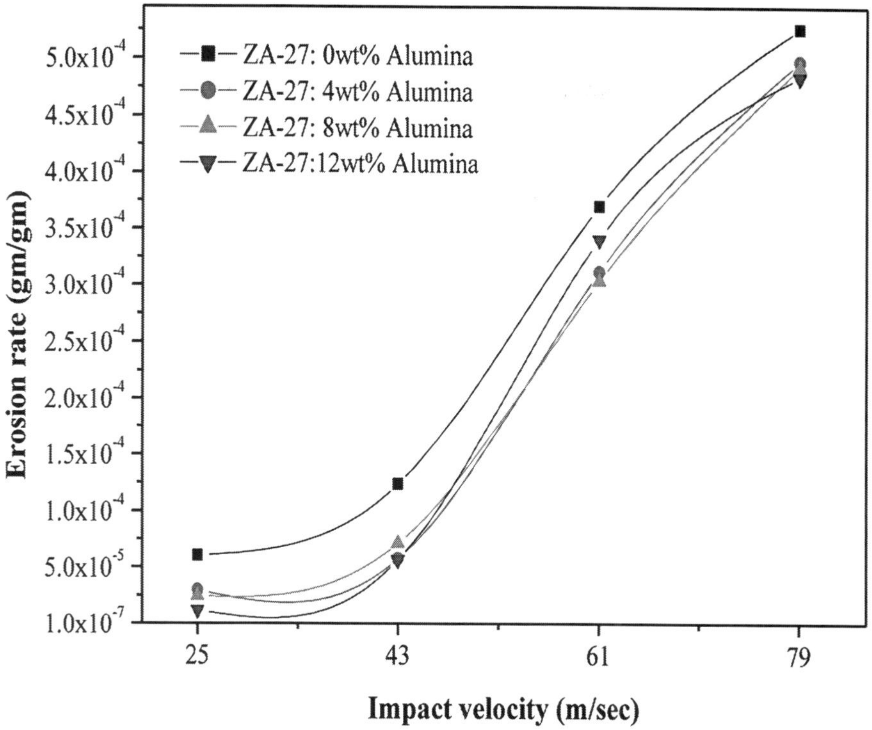

FIGURE 18.4 Variation of erosion rate with impact velocity for the alumina-filled ZA-27 composites.

18.3.1.3 TAGUCHI EXPERIMENTAL ANALYSIS

In Table 18.3, the eight columns represent S/N ratio of the wear rate which is, in fact, the average of two replications of composites. The overall mean for the S/N ratio of the wear rate is found to be 86.56 db. The analysis was made using the popular software specifically used for DoE applications known as MINITAB 15. The analysis of the result leads to the conclusion that factor combination of A_1, B_3, C_3, D_4, and E_2 gives minimum erosion rate for alumina-filled MMCs (Fig. 18.5). This analysis concludes that wear rate also depends upon the types of filler, filler content, and fabrication methodology.

TABLE 18.3 Experimental Design using L_{16} Orthogonal Array.

Sl no.	Impact velocity (m/s)	Filler content (%)	Erodent temperature (°C)	Impingement angle (degree)	E (erodent size) (μm)	Erosion rate (g/g)	S/N ratio (db)
1	25	0	40	30	250	0.0000128	97.856
2	25	4	60	45	350	0.0000105	99.576
3	25	8	80	60	450	0.0000005	126.021
4	25	12	100	90	550	0.0000044	107.131
5	43	0	60	60	550	0.0000725	82.793
6	43	4	40	90	450	0.0000246	92.181
7	43	8	100	30	350	0.0000030	110.458
8	43	12	80	45	250	0.0000721	82.841
9	61	0	80	90	350	0.0000051	105.849
10	61	4	100	60	250	0.0004074	67.800
11	61	8	40	45	550	0.0002661	71.499
12	61	12	60	30	450	0.0002630	71.601
13	79	0	100	45	450	0.0004502	66.932
14	79	4	80	30	550	0.0004919	66.162
15	79	8	60	90	250	0.0002953	70.595
16	79	12	40	60	350	0.0005149	65.766

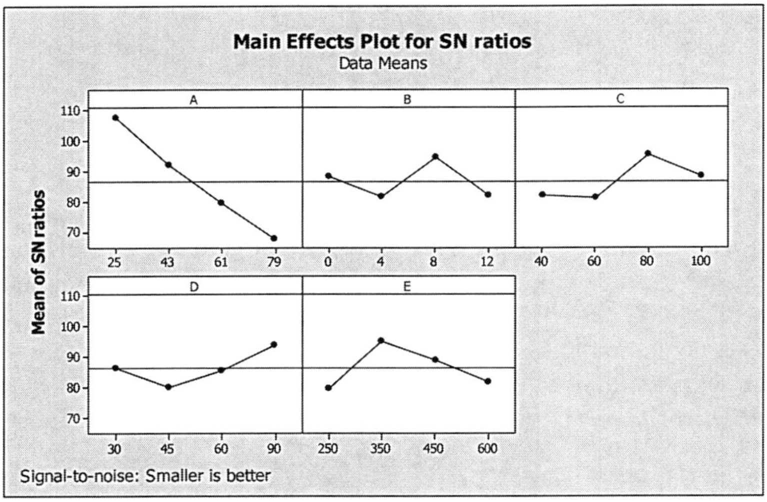

FIGURE 18.5 Effect of control factors on wear rate.

18.3.1.4 FACTOR SETTINGS FOR MINIMUM EROSION RATE

In this study, an attempt is made to derive optimal settings of the control factors for minimization of erosion rate. The single-objective optimization requires quantitative determination of the relationship between erosion rates with a combination of control factors. In order to express erosion rate in terms of mathematical model, the following form is suggested:

$$Er = K_0 + K_1 \times A + K_2 \times B + K_3 \times C + K_4 \times D + K_5 \times E \qquad (18.2)$$

Here, Er is the performance output terms and K_i (i=0,.........0.5) are the model constants. The constants are calculated using nonlinear regression analysis with the help of SYSTAT 7 software and the following relations are obtained.

The correctness of the calculated constants is confirmed as high correlation coefficients (r^2) in the tune of 0.89 are obtained for Equation 18.2 and, therefore, the models are quite suitable to use for further analysis.

18.4 CONCLUSIONS

1. The particle-reinforced composites show good tribological properties, with the erosion becoming more uniform and the erosion rate decreasing appreciably with the increase in filler content. Among all, 8 wt.% alumina-filled composite shows maximum erosion resistance as compared with the rest of the filled and unfilled composites.
2. In the variation of erosion rate with impingement angles, the material loss is dictated mainly more at 60° impingement angles, in the steady-state erosion rate with other control factors are remain constant. However, compared with the computational models, the FE model (ANSYS/AUTODYN) is much closer to the experimental results. The major advantages of simulated results are during the experimental study as it is very difficult to analyze the flow direction, and particularly at low impingement angle, most of the erodent particles are sliding on the target material instead of reback of erodent particles. However, in FE simulated model, the above conditions can be measured easily including the residual stress and the depth of penetration which is difficult to determine by the experimental method.

KEYWORDS

- **composites**
- **finite element model**
- **Taguchi technique**
- **erosion wear**

REFERENCES

1. Pruthviraj, R. D.; Krupakara, P. V. Influence of Sic Additions on Mechanical Properties of the Zn–Al Alloy (Za-27). *Int. J Mater Sci.* **2007**, *2,* 53–57.
2. Babic, M.; Ninkovic, R.; Rac, A. Sliding Wear Behavior of Zn-Al Alloys in Conditions of Boundary Lubrication. *Ann. Univ. "Dunarea De Jos" Galati, Fascicle VIII Tribol.* **2005,** *8,* 60–64.
3. Prasad, B. K.; Modi, O. P.; Khaira, H. K. High-Stress Abrasive Wear Behavior of a Zinc-Based Alloy and Its Composite Compared with a Cast Iron Under Varying Track Radius and Load Conditions. *Mater. Sci. Eng. A* **2004**, *381,* 343–354.
4. Modi, O. P.; Rathod, S.; Prasad, B. K.; Jha, A. K.; Dixit, G. The Influence of Alumina Particle Dispersion and Test Parameters on Dry Sliding Wear Behavior of Zinc-Based Alloy. *Tribol Int.* **2007**, *40,* 1137–1146.
5. Xiandong, S.; Chengping, L.; Zhuoxuan, L.; Liuzhang, O. The Fabrication and Properties of Particle Reinforced Cast Metal Matrix Composites. *J. Mater. Process. Technol.* **1997,** *63,* 426–431.
6. Ling, F. F.; Pan, C. H. T.; Eds. Approaches to Modeling of Friction And Wear. In *Proceedings of the Workshop on the Use of Surface Deformation models to Predict Tribology Behavior,* Springer–Verlag: New York, 1986.
7. Ludema, K. C.; Bayer, R. G., Eds. *Tribological Modeling for Mechanical Designers;* ASTM STP1105, American Society for Testing and Materials: Philadelphia, 1991.
8. Hsu, S. M.; Shen, M. C.; Ruff, A. W. Wear Prediction for Metals. *Tribol. Int.* **1997,** *30*(5), 377–383.
9. Komvopoulos, K.; Choi, D. H. Elastic Finite Element Analysis of Multi-Asperity Contacts. *J. Tribol.* **1992,** *114,* 823–831.
10. Adler, W. F. Waterdrop Impact Modeling. *Wear* **1995,** *186–187,* 341–351.
11. Johnson, G. R.; Cook, W. H. A Constitutive Model and Data for Metals Subjected to Large Strains, High Strain Rates and High Temperatures. In: *Proceedings of the 7th International Symposium on Ballistics,* The Hague, 1983, pp 541–547.
12. Johnson, G. R.; Cook, W. H. Fracture Characteristics of Three Metals Subjected to Various Strains, Strain Rates, Temperatures and Pressures. *Eng. Fract. Mech.* **1985,** *21*(1), 31–48.
13. ElTobgy, M. S.; Ng, E.; Elbestawi, M. A. Finite Element Modeling of Erosive Wear. *Int. J. Mach. Tool Manuf.* **2005,** *45,* 1337–1346.
14. Woytowitz, P. J.; Richman, R. H. Modeling of Damage from Multiple Impacts by Spherical Particles. *Wear* **1999,** *233–235,* 120–133.

CHAPTER 19

EFFECT OF HEAT TREATMENT ON MECHANICAL AND WEAR PROPERTIES OF ALUMINUM–RED MUD–TUNGSTEN CARBIDE METAL MATRIX HYBRID COMPOSITES

CHINTA NEELIMA DEVI[1,*], N. SELVARAJ[2], and V. MAHESH[3]

[1]*Department of Mechanical Engineering, JNTUK University College of Engineering, Vizianagaram, Andhra Pradesh, India*

[2]*Department of Mechanical Engineering, National Institute of Technology, Warangal, Telangana, India, E-mail: nsr14988@yahoo.co.in*

[3]*Department of Mechanical Engineering, SR Engineering College, Warangal, Telangana, India, E-mail: v.mahesh2@gmail.com*

**Corresponding author. E-mail: cneelima.me@jntukucev.ac.in*

ABSTRACT

Aluminum hybrid composites are finding increased demand and usage in new aircraft designs and automotive industries due to their higher strength capabilities. In this work, an attempt has been made to utilize the red mud (bauxite residue) which is an industrial byproduct for value added applications. It is received from Nalco and reinforced aluminum with 2, 4, and 6% weight fractions of 100 μm and 42 nm red mud with 4% tungsten carbide (WC). The aluminum-red mud-tungsten carbide metal matrix hybrid composites are prepared by conventional sintering process. The chapter presents mechanical behavior such as hardness and wear properties of pure Al with red mud and WC at normal condition and heat treatment condition (350, 400, 450, and 500°C). A mathematical model is developed using regression analysis for the hardness and wear rate values of Al, red mud, and WC. The predicted equations are significant in nature with near-unity R-squared values. Nano-red-mud specimens have more hardness and wear resistance when compared with micronatured specimens. The wear resistance is more for 42 nm of 94% Al and 6% weight fraction of bauxite residue with pure aluminum and tungsten carbide at 600 rpm speed.

19.1 INTRODUCTION

Many of the modern technologies require new engineering materials and great interest for potentially capable materials in the field of defense, automobile, underwater, and spacecraft applications, which will possess low densities, are strong, and abrasion, corrosion and impact resistant. Aluminum–red mud–WC hybrid composites can become alternative materials for different engineering applications with their excellent qualities such as lightweight, environment-friendliness, quality, performance with superior mechanical properties, wear characteristics, and low cost. Therefore, the innate ability of aluminum hybrid composites has a wide variety of applications in various research activities all over the world.[1–3] Aluminum-based metal matrix composites (MMCs) containing hard particles offer superior operating performance and resistance to wear.[4–5] Red mud is one of the wastages produced during manufacturing of alumina. For every 2.5 t of alumina production, 1.5 t of red mud is produced.[6–9] The present research work is aimed to find out the value-added applications of utilizing the industrial by-product to form metal matrix hybrid composite and eliminate the large area of land filling with bauxite residue. The application of red mud as a basic catalyst for biodiesel production and manufacturing of ceramic tiles are also studied.[10–11]

19.2 EXPERIMENTAL WORK

19.2.1 GRADATION TEST

Gradation test or sieve is a procedure used to assess the particle size distribution of a granular material. In the present work, the red mud is taken from Nalco, India. Tables 19.1, 19.2, and 19.3 present the chemical compositions of Al, red mud, and WC, respectively.

TABLE 19.1 Chemical Composition of Aluminum.

Element	Fe	Mg	Si	Zn	Cu	Mn	Others
Wt.%	0.17	0.001	0.07	0.003	0.005	0.0008	Balance

TABLE 19.2 Chemical Composition of Red Mud.

Element	Fe_2O_3	TiO_2	Al_2O_3	SiO_2	V_2O_5	Na_2O	Others
Wt.%	53.8	3.9	14.3	8.34	0.38	4.3	Balance

TABLE 19.3 Chemical Composition of Tungsten Carbide.

Element	W	Cr	Si	B	C	Others
Wt.%	8.5	3.52	2.0	1.8	0.5	Balance

The sieve analysis is carried out to obtain the microlevel red mud powder of 100, 150, and 200 μ size in sieve shaker machine.

19.2.2 HIGH-ENERGY BALL MILL

High-energy ball mill, referred to as PULVERISETTE, are used for conducting mechanical attrition experiments. The sample red mud powder is collected after 6, 13, 24, and 30 h of high-energy ball mill processing. Mechanical drier is used to dry the collected red mud sample. The X-ray diffraction (XRD) tests are conducted. Figure 19.1 shows the XRD pattern of red mud at 30 h. The particle size of 42 nm is obtained after 30 h of ball milling.

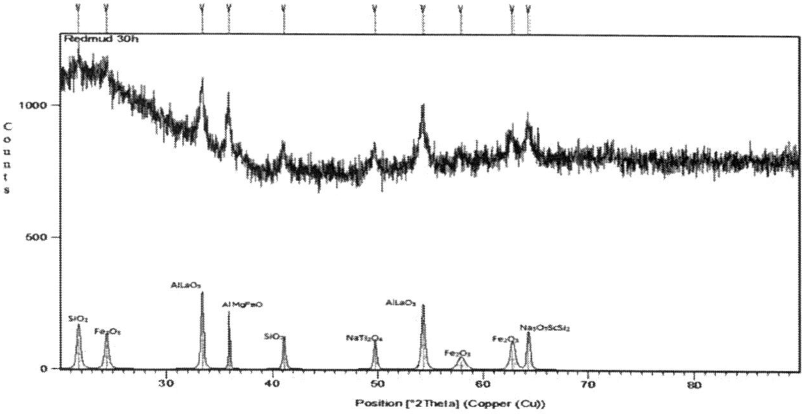

FIGURE 19.1 X-Ray diffraction (XRD) pattern of red mud at 30 h.

19.2.3 MIXING AND COMPACTING

The microlevel red mud powders of 100, 150, and 200 μm and nanolevel red mud powder of 42 nm at 2, 4, and 6% of weight fractions is mixed with pure aluminum powder in a double-cone mixer for 10 h in order to obtain proper mixing of particles with each other. Using the load capacity of 100 t hydraulic press, the different proportions of aluminum–red mud sample materials are compacted.

19.2.4 CONVENTIONAL SINTERING

Sintering is a thermal treatment process. The vacuum sintering is done on Model No ZSJ-25x25x50. The compacted Al, red mud, and WC samples are shown in Figure 19.2.

FIGURE 19.2 Compacted pure Al, red mud, and tungsten carbide (WC) samples in sintering furnace.

19.3 RESULTS AND DISCUSSIONS

19.3.1 HARDNESS

Hardness values are measured using Micro Vickers Hardness tester according to ASTM E32. The Vickers Hardness Number (VHN) is given by $1.854L/d^2$, where L is known as the applied load in kgf and d is known as diagonal length of a square impression in millimeter. The graph between hardness and % weight fraction of 4% WC and red mud with pure aluminum at normal condition is shown in Figure 19.3.

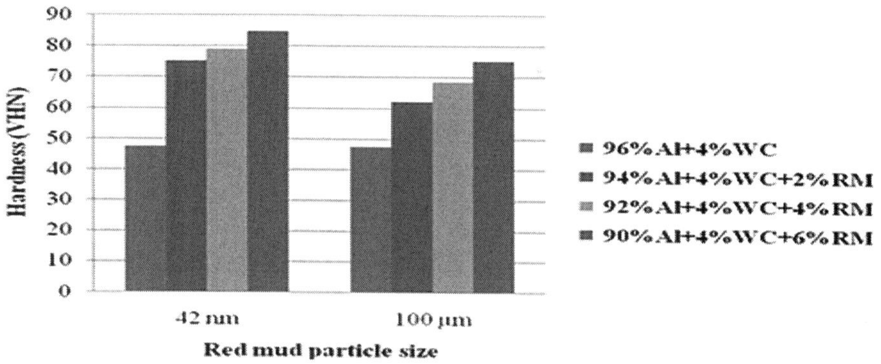

FIGURE 19.3 Hardness versus % weight fraction of 4% WC and red mud with pure aluminum at normal condition.

It is observed that 6% red mud (nano level) with pure Al and 4% WC has a best hardness value of 84.9 VHN compared to remaining proportions. The plots between hardness and percentage (%) weight fraction of Al with red mud and 4% WC at heat treatment conditions of 350, 400, 450, and 500°C are shown in Figures 19.4–19.7.

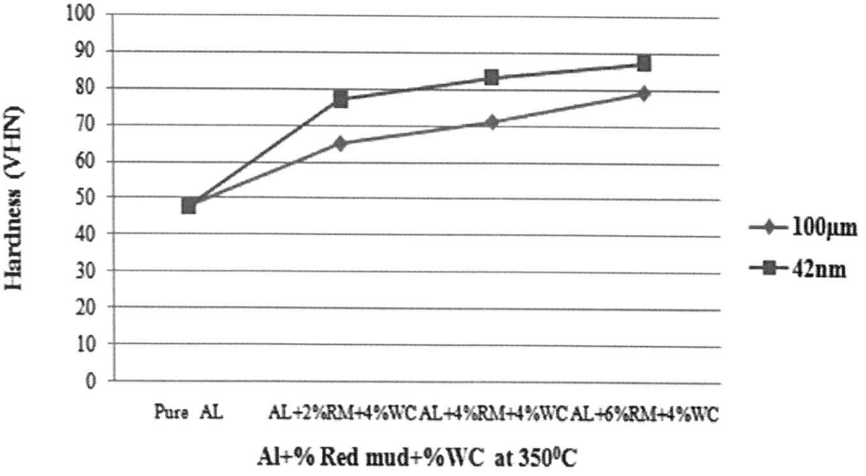

FIGURE 19.4 Hardness versus % weight fraction of 4% WC and red mud with pure aluminum at 350°C.

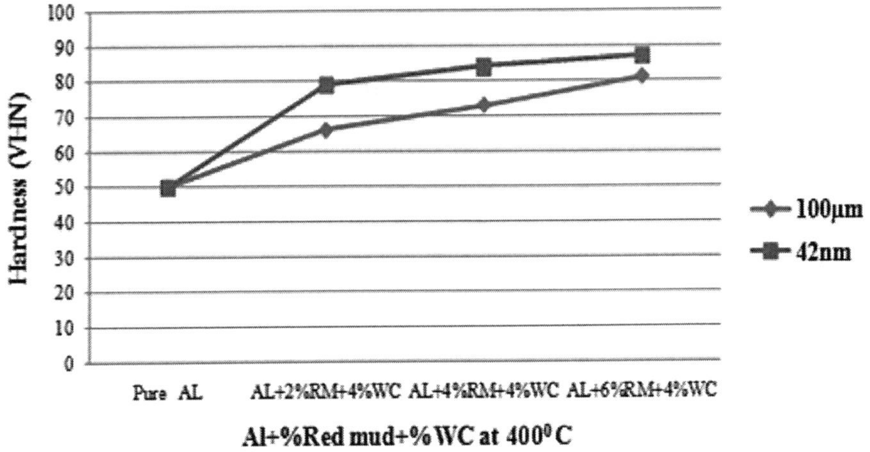

FIGURE 19.5 Hardness versus % weight fraction of 4% WC and red mud with pure aluminum at 400°C.

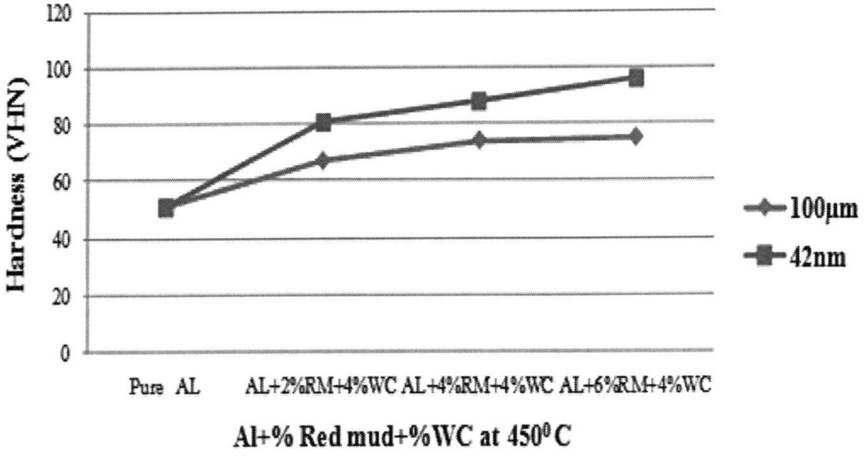

FIGURE 19.6 Hardness versus % weight fraction 4% WC and red mud with pure aluminum at 450°C.

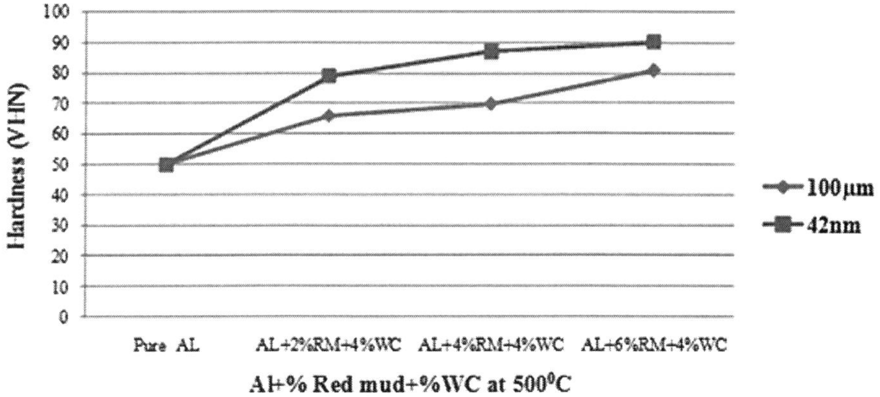

FIGURE 19.7 Hardness versus % weight fraction of 4% WC and red mud with pure aluminum at 500°C.

19.3.2 WEAR TEST

Wear tests are conducted as per ASTM G99 standards on pin-on-disc wear testing machine and the graph between wear rate and % weight fraction of 4% WC and red mud with pure aluminum at normal condition is shown in Figure 19.8.

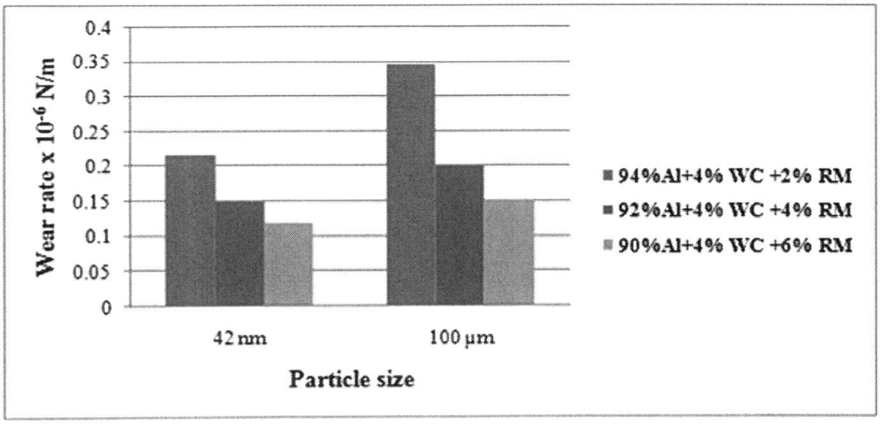

FIGURE 19.8 Plot between wear rate and particle size of Al + %WC + %RM at normal condition.

The plot between wear rate, % weight fraction of red mud, and temperature of 100 μm and 42 nm particle sizes with pure aluminum and 4% tungsten carbide at 10 N load is shown in Figures 19.9 and 19.10, respectively.

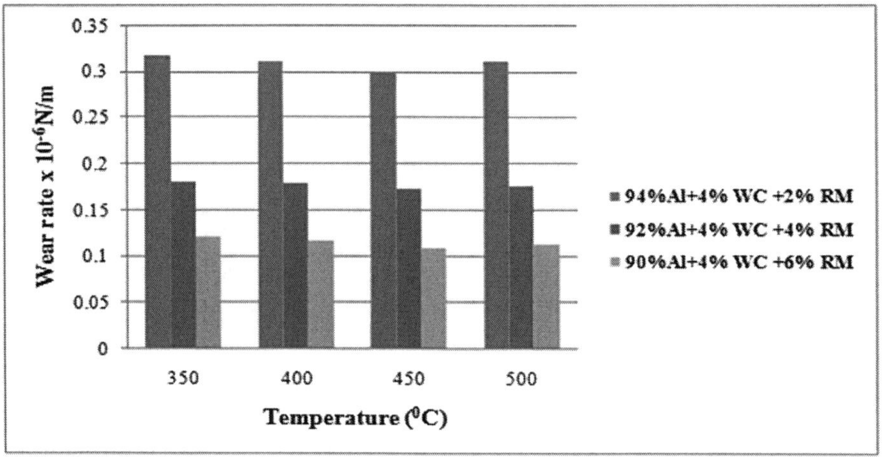

FIGURE 19.9 Wear rate, % weight fraction of red mud (100 μm) with aluminum and tungsten carbide and temperature.

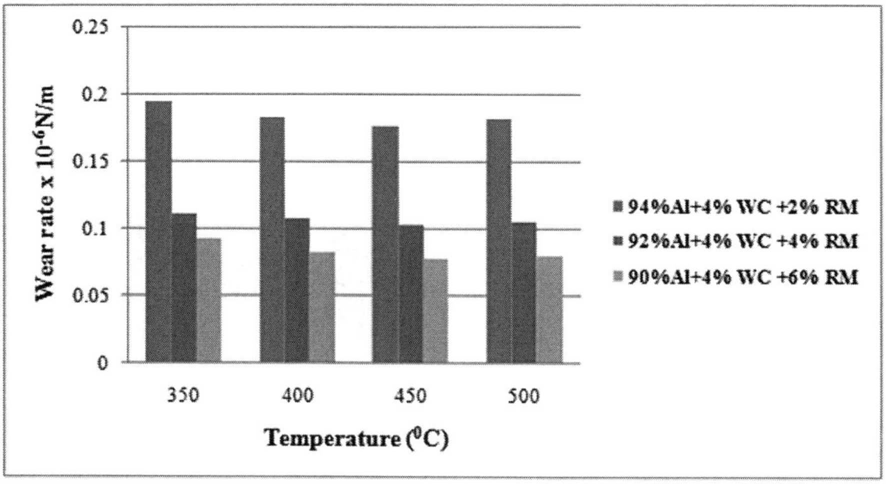

FIGURE 19.10 Wear rate, % weight fraction of red mud (42 nm) with aluminum and tungsten carbide and temperature.

In heat treatment conditions, wear resistance values are better for nanolevel than microlevel reinforcement compared with normal condition due to the increase in surface area of contact and higher bond strengths.

19.3.3 MATHEMATICAL MODELING

A mathematical model is developed using regression analysis for the hardness values of Al, red mud, and WC. The overall equations for normal condition and heat treatment conditions (350, 400, 450, and 500°C) are shown in Equations 19.1 and 19.2, respectively.

$$\text{Hardness} = 59.9167 - ([0.08311] \times \text{particle size}) + ([4.37475] \times \%\text{weight of red mud}). \quad (19.1)$$

$$\text{Hardness} = 55.9024 - ([0.08422] \times \text{particle size}) + ([4.5149] \times \%\text{weight of red mud}) + ([0.02113] \times \text{temperature}). \quad (19.2)$$

It has been observed that the R-squared values for Equations 19.3 and 19.4 are 0.8861 and 0.8883, respectively, from regression statistics. A mathematical model is developed for aluminum, red mud, and tungsten carbide for wear rate values at normal condition and heat treatment conditions are shown in Equations 19.3 and 19.4, respectively. The R-squared values of 0.8757 and 0.8643 are obtained from regression statistics.

$$\text{Wear rate} = (\text{particle size} \times [0.00070693]) - (\text{percentage weight fraction of red mud} \times [0.0365]) + 0.3079739. \quad (19.3)$$

$$\text{Wear rate} = 0.3278566 - (\text{temperature in } ^\circ\text{C} \times [0.000153666]) + (\text{particle size} \times [0.00083451]) - (\%\text{ weight composition of red mud} \times [0.034]). \quad (19.4)$$

As per the regression statistics, the influence of red mud particle size is much higher than temperature and speed properties.

19.4 CONCLUSIONS

The hardness values are improved up to 450°C; after that, there is a decrement in hardness values. In heat treatment conditions, hardness values are better for nanolevel than microlevel reinforcement compared with normal condition. This results because of the increase in surface area of contact and higher bond strengths. Wear resistance is more for nanolevel 6% weight fraction of red mud with 90% Al and 4% WC metal matrix hybrid composite. As % weight composition of red mud with Al and WC increases, there is an improvement in wear resistance. A decrease in wear rate is observed with increase in the amount of temperature up to 450°C and then it starts declining in nature. As the speed of rotation of the specimen increases, the wear resistance is increased. The highest wear resistance is obtained for 6% weight fraction of red mud powder test specimen with 42 nm particle size at 600 rpm speed.

KEYWORDS

- **aluminum**
- **red mud**
- **tungsten carbide**
- **hybrid composites**

REFERENCES

1. Siva Prasad, D.; Shoba, C. Hybrid Composites—A Better Choice for High Wear Resistant Materials. *J. Mater. Res. Technol.* **2014,** *3*(2), 172–178.
2. Slipenyuk, A.; Kuprin, V.; Milman, Y.; Goncharuk, V.; Eckert, J. Properties of P/M Processed Particle Reinforced Metal Matrix Composites Specified by Reinforcement Concentration and Matrix-To-Reinforcement Particle Size Ratio. *J. Acta Mater.* **2006,** *54,* 157–166.
3. Wilson, S.; Alpas, A. T. Wear Mechanism Maps for Metal Matrix Composites. *J. Wear* **1997,** *212,* 41–49.
4. Pramila Bai, B. N.; Ramashesh, B. S.; Surappa, M. K. Dry Sliding Wear of A356-Al–Sic Composites. *J. Wear* **1992,** *157,* 295–304.
5. Prasad, S. D.; Krishna, R. A. Production and Mechanical Properties of A356.2/Rha Composites *Int. J. Adv. Sci. Technol.* **2011,** *33,* 51–58.

6. Liu, Y.; Lin, C. X.; Wu, Y. G. Characterization of Red Mud Derived from a Combined Bayer Process and Bauxite Calcinations Method. *J. Hazard. Mater.* **2007,** *146,* 255–261.
7. Paramguru, R. K.; Rath, P. C.; Misra, V. N. Trends in Red Mud Utilization—A Review. *J. Miner. Process. Extr. Metall.* **2005,** *26,* 1–29.
8. Singh, M.; Upadhaya, S. N.; Prasad, P. M. Preparation of Iron Rich Cement from Red Mud. *J. Cem. Concr. Residue* **1997,** *27*(7), 1037–1046.
9. Samal, S.; Ray, A. K.; Bandopadhyay, A. Proposal for Resources, Utilization and Processes of Red Mud in India—A Review. *J. Miner. Process.* **2013,** *118,* 43–55.
10. Liu, Q.; Xin, R.; Chengcheng, L.; Xu, C.; Yang, J. Application of Red Mud as a Basic Catalyst for Bio-Diesel Production. *J. Environ. Sci.* **2013,** *25,* 823–829.
11. Youssef, N. F.; Shater, M. O.; Abadir, M. F.; Ibrahim, O. A. Utilization of Red Mud in the Manufacture of Ceramic Tiles. *J. Eng. Mater.* **2002,** *206,* 1775–1778.

CHAPTER 20

EFFECT OF GRAIN SIZE ON HARDNESS OF IS2062 METAL INERT GAS WELDMENT BEFORE AND AFTER HEAT TREATMENT PROCESS

M. BALA CHENNAIAH[1,*], P. NANDA KUMAR[2], and K. PRAHLADA RAO[3]

[1]*Department of Mechanical Engineering, V. R. Siddhartha Engineering College, Kanuru, Vijayawada, Andhra Pradesh, India*

[2]*Department of Mechanical Engineering, NBKR Institute of Science and Technology, Vidhyanagar, Nellore, Andhra Pradesh, India, E-mail: podaralla.nandakumar@gmail.com*

[3]*Department of Mechanical Engineering, JNT University, Anatapur, Andhra Pradesh, India, E-mail: drkprao@gmail.com*

**Corresponding author. E-mail: Chennai303.mech@vrsiddhartha.ac.in*

ABSTRACT

In the present work, application of optical metallography was implemented to study the microstructure of the IS2062 which is a low carbon steel specimen subjected to metal inert gas (MIG) welding. Specimens before and after heat treatment process are considered. The optical inspection was carried out at the base metal (BM) zone, heat treatment zone, and welded zone using an optical microscope. Using the test reports obtained from the analysis, we established relations between various microstructural and mechanical properties such as grain size, volume fraction, strength, and hardness.

The present work is the experimental investigation of the effect of grain size on hardness of IS2062 MIG weldment before and after heat treatment process. The butt joint of IS2062 steel is welded by MIG welding process.

The two similar butt joint welded structures are made, one without heat treatment and the other with heat treatment to investigate the effect of heat treatment. The optical metallography (OM) was implemented to study the microstructure of the welded structure. The optical inspection was carried out at the BM area, heat affected area, and welded zone using an optical microscope in both structures to obtain grain size and volumetric phase analysis. It is observed that the grain was indirectly proportional to hardness and strength of materials.

20.1 INTRODUCTION

Steel is an important material because of its tremendous flexibility in fabrication processes and metal working. During the production of steel, several elements are added to impart special properties and avoid cold cracking. Some methods are used to improve the mechanical and metallurgical properties of steels. Low carbon steel is the most common form of steel due to the fact that its material properties are acceptable for automobile, structural, and other engineering applications. Generally, the carbon content is below 0.22% and remaining other elements such as 1.5% of Si, 0.4% of S, and 0.045% of P. Because of these, metals are readily available in various size, shape, and high weldability and formability characteristics in nature. Welding is a joining process, in which similar and dissimilar metals are joined permanently with or without filler metal. During welding, the work pieces to be joined are melted and solidified at suitable working medium to obtain a permanent joint. These joints give higher strength than the BMs. It can be used as a monolithic structure for structural applications. Keehan[1] studied the various alloying elements, their effect on the BM micro structure and its properties such as tensile, hardness, and impact strength. The selection of welding process parameters[2] also affects the mechanical properties of weldments. Joseph and Alo[3] conducted different heat treatment processes on low carbon steel weldments. These weldments are compared with their mechanical and metallurgical properties. Finally, they concluded that the heat-treated weldment hardness is decreased as compared to the without heat-treated weldments. Especially, in normalizing processes, weldments give better-quality microstructure and higher fatigue strength is obtained. Schastivtsev and Orlo[4] studied the microstructure and properties of low carbon weld steel after thermos mechanical strengthening. Finally, they concluded that there is an improvement in rolled sheets of KH90 with direct quenching processes. Analysis of a materials microstructure aid in determining if the material has been processed correctly and it is therefore a critical step for determining product reliability and for determining why a material failed. The basic steps for proper metallographic specimen preparation include documentation, sectioning and cutting, mounting, planar grinding, rough polishing, final polishing, etching, microscopic analysis, and hardness are testing.[5] Ahmad and Channa[6] studied the mathematical relationship between ferrite, pearlite, and average grain size of steel. Optical microscopy is used to observe average grain size, % of ferrite, and % of pearlite by intercept method. Finally, they concluded that the average grain size of hot-rolled AISI-1060 steel is equal to the volume fraction of pearlite and ferrite grain size. Herring[7] studied the grain size and its influence on materials properties. Finally he concluded that grain size has a measurable effect on most mechanical properties. Machinability is also affected; rough machining favors coarse grain size, while finish machining favors fine grain size. The effect of grain size is greatest on properties that are related to the early stages of deformation. Fine-grain steels do not harden quite as deeply and have less tendency to crack than coarse-grain steels of similar analysis. Also, fine-grain steels have greater fatigue resistance, and a fine grain size promotes a somewhat greater toughness and shock resistance.[8] Increasing of the arc voltage and welding current increases the welding heat input: accordingly the chance of defects formation such burn through in weld metal also increases. Besides, the high welding current reduces the hardness, yield strength, and ultimate tensile strength (UTS) of weld metal. However, increasing in welding speed decreases the welding heat input and chance of defects formation in weld metal. Thus, increasing the welding speed increases the hardness, yield strength, and UTS of weld metal.

20.2 EXPERIMENTATION

The two IS2062 butt joint welded structures are prepared using metal inert gas welding process. The one welded structure has undergone heat treatment. These weldments are undergone with nondestructive testing methods to evaluate the properties and microstructural analysis. Test specimens are obtained from both welded structures with and without heat treatment at three different zones to be examined (Fig. 20.1). Optical microscopy is used to analyze the microstructure of a low carbon steel specimen (IS2062), BM, heat affected area, and welded zone.

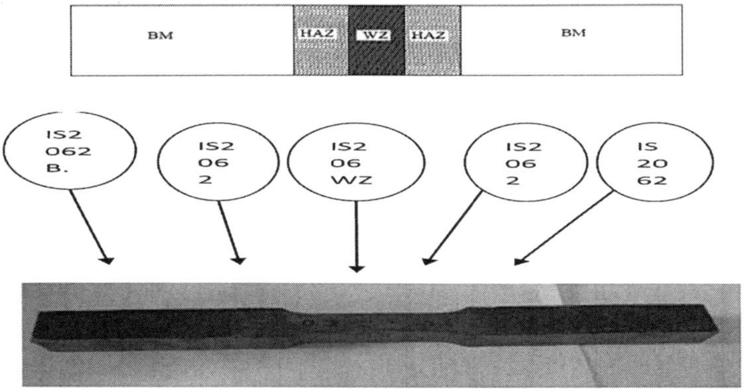

FIGURE 20.1 Specimen with zones.

Specimen preparation is done by sectioning the sample from the required metal, undergoes polishing with different graded grit papers. Finally polishing of the test piece by diamond abrasive paste is done to polish the surface to be inspected using the microscope. These samples are examined under the optical microscope which is provided with a system linked software MP foundry plus, by this software application different areas of interest have been examined and photos of the examined spots are taken by the embedded camera of the microscope as shown in Figure 20.2. These results obtained from the microscope are analyzed using the software linked with the microscope. Microscopic parameters such as grain size, the volume fraction of pearlite, and ferrite are calculated. These outputs are studied to understand the relationship between the mechanical and microscopic properties.

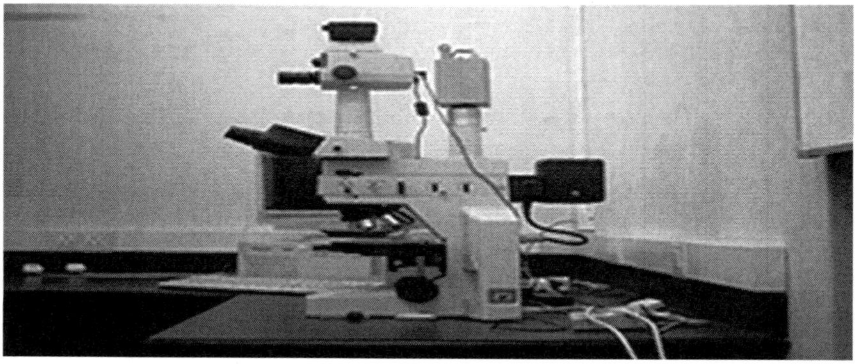

FIGURE 20.2 Optical microscope.

20.3 RESULT ANALYSIS

20.3.1 HARDNESS

Mechanical properties of the treated and untreated samples were determined using standard methods. For hardness testing, oxide layers formed during heat treatment were removed by stage grinding and then polished. Average Rockwell hardness number (HRB) readings were determined by taking hardness readings at BM, heat-affected zone (HAZ), weld zone (WZ) positions on the samples, Rockwell hardness test results are shown in Table 20.1.The histogram plots of hardness values before and after heat treatments of welded specimens are presented in Fig. 20.3, which shows that the hardness values after heat treatment are reduced due to recrystallization of microstructure during heat treatment.

TABLE 20.1 Hardness at Different Zones.

Sr. no.	Before heat treatment			After heat treatment		
	Base metal (BM)	Heat-affected zone (HAZ)	Weld zone (WZ)	BM	HAZ	WZ
1	132	140	119	105	120	119
2	131	133	129	108	103	118
3	139	135	149	108	108	116
4	148	130	152	109	105	114
Average	137	134	137	107	109	116

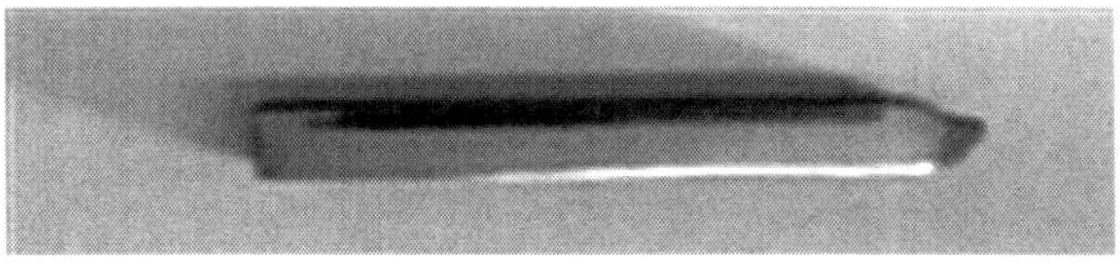

FIGURE 20.3 HT-IS2062.

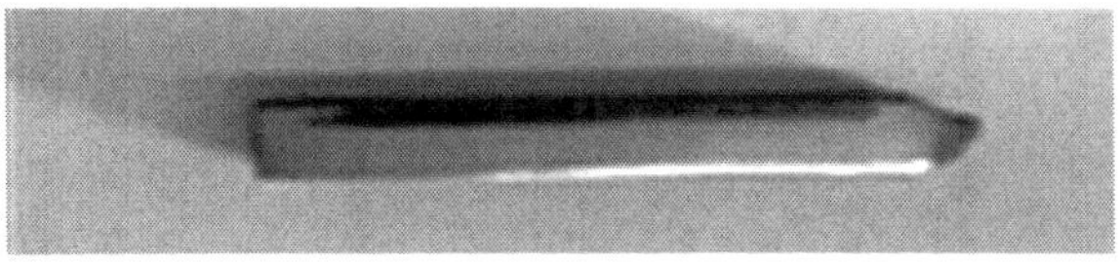

FIGURE 20.4 NHT-IS2062.

20.3.2 MICROSTRUCTURE

Microstructure is the image obtained from a substance when undergone magnification revealing its details by use of optical microscope. By the microstructure examination, its composition can be studied. Any substance's microstructure (grain size) and its volume fraction analysis can be observed by using optical microscope which has undergone suitable magnification. The intercept method is used to study grain size and volume fraction analysis of substance. The grain size of BM and alloy steels is generally understood to mean prior austenitic grain size. During the heating and subsequence cooling of weldments, transformation to austenite takes place. When first formed, these grains are fine grains, but grow in large size as the time and temperature are increased. The size will remain fine grain for steels held at temperatures just above the (upper) critical. At higher range of temperature, steels exhibits different ranges of grain structures, depending on the chemical composition of steels. Addition of filler material elements at slow rate of cooling gives better mechanical properties of the grain growth as compared to the faster rate cooling. The grain size of steel can be measured before heat treatment using an optical microscope by intercepts method the number of grains within a given area range from 0 to 7.5 by determining the number of grains that intersect a given length of 25.31 μm random lines. Figure 20.5a shows the microstructure of BM which has a uniform structure with uniformly distributed very fine strengthening precipitates. With ASTM grain size of 7.5, the light-etched structures which are observed in white color are the ferrite phase and dark-etched structure which is pink in color represents pearlite phase. Figure 20.5b indicates the microstructure of BM IS2062 after heat treatment. The microstructure observation also shows that the grain size increases to 7 after heat treatment compared with a value of 7.5 before the heat treatment. Figure 20.6a shows the microstructure observed in the HAZ without heat treatment. A grain size of 7.5 was observed. Figure 20.6b shows the microstructure observed in the HAZ with heat treatment. It is clearly evident that the grain size increases from 7.5 to 7 with heat treatment. Figure 20.7a shows the microstructures observed in the WZ without heat treatment and with heat treatment, respectively. In case of specimens without heat treatment, grain size of 7 has been observed as shown in Figure 20.7b. In the WZ adjacent faying surfaces fine equiaxed grain are developed, gradually toward the weld axis coarse grains are observed. The HAZ is adjacent to the fusion zone containing grains larger than those in the BM. The irregularities observed in the pattern of microstructure for different welded joints were solely due to the level of diffusion take place during welding. In the BM, coarse grains are observed. In all the zones, after heat treatment, the appreciable grain refinement was observed.

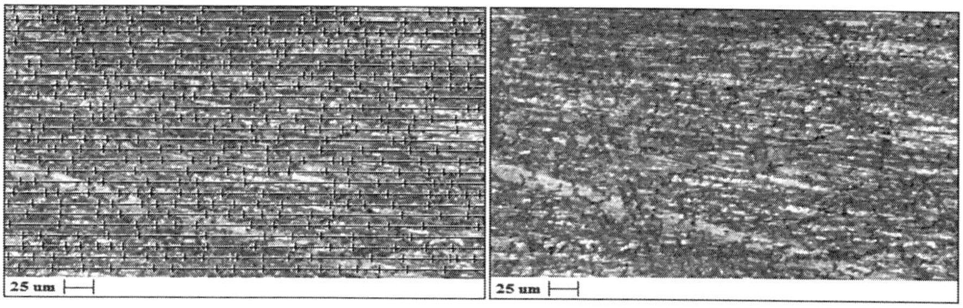

FIGURE 20.5 Microstructure at base metal (a) Before heat treatment and (b) after heat treatment.

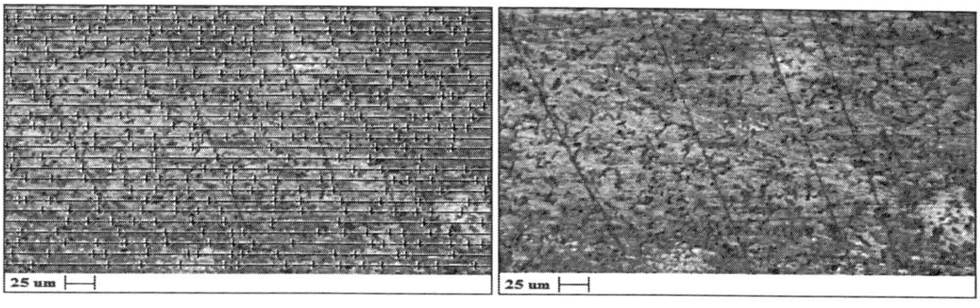

FIGURE 20.6 Microstructure at Heat Affected Zone (a) Before heat treatment and (b) after heat treatment.

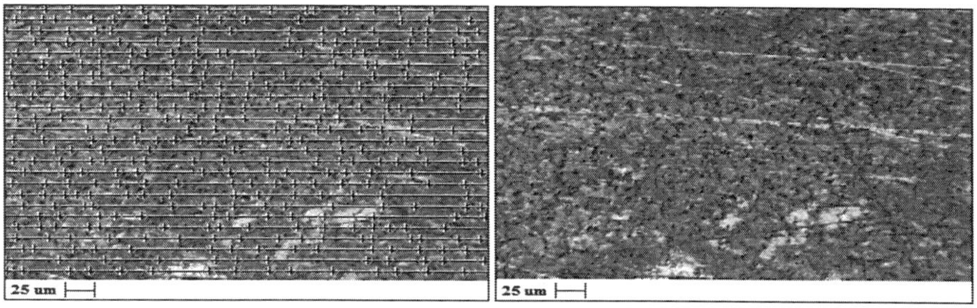

FIGURE 20.7 Microstructure at welded zone (a) before heat treatment and (b) after heat treatment.

20.4 DISCUSSIONS

20.4.1 GRAIN SIZE EFFECT ON PROPERTIES

Grain size has a measurable effect on most of the mechanical properties. For example, at room temperature, hardness, yield strength, tensile strength, fatigue strength, and impact strength all increase with decreasing grain size. Machinability is also affected; rough machining favors coarse-grain size, while finish machining favors fine-grain size. The effect of grain size is greatest on properties that are related to the early stages of deformation. Fine-grain steels do not harden quite as deeply and have less tendency to crack than coarse-grain steels of similar analysis. Moreover, fine-grain steels have greater fatigue resistance, and a fine-grain size promotes a somewhat greater toughness and shock resistance. The data obtained from the optical microscope, found that the ASTM grain size of specimens without heat treatment is 7.5, whereas the test specimen subjected to heat treatment is 7. A lower number means greater number of grains as per ASTM. As number of grains/mm^2 increases, it denotes that grain size is gradually reduced which indicates the increase in hardness of the test sample subjected to heat treatment. Since the grain size is less when the heat treatment is done, it proves that the heat treatment process enhances the mechanical properties of the specimen as shown in Table 20.2. A graph is plotted between hardness and zones of the specimen. We found that the hardness of the nonheat-treated specimen increased from BM to the weld metal zone. The welded zone in the without heat treatment specimen have higher hardness. In the heat-treated specimen, the hardness does not change—it almost remains the same. From Figure 20.3 of Rockwell hardness and zones of the specimen, it is found that the welded zone have higher hardness followed by heat-affected area and BM of a nonheat-treated specimen, whereas the heat-treated specimen shows almost uniform hardness throughout—this might be the hardness gained in heat treatment.

20.4 CONCLUSION

From the present work, we found that increase in grain size decreases the hardness of the specimen and strength. Grain size is inversely proportional to the hardness and strength. Hardness is found to be more in the welded zone, which might be due to the finer particles formed during the welding process. For a heat-treated element, the hardness of the grain size at various zones is almost the same. Some deviations in the metallurgical principles are observed due to the improper cooling process of the weldment and rough machining process employed in the preparation of metallographic specimen.

TABLE 20.2 Average Grain Size at Different Zones with Heat Treatment and Without Heat Treatment.

Field measured	2
Analyzed area	0.2344 mm^2
Standard used	ASTM E 1382
Before heat treatment	
ASTM grain size#	7.5
Intercepts	781
Mean internal length (μm)	25.31686
Standard deviation	1.971
95% Confidence interval	5.998
%RA	23.691
After heat treatment	
ASTM grain size#	7
Intercepts	760
Mean internal length (μm)	25.94041
Standard deviation	0.386
95% Confidence interval	4.529
%RA	4.529

ACKNOWLEDGMENT

I express my sincere thanks to UGC for assisting my work through funding Minor Research Project, MRP-6188/15 (SERO/UGC) for the purchasing the Optical Microscope which helped me to carry out my work successfully.

KEYWORDS

- **MIG welding**
- **microstructure**
- **grain size measurement**

- **Rockwell hardness**
- **base metal**
- **weld zone**
- **heat-affected zone**
- **American Society for Testing and Materials**

REFERENCES

1. Keehan, E. Effect of Microstructure on Mechanical Properties of High Strength Steel Weld Metals. Doctoral Thesis, University of Cambridge, 2004.
2. Bin Abdillah, M. Q. Effect of Welding Parameters on Mechanical Properties of Welded Carbon Steel. Thesis, Universiti Malaysia Pahang, May 2012.
3. Joseph, O. O.; Alo, F. I. An Assessment of the Microstructure and Mechanical Properties of 0.26% Low Carbon Steel under Different Cooling Media. *Ind. Eng. Lett.* **2014,** *4*(7), 39–45.
4. Schastlivtse, V. M.; Orlo, V. V. Microstructure and Properties of Low-Carbon Weld Steel After Thermo Mechanical Strengthening. *Phys. Met. Metallogr.* **2012,** *113*(5), 480–488.
5. Zipperian, D. C. Metallographic Specimen Preparation Basics. PhD thesis.
6. Ahmad, S.; Channa I. A. Mathematical Relationship between Ferritic, Pearlitic and Average Grain Size of Steel. *J. Mod. Sci. Tech.* **2013,** *1*(1), 1–18.
7. Herring, D. H. Grain Size and Its Influence on Materials Properties.
8. Bahman, A. R.; Alialhossein, E. Change in Hardness, Yield Strength and UTS of Welded Joints Produced in St37 Grade Steel. *Indian J. Sci. Technol.* **2010,** *3*(12), 1162–1164.

CHAPTER 21

OPTIMIZATION OF CUTTING PARAMETERS IN TURNING OF MARAGING STEEL

K. ARAVIND SANKEERTH[1,*], K. VENKATASUBBAIAH[1], G. VENKATESWARA RAO[2], and AJAY KUMAR SINGH[3]

[1]*AUCE(A), Andhra University, Visakhapatnam, Andhra Pradesh, India*

[2]*NIT, Warangal, Telangana, India*

[3]*DRDO, Hyderabad, Telangana, India*

Corresponding author. E-mail: sunny.aravind@gmail.com

ABSTRACT

The present work is based on turning experiments conducted on a lathe using uncoated tungsten carbide cutting tools for the machining of maraging steel. The L9 orthogonal array based on design of experiments was applied to plan the experiments, by selecting three controlling factors, namely the cutting speed (v), feed (f) and depth of cut (d). The Taguchi analysis was applied to examine how these cutting factors influence the cutting force (Fz). An optimal parameter combination was then obtained using the experimental results. Analysis of variance was also carried out to examine the most significant factors for the Fz in the turning process.

21.1 INTRODUCTION

Maraging steels are widely used in many technological sectors such as aerospace, military as well as for tools and dies. For most of the applications, these steels require high strength in combination with good toughness. Moreover good weld ability, high strength to weight ratio, and dimensional stability during aging are attractive features for applications. Machining of these steels require tool materials that can withstand the extreme forces. Cutting forces are important in machining as they provide unique signature of the mechanics of machining while determining the energy consumed and machining power required for the process, tool, and workpiece deflections. There have been many studies concerning the effect of cutting parameters such as speed, feed, depth of cut, tool geometry, and so forth, on cutting forces while machining different materials.[1–4] Yang and Tarng[5] investigated cutting characteristics of S45c steel bars using tungsten carbide cutting tools using orthogonal array, the signal to noise (S/N), ratios, and the analysis of variance (ANOVA). Lalwani et al.[6] studied the effect of cutting parameters (cutting speed, feed rate, and depth of cut) on cutting forces and surface roughness in finish hard turning of MDN250 steel (equivalent to 18Ni (250) maraging steel) using coated ceramic tool. The results show that cutting forces and surface roughness do not vary much with cutting speed in the range of 55–93 m/min. A nonlinear quadratic model best describes the variation of surface roughness with major contribution of feed rate and secondary contributions of interaction effect between feed rate and depth of cut. Bartarya and Choudhury[7] developed a force prediction model during finish machining of EN31 steel (equivalent to AISI 52100 steel) hardened to 60 ± 2 hardness on Rockwell C using hone edge uncoated CBN tool and to analyze the combination of the machining parameters for better performance within a selected range of machining parameters. The predictions from the developed models were compared with the measured force and surface roughness values to propose the favorable range of the machining parameter values for energy-efficient machining.

Though many studies were carried out on the machining performance in terms of cutting force in the field of turning operations, only a few studies were reported on bending analysis of maraging steel. In the present work, turning experiments were conducted on a lathe with uncoated tungsten carbide cutting tools for the machining of maraging steel. The L9 orthogonal array based on the design of experiments was applied to plan the experiments, by selecting three controlling factors, namely the cutting speed (v), feed (f), and depth of cut (d). The Taguchi analysis is applied to examine how these cutting factors influence the cutting force (Fz). An optimal parameter combination was then obtained using the experimental results. Furthermore, ANOVA was also carried out to examine the most significant factors for the Fz in the turning process.

21.2 EXPERIMENTAL DETAILS AND RESULTS

The turning experiments were carried out on a precision lathe setup using tungsten carbide cutting tools for the machining of maraging steel bar which is 50 mm in diameter and 100 mm in length. The chemical composition and mechanical properties of the workpiece material are listed in Table 21.1.

TABLE 21.1 Chemical and Mechanical Properties of Maraging Steel.

Chemical composition	Nickel	Chromium	Molybdenum	Titanium
Wt.%	18	8	5	0.4
Mechanical properties	Density (kg/m^3)	Thermal conductivity (W/m·K)	Shear modulus (GPa)	Hardness (hardness on Rockwell C)
	1.8	25.5	77	50

Carbide tool materials are employed in the machining of maraging steels due to their improved performance in terms of tool life and surface finish. Carbide tool inserts CNMG 120408, along with the tool holder PCLNR 2020 M12 (tool geometry: approach angle: 95°, back rake angle: 6° and inclination angle: 6°) were used in the present investigation.

In full factorial design, the number of experimental runs exponentially increases with the increase in the number of factors as well as their levels resulting in a huge experimentation cost and considerable time period.[8] In order to compromise these two adverse factors and also to find the optimal process condition through a limited number of experimental runs, Taguchi's L9 orthogonal array consisting of nine sets of data was selected. Experiments were conducted with the process parameters, given in Table 21.2.

TABLE 21.2 Cutting Parameters and Their Levels.

Cutting parameter	Notation	Units	Level of factors		
			1	2	3
Speed	v	m/min	30	50	70
Feed	f	mm/rev	0.2	0.4	0.6
Depth of cut	d	mm	0.05	0.10	0.15

The response variable measured is cutting force. The cutting forces generated during machining trials were measured using piezoelectric tool postdynamometer (Kistler, 9272). The force signals generated during machining were fed into a charge amplifier (Kistler, 5070) connected to the dynamometer. This amplifier converts the analog signal to digital signal that was continuously recorded by the data acquisition system connected to the charge amplifier. One of the cutting force signals obtained for v=30 m/min, f=0.60 mm/rev, and d=0.15 mm using DynoWare software is shown in Figure 21.1. Based on Taguchi L9 orthogonal array consisting nine sets of coded conditions and the experimental results for the responses of Fz are shown in Table 21.3.

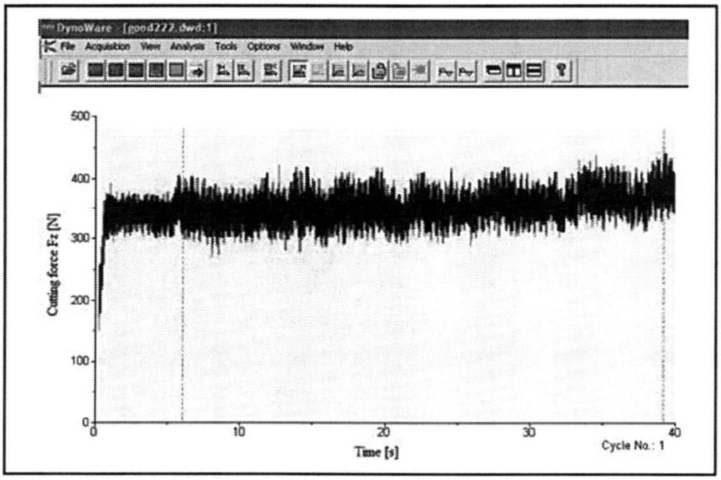

FIGURE 21.1 Cutting force signal from dynamometer.

TABLE 21.3 Experimental Runs, Results, and Corresponding S/N Ratios.

Run no.	Parameters levels			Experimental result	S/N ratio
	v	f	d	Fz (N)	Fz (N)
1	30	0.2	0.05	270	−48.63
2	30	0.4	0.10	314	−49.94
3	30	0.6	0.15	348	−50.83
4	50	0.2	0.10	203	−46.15
5	50	0.4	0.15	331	−50.40
6	50	0.6	0.05	211	−46.49
7	70	0.2	0.15	198	−45.93
8	70	0.4	0.05	161	−42.41
9	70	0.6	0.10	191	−45.62

21.3 ANALYSIS OF RESULTS

21.3.1 ANALYSIS OF S/N RATIO

In Taguchi method, the term "signal" represents the desirable value (mean) for the output characteristic and the term "noise" represents the undesirable value for the output characteristic. Taguchi uses S/N ratio to measure the quality characteristic deviating from the desired value. There are basically three types of characteristics in S/N ratios, that is, higher-the-better, lower-the-better, and nominal-the-better. The S/N ratio η is defined as

$$\eta = -10 \log (\text{MSD}) \tag{21.1}$$

where MSD is the mean-square deviation for the output characteristic.

Smaller values of Fz are desirable for any machining operation. Thus, smaller-the-better criteria for Fz were selected during the present work. The MSD for smaller-the-better quality characteristic can be expressed as:

$$MSD = \frac{1}{m} \sum T_i^2 \tag{21.2}$$

where m is the number of tests and T_i is the value of Fz of the ith test.

Table 21.3 shows the experimental results for Fz and the corresponding S/N ratio using Equations 21.1 and 21.2. The mean S/N ratio for the cutting speed at levels 1, 2, and 3 can be calculated by averaging the S/N ratios for the experiments 1–3, 4–6, and 7–9, respectively. The mean S/N ratio for each level of the other cutting parameters can be computed in a similar manner and are listed in Table 21.4. Regardless of the lower-the-better or the higher-the-better quality characteristic, the greater the S/N ratio corresponds to the smaller variance of the output characteristics around the desired value.

TABLE 21.4 S/N Response Table for F_Z (N).

Cutting parameters	S/N response to force Fz (N)				Rank
	Level 1	Level 2	Level 3	Max–min	
Speed	−49.799	−47.677	−45.230	4.569	1
Feed	−46.903	−48.157	−47.646	1.254	3
Depth of cut	−46.416	−47.236	−49.054	2.637	2
Average			−47.56		

21.3.2 ANALYSIS OF VARIANCE

ANOVA was introduced by Sir Ronald Fisher.[9] This analysis was carried out for a level of significance of 5%, that is, for 95% level of confidence. The purpose of ANOVA is to investigate which turning parameter significantly affects the performance characteristics.[10] Table 21.5 shows the results of ANOVA for Fz. It is found that cutting speed and depth of cut are significant cutting parameters affecting cutting force. Therefore, based on S/N ratio and ANOVA analyses, the optimal cutting parameters for cutting force are the cutting speed at level 3, the feed at level 1, and the depth of cut at level 1.

TABLE 21.5 Results of Analysis of Variance (ANOVA) for F_Z (N).

Symbol	Parameter	DOF	SS	MS	F	P (%)
A	Speed	2	31.366	15.683	14.701	67.000
B	Feed	2	2.384	1.192	1.118	5.093
C	DOC	2	10.931	5.465	5.123	23.349
Error		2	2.134	1.067		4.558
Total		8	46.815			100.000

21.3.4 CONFORMATION TEST

Once the optimal level of the design parameters has been selected, the final step is to predict and verify the improvement of the quality characteristic using the optimal level of the design parameters. The estimated S/N ratio $\tilde{\eta}$ using the optimal level of the design parameters can be calculated as:

$$\tilde{\eta} = \eta_m + \sum_{i=1}^{o} (\bar{\eta}_i - \eta_m) \tag{21.3}$$

where η_m is the total mean S/N ratio, $\bar{\eta}_i$ is the mean S/N ratio at the optimal level, and o is the number of the main design parameters that affect the quality characteristic.

Table 21.6 shows the comparison of obtained cutting forces with initial set of parameter for machining and the optimal set of parameters obtained from Taguchi analysis. The increase of the S/N ratio from initial cutting parameters to the optimal cutting parameters is 5.598 dB, which indicates that the cutting forces can be reduced by machining at optimum process parameters. In other words, the experimental results conform the design and analysis for optimizing the cutting parameters through this approach.

TABLE 21.6 Result of Conforming Experiment for F_Z.

	Initial cutting parameter	Optimal cutting parameters	
		Prediction	Experiment
Level	$v_2f_2d_2$	$v_3f_1d_1$	$v_3f_1d_1$
Cutting force (N)	261	135.61	137
S/N ratio	−50.306	−43.412	−42.734

Note: Improvement of S/N ratio = 5.598 dB.

21.4 CONCLUSIONS

This chapter discussed an application of the Taguchi method for optimizing the cutting parameters in turning operation. The Taguchi method provides a systematic and efficient methodology for the design and optimization of cutting parameters with far less effort than would be required for most optimization techniques. It has been shown that cutting force will be reduced significantly for turning operation by conducting experiments at the optimal parameter combination and also by analyzing S/N ratio. The conformation experiments were also conducted to verify the optimal combination of parameters obtained. Good agreement between the predicted and actual values for forces is observed.

KEYWORDS

- **machining**
- **turning**
- **maragingsteel**
- **Taguchi optimization methods**
- **ANOVA**

REFERENCES

1. Qian, L.; Hossan, M. R. Effect on Cutting Force in Turning Hardened Tool Steels with Cubic Boron Nitride Inserts. *J. Mater. Process. Technol.* **2007,** *191,* 274–278.

2. Thandra, S. K.; Choudhury, S. K. Effect of Cutting Parameters on Cutting Force, Surface Finish and Tool Wear in Hot Machining. *Int. J. Mach. Mach. Mater.* **2010,** *7*(3/4), 260–273.
3. Lin, Z.-C.; Lo, S.-P. Effect of Different Tool Flank Wear Lengths on the Deformations of an Elastic Cutting Tool and The Machined Workpiece. *Int. J. Comput. Appl. Technol.* **2006,** *25*(1), 30–39.
4. Saglam, H.; Unsacar, F.; Yaldiz, S. Investigation of the Effect of Rake Angle and Approaching Angle on Main Cutting Force and Tooltip Temperature. *Int. J. Mach. Tools Manuf.* **2006,** *46*(2), 132–141.
5. Yang, W. H.; Tarng, Y. S. Design Optimization of Cutting Parameters for Turning Operations Based on Taguchi Method. *J. Mater Technol.* **1998,** *84,* 122–129.
6. Lalwani, D. I.; Mehta, N. K.; Jain, P. K. Experimental Investigations of Cutting Parameters Influence on Cutting Forces and Surface Roughness in Finish Hard Turning of MDN250 Steel. *J. Mater. Process. Technol.* **2008,** *206*(1–3), 167–179.
7. Bartarya, G.; Choudhury, S. K. Effect of Cutting Parameters on Cutting Force and Surface Roughness During Finish Hard Turning AISI52100 Grade Steel. *Procedia CIRP* **2012,** *1,* 651–656.
8. Kosaraju, S.; Anne, V. G.; Popuri, B. B. Taguchi Analysis on Cutting Forces and Temperature in Turning Titanium Ti-6Al-4V. *Int. J. Mech. Ind. Eng. (IJMIE)* **2012,** *1*(4) 55–59.
9. Fisher, R. A. *Statistical Methods for Research Workers;* Oliver and Boyd: London, 1925.
10. Kosaraju, S.; Gopal, A. V.; Popuri, B. B. Analysis for Optimal Decisions on Turning Ti–6Al–4V with Taguchi–Grey Method. *Proc. Inst. Mech. Eng. Part C* **2014,** *228*(1), 152–157.

CHAPTER 22

OPTIMIZATION OF POWDER-MIXED ELECTRIC DISCHARGE MACHINING PARAMETERS USING TAGUCHI AND GREY RELATIONAL ANALYSIS BASED ON MATERIAL REMOVAL RATE AND SURFACE ROUGHNESS

BSV RAMARAO[1,*] and P SAILESH[2]

[1]*Department of Mechanical Engineering, Aurora's Scientific & Technological Institute, Ghatkesar, Telangana, India*

[2]*Department of Mechanical Engineering, Methodist College of Engineering, Hyderabad, Telangana, India, E-mail: palapartyshailesh@gmail.com*

[*]*Corresponding author. E-mail: svramaraoboda@gmail.com*

ABSTRACT

Survival of an industry is dependent on the task that how best you are producing/selling for lesser price. Industries will always work on optimizing the inputs to get the better results or outputs. Optimization is the best tool used on the production side to obtain the best manufacturing conditions. There are many optimization techniques that are available to apply to the required task. Grey relational analysis (GRA) is one of best technique to get the best results. The main aim of this chapter is to get the best set of input values, that is, discharge current, pulse ON time, pulse OFF time, and powder concentration to determine the variation in the performance measures such as material removal rate, surface roughness, and tool wear rate when working on titanium Ti-6Al-4V. The L27 orthogonal array has been taken to design the experiments on powder-mixed electric discharge machining by considering the B4C as the powder, which is mixed in dielectric fluid. The analysis has been made using GRA and Taguchi method. Finally, the results are reverified and percentage of error is calculated.

22.1 INTRODUCTION AND LITERATURE

Titanium (Ti) alloys are popular in using many applications of manufacturing industries including aerospace due to their advantages that the alloy is having in terms of corrosion resistance, fracture resistance, and very high strength to weight ratio. Especially, Titanium grade 5 material, that is, Ti-6Al-4V alloys are very important and popular among titanium alloys/material which is frequently used for the purpose mentioned earlier as this particular alloy is having the similar characteristics. Along with the aerospace industry, this alloy is also used in automobile industry, biomedical industry, power generation industry, and even in marine applications. Ezugwu andWang[5] stated that titanium alloy is considered as very difficult-to-cut material due to its high cutting temperature and the high stresses at and/or close to the cutting edge during machining.

Dewangan et al.[4] conducted experiments designed using response surface methodology and obtained white layer thickness, surface crack density, and surface roughness (SR). They utilized grey-fuzzy logic-based hybrid approach to obtain the grey fuzzy reasoning grade (GFRG) and concluded with the improvement in the GFRG for the predicted and experimental optimum machining parameters. Talla et al.[13] have conducted multiresponse optimization of powder-mixed electric discharge machining (PMEDM) of al/Al MMC using GRA. He has taken material removal rate (MRR) and SR as output parameters and applied analysis of variance. He has concluded the recommended level of process parameters.

Kolli and Kumar[9] have studied the effect of the addition of different B4C concentrations mixed in commercial EDM oil grade SAE 450 working on Ti-6Al-4v which has influenced the MRR and tool wear rate (TWR) and SR.

Gu et al.[7] investigated the performance of the EDM for multihole using the bundled electrodes. Azad and Puri[1] have experimented on PMEDM using silicon carbide. They have obtained better results by using response surface methodology.

Chow et al.[2] performed experiments using Sic powder in water for microslit. Here, a rotating copper disk is used as electrode. They have concluded that the combination of Sic and water is capable to attain better results than using water alone.

Kirbia et al.[8] have performed experiments by considering the kerosene and deionized water as dielectric fluids in which B4C was mixed. He compared the results in both the cases and noticed that MRR and TWR are more in case of deionized water comparatively with kerosene.

Kolli and Kumar [10] have conducted experiments on electric discharge machining considering the B4C as the mixed powder and reported that the output parameters of MRR, TWR, and SR are increased with the increase in the level/value of discharge current.

Srinivasan et al.[12] have investigated the optimal set of process parameters such as current, pulse ON, and OFF time to identify the variations in rate of material removal, wear rate on tool, and SR value on the work material for machining mild steel IS 2026 using copper electrode. Analysis has been made using grey relational analysis (GRA), which is a Taguchi method. The obtained results show that the Taguchi GRA is being effective technique to optimize the machining parameters for EDM process.

Ganachari[6] has demonstrated that the presence of suspended particle in dielectric fluid significantly increases the surface finish. They have conducted experiments on a ZNC (a machine can work in Z-axis along with X- and Y-axis, numerical controlled) control EDM machine manufactured by Electronica Machine Tools Ltd. India. Taguchi method with GRA optimization is used to study the effect of independent variables on responses and develop predictive models. They have targeted to find the range of parameters to get the good surface finish.

Rodic et al.[11] have worked on adaptive-network-based fuzzy inference system, artificial neural network, and estimated SR. One of the most important advantages of genetic programming modeling is that specific equations are obtained and models can be used independently. Because of the scarcity of space and slight complexity of generated membership functions, they are not shown in their paper.

Das et al.[3] have applied the Taguchi method and GRA to improve the multiple performance characteristics of the electrode wear ratio, MRR, and SR. As a result, this method greatly simplifies the optimization of complicated multiple performance characteristics.

Extensive literature survey expresses that the all the researchers have worked for the betterment of output results such as MRR, TWR, and SR by using many techniques. They have used different powders in different dielectric fluids when working on different alloys. The present work is mainly focused to achieve the optimized set of input parameters to obtain the best values of MRR and SR when working on Ti-6Al-4V alloy by using B4C powder added to drinking water as the dielectric fluid. The methodology of Taguchi with GRA is applied to obtain the same.

22.2 EXPERIMENTATION

22.2.1 MATERIAL USED

Ti-6Al-4V and copper electrode is selected as the material to work on. The material properties of Ti-6Al-4V are tabulated below in Table 22.1.

TABLE 22.1 Material Properties of Ti-6Al-4V

Property	Values
Hardness (HRC)	32–34
Melting point (°C)	1649–1660
Density (g/cm^3)	4.43
Ultimate tensile strength (MPa)	897–950
Thermal conductivity (W/m · k)	6.7–6.9
Specific heat (J/kg · K)	560
Mean coefficient of thermal expansion (0–100°C/°C)	8.6×10^{-6}
Volume electrical resistivity (ohm-cm)	170
Elastic modulus (GPa)	113–114

And its chemical composition is given under in Table 22.2

TABLE 22.2 Chemical Composition of Ti-6Al-4V.

Element	Percentage (max)
C	0.014
Al	6.07
V	4.02
N	0.0036
O	0.1497
Fe	0.03
Ti	Remaining (Apx 89.7012)

The B4C powder is used to mix with the dielectric fluid, that is, drinking water and its properties are given below in Table 22.3.

TABLE 22.3 Properties of B4C.

Property	Value
Density (g/cm^3)	2.51
Particle size (μm)	20–30
Melting point (°C)	2450–2630
Hardness (kg/mm^2)	2900–3580
Fracture toughness (MPa)	2.9–3.7
Thermal conductivity (W/m · K at 25°C)	140
Thermal expansion 0–100°C	4.6×10^{-6}
Compressive strength (MPa)	2800
Young's modulus (GPa)	450–470

22.2.2 PARAMETER TABLE

Discharge current, pulse ON time, pulse OFF time, and powder concentration are considered as the input control parameters with three levels of each parameter and their range is tabulated below in Table 22.4.

TABLE 22.4 Parameter Table.

Parameter	Level 1	Level 2	Level 3
Discharge current (A)	10	15	20
Pulse ON time (B)	25	45	65
Pulse OFF time (C)	24	36	48
Powder concentration (D)	5	10	15

22.3.3 DESIGN OF EXPERIMENTS/ORTHOGONAL ARRAY

During the availability of minimum resources, design of experiment is a best suitable tool to design and perform those experiments. The L27 orthogonal array (OA) is selected as the suitable OA and using Minitab, it is obtained that the following set of values to perform the required experiments, which are given in Table 22.5.

22.3.3.1 MATERIAL REMOVAL RATE

It is calculated based on the change of the weight before and after the machining process. It is formulated as shown below in the Equation 22.1.

$$MRR = [(W_i - W_f) / (D \text{ X } t)] \times 1000 \tag{22.1}$$

W_i = Initial weight in grams
W_f = Final weight in grams
D = Density of the workpiece material in g/cm^3
t = Time period of the experiment in minutes

22.3.3.2 SURFACE ROUGHNESS

It is calculated based on the average roughness by comparing all the peak and valley points to the center line. Here, in this experimentation, the SR may be calculated using Talysurf.

22.4 ANALYSIS

By varying the process parameters, the predefined experiments are being conducted to obtain the quality characteristics. To calculate the quality characteristics, "larger is better and smaller is better" are used generally through S/N ratios, depending upon the output parameter that has been taken. For example, Larger is Better has been taken for the case of MRR and Smaller is Better is taken for SR. Experiments were conducted based on the OA.

The transformation of original values into S/N ratios can be obtained using Minitab 17.0, in which Larger is Better and Smaller is Better were used and proceeded for subsequent calculations and analysis. Then the corresponding S/N ratios were tabulated in the following Table 22.5.

Then the normalization of the data can be done for all the obtained S/N values which are obtained for output parameters again by considering the Larger the better for MRR and Smaller the better for SR and resultant data is tabulated in Table 22.5.

The grey relational coefficient (GRC), through the calculation of deviation sequences, is found out to view the relation between the best and actual normalized values which were calculated above. Based on the values of the

GRC for each set of experiment, the average is found out and is named as the grey relational grade (GRG). Those are listed out in Table 22.6.

TABLE 22.5 S/N Ratios and Normalized Values of Output Parameters.

Experiment no.	A	B	C	D	S/N ratios		Normalized values	
					Material removal rate (MRR)	Surface roughness (SR)	MRR	SR
1	1	1	1	1	10.56891	−10.96	0.3040	0.4872
2	1	1	2	2	7.708669	−13.602	0.0643	0.7268
3	1	1	2	3	9.675548	−7.7026	0.2291	0.1917
4	1	2	1	2	11.27326	−14.461	0.3630	0.8047
5	1	2	2	3	7.85265	−13.45	0.0764	0.7130
6	1	2	2	1	9.457273	−10.663	0.2108	0.4602
7	1	3	1	3	9.483839	−10.305	0.2131	0.4278
8	1	3	2	1	6.940873	−10.729	0.0000	0.4662
9	1	3	2	2	9.454057	−8.8944	0.2106	0.2998
10	2	1	1	1	14.93476	−13.749	0.6698	0.7402
11	2	1	2	2	12.11457	−12.076	0.4335	0.5884
12	2	1	2	3	13.62742	−8.3826	0.5603	0.2534
13	2	2	1	2	14.24008	−13.674	0.6116	0.7333
14	2	2	2	3	12.69727	−15.047	0.4823	0.8579
15	2	2	2	1	14.36693	−12.759	0.6222	0.6503
16	2	3	1	3	14.63827	−10.797	0.6450	0.4723
17	2	3	2	1	11.72957	−12.999	0.4012	0.6721
18	2	3	2	2	12.77644	−5.5888	0.4890	0.0000
19	3	1	1	1	18.49224	−12.954	0.9679	0.6680
20	3	1	2	2	17.61757	−14.003	0.8946	0.7631
21	3	1	2	3	18.44883	−11.087	0.9643	0.4987
22	3	2	1	2	18.45952	−14.58	0.9651	0.8155
23	3	2	2	3	17.0983	−16.614	0.8511	1.0000
24	3	2	2	1	17.73605	−12.51	0.9045	0.6278
25	3	3	1	3	18.87546	−13.62	1.0000	0.7284
26	3	3	2	1	16.91304	−13.282	0.8356	0.6978
27	3	3	2	2	17.69175	−8.075	0.9008	0.2255

TABLE 22.6 Grey Relational Grades.

Experiment no.	A	B	C	D	Deviation sequence		Grey relational coefficient		Average GRG	Rank
					MRR	SR	MRR	SR		
1	1	1	1	1	0.6960	0.5128	0.4181	0.4937	0.4559	21
2	1	1	2	2	0.9357	0.2732	0.3483	0.6466	0.4975	18
3	1	1	2	3	0.7709	0.8083	0.3934	0.3822	0.3878	27
4	1	2	1	2	0.6370	0.1953	0.4398	0.7191	0.5794	13
5	1	2	2	3	0.9236	0.2870	0.3512	0.6353	0.4933	19
6	1	2	2	1	0.7892	0.5398	0.3879	0.4809	0.4344	22
7	1	3	1	3	0.7869	0.5722	0.3885	0.4663	0.4274	23
8	1	3	2	1	1.0000	0.5338	0.3333	0.4836	0.4085	25
9	1	3	2	2	0.7894	0.7002	0.3878	0.4166	0.4022	26
10	2	1	1	1	0.3302	0.2598	0.6023	0.6580	0.6301	10
11	2	1	2	2	0.5665	0.4116	0.4688	0.5485	0.5086	17
12	2	1	2	3	0.4397	0.7466	0.5321	0.4011	0.4666	20
13	2	2	1	2	0.3884	0.2667	0.5628	0.6521	0.6075	12
14	2	2	2	3	0.5177	0.1421	0.4913	0.7786	0.6350	9
15	2	2	2	1	0.3778	0.3497	0.5696	0.5885	0.5790	14
16	2	3	1	3	0.3550	0.5277	0.5848	0.4865	0.5357	15
17	2	3	2	1	0.5988	0.3279	0.4551	0.6039	0.5295	16
18	2	3	2	2	0.5110	1.0000	0.4945	0.3333	0.4139	24
19	3	1	1	1	0.0321	0.3320	0.9397	0.6010	0.7703	4
20	3	1	2	2	0.1054	0.2369	0.8259	0.6785	0.7522	5
21	3	1	2	3	0.0357	0.5013	0.9333	0.4994	0.7163	6
22	3	2	1	2	0.0349	0.1845	0.9348	0.7305	0.8326	2
23	3	2	2	3	0.1489	0.0000	0.7705	1.0000	0.8853	1
24	3	2	2	1	0.0955	0.3722	0.8397	0.5732	0.7065	7
25	3	3	1	3	0.0000	0.2716	1.0000	0.6480	0.8240	3
26	3	3	2	1	0.1644	0.3022	0.7525	0.6233	0.6879	8
27	3	3	2	2	0.0992	0.7745	0.8345	0.3923	0.6134	11

As the OA is used in the design of the experiment, it is easy to separate out the effect of each input parameter at different pre-assumed levels. The mean of the GRG of all the levels of the input parameter is shown in Table 22.7.

TABLE 22.7 Influence of Input Parameters.

Parameter	Level 1	Level 2	Level 3	Max–min	Rank
Discharge Current	4.0863	4.906	6.7885	2.7022	1
Pulse ON Time	5.1854	5.7529	4.8423	0.5675	2
Pulse OFF Time	5.663	5.3977	4.7201	0.2653	3
Powder Concentration	5.2021	5.2074	5.3713	0.1692	4

The next is to determine the optimal factor and it's level combination which will be expressed in Figure 22.1. Figures 22.2–22.5 show the GRG with maximum MRR and minimum SR. The graphs and Table 22.7 are showing that the process is maximum influenced by discharge current.

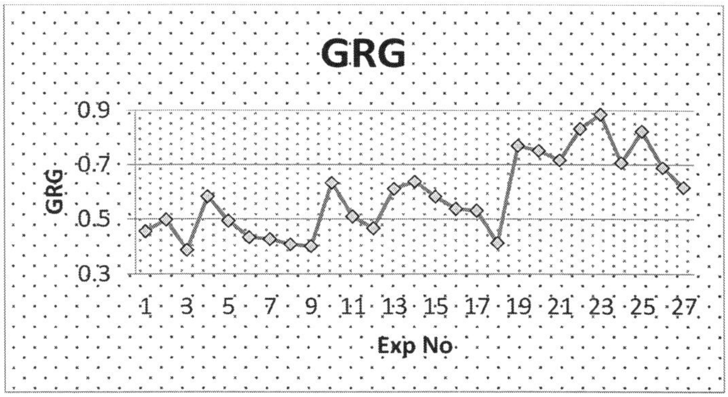

FIGURE 22.1 Variations of GRG.

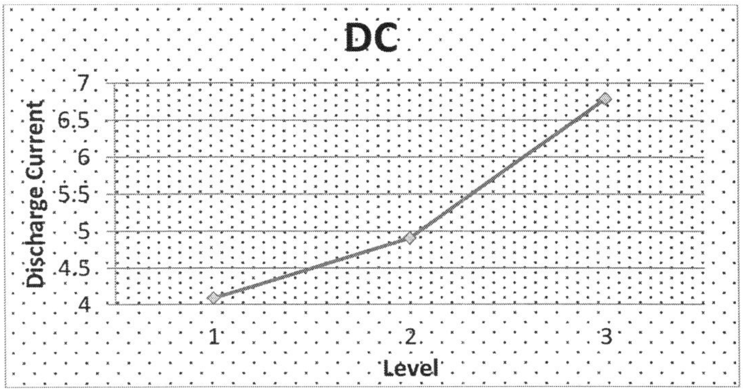

FIGURE 22.2 Variations of DC.

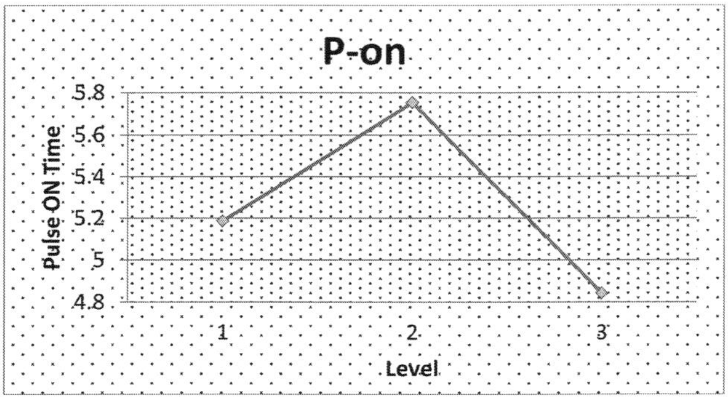

FIGURE 22.3 Variations of pulse ON time.

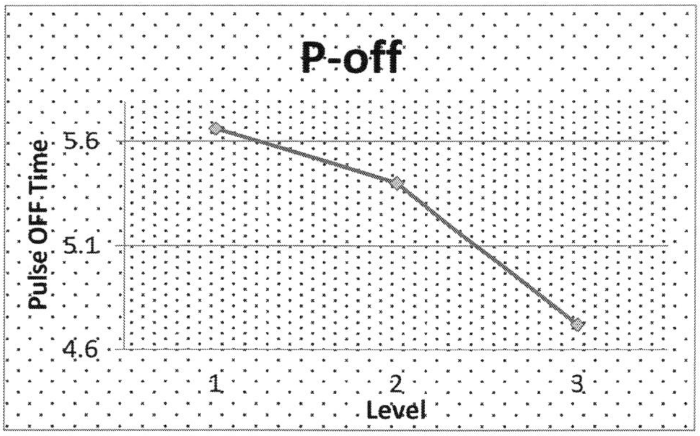

FIGURE 22.4 Variations of pulse OFF time.

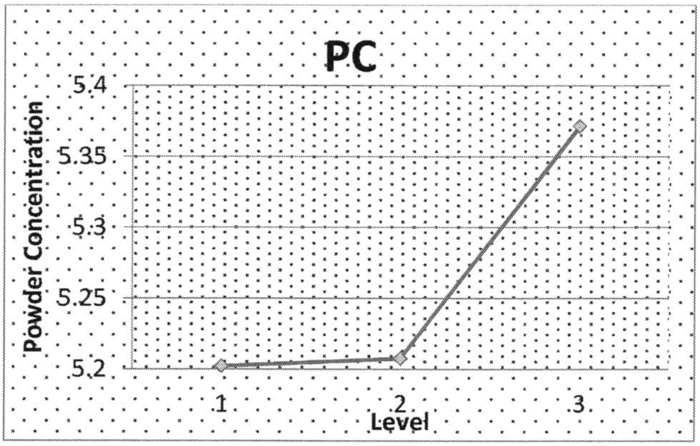

FIGURE 22.5 Variations of powder concentration.

22.4.1 REVERIFICATION OF RESULTS

The reverification of results for optimal parameters with its levels are done to identify the quality characteristics of PMEDM with Table 22.4, indicating the initial process parameter set of A3B2C2D3 for the best output among the 27 experiments which are performed. Table 22.8 shows the comparison of the predicted and experimental values and then the percentage of error may be calculated as below

TABLE 22.8 Comparison of Predicted and Experimented Values.

Output parameters	Optimized set		% of Deviation
	Experimented values	Predicted value	
Material removal rate	7.160	7.4008	3.24%
Surface roughness	6.772	6.1702	8.88%

The percentage of error may be calculated using the following Equation 22.1

$$\% \text{ of Deviation} = [(\text{Experimental value} - \text{Predicted value}) / \text{Experimental Value}] \times 100 \quad (22.1)$$

The calculated average percentage of error of the two output parameters is 6.06%.

22.5 CONCLUSIONS

To improve the multiresponse characteristics such as MRR and SR, the methods Taguchi and GRA were applied. It is determined that the best combination is A3B2C2D3 and then the predicted values of the set have been found with the average percentage of positive deviation as 6.06%.

KEYWORDS

- **powder-mixed electric discharge machining**
- **grey relational analysis**
- **grey relational coefficient**
- **grey relational grade**
- **Taguchi**

REFERENCES

1. Azad, M. S.; Puri, A. B. Simultaneous Optimization of Multiple Performance Characteristics in Micro-EDM Drilling of Titanium Alloy. *Int. J. Adv. Manuf. Technol.* **2012,** *61,* 1231–1239.
2. Chow, H.-M.; Yang, L.-D.; Lin, C.-T.; Chen, Y.-F. The Use of SiC Powder in Water as Dielectric for Micro-Slit EDM Machining. *J. Mater. Process. Technol.* **2008,** *195,* 160–170.
3. Das, V. C.; Srinivas, C. Optimization of Multiple Response Characteristics on EDM Using the Taguchi Method and Grey Relational Analysis. *Int. J. Sci. Res. Rev.* **2014,** *3,* 25–41.
4. Dewangan, S.; Gangopadhyay, S.; Biswas, C. K. Multi-Response Optimization of Surface Integrity Characteristics of EDM Process Using Grey-Fuzzy Logic-Based Hybrid Approach. *Eng. Sci. Technol. Int. J.* **2015,** *18*(3), 361–368.
5. Ezugwu, E. O.; Wang, Z. M. Titanium Alloys and their Machinability—A Review. *J. Mater. Process. Technol.* **1997,** *68*(3), 262–274.
6. Ganachari, V. S.; Kavade, M. V.; Mohite, S. S. Effect of Mixture of Al and SiC Powder on Surface Roughness in PMEDM Using Taguchi Method with GRA Optimization. *Int. J. Adv. Eng. Res. Stud.* **2013,** E-ISSN: 2249–8974.
7. Gu, L.; Li, L.; Zhao, W.; Rajurkar, K. P. Electrical Discharge Machining of Ti-6Al-4V with a Bundled Electrode. *Int. J. Mach. Tools Manuf.* **2012,** *53,* 100–106.
8. Kirbia, G.; Sarkar, B. R.; Pradhan, B.; Bhattacharyya, B. Comparative Study of Different Dielectrics for Micro-EDM Performance During Micro Hole Machining of Ti-6Al-4V Alloy. *Int. J. Adv. Manuf. Technol.* **2009.** DOI: 10.1007/s00170-009-2298-y.
9. Kolli, M.; Kumar, A. Effect of Boron Carbide Powder Mixed into Dielectric Fluid on Electrical Discharge Machining of Titanium Alloy. International Conference on Advances in Manufacturing and Materials Engineering, AMME 2014. *Procedia Mater. Sci.* **2014,** *5,* 1957–1965.
10. Kolli, M.; Kumar, A. *Effect of Additives Added in Dielectric Fluid on EDM of Titanium Alloy,* ICARMMIEM, India, 2014, paper 52.
11. Rodic, D.; Gostimirovic, M. Predicting of Machining Quality in Electric Discharge Machining Using Intelligent Optimization Techniques. *Int. J. Recent Adv. Mech. Eng. (IJMECH)* **2014,** *3,* 1–9.
12. Srinivasan, R.; Kothandapani, T.; Panneerselvam, T.; Santosh S. Optimization Of EDM Parameters Using Taguchi Method and Grey Relational Analysis for Mild Steel is 2026. *Int. J. Innovative Res. Sci. Eng. Technol.* **2013,** *2*(7), ISSN: 2319–8753.
13. Talla, G.; Gangopadhyay, S.; Biswas, C. K. Multi Response Optimization of Powder Mixed Electric Discharge Machining of Aluminum/Alumina Metal Matrix Composite Using Grey Relation Analysis. International Conference on Advances in Manufacturing and Materials Engineering. *Procedia Mater. Sci.* **2014,** *5,* 1633–1639.

CHAPTER 23

GEOMETRY-BASED DYNAMIC BEHAVIOR OF FUNCTIONALLY GRADED MATERIAL UNDER FLUCTUATING LOADS

S. VIJAY[1,*] and CH. SRINIVASA RAO[2,*]

[1]*Department Bapatla Engineering College Bapatla, Andhra Pradesh, India*

[2]*Department AU College of Engineering (A), Visakhapatnam, Andhra Pradesh, India*

[*]*Corresponding author. E-mail: vizzu515@gmail.com, csr_auce@yahoo.co.in*

ABSTRACT

Material selection is one of the key points in the optimization of design process. Main challenge is, when the component is in hazardous working conditions. Pure metals hardly find any applications in present days because of the requirement of varying properties in both hardness and ductility. Over the years; metal matrix composites (MMC) have been developed to meet the functional requirement. Due to the varying properties of the MMCs they are allowed to use in wide variety of applications with changed or controlled volume fraction in the reinforcement phase. A low to moderate toughness of conventional MMCs is the main disadvantageous to some of the applications in which additional requirement of high-wear resistance and toughness. The need for a different property at various locations within a single engineering component is becoming greater as advancement of technology. This leads to the development of functionally graded materials (FGM).

In this work, an attempt is made to investigate the material behavior considering different basic geometries under fluctuating loads. The wind pressure acting on vehicle surface from starting to steady state speed of the vehicle on the straight road is taken as dynamic loads for different geometries. Transient structural analysis using finite element method is carried out with ANSYS Workbench 14.0 software. Five-layered FGM with titanium (Ti) as the metal and ZrO_2 as the ceramic are used for the analysis. Engineering properties such as stresses, strains, strain energy and deformations in structural problems are evaluated for FGMs with three different domains. Comparative graphs were developed for different properties.

23.1 INTRODUCTION

Functionally graded material (FGM), comes under advanced materials with change in properties in the thickness direction. As the human always try to learn from the nature, FGMs are also gifted by nature in terms of bones, teeth, trunk, and so forth. FGMs eliminate the sharp changes of the properties that are common in general composites that cause initial failure. FGMs have the facility to change the material properties from one portion to the other depending on the requirement. Depending on the application different kinds of fabrication processes are evolved such as electrophoretic deposition, in situ centrifugal casting, spraying, powder metallurgy, and laser cladding. The easy and effective process in which microstructure, composition, and sudden gradients can be controlled is the powder metallurgy.

Continuous variation of composition of the metal-ceramic system is the key to accommodate sudden changes of stresses in FGM. Powder metallurgy technique is the most suitable method to fabricate the proposed FGMs because of the domain thickness. As any scientific procedure, powder metallurgy technique also involved step by step procedure such as (a) weighing and blending of metal and ceramic powder based on the required spatial distribution, (b) piling and compressing of the homogeneous mixture, (c) sintering to transmit the green product to final product. The present work is dealing with analyzing the dynamic behavior of the FGM under fluctuating loads. The domain taken is of three types each from basic geometries called plate (cartesian), cylindrical (polar), and hollow sphere (spherical)-shaped FGM with five layers.[5] Optimization of the layers is adopted from literature. Static behavior of the functionally graded materials has been investigated by many authors. Zhao et al.[8] evaluated transient temperature fields in FGM under convective boundary conditions. Kim et al. [6] proposed volume fraction optimization of FGMs for critical temperatures. They used 3D FE model with brick element having 18 nodes to analyze the variation of FGM properties, for example temperature field, accurately in the direction of thickness. Li and Weng et al.[4] investigated magnitude of dynamic stress intensity factor of FGM under anti-plane shear load including finite crack.

Nemat-Alla[7] proposed an optimum composition variation parameter of ZrO2/Ti 2D-FGM under severe thermal loading cycles. Dynamic behaviors of FGMs have attempted by few authors and their work is limited to particular geometry and particular engineering property. In this wok, dynamic behavior of FGMs with different basic is reviewed.

23.2 FORMULATION AND MODELING OF DOMAIN

Shapes of most of the engineering components fall under three basic geometries called planar, cylindrical, and spherical shapes. Domain considered for this analysis is small portion of vehicle body subjected to wind pressure with basic geometries. Dimension of the planar domain is taken as $100 \times 100 \times 5$ mm for each layer, $100 \times 90° \times 5$ mm for polar domain, and $100 \times 90° \times 45°$ for spherical domain are considered (Figure 23.1). For each domain, five layers of 5 mm thickness are used.

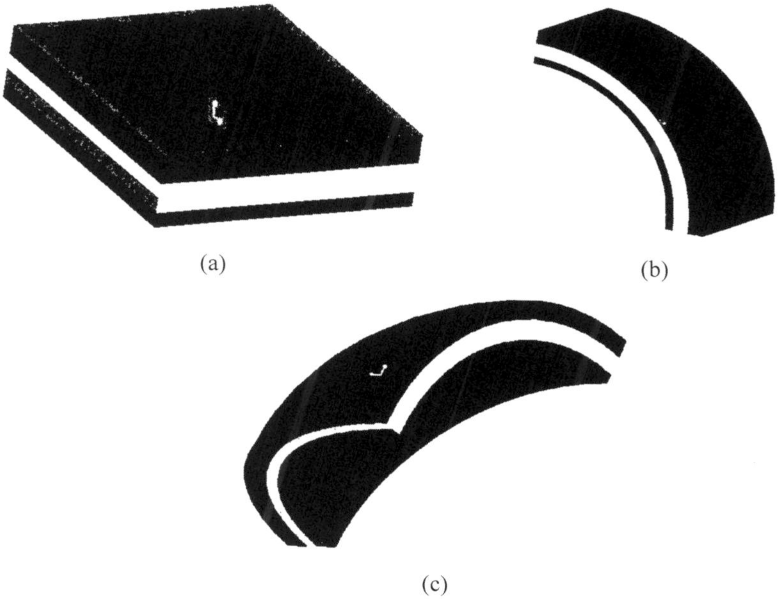

FIGURE 23.1 (a) Planar domain, (b) cylindrical domain, and (c) spherical domain.

Each of the models is having the five layers in which top layer are purely ZrO_2 (ceramic) and the bottom layer is purely Ti (metal). Interior layers have the gradients of metal properties which follow the rule of mixture, in which the properties of individual layers are considered.

$$V_m(t) = \left(1 - \frac{t}{h}\right)^n, \quad V_c(t) = 1 - V_m(t) \tag{23.1}$$

where *Vm* is the volume fraction of metal and *Vc* is the volume fraction of ceramic. *n* is the volume fraction index indicating the material variation profile through the thickness direction and a nonnegative real number. Effective material properties of different layers can be expressed as follows[6]

$$MP_{eff}(t) = MP_m * V_m(t) + MP_c * V_c(t) \tag{23.2}$$

where MP_m and MP_c are any material properties of metal and ceramic. Different material properties of metal (Ti) and ceramic (ZrO_2) are adopted from the literature[6] and are listed in Table 23.1 Properties for different layers were calculated using Equation 23.2 and tabulated in Table 23.2.

TABLE 23.1 Material Properties of Titanium and Zirconium Oxide.[2]

Property	Titanium (Ti)	Zirconium oxide (ZrO_2)
Young's modulus E(GPa)	$122.7-0.0565T$	$132.2-50.3\times10^{-3}T-8.1\times10^{-6}T^2$
Conductivity k(W/m^{0}K)	$1.1+0.017T$	$1.71+0.21\times10^{-3}T+0.116\times10^{-6}T^2$
Poisson's ratio (υ)	$0.2888+32\times10^{-6}T$	0.333
Thermal Expansion α (1/^{0}K)	$7.43\times10^{-6}+5.56\times10^{-9}T-2.69\times10^{-12}T^2$	$13.31\times10^{-6}-18.9\times10^{-9}T+12.7\times10^{-12}T^2$
Yield strength σ_y (MPa)	$1252.0-0.8486T$	–
Ultimate tensile σ_{ut} (MPa)	–	$148.1+1.184\times10^{-3}T-31.4\times10^{-6}T^2$
Ultimate compress. σ_{uc}(MPa)	–	$3181.2+25.43\times10^{-3}T-0.675\times10^{-3}T^2$
Density ρ (kg/m^3)	4500	6505

For the dynamic load calculation, pressure acting on the vehicle surface from the starting of the vehicle to its steady state in the period of 1 h on a smooth road is considered. Pressure variation depends on the speed of the vehicle and the relation is adopted as:

$$p(Pa) = 0.3073 * V^2 \tag{23.3}$$

where V is speed of the vehicle (kmph). Average speeds of the vehicle at different gear changes are assumed according to the code rule of road transport and highways by Government of India,[3] and calculated values of pressure are listed in Table 23.3 and Figure 23.2.

TABLE 23.2 Properties of Materials in Different Layers at 20°C.

Property	Layer 1	Layer 2	Layer 3	Layer 4	Layer 5
Young's modulus E(GPa)	131.19	128.78	127.38	125.06	121.57
Conductivity k(W/m0K)	1.7142	1.6456	1.5771	1.5085	1.44
Poisson's ratio (υ)	0.33	0.3221	0.3193	0.2998	0.2894
Thermal Expansion α (1/0K)	1.2937XE-05	1.1587XE-05	1.0240XE-05	8.8893XE-06	7.3177XE-06
Density ρ (kg/m3)	6505	6003.7	5502.5	5001.2	4500

TABLE 23.3 Wind Pressure at Different Gears.

Gear change	Avg. speed (kmph)	Wind pressure(Pa)
1	5	7.9325
2	15	71.3925
3	30	285.57
4	40	507.68
5	80	2030.72

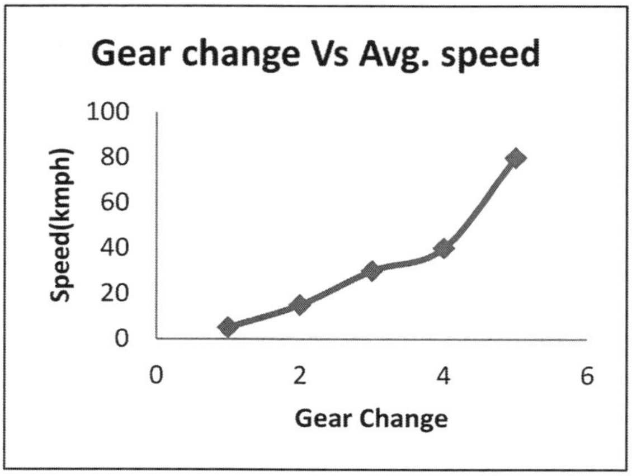

FIGURE 23.2 Different gear changes and its speed.

23.3 TREATMENT OF BOUNDARY CONDITIONS AND APPLICATION OF LOADS

As the domains are small portions of large plates of the vehicle body, the lateral surfaces of the domain are fully constrained so that the problem becomes plane strain problem. Bottom surface of the domain is free to expand and the upper surface is exposed to the variation of pressure according to the speed variation of the vehicle (Table 23.3) (Figure 23.3).

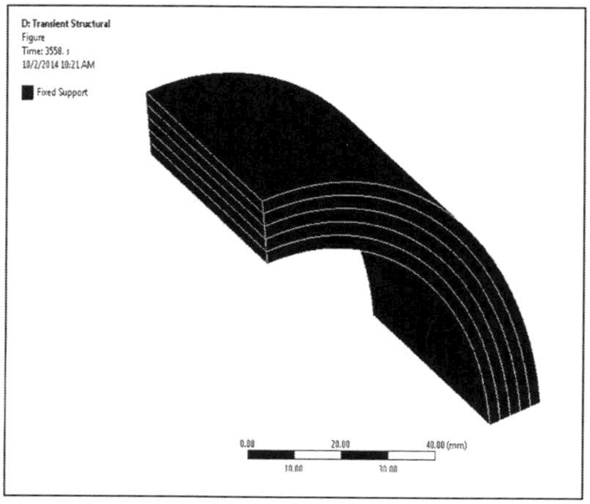

FIGURE 23.3 Boundary conditions for cylindrical domain.

23.4 THREE-DIMENSIONATIONAL FINITE ELEMENT ANALYSIS WITH DYNAMIC LOADING

For analyzing the dynamic behavior of FGM, three-dimensional finite element analysis is adopted and the following *assumptions* are considered for analysis.

i. Material properties within the layer are not changing
ii. Strength of the interface of the layer is same as the layer strength
iii. Path of the vehicle is smooth, vibrations of the body are neglected
iv. Vehicle is running continuously for 1 h
v. Relative velocity of the wind is neglected
vi. Speed accelerates for every 5 km

The graph shown in Figure 23.4 indicates the cyclic loads up to 1488 s; it is continued up to 3600 s in the analysis. Eight noded brick element is used for meshing and h-type convergence criterion is used to avoid error in the descritization process. Height of the element is changed from 5 to 2 mm in a regular interval for checking the convergence of the analysis. Figure 23.5 shows the convergence of total deformation of the domain. The results are adopted for the remaining analysis. Convergence test is also conducted in terms of other solutions like equivalent stresses induced in the elements, and strain energy, and so forth, and obtain same results as the deformation. From the results of the convergence test 2 mm is taken as the optimum element size.

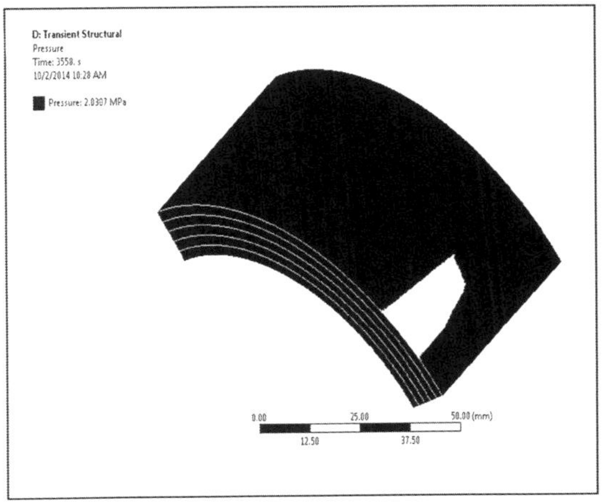

FIGURE 23.4 Load application on the outer surface.

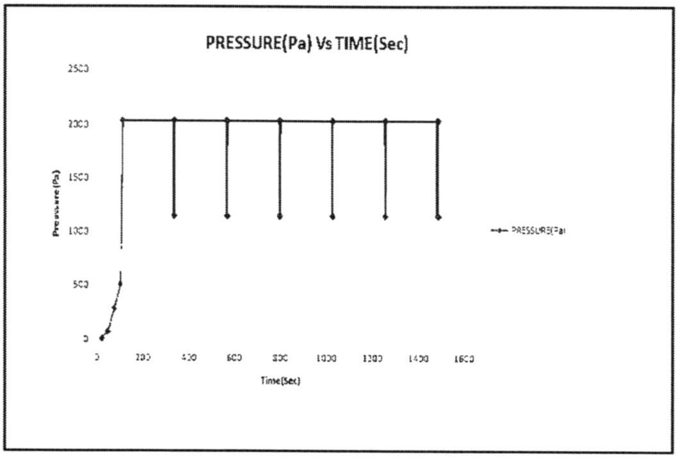

FIGURE 23.5 Cyclic loads for dynamic analysis.

23.5 RESULTS AND DISCUSSION

Dynamic loading shown in Figure 23.5 is applied on the upper surface of each domain and transient structural analysis has conducted using ANSYS Workbench 14.0 software. Total deformation, equivalent stresses, equivalent strains, and strain energy results for each domain and each layer have been observed. From the deformation results, it is observed that spherical domain has less deformation for the same load application when compared to the planar and cylindrical domain. Figure 23.6 shows total deformation of the domain after the application of dynamic loads, from the analysis it is observed that spherical domain is having less value of 0.0035035 mm for the top layer, which is made up of pure ceramic ZrO_2. In all the domains, deformation increases from top layer, which is ceramic to bottom layer, which is pure metal *Ti*.

Figure 23.7 gives the equivalent iso-stress curve patterns for cylindrical domain. Same patterns were obtained for the planar and spherical domains. From the comparative graphs in Figure 23.8, it is observed that pure metals have larger deformation and higher stresses for all the domains. Whereas, FGM and ceramics have smaller deformation and lesser stresses compared to pure metal. Mixed results are observed in FGMs and ceramics with different domains. In planar and spherical domains, FGM has lesser values of both deformation and stresses but in cylindrical domain ceramic plates are having lesser values comparatively with minute difference in FGMs.

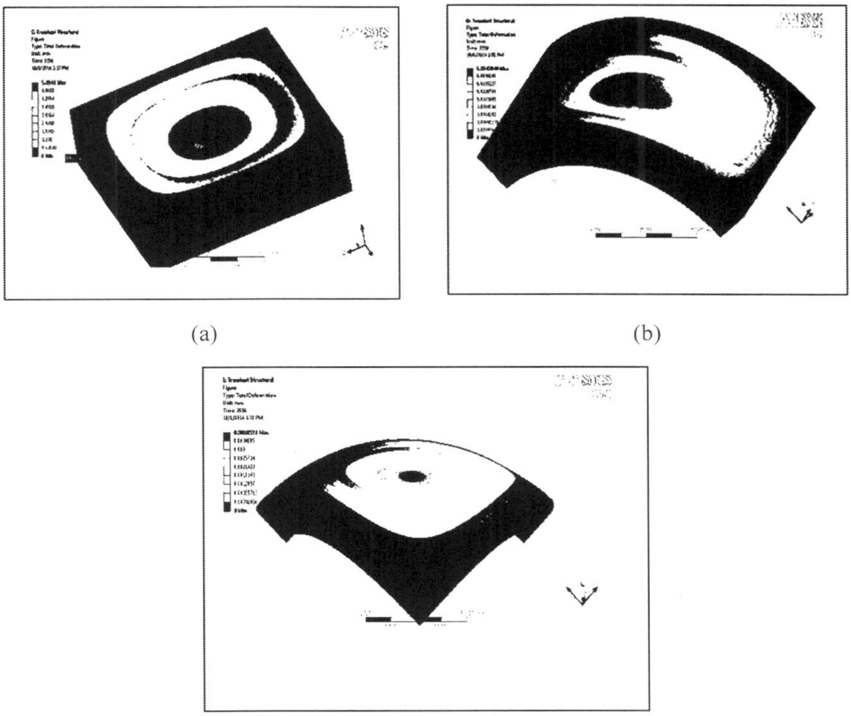

(a) (b)

(c)

FIGURE 23.6 Total deformation in (a) planar, (b) cylindrical, and (c) spherical domain.

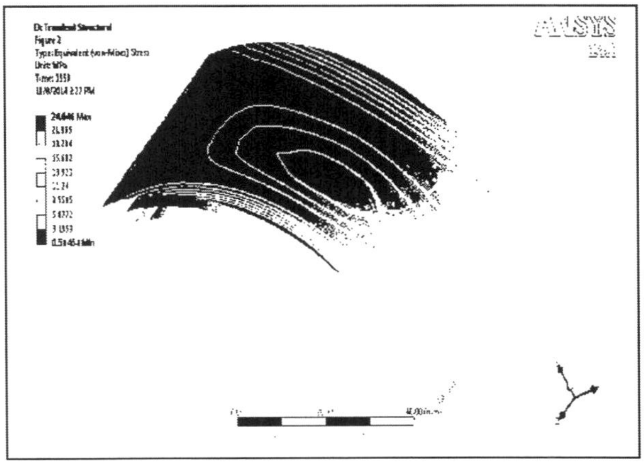

FIGURE 23.7 Equivalent iso-stress curves for cylindrical domain.

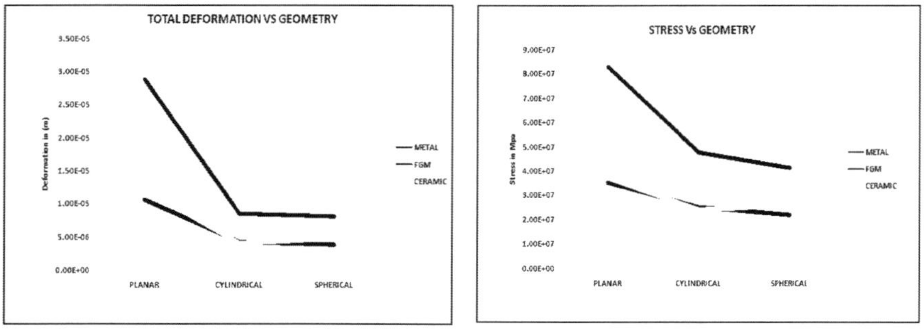

(a) (b)

FIGURE 23.8 Comparative graphs (a) deformation (b) stress.

23.6 CONCLUSION

Dynamic analysis using fluctuating loads on three basic geometrical domains of the functionally graded material was carried out. 3D finite element analysis is used for evaluating total deformation, equivalent stresses, equivalent strain, and strain energy of the domain and in individual layers. It is concluded from this analysis that due to the normal surface of the planar domain to the applied load total displacement as well as the stresses are too high when compared to cylindrical and spherical domains. The deformations are high in the top layers and less in the bottom layers for the planar domain, and it is reverse in the case of other domains. Smaller values of deformations in cylindrical and spherical domains are mainly due to the curved surface. Stress variation pattern is same in all domains. Minimum equivalent stress is observed in the middle layer. Due to continuous variation of material properties, the uniform distribution of deformation and stresses are observed in FGM.

Equivalent iso-stress curves of FGM reveal that the stress concentration intensity is reduced and distributed in all directions. From the path operations, it is concluded that inter-laminar rises of stress intensities can be avoided by using FGMs. From this work, it is suggested that components subjected to fluctuating sudden loads, functionally graded materials are preferable to avoid fatigue failures.

KEYWORDS

- **functionally graded materials**
- **fluctuating loads**
- **geometry**

REFERENCES

1. CAN-BEST window & IG Testing Laboratories wind Speed and Pressure Conversion Data. Google Open Source. http://can-best.com/index_files/Wind%20Speed%20Pressure%20Conversion.pdf
2. Chiba, R.; Sugano, Y. Optimization of Material Composition of Functionally Graded Materials Based on Multiscale Thermoelastic Analysis. *Acta Mech.* **2012,** *223*, 891–909.
3. Document on Test Method, Testing Equipment and Related Procedures, Ministry of Road Transport and Highways, Government of INDIA march-2010.
4. Li, C.; Weng, G. J. Dynamic Stress Intensity Factor of a Functionally Graded Material Under Antiplane Shear Loading. *Acta Mech.* **2001,** *149*, 1–10.
5. Michalak, B.; Wirowski, A. Dynamic Modeling of Thin Plate Intensity Factor of a Functionally Graded Material under Antiplane Shear Made of Certain Functionally Graded Materials. *Meccanica* **2012,** *47*, 1487–1498.
6. Na, K.-S.; Kim, J.-H. Optimization of Volume Fractions for Functionally Graded Panels Considering Stress and Critical Temperature. *Int. J. Compos. Struct.* **2009,** 509–516.
7. Nemat-Alla, M. Reduction of Thermal Stresses by Composition Optimization of Two-Dimensional Functionally Graded Materials. *Acta Mech.* **2009,** *208*, 147–161.
8. Zhao, J.; Ai, X.; Li, Y. Z. Transient Temperature Fields in Functionally Graded Materials with Different Shapes under Convective Boundary Conditions. *Heat Mass Transfer* **2007,** *43*, 1227–1232.

CHAPTER 24

A COST AND ENERGY-EFFECTIVE AUTOMATIC MATERIAL HANDLING SYSTEM FOR SMALL-SCALE INDUSTRY

S. G. KUMBHAR[1,*], S. M. TELI[2,3], T. U. MALI[2,4], S. M. SHINDE[2,5], A. K. KUMBHAR[2,6], and P. M. ZENDE[2,7]

[1]*Department of Automobile Engineering, Rajarambapu Institute of Technology,Rajaramnagar, Sakharale, Sangli, Maharashtra, India*

[2]*Department of Automobile Engineering, Annasaheb Dange College of Engineering and Technology, Ashta, Maharashtra, India*

[3]*satishteli1221@gmail.com*

[4]*tusharmali2020@gmail.com*

[5]*shreeshinde6514@gmail.com*

[6]*avadhut.k94@gmail.com*

[7]*pandhrinathzende@gmail.com*

[*]*Corresponding author. E-mail: surajkumar.kumbhar@ritindia.edu*

ABSTRACT

High-end automation in material handling has been a complex and expensive marvel to small-scale industries till date. Labor intrinsic transportation, the constraint of an external energy source, manual operation modes make a high proportion of total operating costs of the plant. These depict a holistic room to develop a cost and energy effective automatic material handling system for small scale industries. The present work aims to design and develop an automatic material handling system which can accomplish maximum objectives of material handling. The new system is also compared with some old design of the similar approach to ascertain the magnificence of the design and found to be better in all aspects. The implementation and utilization of present eco-friendly material handling system found to be a feasible and profitable way to handle and transport milk products in Sadguru Dudh Sangh Pvt. Ltd. Empirical results specify that the present system is cost-effective and time saving due to the automatic handling without using any external energy source.

24.1 INTRODUCTION

A material handling does not add to the value of the product but helps in the production. An unavoidable, depending on the weight, volume, and throughput of materials became the key features of the material handling devices. Nevertheless, it costs money, huge demand is to eliminate or reduce this as much as possible. Though some of anticipated mechanical/ mechatronics handling, reduce the labor costs of manual handling (MH) of materials, but the suitability of the same with respect to affordability should be taken into consideration especially in case of small manufacturing enterprises (SME). Hence, SME's puts a huge demand for cost and energy effective materials handling that should be carefully designed to suit the application.

The requirement of safe handling, maintaining quality, and condition of material to be handled still made the material handling system vital in industry field. In a typical manufacturing plant, many researchers pinpointed the accounts of material handling for 25% of all employees, 55% of all company space, 87% of the production time, and 15–75% of the total cost of a product.[1] Instead of enhancing the level of automation, SME's demands material handling systems with cost-effectiveness, flexibility, reconfigurability, and high availability. To weigh the benefits against the limitations or disadvantages of material handling systems as additional investment, lack of flexibility, vulnerable to downtime whenever there is a breakdown, additional maintenance staff and cost, cost of auxiliary equipment, space and other requirements, and so forth have been identified to make a judicious balance of the total benefits and limitations which is required before an economically sound decision is made.

To select appropriate material handling equipment, the well-known multicriteria decision-making methods have been used by many researchers due to its potential to convert a complex problem to a paired comparison. The prominent significance of analytic hierarchy process has been focused generally for solving the problem of selection of material handling equipment on the basis of some criteria (material, move, and method) are selected. SMEs with highly flexible product mixtures rarely considers automatic-guided vehicles (AGVs) as a potential solution.[1]

For fluctuating material handling environments, the fixed wire guidance systems were found inadequate due to their insensitivity and persistence. The brilliance of an AGV has been demonstrated in recent times due to the capability, quicker, and safer throughout operations. Despite the high initial cost unified with the inflexibility of navigable paths, it becomes a hurdle to the use AGV systems in SMEs. The development of the SmartCaddy, a low-cost, modular, and flexible AGV system are boosted nowadays, whereas low cost was achieved by using simple, off-the-shelf components, and by avoiding large software and hardware development costs.[2]

Many of the techniques and methodologies to design, implement, and control AGV systems has been limited in scale due to the complex nature of systems.[3–7] The requirements of the design effort in routing and scheduling of AGV's have been revised due to the increased flexibility and complexity.[8] Throughput, unit load, flow path design, and fleet size have been made a significant influence on the design and operation performance of AGV in a small manufacturing unit.[9] The formulation to control the traffic inside an industrial workspace has been found an interested scope in the research of AGV's. Especially, the possibility of improvement in the guided tape type AGV utilizing better navigation techniques and the ability to adapt to any environment easily which is economical among autonomous robot has been denoted which highlight the use of fussy brain of the autonomous robot integrated with both central and sensor system for detection of the obstacles in the warehouse.[10] The selection of appropriate material handling systems is extremely influenced by consideration of initial investment cost rather than the operational cost and prolonged effects, particularly in SME's.[11] In the area of research, published work is mostly limited to complicated and costly material handling system uses external energy. Therefore, need of SME's depicts a scope for a more comprehensive and holistic development of cost-effective automatic material handling system.

24.2 MATERIAL HANDLING SYSTEM

This project work has been executed to accomplish the aim of effective and efficient material handling system for small-scale industry. First, several present systems and practices of the material handling in small-scale industry were analyzed theoretically to gain the knowledge and difficulties occurring while designing new methodology in desired constraints. Additionally, an industrial training of 2 weeks at Sadguru Dudh Sangh, MIDC Tasvade, Karad had been undergone to support the theoretical study. The functioning in small-scale industry environment can be supportive to understand and analyze the present and unknown problems which may occur during the actual operation of the system in the industry. Along with these, our team has been consulting with some manufacturers and fabricators to clear out some uncertainties and feasibility problems that may be upraised during manufacturing of the system. The entire discussion with these consultants and MD of Sadguru Dudh Sangh Pvt. Ltd. has gone through various issues such as the total cost of the system, fabrication hurdles, implementation strategies, safety and the environmental concerns, and sponsorship details of the present approach. One possible and suitable solution to design and develop an eco-friendly and effective AGV system which requires minimal external power to operate, having cost and time-saving capability than manual mode considering safety and the environment perspective has been opted.

24.3 DESIGN AND DEVELOPMENT OF AUTOMATED-GUIDED VEHICLE

With some fundamental principles and theories, a new approach based upon of gravitation and rack and pinion mechanism has been used to design an energy effective AGV system which can accomplish maximum objectives of material handling. Rack and pinion offers a maximum advantage of smooth and effective travel. The entire arrangement of system has been made in such a way that it befitted a composite of conventional and automated system. As decided to use the principle of self-weight for the final movement of the system, neither manual efforts nor external power source is required to operate. The correlation between load, speed, and time taken to travel during material handling has been taken into consideration while designing the entire system.

Some of the old approaches had inadequacy in material loading–unloading, balancing of the system as well as the absence of traveling distance adjustment and start/stop facility. The present system has been designed for adequate loading–unloading and diminishes the problem of balancing. Old approaches were found unsuitable to load and unload high-weighted material due to the position of weight carrier which was fixed at the higher level from the ground. In this design, the height of weight carrier is fixed nearly half of the previous designs which make it user friendly than earlier designs. Moreover, the problem of balancing of the old system due to the central position of weight carrier on the top of the floating pipe was resolved by fixing the weight carrier on the side of floating pipe., whereas the problem of balancing in this design also has been analyzed and adjusted at the design stage only by varying some design variables.

Along with this traveling, distance could not be adjusted in older versions which have been overwhelmed beautifully with the present design by providing flexibility to adjust the spring deflection. The travel distance adjuster is free to slide along the length of rigid pipe and can be fixed in a specific range, which can change the total travel distance of the AGV. The absence of start/stop facility in old designs is the more inconvenient problem for the user. The present design has an arrangement of start/stop of the system similar to travel distance adjustment. At the end, detailed drawings of each part were finalized by departmental design expert faculties and forwarded to fabrication.

24.4 EXPERIMENTATION AND TESTING

The entire system has been fabricated as per the predefined plan and strategies. Various problems and difficulties occurred during manufacturing were removed or adjusted as per the decision taken by fabrication experts and design team.

Figure 24.1 shows the schematic view of the new AGV. The whole system has been tested as per the text matrix in the laboratory environment to assess the performance as per design constraints and features. Some limitations and problems regarding balancing and vibrations were removed from trial and error method regularly. Finally, the system has been implemented and utilized in Sadguru Dudh Sangh Pvt. Ltd. for handling and transportation of milk cans and its performance were evaluated on the basis of considering desired constraints in industrial environment. Hence, compared to older designs of the similar approaches in different operating conditions of load, time to travel, and travel distance; new design has been demonstrated its magnificence in design and found to be better in all aspects.

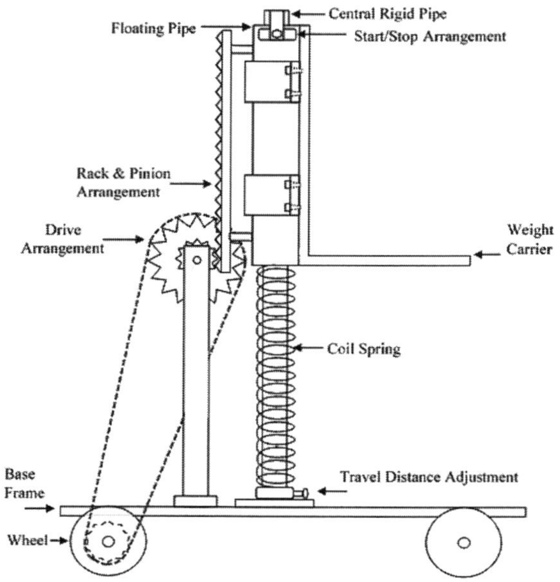

FIGURE 24.1 Schematic view of new designed automated-guided vehicle.

24.5 RESULTS AND DISCUSSION

The analysis is done for the time required to transport a batch of milk and numbers of labors required. Table 24.1 shows an average time taken by each MH (MH), roller conveyor (RC), and AGV over a period of 3 months for a respective batch of milk transported. To determine the significant impact of the AGV and RC system compared to MH, a comparative analysis of time taken by this mode of the transportation system is performed.

TABLE 24.1 Average Time Required to Transport A Batch of Milk/Day and Number of Labors Required.

Mode of transportation	Number of labors required	The average time required to transport (min)				
		1000 l	5000 l	10,000 l	15,000 l	20,000 l
Manual handling (MH)	06	43	72	101	128	157
Roller conveyor (RC)	03	19	37	61	78	99
Automatic-guided vehicle (AGV)	02	5	19	38	56	75

It was observed that there is, on average a large number of cans waiting for transport by the AGV in critical conditions of transport of milk cans. This indicates that the AGV is a bottleneck for a milk batch of more than 20,000 l. It was observed that the transport of more numbers of milk cans in prescribed time limit cannot be achieved with only one AGV in such condition. It is due to the time cycle limitation which is based on the fundamental principle of the system. Hence, an AGV fleet of the single machine has been found a critical alternative to the current problem of cost-effective transport of milk cans.

It was also monitored that conventional approaches perform more efficiently in such situation where more number of cans to be transported in minimal time due to the flexibility in the transport of cans. Wherever, such condition requires more human effort which is not possible in AGV approach as it cannot be considered as a sustainable alternative due to cost and effort inadequacies. Henceforth, analysis with single AGV utilization, and milk batch more than 20,000 l is not included in the current analysis. A reasonable solution for such problem is nothing but increase the number of AGV's with proper management of fleet to transport milk batches lesser than 20,000 l.

Feet of AGV containing two AGV's is used in such a way that the loading time of the one equals with unloading time of others. This contrast or multiplicity is used to utilize the same number of workers for fleet and reduce the overlapping and labor involvement. Figure 24.2 shows that AGV reduces nearly half of the meantime consumed by MH for each batch of the milk.

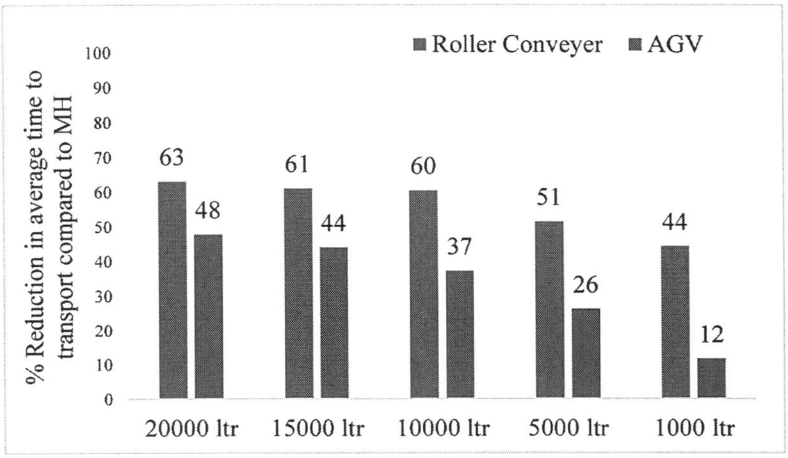

FIGURE 24.2 Percentage reduction of average time to transport.

The current approach of fleet of two AGV's implemented for 3 months in 60 production days' results in the average total time required to transport of milk cans to be reduced to 5.15 min/day as analyzed in Table 24.2. Compared to RC approach, this results in a drastic reduction in time and labor required to transport a similar batch of milk cans.

TABLE 24.2 Average Time Reduced to Transport.

Parameter	Average time required to transport (min)					Total time (min)
	1000 l	5000 l	10,000 l	15,000 l	20,000 l	
MH versus RC	24	35	40	50	58	207
MH versus AGV	38	53	64	72	82	309
Average time reduced by RC/day						3.45
Average time reduced by AGV/day						5.15

The one-way analysis of variance (ANOVA) has been performed to determine whether there are any statistically significant differences between the means for three modes of transport.[12] Table 24.3 provides descriptive statistics, including the mean, standard deviation, and 95% confidence intervals for the dependent variable (time) for each separate mode of transportation, as well as when all are combined (total). Table 24.4 shows the output of the ANOVA and whether there is a statistically significant difference between means.

TABLE 24.3 Mean, Standard Deviation, and 95% Confidence Interval for Different Modes of Transportation.

Mode of transportation	Mean	Standard deviation	Standard error	Margin of error	95% Confidence interval for mean	
					Lower bound	Upper bound
MH	100	44.91	20.08	17.60	117.80	82.60
RC	59	31.82	14.23	12.47	71.27	46.33
AGV	39	28.03	12.53	10.99	49.49	27.51

The p-value corresponding to the F-statistic of one-way ANOVA is lower than 0.05, suggesting that the one or more modes are significantly different as determined by one-way ANOVA ($F\ [2, 12]=3.8893$, $p=0.0499$). The significance of the time found by an ANOVA at 5% level of significance and a 95% confidence level. Moreover, for each of the three modes of transportation, Tukey's honest significant difference (HSD) test has been performed to pinpoint which of them exhibits statistically significant difference. The critical value of the Tukey-Kramer HSD Q statistic based on the N=3 modes and 12° of freedom for the error term, in the studentized range distribution, are 5.0430 and 3.7711 for significance level $p=0.01$ and 0.05, respectively. These have been evaluated to know whether $Q_{i,j} > Q_{critical}$.

TABLE 24.4 Analysis of Variance for Time to Transport.

Source	Sum of Squares	Degrees of Freedom DF	Mean Square	F-statistic	p-value
Between	09915.6000	2	4957.8000	3.8893	0.0499
Within	15296.8000	12	1274.7333		
Total	25212.4000	14			

A Tukey HSD test in Table 24.5 revealed that the average time to transport was statistically significantly lower for AGV compared to the MH. There was no statistically significant difference found between the other combinations.

TABLE 24.5 Tukey's Honest Significant Difference Results.

Mode of transportation	Tukey honest significant difference (HSD) Q statistic	Tukey HSD p-value	Tukey HSD inference
MH versus RC	2.5928	0.2004243	Insignificant
MH versus AGV	3.8705	0.0441609	$p<0.05$
RC versus AGV	1.2776	0.6428486	Insignificant

Figure 24.3 shows 95% confidence interval for the means where mean time reduction of AGV system is somewhere between 49 and 28 min, which is more precise than MH as well as RC.

Table 24.6 shows the cost analysis of expenditure on material handling for 3 months in Sadguru Dudh Sangh Pvt. Ltd. The humanitarian efforts to drive this new AGV is significantly less than manual as well as RC approach. The total labor expended by Sadguru Dudh Sangh over 3 months for AGV is ₹ 50,000 which is nearly 31% lesser than MH. Along with this, the designed AGV do not require any external power source which extremely reduces the energy expenditure to zero.

In all above aspects, AGV has been demonstrated the profitability to handle and transport milk cans in small-scale enterprises like Sadguru Dudh

Sangh Pvt. Ltd. Experimental data analysis denotes the physical agreement of the present system. Overall results reveal the user compatibility and cost-effectiveness of the system.

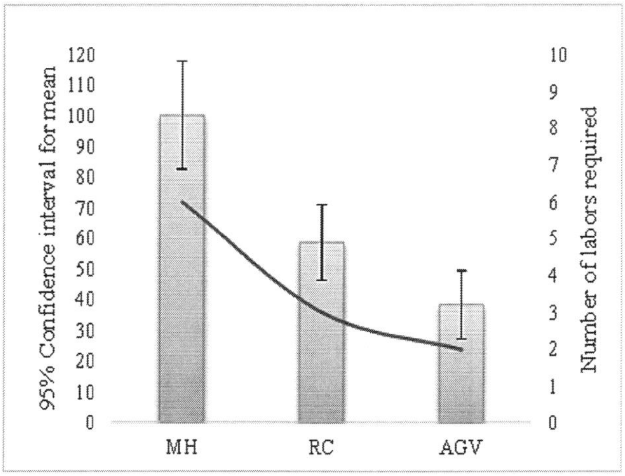

FIGURE 24.3 Comparison of 95% confidence intervals for means and number of labors required in respective mode of transportation.

TABLE 24.6 Cost Analysis.

Mode of transportation	Labor cost (₹)	Other cost (Overtime/ maintenance, repair, etc.) (₹)	Total cost (₹)
MH	6*8000*3=144000	18000	162000
RC	3*8000*3=72000	5000	77000
AGV	2*8000*3=48000	2000	50000

24.6 CONCLUSION

In the present work, a cost and energy effective automatic AGV system have been developed for Sadguru Dudh Sangh Pvt. Ltd. to handle and transport milk cans in desired constraints. The system has been designed and fabricated to accomplish maximum objectives of the material handling system. The entire system has been found user friendly as compared to older designs. The system is implemented and utilized in Sadguru Dudh Sangh Pvt. Ltd. and its performance are evaluated and validated in industrial environment.

The new system has been compared with some old design of the similar approach in cost, energy, compatibility to the user, and time-saving aspects to ascertain the magnificence of the design and found to be better. From the obtained responses, the following conclusions are drawn:

- AGV reduces the average total time of material transportation by 5.15 min/day compared to the MH. This increase the production time, which has been found profitable to the industry.
- Start/stop phenomena of the system as well as total travel distance can be easily controlled by the user easily as per constraint or application.
- The total cost expenditure on designed material handling systems is found to be 31% lesser than MH.
- AGV found to be a user-friendly material handling system than old approached for small-scale industry to transfer material within desired constraints.
- The results obtained are agreeing with physical reasoning and expectations.

ACKNOWLEDGMENT

The authors would like to thank Mr. Ranjeet Jadhav, Managing Director, Sadguru Dudh Sangh Pvt. Ltd., MIDC Tasvade, Karad, Maharashtra, India and the entire team for valuable contribution and suggestions on this work. The authors are also thankful to Dr. S. R. Kumbhar and Dr. R. G. Desavele, Rajarambapu Institute of Technology, Rajaramnagar, Sakharale, Sangli, Maharashtra, India, for inspiration and guidance. The authors also would like to acknowledge the efforts of students and faculties from Automobile Engineering Department, Annasaheb Dange College of Engineering and Technology, Ashta, Sangli, Maharashtra, India.

KEYWORDS

- **material handling**
- **automatic guided vehicles**
- **small manufacturing enterprises**

REFERENCES

1. Eason, G.; Noble, B.; Sneddon, I.N. On Certain Integrals of Lipschitz-Hankel Type Involving Products of Bessel Functions. *Phil. Trans. Roy. Soc. London* **1955**, *247*, 529–551.
2. Kumar Sharma, P.; Soni, M.; Patidar, S. Selection of Material Handling Equipment by using Analytical Hierarchy Process. *Int. J. Manage. Soc. Sci. Res. (IJMSSR)* **2014**, *4*(2).
3. Rosandich Ryan, G.; Lindeke Richard, R. Development and Automated Guided Vehicles for Small to Medium Sized Enterprises. May 2007.
4. Le-Anh, T.; De Koster M. B. M. Review of Design and Control of Automated Guided Vehicle System. *ERS* **2004**, 01–34.
5. Bülent Sezen, "Modeling, Automated Guided Vehicle Systems in Material Handling. *Dogus Üniversitesi Dergisi* **2003**, *4*(2), 207–216.
6. Tanchoco, J. M. A; Egbelu, P. J.; Taghaboni, F. Determination of the Total Number of Vehicles in an AGV-Based Material Transport System. *Mater. Flow* **1987**, *4*, 33–51.
7. Egbelu, P. J.; Tanchoco, J. M. A. AGVSim User's Manual. *Department of Industrial Engineering and Operations Research*; Technical Report No. 8204, Virginia Polytechnic Institute and State University: Blacksburg, VA. 1982.
8. Bozer, Y. A.; Srinivasan, M. M. Tandem Configurations for AGVS and the Analysis of Single Vehicle Loops. *IIE Transactions* **1991**, *23*(1), 72–82.
9. Russell, E. K.; Carl, W. A Review of Automated Guided-Vehicle Systems Design and Scheduling, Production Planning and Control. *Managem. Oper.* **2007**, 2(1), 44–51.
10. Gaur, A. V.; Pawar, M. S. AGV Based Material Handling System: A Literature Review. *IJRSI* **2016**, *III*(IA), 33–36.
11. Das, S. K.; Pasan, M. K. Design and Methodology of Automated Guided Vehicle-A Review. *IOSR-JMCE*, e-ISSN: 2278-1684, p-ISSN: 2320–334X.
12. www.astatsa.com/OneWay_Anova_with_TukeyHSD/@2013 Navendu Vasavada (accessed Jan 17, 2017).

CHAPTER 25

A SURVEY ON APPLICATION OF A SYSTEM DYNAMIC APPROACH IN SUPPLY CHAIN PERFORMANCE MODELING

K. JAGAN MOHAN REDDY*, A. NEELAKANTESWARA RAO, and LANKA KRISHNANAND

[1]*Department of Mechanical Engineering, National Institute of Technology Warangal, Warangal, Telangana*

**Corresponding author. E-mail: jaganreddyie@gmail.com*

ABSTRACT

In today's volatile business environment, supply chain (SC) performance measurement (PM) plays an important role in the success of industries. It is proved that efficient PM method is required to measure the performance of whole SC. Over the past two decades, a number of authors have used various methods and techniques for different strategies. However, little literature is available in the field of system dynamics (SD) approach to evaluating the dynamic behavior of SC. Hence, this paper reviews the SC performance modeling using SD approach. Research articles available on Scopus and Institute for Scientific Information (ISI) databases are reviewed. This paper gives a bird's eye view of SC metrics, taxonomy of research and developments, simulation and modeling techniques of SC and recent research activities. It is observed that the application of SD in integrated SC performance modeling is a fruitful area to do research.

25.1 INTRODUCTION

In present business, technoeconomic scenario, huge competition is there between one supply chain (SC) and another. Every organization is trying to produce their goods and services at minimal cost, at right quantity and right place with proper quality to satisfy the customer. In this context, emerging tools like SC Management (SCM), Just-In-Time (JIT), Total Quality Management (TQM) and Six Sigma were evolved, but SCM is a versatile technique.[1] SCM is an approach whereby the entire network from the supplier to the ultimate customer is analyzed and managed in order to achieve the best outcome for the whole system.[37] An SC consists of all parties involved, directly or indirectly, in fulfilling a customer request. The SC as shown in Figure 25.1 includes not only the manufacturer and suppliers but also transporters, warehouses, retailers, and even customers themselves.[12] So, SCM plays a vital role in today's business transactions like fund flow and product flow. Implementation of the SCM is not sufficient, to withstand in today's fluctuating market, so there needs to be a continuous improvement in it. SC performance measurement (PM) approaches and techniques play an important role in the SC performance improvement process. However, many companies have failed in the improvement of SC performance, due to lack of proper approaches and PM frameworks.[20] PM systems are described as the overall set of metrics used to quantify both the effectiveness and efficiency of action.[30]

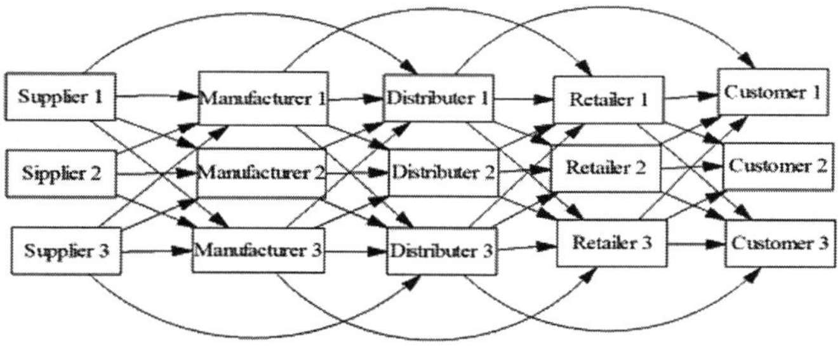

FIGURE 25.1 Schematic of supply chain.

Over the last few decades in SCM evolution, wide steady streams of research papers dealing with SCPM have been published. Several authors have used various approaches and techniques for PM like process-based approach,[4,6,9–11,16,26,27,28,29,33,34] perceptive-based approach,[8,33,36,38,39] hierarchicy-based approach,[6,7,14,20,33,36,41] six-sigma-based approach,[14,26] techniques like analytic hierarchy process,[7,8,10,36] data envelopment analysis,[2,28,38,40,41] Delphi[8]–Heuristic techniques-based model,[3] hybrid techniques-based model,[6,11,31] and simulation-based model.[29] Measuring SC performance of a single business organization is comparatively easier than multi-tier business organization. Measuring and setting benchmarks for various strategies in multi-tier SC is highly complicated. Identifying the key performance metrics and linking them with the operational strategies in this dynamic nature is too difficult to do. So, in this context simulation plays a vital role in the modeling of performance metrics in dynamic nature. However, in the last couple of decades, few authors have used the simulation techniques for SCPM. Moreover, Simulation is a powerful tool to simulate the entire SC performance in present and as well as in future. Simulation techniques such as system dynamics (SD) approach plays a novel role in modeling the SCP metrics in the dynamic environment. This paper reports a comprehensive review of SCPM using SD approach and shows a fruitful path for feature researchers in this area.

The rest of the paper is organized as follows: the next section discusses the historical introduction of SD to the SCPM. Section 25.3 reports the literature search procedure and methodology followed. Section 25.4 discusses the diffusion of SD in SCPM And finally, Section 25.5 gives the conclusions of this paper.

25.2 A HISTORICAL INTRODUCTION OF SYSTEM DYNAMICS (SD) TO THE SUPPLY CHAIN PERFORMANCE MEASUREMENT (SCPM)

SD is a computer-aided approach for examining and taking care of complex issues with an attention to policy analysis and design. In the beginning, it was called as industrial dynamics (ID),[18] a field developed by Jay W. Forrester at the Massachusetts Institute of Technology (MIT). Initially, SD was developed in control engineering and management; later its applications have been spread to other fields over last two decades. Forrester (1961) characterizes ID as "the investigation of the data criticism qualities of modern action to show how authoritative structure, intensification (in arrangements), and time delays (in choice and activities) cooperate to impact the achievement of the undertaking. It treats the connections between the streams of data, cash, orders, materials, work force, and capital hardware in an organization, an industry, or a national economy". SD models are useful in solving the problems by updating all the variables in small time intervals with positive and negative feedbacks and also updating time delays in structuring the interactions among the variables. In the SD methodology, a problem or a system is first represented as a causal loop diagram, which is a simple map of a system with all its constituent components and their interactions. Later, a stock and flow diagram is created from this causal loop diagram, this diagram is used to study and quantitative analysis of the system using computer simulation.

For last two decades, SD was applied to many disciplines. It includes work in corporate planning and policy design,[18,35] public management and policy,[17] economic behavior,[32] energy and the environment,[17] (biological and medical modeling[21]) dynamic decision-making,[15] theory development in the natural and social sciences,[35] software engineering,[5] complex nonlinear dynamics,[19] and SC management.[3,5,35]

SCPM and simulation using SD approach has become an innovative research area in the history of SD era for last few years. Besides, the utilization of SD modeling to SCM has its underlying foundations in ID.[18] According to Ackerman et al.,[23] research in theory building include the uses of SD to study the interrelationships among the different components of an SC system. Towill[35] utilizes a three-echelon production system as an SC reference model for comparing various methods of improving total dynamic performance. Towill[35] in his paper presents different routes, in which ID may have to be built and exploited in SC reengineering. He had utilized observation-based, system knowledge-based, and people-based sources to develop a real-life model for an electronics product's SC. Towill used SD as a methodology to solve difficult problems such as inventory oscillations, SC reengineering, and SC design. Barlas and Aksogan[5] developed an SD model to reduce the cost of inventory policies at four levels of SC of an apparel industry. Sterman[32] built an SD model for managing stocks that can be applicable in many situations such as raw material handling and production control at the microeconomic level. After a long slack period, Ge et al.[19] developed an SD model for the supermarket chain in the UK using MATLAB. This model simulates the effect of demand amplification on other variables. The simulation results of this model show that mutually sharing of necessary information among suppliers and buyers within every tier relationship in an SC is very important to enhance SC performance. De Souza et al.[25] explored the dynamics of the SC both qualitatively and quantitatively. Their clarifications provided some guidelines for SC reengineering. Angerhofer and Angelides[3] proposed taxonomy of research work and related development in SD modeling in SCM. Lai et al.[24] argue that SD is a practical approach to identify the relationship between different service processes and improve the operational efficiency.

Over the past few decades, many authors have used the SD approach for various SC problems like demand amplification, inventory osculation, supplier selection, logistics, and performance evaluation problems. This paper attempted to give an overview of the literature related to the SCPM using SD approach.

25.3 METHODOLOGY

The extensive literature review of SD application to SCPM has been conducted by following a systematic approach. This paper was prepared based on the articles that have been published in selected peer-reviewed international journals in the last 20 years from databases resources such as Scopus, Google scholar, Emerald, and ISI Web of Knowledge. Initially, we set the conceptual boundaries to help us in searching literature contributions, which may clear our review questions like how much SD approach has been used to model the SC performance metrics as shown in Figure 25.2.

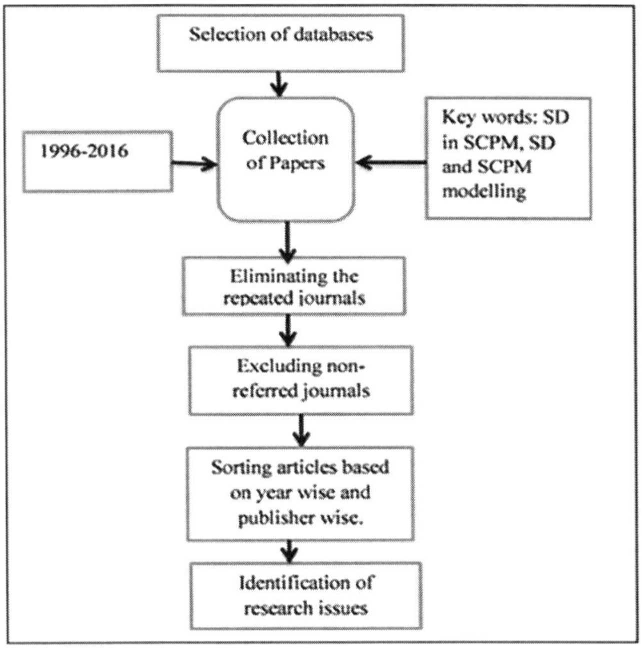

FIGURE 25.2 Systematic review methodology.

We have searched the articles with the keywords application of SD approach in SCPM and SCPM. We have reviewed more than 200 articles, and finally, we were able to scrutinize 50 articles based on authors' perspective. We have excluded the articles that contain the key search themes but do not focus on SD application to SCPM.

Less number of authors have reviewed the application of SD in the SCM field, so this paper tried to present a taxonomic review on application of SD in SCPM. In this point, initially, we have analysed the extensive research has been done on SCPM using SD approach. In this paper, articles were highlighted based on the author(s), year of publication, country, and journal. Decision hierarchy (strategic, tactical, or operational), and research method (quantitative, qualitative) and whether single echelon or multi-echelon. The whole classification of the articles is shown in Appendix 25.1.

25.4 THE DIFFUSION OF SD IN SCPM FIELD

To understand the amount of work that have been done in the field of SCPM using SD approach, the graph showing year wise representation of articles published in the academic journals have been plotted in Figure 25.3, and shows the adaption rate of SD application to SCPM field. Noticeably, the graph shows that the application of SD methodology to SCPM started in the late 1970s. But well ahead; a few authors have been used SD methodology for modeling and simulation in SC context because the lack of computing power limited the applicability of SD. The use of SD modeling in SCM has only recently re-emerged after a long slack period.

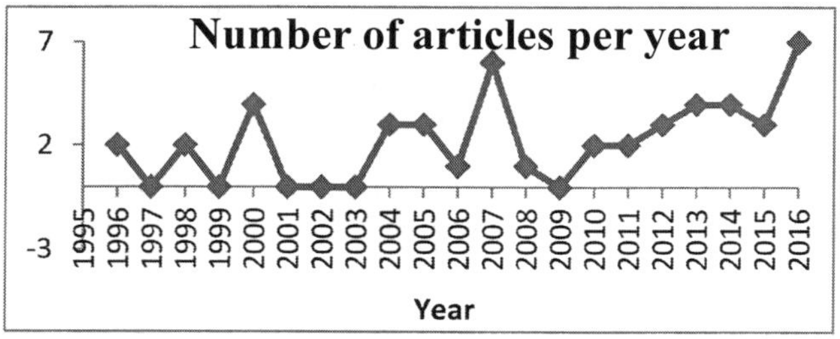

FIGURE 25.3 Diffusion of system dynamics (SD) to supply chain performance measurement (SCPM) per year.

25.4.1 APPLICATION SD TO SCPM BASED ON DECISION HIERARCHY LEVEL

In any industry or integrated SC, managers have to take decisions to meet the requirements of the customer. In SC decisions are taking at each stage due to dynamic changes in demand of the customer. According to Rushton and Oxley (1989), these decisions are classified based on the time frame, as strategic level, tactical level and operational level. This paper has shown a pictorial view of literature at each decision level. Figure 25.4 shows that the majority of literature fall into operational level moreover, a few authors only have been applied SD to SCPM at the strategic level. Noticeably, this graph shows a fruitful path for academicians to carry out the research on the application of SD to SCPM at a strategic level.

Strategic decisions play a vital role in meeting the objectives of focal SC.[20] These decisions are taken by top-level management through involving each level in SC. The strategic level measures influence the top level management decisions, very often reflecting investigation of broad based policies, corporate financial plans, competitiveness and level of adherence to organizational goals. This paper attempted to show the related literature at each decision level based on the performance metrics which have been used by researchers and academics. For last 20 years, a less number (16%) of authors have used the SD to model the SCP based on strategic metrics. Gunasekaran et al.[20] have given the metrics at the strategic level, such as level of customer perceived value of the product, variances against budget, order lead time, information processing cost, net profit verses productivity ratio, total cycle time, total cash flow time, product development cycle time, and so forth. This paper has attempted to show the number of literature have used these metrics to model SCP through SD approach. For example Bolarin et al.[5] used the dynamic model to predict the inventory costs consequences of variability demand process within a multi-stage SC consisting of supplier-manufacturer-distributor-retailer. They used Vensim software to relate inventory carrying cost for all the SC stages. Debabrata Das and Pankaj Dutta[13] proposed framework can be used to understand the long-term behavior of a system under various managerial issues.

The tactical level deals with resource allocation and measuring performance against targets to be met in order to achieve results specified at the strategic level. Measurement of performance at this level provides valuable feedback on mid-level management decisions. Gunasekaran et al.[20] have proposed some tactical decision level metrics such as supplier delivery performance, supplier lead time against industry norm, supplier pricing against the market, an efficiency of purchase order cycle time, efficiency of cash flow method, percentage of finished goods in transit and delivery reliability performance, etc. Figure 25.4 shows the number of literature (24%) related to SCPM using SD approach at tactical level for the last couple of decades. Comparing with strategic level, many authors have worked at the tactical level. For example, Narasimha[23] has done some work related to capacity augmentation of an SC for a short lifecycle product using SD approach. Janamanchi and Burns[22] developed insights for inventory management to prevent stock-outs and unfilled orders and to fill customer orders at the lowest possible cost to SC partners under different scenarios, in a two-player supplier–retailer SC. Moderate levels of inventory, defining appropriate performance functional, appear to be crucial in choosing the right policies for managing retail SC systems.

Operational level measurements and metrics require accurate data and assess the results of decisions of low level managers. Supervisors and workers set operational objectives that, if met, will lead to the achievement of tactical objectives. As per Figure 25.4, most of the authors (60%) have used the SD approach at the operational level only in SCPM context. For example, Towill,[35] had used SD approach for demand application and inventory management. Lia[25] also used the SD approach to model the operational performance metrics such as inventory on hand, quality of delivered goods, on-time delivery of goods, the effectiveness of delivery invoice methods, number of faultless delivery notes invoiced and percentage of urgent deliveries and, so forth.

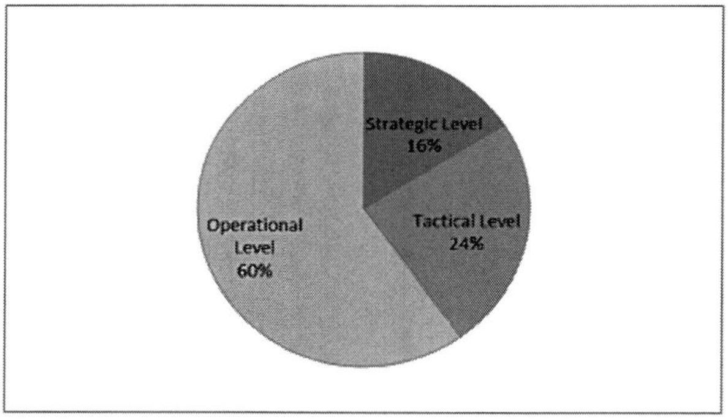

FIGURE 25.4 Application of SD to SCPM at each decision level.

25.4.2 APPLICATION OF SD AT VARIOUS SC PERFORMANCE METRICS

Chan[11] presents SCMPM approach which consists of qualitative and quantitative measures. Quantitative measures are cost and resource utilization, and qualitative measures are quality, flexibility, visibility, trust, and innovativeness. Figure 25.5 shows that the majority (64%) of literature are related to quantitating the performance measures only due to complexity in the modeling of qualitative measures, but these days many authors are using SD to model or simulate the qualitative SC performance measures.

This graph displays the application of SD approach to model the qualitative and quantitative performance metrics in the SC.

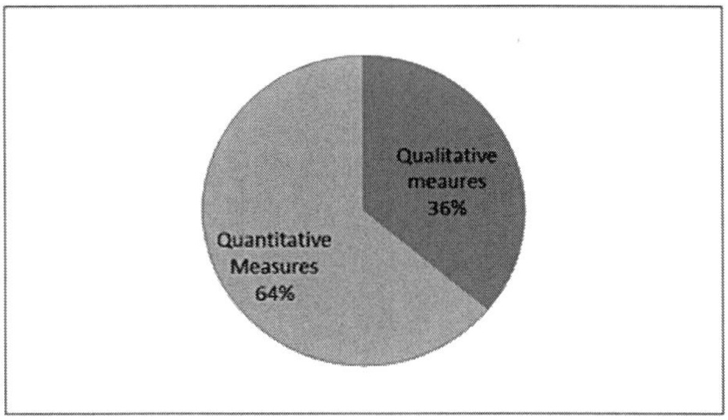

FIGURE 25.5 Application of SD at various types SCP metrics.

25.4.3 APPLICATION OF SD TO SCPM AT ECHELON LEVEL

Eventually, this paper focuses on the unit of echelons considered by the various publications. Since Forrester ID, majority (72%) of authors have used the SD at single echelon (either manufacturing–retailer, supplier–manufacturer and retailer–customer level). Very few authors (28%) only used SD approach at integrated SC or multi-echelon SC (supplier–manufacturer–distributor–retailer–customer).

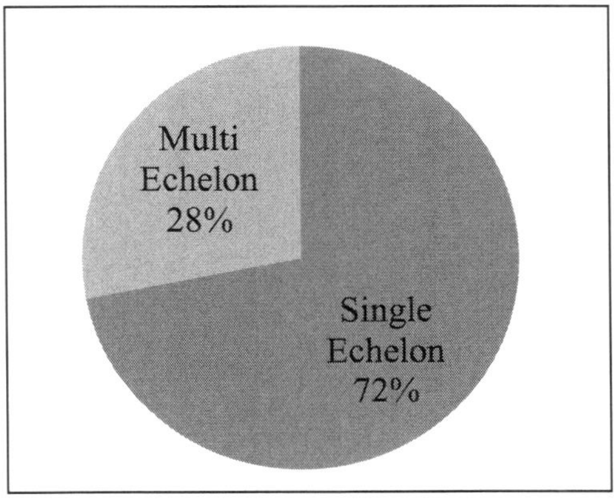

FIGURE 25.6 Application of SD to SCPM at echelon level

25.5 CONCLUSIONS AND FEATURED CHALLENGES

SD is a methodology that has been applied to different fields of research. This paper analyzed its applications to SCPM field through an extensive literature review. This paper has identified a critical mass of SD application to SCPM field. This literature review has put an evidence of how SD approach has to be considered in a flexible research methodology that can be soundly integrated with SCPM, simulation, and frameworks. This paper has classified the literature related to the SCPM using SD approach based on decision level, methods, and level of echelon. Based on the decision level, 60% of the articles have used SD approach at the operational level, 24% literature have used at the tactical level, and remaining 16% of the articles have used at the strategic level for last two decades. This paper separated the articles based on the research method. The majority (64%) of literature deals with the quantitative measures and 36% of articles are related to the quantitative measures. Finally, articles are separated based on the application level of SC. Number (72%) of articles have used SD at single-echelon SC by comparing with the multi-echelon SC.

Despite the current value of this paper, future research in support of the different application of SD modeling to provide a greater insight into the SCPM is clearly necessary. Moreover, vast research scope is there to carry out the research in the field of focal SC performance metrics modeling using SD.

KEYWORDS

- **supply chain performance modeling**
- **system dynamics**
- **supply chain measurement and supply chain modeling**

REFERENCES

1. Abdel-Hamid, T. K. The Dynamics of Software Development Project Management: An Integrative System Dynamics Perspective, Ph.D. Thesis, Massachusetts Institute of Technology, 1984.
2. Abu Bakar, A. H.; Lukman Hakim, I.; Chong, S. C.; Lin, B. Measuring Supply Chain Performance Among Public Hospital Laboratories. *Int. J. Prod. Perform. Manage.* **2010**, *59*(1), 75–97.

3. Angerhofer, B. J.; Angelides, M. C. A Model and a Performance Measurement System for Collaborative Supply Chains. *Decis. Support Syst.* **2006**, *42*, 283–301.
4. Askariazad, M.; Wanous, M. A Proposed Value Model for Prioritising Supply Chain Performance Measures. *Int. J. Bus. Perform. Supply Chain Modell.* **2009**, *1*(2/3), 115–128.
5. Barlas, Y.; Aksogan, A. Product Diversification and Quick Response Order Strategies in Supply Chain Management. Bogazici University.**1997**, http://ieiris.cc.boun.edu.tr/faculty/barlas (accessed Aug 27, 1999).
6. Berrah, L.; Cliville′, V. Powards an Aggregation Performance Measurement System Model in a Supply Chain Context. *Comput. Ind.* **2007**, *58*, 709–719.
7. Bhagwat, R.; Sharma, M. K. An Application of the Integrated AHP-PGP Model for Performance Measurement of Supply Chain Management. *Prod. Plann. Control* **2009**, *20*(8), 678–690.
8. Bigliardi, B.; Bottani, E. Performance Measurement in the Food Supply Chain: a Balanced Scorecard Approach. *Fa-cilities* **2010**, *28*, (5/6), 249–260.
9. Bullinger, H. G.; Kuhner, M.; Van Hoof, A. Analysing Supply Chain Performance Using a Balanced Measurement Method. *Int. J. Prod. Res.* **2002**, *40*(15), 3533–3543.
10. Chan, F. T. S. Performance Measurement in a Supply Chain. Int. *J. Adv. Manuf. Technol.* **2003**, *21*, 534–548.
11. Chan, F.; Qi, H.; Chan, H.; Lau, H.; Ip, R. A Conceptual Model of Performance Measurement for Supply Chains. *Manage. Decis.* **2003**, *41*(7), 635–642.
12. Chopra, S.; Meindl, P. *Supply Chain Management*, 3rd ed.; Pearson Education: India, 2007.
13. Das, D.; Dutta, P. A System Dynamics Framework for Integrated Reverse Supply Chain with Three-Way Recovery and Product Exchange Policy. *Comput. Ind. Eng.* **2013**, *66*(4), 720–733.
14. Dasgupta, T. Using the Six-Sigma Metric to Measure and Improve the Performance of a Supply Chain. *Total Qual. Manage. Bus. Excellence* **2003**, *14*(3), 355–366.
15. Dill, M. *Capital Investment Cycles: A System Dynamics Modelling Approach to Social Theory Development*. International System Dynamics Conference, Istanbul, Turkey, 1997.
16. Drzymalski, J.; Odrey, N. G.; Wilson, G. R. Aggregating Performance Measures of a Multi-Echelon Supply Chain Using the Analytical Network and Analytical Hierarchy Process. *Int. J. Serv. Econ. Manage.* **2010**, *2*, (3/4), 286–306.
17. Ford, A.; Lorber, H. W. Methodology for Analysis of Impacts of Electric Power Production in the West. Environmental Protection Agency Conference 1997.
18. Forrester, J. W. Industrial Dynamics: A Major Breakthrough for Decision Makers. *Harv. Bus. Rev.* **1961**, *36*(4), 37–66.
19. Ge, Y.; Yang, J.-B.; Proudlove, N.; Spring, M. System Dynamics Modelling for Supply-Chain Management: A Case Study on a Supermarket Chain in the UK. *Int. Trans. Oper. Res.* **2014**, 495–509.
20. Gunasekaran, A.; Patel, C.; McGaughey, R. A. A Framework for Supply Chain Performance Measurement. *Int. J. Prod. Econ.* **2004**, *87*(3), 333–347.
21. Hansen, J. E.; Bie, P. Distribution of Body Fluids, Plasma Protein, and Sodium in Dogs: a System Dynamics Model. *Syst. Dyn. Rev.* **1987**, *3*(21), 116–135.
22. Janamanchi, B.; Burns, J. R. Control Theory Concepts Applied to Retail Supply Chain: A System Dynamics Modelling Environment Study. Hindawi Publishing Corporation

Modelling and Simulation in Engineering. 2013, *2013*, Article ID: 421350, 14 pages. http://dx.doi.org/10.1155/2013/421350.
23. Kamath, N. B.; Roy, R. Capacity Augmentation of a Supply Chain for a Short Lifecycle Product: A System Dynamics Framework. *Eur. J. Oper. Res.* **2007**, *179*, 334–351.
24. Lai, K. H.; Ngai, E. W. T.; Cheng, T. C. E. Measures for Evaluating Supply Chain Performance in Transport Logistics. *Transp. Res. Part E*, **2002**, *38*, 439–456.
25. Lia, C.; Rena, J.; Wangb, H. A System Dynamics Simulation Model of Chemical Supply Chain Transportation Risk Management Systems. *Comput. Chem. Eng.* **2016**, *89*, 71–83.
26. Lin, L. C.; Li, T. S. An Integrated Framework for Supply Chain Performance Measurement Using Six-Sigma Metrics. *Software Qual. J.* **2010**, *18*, 387–406.
27. Mosekilde, E.; Larson, E. R.; Sterman, J. D. *Coping with Complexity: Deterministic Chaos in Human Decisionmaking Behaviour*; CRC Press: Boston, 1991; 199–229.
28. Parkan, C.; Wang, J. Gauging the Performance of a Supply Chain. *Int. J. Prod. Qual. Manage.* **2007**, *2*(2), 141–176.
29. Persson, F.; Olhager, J. Performance Simulation of Supply Chain Designs. *Int. J. Prod. Econ.* **2002**, *77*(3), 231–245.
30. Shepherd, C.; Günter, H. Measuring Supply Chain Performance: Current Research and Future Directions. *Int. J. Prod. Perform. Manage.* **2006**, *55*(3), 242–258.
31. Soni, G.; Kodali, R. Internal Benchmarking for Assessment of Supply Chain Performance. *Benchmarking: Int. J.* **2010**, *17*(1), 44–76.
32. Sterman, J. D. *Business Dynamics: Systems Thinking and Modelling for a Complex World*. McGraw-Hill: Burr Ridge, Illinois, 2000.
33. Thakkar, J.; Kanda, A.; Deshmukh, S. G. Supply Chain Performance Measurement Framework for Small and Medium Scale Enterprises. *Benchmarking: Int. J.* **2009**, *16*(5), 702–723.
34. Theeranuphattana, A.; Tang John, C. S. A Conceptual Model of Performance Measurement for Supply Chains Alternative Considerations. *J. Manuf. Technol. Manage.* **2008**, *19*(1), 125–148.
35. Towill, D. R. Industrial Dynamics Modeling of Supply Chains. *Logistics Inf. Manage.* **1996a** *9*(4), 53–56.
36. Varma, S.; Wadhwa, S.; Deshmukh, S. G. Evaluating Petroleum Supply Chain Performance Application of Analytical Hierarchy Process to Balanced Scorecard. *Asia Pac. J. Mark. Logistics* **2008**, *20*(3), 343–356.
37. Wolf-Rudiger, B.; Karim, B. Sustainable Logistics: Responses to a Global Challenge. Springer Science & Business Media, 2012.
38. Wong, W. P. Performance Evaluation of Supply Chain in Stochastic Environment: Using a Simulation Based DEA Framework. *Int. J. Bus. Perform. Supply Chain Modell.* **2009**, 1(2/3), 203–228.
39. Wong, W. P.; Wong, K. Y. A Review on Benchmarking of Supply Chain Performance Measures. Benchmarking: Int. J. **2008**, 15(1), 25–51.
40. Xu, J.; Li, B.; Wu, D. Rough Data Envelopment Analysis and its Application to Supply Chain Performance Evaluation. *Int. J. Prod. Econ.* **2009**, *122*, 628–638.
41. Yang, F.; Wu, D.; Liang, L.; Bi, G.; Wu, D. D. Supply Chain DEA: Production Possibility Set and Performance Evaluation Model. *Ann. Oper. Res.* **2009,** DOI: 10.1007/s10479-008-0511-2.

APPENDIX

APPENDIX.1

S.No	Author(s)	Year	Journal	Country	Strategic level	Tactical leve	erational le	Qualitative	Quantitative	Single echlo	Multi echlon	Review
1	Denis R. Towill	1996	International Journal of Physical Distribution & Logistics Management	UK		✓			✓		✓	
2	K. Hafeeza, M. Griffithsb, J. Griffithsc and M.M. Naimd	1996	Int. J. Production Economics	UK			✓		✓		✓	
3	C Lin, T S Baines, J and O'Kane, D Link	1998	International Conference on Simulation	UK			✓		✓	✓		
4	Benita M. Beamon	1998	Int. J. Production Economics	USA		✓		✓		✓		✓
5	Bernhard J. Angerhofer and Marios C. Angelides	2000	Proceedings of the 2000 Winter Simulation Conference	UK			✓		✓	✓		✓
6	Shotaro Minegishi and Daniel Thiel	2000	Simulation Practice and Theory	France			✓		✓		✓	
7	Jack G.A.J. van der Vorst , Adrie J.M. Beulens and Paul van Beek	2000	European Journal of Operational Research	The Netherlands	✓			✓		✓		
8	B. Johansson, S. Jain, J. Montoya-Torres, J. Hugan, and E. Yücesan, eds	2000	Proceedings of the 2010 Winter Simulation Conference	Italy			✓	✓			✓	
9	Y. Ge, J.-B. Yang, N. Proudlove and M. Spring	2004	International transtions on operational research	UK			✓	✓			✓	
10	U.M.Bhushi and C.M.Javalagi	2004	International Engineering Management Conference	India			✓	✓		✓		✓
11	Yong Zhang and David Dilts	2004	Information systems and e- Business Management	USA			✓		✓		✓	
12	Ashish Agarwal and Ravi Shankar	2005	Asian Academy of Management Journal	UK			✓		✓	✓		
13	Patroklos Georgiadis, Dimitrios Vlachos and Eleftherios Iakovou	2005	Journal of Food Engineering	Greece			✓		✓	✓		
14	Henk Akkermans and Nico Dellaert	2005	System Dynamics Review	The Netherlands	✓			✓		✓		✓
15	Thi Le Hoa Vo and Daniel Thiel	2006	System Dynamics conference	France		✓			✓		✓	
16	Bernhard J. Angerhofer, Marios C. Angelides	2006	Decision Support Systems	UK			✓		✓	✓		
17	D. R. Towill	2007	International Jouranl of computer intigrated manufacturing	UK			✓		✓	✓		
18	Mustafa O¨ zbayrak , Theopisti C. Papadopoulou , and Melek Akgun	2007	Simulation Modelling Practice and Theory	UK			✓		✓	✓		
19	Mustafa o zbayrak,Theopisti C.Papadopoulou and Melek Akgun	2007	Simulation Modelling Practice and Theory	Turkey			✓		✓		✓	
20	Narasimha B. Kamath and Rahul Roy	2007	European Journal of Operational Research	India	✓				✓	✓		
21	Dimitrios Vlachos, Patroklos Georgiadis and Eleftherios Iakovou	2007	Computers & Operations Research	Greece		✓			✓	✓		
22	T. S. Baines & D. K. Harrison	2007	Production Planning & Control: The Management of Operations	UK		✓		✓		✓		✓
23	L. Rabelo, M. Helaly, C. Lertpattarapongz, R. MoragaxAnd A. Sarmientoy	2008	International Journal of Production Research	USA		✓		✓		✓		
24	Jafar Mahmod & Mohamad Hosein Minaee	2010	International Journal of Industrial Engineering & Production Research	Iran			✓		✓	✓		
25	Ahmad Norang, Mohammad Ali Eghbali andAmir Hajian	2010	Logistics Systems and Intelligent Management Conference	Iran	✓			✓		✓		
26	Antuela A. Tako and Stewart Robinson	2011	Decision support systems	UK			✓		✓	✓		✓
27	Anna Corinna Cagliano Alberto DeMarco Carlo Rafele Sergio Volpe	2011	Journal of Manufacturing Technology Management	Italy		✓			✓		✓	
28	Yang Feng	2012	Physics Procedia	China			✓		✓		✓	
29	Rajbir Singh Bhatti, Pradeep Kumar and Dinesh Kumar	2012	Int. J. Indian Culture and Business Management	India		✓		✓		✓		
30	Shidi Miao and Chunxian Teng, Lu Zhang	2012	Journal of software	China			✓		✓		✓	
31	Josefa Mula , Francisco Campuzano-Bolarin & Katerine M. Carpio	2013	International Journal of Production Research	spain			✓	✓		✓		
32	Behrooz Asgari and Md Aynul Hoque	2013	Ritsumeikan Journal of Asia Pacific Studies	Bangladesh	✓			✓		✓		
33	Wen Wang, Weiping Fu, Hanlin Zhang and Yufei Wang	2013	Research Journal of Applied Sciences, Engineering and Technology	China	✓			✓		✓		
34	Roberto Poles	2013	Int. J. Production Economics	Australia			✓		✓	✓		
35	Yihui Tian a, Kannan Govindan b and Qinghua Zhu	2014	Journal of Cleaner Production	China			✓		✓	✓		
36	Mohammed G. Sayed	2014	International Journal of Computer and Information Technology	France			✓		✓	✓		
37	Ghada Elkady, Jonathan Moizer, and Shaofeng Liu	2014	International Journal of Innovation, Management and Technology	UK			✓		✓	✓		
38	Yanxin Wang, Minfang Huang and Jianbin Chen	2014	International Journal of u- and e- Service, Science and Technology	China			✓		✓	✓		
39	Salvatore Cannella, Jose M. Framinan and Manfredi Bruccoleri	2014	European Journal of Operational Research	spain	✓				✓	✓		
40	Sujit Singh and Puja Chhabra Sharma	2015	International Journal of Humanities and Social Science	Malaysia			✓	✓			✓	
41	Golman Rahmanifar1, Babak Shirazi and Hamed Fazlollahtabar	2015	Int. J. Sensing, Computing & Control	Iran			✓	✓		✓	✓	
42	Rafi Rahanandeh Poor Langroodi and Maghsoud Amiri	2015	Expert Systems With Applications	Iran		✓			✓	✓		
43	Saeed Rahimpour Golroudbary and Seyed Mojib Zahraee	2015	Simulation Modelling Practice and Theory	Malaysia		✓		✓			✓	
44	Balaji Janamanchi and James R. Burns	2016	Cogent Business & Management	UK			✓		✓		✓	
45	Chihhao Fan, Shu-Kai S. Fanb, Chen-Shu Wang and Wen-Pin Tsai	2016	Resources, Conservation and Recycling	Taiwan			✓		✓	✓		
46	Virginia L.M. Spiegler, Mohamed M. Naim, Denis R. Towill and Joakim Wikner,	2016	European Journal of Operational Research	UK			✓		✓	✓		
47	M.D Aynul Hoque and Qantara Khaleda Khan	2016	British Journal of Business Design & Education	Bangladesh		✓		✓		✓		✓
48	Esra Ekinci Adil Baykasoglu	2016	Kybernetes	Turkey	✓				✓	✓		
49	An-quan Zou, Xing-ling Luo and Wan-tong Zou	2016	Proceedings of the 22nd International Conference on IEM	China			✓		✓	✓		
50	Chaoyu Li, Jun Ren and Haiyan Wangb	2016	Computers and Chemical Engineering	UK		✓		✓		✓		

CHAPTER 26

GEL CASTING OF Si_3N_4–SiO_2 CERAMIC COMPOSITES AND EVALUATION CHARACTERISTICS

NAGAVENI THALLAPALLI[1,*] and C. S. P. RAO[2,3]

[1]*Department of Mechanical Engineering, University College of Technology, Osmania University, Hyderabad, 500007, Hyderabad, India*

[2]*Department of Mechanical Engineering, National Institute of Technology, Warangal, 506004, Warangal, India*

[3]*cspr@nitw.ac.in*

[*]*Corresponding author. E-mail: tnagaveni@gmail.com*

ABSTRACT

Si_3N_4–SiO_2 ceramic composites were prepared by gel-casting method and pressure-less sintering using Si_3N_4 and SiO_2 as starting powders, and Al_2O_3 and Y_2O_3 as sintering additives. The effects of SiO_2 on the microstructure and dielectric properties of Si_3N_4–SiO_2 ceramic were investigated. After sintering for 2 h at 1500°C, Si_3N_4–SiO_2 ceramic composites, having a bending strength of 60.59~178.47 MPa and a dielectric constant of 3.2~6.8 at 1~30 MHz frequency were obtained. β-Si_3N_4 grains and Si_2N_2O phase make the sintered body. The formation of β-Si_3N_4 grains was demonstrated by X-ray diffraction (XRD) and scanning electron microscopy (SEM). The influence of SiO_2 content on the dielectric and mechanical properties has also been discussed in this work. The mechanical properties of Si_3N_4-based ceramics are found to decrease with the increase of SiO_2 content, while the dielectric properties are improved. The low dielectric properties of porous Si_3N_4–SiO_2 ceramic composites made them one of the most candidate materials for wave transparent applications.

26.1 INTRODUCTION

Silicon nitride (Si_3N_4) ceramics are known for its high strength, fracture toughness, strength at elevated temperatures, chemical resistance, and excellent ablation resistance, which make them suitable as radome material, but a relatively high dielectric constant and dielectric loss tangent at elevated temperatures limit its applications as a wave transparent materials.[1–5] SiO_2 ceramics have rather low-dielectric constant, high-chemical stability, and extremely low coefficient of thermal expansion and thermal conductivity than Si_3N_4 ceramics.[6–8] However, it has low flexural strength (less than 80 MPa) and poor rain erosion resistance, which cannot meet the requirements of radome used in high speed missiles. Thus, the Si_3N_4–SiO_2 ceramic composites may have excellent properties because of combined benefits of both the materials.

Gel casting is a near-net-shape technique to fabricate homogeneous and complex-shaped ceramic parts[9–10] and is one of the most effective ways to increase the ceramic reliability and to reduce the production cost.[11]

Different techniques have been developed to fabricate Si_3N_4–SiO_2 composites for structural and functional applications. Qi et al. developed three-dimensional quartz fiber-reinforced silicon nitride composites using infiltration and pyrolysis method by polyhydridomethylsilazane[12] and perhydropolysilazane.[13] They observed the formation of silicon oxynitride in the fiber/matrix interfaces showing a high flexural strength of 114.5 MPa and non-brittle failure behavior. Liu et al.[14] prepared silicon dioxide fiber-reinforced silicon nitride matrix (SiO_2/Si_3N_4) composite by chemical vapor infiltration (CVI) process using the $SiCl_4$-NH_3-H_2 system. Li et al. developed a process combining oxidation-bonding and solgel infiltration-sintering to fabricate Si_3N_4–SiO_2 ceramics. The ceramic with a porosity of 23.9% attained a flexural strength of 120 MPa, a Vickers hardness of 4.1 GPa, a fracture toughness of 1.4 $MPam^{1/2}$, and a dielectric constant of 3.80 with a tangent loss of 3.11×10^{-3} at a resonant frequency of 14 GHz.[15] Recently, they also studied the effect of infiltration time on mechanical and dielectric properties.[16] Cai et al.[17] fabricated porous Si_3N_4–SiO_2 ceramics at low cost by combining carbothermal reduction and solgel infiltration-sintering by using nano-sized diatomite grains as a silica source and dextrin as a carbon source. Zhang et al.[18] prepared a coating on porous Si_3N_4-based substrate via dip-coating from silicon slurry, which was prepared by adding pre-treated silicon powder into alumina sol. The chemical coating was synthesized by nitridation of silicon and the solid solution reaction of β-Si, Al_2O_3, and Si_2N_2O. Zou et al.[19]

prepared initially porous Si_3N_4 ceramics by gel casting, and amorphous SiO_2 was introduced into porous Si_3N_4 frames by solgel infiltration. The increase of density and the formation of well-distributed micropores with both uniform pore size and smooth pore wall, they achieved a high mechanical strength and low dielectric constant. Yuchen et al. prepared Si_2N_2O–Si_3N_4 in situ composites by gel-casting process, the composite attained flexural strength 230.46 ± 13.24 MPa and dielectric constant varied from 4.34 to 4.59 from room temperature to elevated temperature (1150°C).[20,21]

In this study, the authors have proposed and conducted tests to study the effects of SiO_2 in Si_3N_4–SiO_2 ceramic composites for a most suitable material having low dielectric constant, low coefficient of thermal expansion, and high flexural strength. The effect of SiO_2 content on the phase composition, microstructures, mechanical properties, dielectric properties of the Si_3N_4–SiO_2 ceramic composites are discussed in the present work.

26.2 EXPERIMENTAL PROCEDURE

All samples used in this experiment were fabricated by gel casting from Si_3N_4 and SiO_2. Premix solution was prepared by mixing methacrylamide (MAM) and *N*, *N'*-methylenebisacrylamide (MBAM) in distilled water with a 0.8 wt.% of Dolapix A88.[22]The premixed aqueous solution contained 10 wt.% monomer content, and the MAM to MBAM ratio was set as 5:1. Ceramic powders were added to premix solution maintaining the solid loading of the suspensions as 50 vol.%. The raw materials required are listed in Table 26.1.

The material codes are listed in Table 26.2; S0, S5, S10, and S15 denote the SiO_2 is added at 0, 5, 10, and 15 vol.%. In each experiment, 3% of Y_2O_3 and 2% of Al_2O_3 were used as sintering aids. In order to break down agglomerates and to achieve good homogeneity the suspensions were put for 24 h in a plastic bottle (ball mill). Hydrochloric acid and ammonia were used to adjust the pH value to approximately 10.5. The resultant suspension was de-aired for 5–10 min, and an initiator (10 wt.% aqueous solution of ammonium persulfate, APS) and catalyst (tetramethylethylenediammine, TEMED) were introduced into it. Immediately, the suspension was cast into nonporous mold and dried under controlled humidity. For the binder burnout, a temperature of 550°C was chosen, with a heating rate of 2°C/min for 2 h and then, the samples were sintered at 1500°C for 2 h in N_2 atmosphere, followed by natural cooling.

TABLE 26.1 Raw Materials Used in the Experiments Materials.

Materials	Function	Supplier
Silicon nitride (Si_3N_4)	Starting powder	Denka Kogaku, Japan, d_{50}~4.2 µm
Silicon dioxide (SiO_2)	Starting powder	Sigma Aldrich, USA, 99.5%; purity, particle size 1-5 µm
Yttriumoxide (Y_2O_3)	Sintering aid	Alfa Aesar, USA, average particle size 50 nm
Alumina (Al_2O_3)	Sintering aid	Alfa Aesar, USA, average particle size100 nm
Methabisacrylamide (MAM)	Monomer	MAM, Sigma Chem. Co., USA
N,N′-Methylenebisacrylamide (MBAM)	Cross-linker	MBAM, Sigma Chem. Co., USA
Darvan A88(amino alcohol)	Dispersant	Zschimmer & Schwarz GmbH
Distilled water	Solvent	
Ammonium persulfate (APS)	Initiator	APS, Sigma Aldrich, USA
N,N,N′,N′-tetramethylethylenediamine (TEMED)	Catalyst	Sigma Aldrich, USA

TABLE 26.2 The Composition Ratio of the Experimental Material.

Sample codes	Si_3N_4 (vol.%)	SiO_2 (vol.%)	Y_2O_3 (vol.%)	Al_2O_3 (vol.%)
S0	95	0	3	2
S5	90	5	3	2
S10	85	10	3	2
S15	80	15	3	2

26.3 TERIALS CHARACTERIZATION

The densities of green and sintered body were determined by Archimedes' displacement method and X-ray diffraction (XRD); Rigaku, D/max2550HB +/PC was used for determining the composition of sintered samples and scanning electron microscopy (SEM); JEOL, JSM-6460LV was used for the study of microstructure. In order to determine flexural strength of the samples, the three-point bending test at room temperature was carried out following the ASTM D-143 (1996) standard on specimen with dimensions of $40 \times 4 \times 3$ mm at a crosshead speed of 0.5 mm/min on an electronic Universal Testing Machine (Dak Systems, India). The fracture toughness (KIC) was tested by single edge notched beam (SENB) method for specimens with dimension of $2.5 \times 5 \times 30$ mm and notch length of 2.5 mm. The loss tangent and dielectric constant were measured using an impedance analyzer (MTZ-35, Biologic Science Instruments, France) in the frequency range of 0.1~35 MHz.

26.4 RESULTS AND DISCUSSION

The XRD patterns of Si_3N_4–SiO_2 ceramics are shown in Figure 26.1, and it is revealed that the main crystal phase of the composites is β-Si_3N_4, Si_2N_2O and small amounts of SiO_2, cristobalite and $Y_2Si_2O_7$ are present. The diffraction intensities of Si_2N_2O increase with an increase in SiO_2 content. The relative content of α-Si_3N_4 and β-Si_3N_4 is listed in Table 26.2. The formation of micro cracks due to the presence of cristobalite can be the major reason for decreased strength of Si_3N_4–SiO_2 ceramics compared to other Si_3N_4 matrix composites. The presence of Si_2N_2O in the monolithic ceramic (Fig. 26.1a), can be explained from the following reaction which occurs during the sintering process between Si_3N_4 and SiO_2 as[18]

$$SiO_2\ (l) + Si_3N_5\ (s) \rightarrow 2Si_2N_2O \qquad (26.1)$$

Actually, amorphous SiO_2 starts to crystallize and forms cristobalite at temperature more than 1300°C, and its nucleation is heterogeneous,[23,24] it may appear from the surface of SiO_2 and extends inwards.[25] From Figure 26.1, we can also see the peak of cristobalite when SiO_2 is added and becomes strong as the SiO_2 content increases, and it means that an extensive crystallization of SiO_2 occurs.[26]

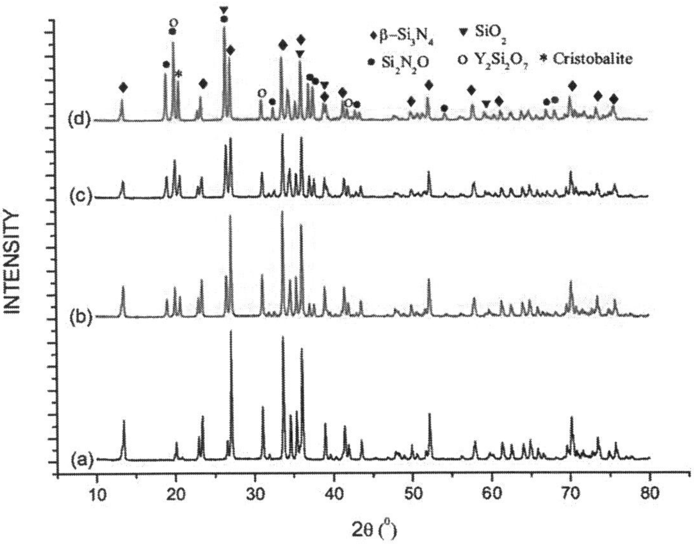

FIGURE 26.1 X-ray diffraction patterns for different SiO_2 content (a) S0, (b) S5, (c) S10, (d) S15.

26.4.1 DENSITY, POROSITY, AND SINTERING SHRINKAGE

The variation of bulk density and porosity of Si_3N_4–SiO_2 composite ceramics with SiO_2 content is shown in Figure 26.2. The density is found to decrease while the porosity increases with the increase in SiO_2 content. The transformation of α-Si_3N_4 into β-Si_3N_4 and its densification was obstructed by the formation of crystalline Si_2N_2O during the liquid phase. From the rule of mixture, the density of SiO_2 was lower than Si_3N_4.This also might be the reason for decreasing the bulk density as the SiO_2 content increases.

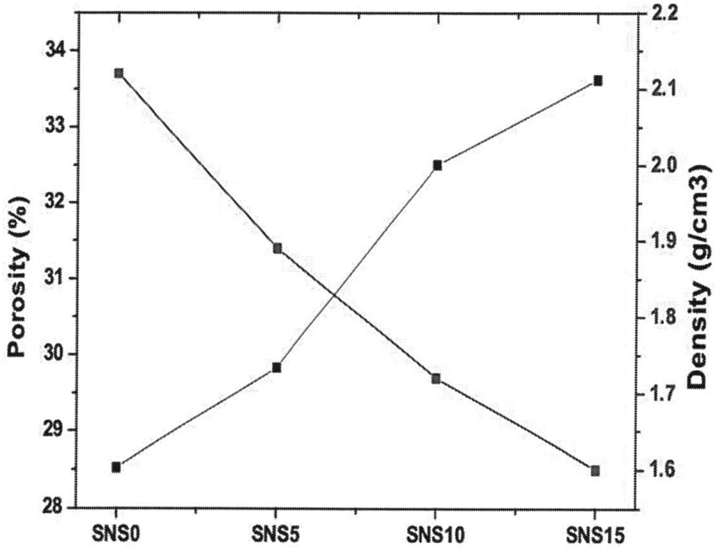

FIGURE 26.2 Porosity and bulk density of ceramics as a function of SiO_2 content.

26.4.2 MICROSTRUCTURE

The fracture morphology as observed by SEM for Si_3N_4–SiO_2 composites are shown in Figure 26.3 for various SiO_2 content. It can be seen that SiO_2 and Si_3N_4 are arranged homogenously in the composite and also in the pores. It is also observed that the pores are evenly distributed in the matrix and with an average size of 4 μm. A uniform microstructure and small amount of β-Si_3N_4, due to its elongated grains, are also observed in SEM micrograph, in Figure 26.3a–d.

In the microstructure analysis, at high aspect ratios, minor columnar grains and bonded particles can be observed, as α/β phase transformations. Stacking of Si_3N_4 particles due to the uniform distribution and connection of open pores in the matrix structure were also observed.

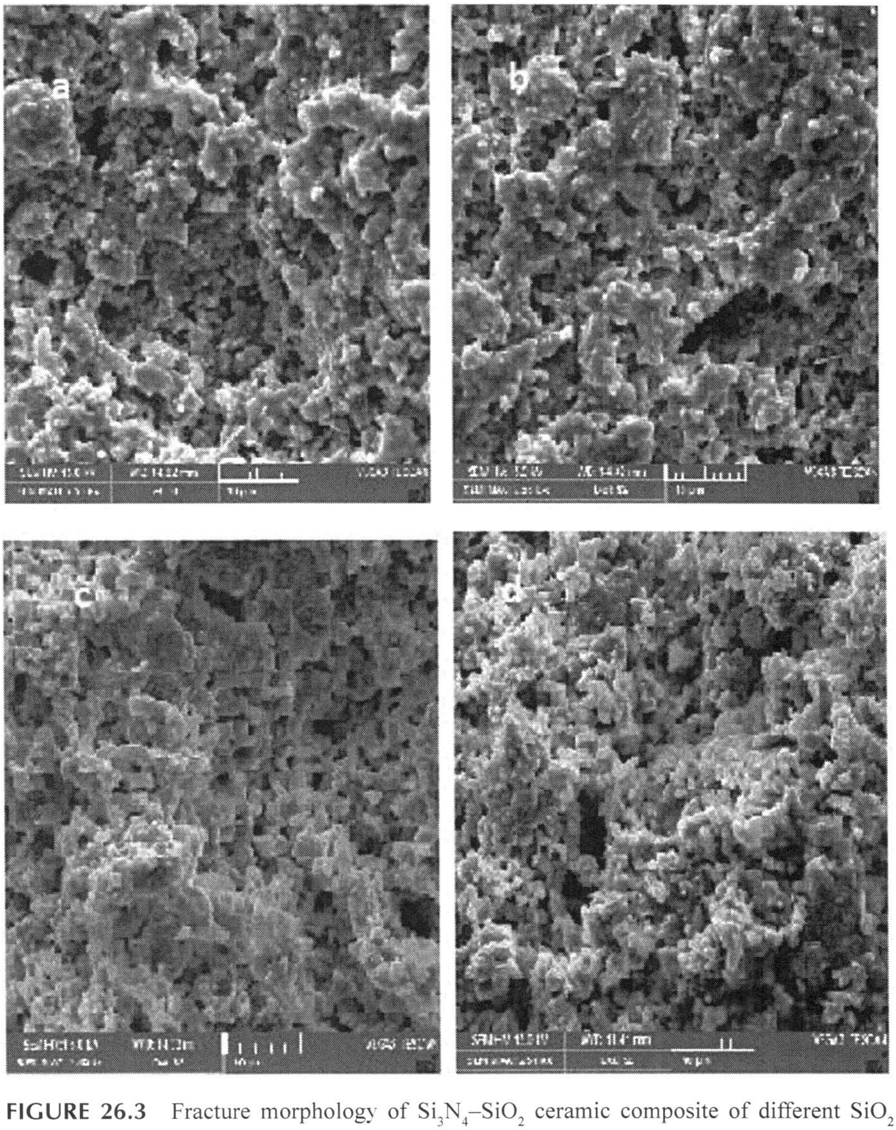

FIGURE 26.3 Fracture morphology of Si_3N_4–SiO_2 ceramic composite of different SiO_2 content. (a) S0, (b) S5, (c) S10, (d) S15.

Some micropores were also observed in SEM analysis, and these may be due to the distribution of organic binder. It is thermodynamically possible to form Si_2N_2O in liquid phase from SiO_2 and α-Si_3N_4 or β-Si_3N_4. A distorted (SiN_3O) tetrahedral structure builds the Si_2N_2O crystal, whereas an oxygen atom was replaced by (SiN_4) tetrahedron in α- and β-Si_3N_4; thus, a tetrahedral (SiN_3O) structural unit of Si_2N_2O was formed. Therefore, the Si_2N_2O nuclei are formed in the liquid phase at the sintering temperature[27]. The grain growth rate of Si_2N_2O is found to be raised after the Si_2N_2O nucleation. The formation of Si_2N_2O enhances due to precipitation process in solution phase in the vicinity of rich liquid phase of SiO_2. The formation of the crystalline Si_2N_2O largely consumed the liquid phases and thus the transformation of α-Si_3N_4 into β-Si_3N_4 and densification are obstructed. This also may be the reason for increasing the porosity due to increase in SiO_2 content.

26.4.3 MECHANICAL PROPERTIES

The sintered materials exhibit a brittle fracture and the fracture surface has not obvious grain pull-out. Micropores can be seen when the amount of SiO_2 reaches 15 wt.% (Fig. 26.3). This may be due to the obstruction of densification and; hence, decrease in flexural strength. Hardness and flexural strength are tested on all the prepared samples differently with respect to the size of particle. The effect of SiO_2 powders on the flexural strength is shown in Figure 26.4.

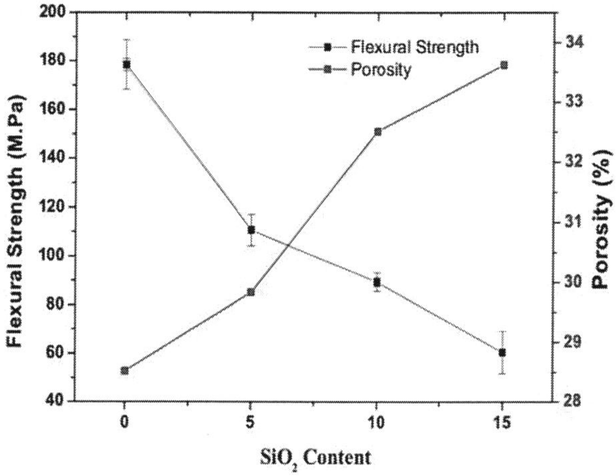

FIGURE 26.4 Flexural strength and porosity of Si_3N_4–SiO_2 ceramic composite as a function of Sio_2 content.

The flexural strength decreases with the increase of SiO_2 powder as SiO_2 is softer than Si_3N_4 matrix. Figure 26.4 shows the flexural strength of the Si_3N_4–SiO_2 composite that ranges from 60.59 to 178.47 MPa. The effective loading across the section of the material has been found reduced due to the presence of pores in the ceramic composite. Therefore, the more is the porosity, the less is the strength. Due to large mismatch of coefficients of thermal expansion between Si_3N_4 and cristobalite formed, this induces the formation of micro cracks, which may also lead to decrease in flexural strength.

26.4.4 DIELECTRIC PROPERTIES

The dielectric loss and dielectric constant (C_{de}) are two important and complimentary properties of a dielectric material. If a material possesses a low dielectric loss then, it is an excellent material for radome as it has the ability to convert the absorbed electromagnetic energy into heat. Figures 26.5 and 26.6 show dielectric constant and loss tangent of the Si_3N_4–SiO_2 ceramic composites with the various wt.% SiO_2 content with respect to frequency.

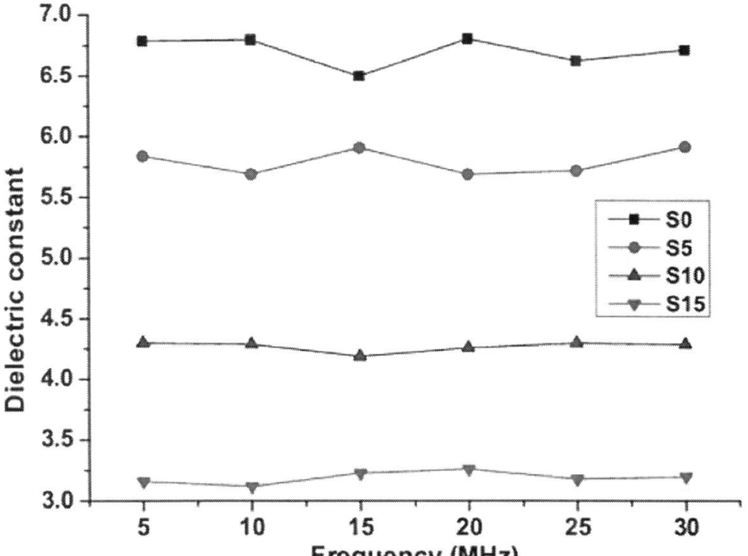

FIGURE 26.5 Dielectric constant of Si_3N_4–SiO_2 ceramic composite as a function of SiO_2 conten8t.

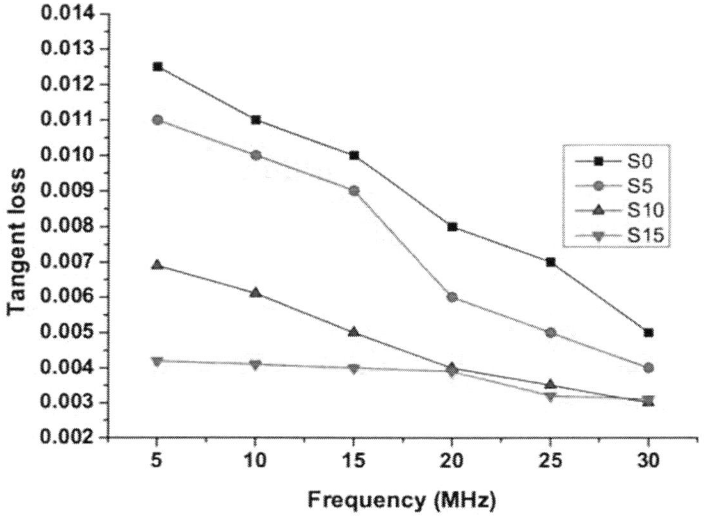

FIGURE 26.6 Tangent loss of Si_3N_4–SiO_2 ceramic composite as a function of SiO_2 content.

The C_{de} value of porous Si_3N_4 is highly influenced by the porosity, Si_3N_4, and phase of Si_2N_2O and cristobalite. It can also be seen in Figure 26.5 that the C_{de} value of composite decreases to the max extent when the porosity is 1. When SiO_2 content increases from 0 to 15 wt.%, the C_{de} of porous Si_3N_4–SiO_2 ceramic composite decreases from 6.8 to 3.2.

Porosity is the major influencing factor for the C_{de} of porous ceramics. SiO_2 ceramics are having lower dielectric properties compared to Si_3N_4 ceramics. Thus, the Si_3N_4–SiO_2 ceramic has a relatively lower C_{de} than that of Si_3N_4 ceramics.

Also, the C_{de} could be lower due to the increase of porosity, because of which we can treat pore as phase with a unit dielectric constant. It is also observed that the polarization mechanism is unchanged with frequency and; hence, the C_{de} does not vary with increasing frequency.

26.5 CONCLUSION

Si_3N_4–SiO_2 ceramic composites were prepared using gel-casting approach followed by pressure-less sintering. The contribution of the SiO_2 content for the mechanical and dielectric properties of the porous Si_3N_4–SiO_2 ceramic composites were studied and the following conclusions were drawn:

1. There is an increase in porosity and decrease in density with the increase of SiO_2 content. The formation of the crystalline Si_2N_2O is seen in liquid phase, so the transformation of α-Si_3N_4 into β-Si_3N_4 and densification are obstructed. This also may be the reason for increase in the porosity as the SiO_2 content increases.
2. The mechanical properties are found to be lower with the increase of SiO_2 content.
3. The dielectric properties may be improved by addition of SiO_2 powder.

The microstructures of the sample are uniform, the pores are distributed evenly. The wt.% of SiO_2 content was in the range of 0–15%, it is found that the flexural strength and dielectric constant of Si_3N_4–SiO_2 composites are in a range of 60.6–178.5 MPa and 3.2–6.8, respectively.

KEYWORDS

- **gelcasting**
- **silicon nitride**
- **ceramic composite**
- **Si_2N_2O**

REFERENCES

1. Greil, P. Processing of Silicon Nitride Ceramics. *Mater. Sci. Engg.* **1989,** *A109*, 27–35.
2. Boberski, C.; Hamminger, R.; Peuckert, M.; Aldinger, F.; Dillinger, R.; Heinrich, J.; Huber, J. High-Performance Silicon Nitride Materials. *Angew. Chem.* **1989**, *101*, 1592–1601.
3. Malghan, S.; Wang, P.; Sivakumar, A.; Somasundaran, P. Deposition of Colloidal Sintering-Aid Particles on Silicon Nitride. *Comps. Interfaces* **1993**, *1*, 193–210.
4. Ganesh, I. Near-Net Shape β-$Si_4Al_2O_2N_6$ Parts by Hydrolysis Induced Aqueous Gelcasting Process. *Int. Jour. App. Cer. Tech.* **2009**, *6*, 89–101.
5. Mishra, S.; Mallika, C.; Das, P.; Mudali, U. K.; Natarajan, R. Development and Characterization of Porous Silicon Nitride Tubes. *Trans. Indian Ceram. Soc.* **2013**, *72*, 52–55.
6. Bergman, B.; Heping, H. The Influence of Different Oxides on the Formation of Si_2N_2O from SiO, and S&N. *J. Eur. Ceram. Soc.* **1990**, *6*, 3–8.
7. Karakuş, N.; Kurt, A.O.; Toplan, H.O. Production of Sinterable Si3N4 from SiO2-Li2O-Y2O3 Mixture. *Mater. Manuf. Processes* **2012**, *27*, 797–801.

8. Jia, D. C.; Zhou, Y.; Lei, T. C. Ambient and Elevated Temperature Mechanical Properties of Hot-Pressed Fused Silica Matrix Composite. *J. Eur. Ceram. Soc.* **2003**, *23*, 801–808.
9. Janney, M. A.; Ren, W.; Kirby, G. H.; Nunn, S. D.; Viswanathan, S. Gelcast tTooling: nNet Shape Casting and Green Machining. *Mater. Manuf. Processes* **1998**, *13*, 389–403.
10. Nie, L.; Wang, Q.; Liu, J. Gelcasting of NiO=YSZ Tubular. Anode-Supports for Solid Oxide Fuel Cells. *Mater. Manuf. Processes* **2014**, *29*, 1153–1156.
11. Kandi, K. K.; Thallapalli, N.; Chilakalapalli, S. P. R. Development of Silicon Nitride-Based Ceramic Radomes—A Review. *Int. J. App. Ceram. Tech.* **2014**. DOI: 10.1111/ijac.12305.
12. Qi, G.; Zhang, C.; Hu, H. High Strength Three-Dimensional Silica Fiber Reinforced Silicon Nitride-Based Composites Via Polyhydridomethylsilzane Pyrolysis. *Ceram. Int.* **2007**, *33*, 891–894.
13. Qi, G. J. Interfacial Reaction Mechanisms of 3D SiO_{2f}/Si_3N_4 Composites Prepared by Perhydropolysilazane. *J. of Comp. Mater*. **2011**, 45, 1621–1626.
14. Liu, Y.; Cheng, L.; Zhang, L.; Xu, Y.; Liu, Y. Microstructure and Properties of Particle Reinforced Silicon Carbide and Silicon Nitride Ceramic Matrix Composites Prepared by Chemical Vapor Infiltration. *J. Univ. Sci. Tech. Beijing* **2007**, *14*, 454–459.
15. Li, X.; Yin, X.; Zhang, L.; Cheng, L.; Qi, Y. Mechanical and Dielectric Properties of Porous. Si_3N_4-SiO_2 Composite Ceramics. *Mater. Sci. Eng. A*, **2009**, *500*, 63–69.
16. Li, X.; Wu, P.; Zhu, D. Fabrication and Properties of Porous Si3N4–SiO2 Ceramics with Dense Surface and Gradient Pore Distribution. *Ceram. Int.* **2014**, *40*, 5079–5084.
17. Cai, Y.; Li, X.; Dong, J. Microstructure and Mechanical Properties of Porous Si3N4-SiO2 Ceramics Fabricated by a Process Combining Carbothermal Reduction and Sol-Gel Infiltration-Sintering. *Mater. Sci. Eng. A* **2014**, *601*, 111–115.
18. Zhang, C. C.; Li, X. L.; Ji, H. M.; Sun, X. Coating on Porous Si3N4 Based Substrate with Solgel Slurry. *Integr. Ferroelectr. Int. J.* **2012**, *138*, 111–116.
19. Zou, C.; Zhang, C.; Li, B.; Cao, F.; Wang, S. Improved Properties and Microstructure of Porous Silicon Nitride/Silicon Oxide Composites Prepared by Sol-Gel Route. *Mater. Sci. Eng. A* **2012**, *556*, 648–652.
20. ShuQin, L.; YuChen, P.; ChangQing, Y.; Jia Lu, L. Mechanical and Dielectric Properties of Porous Si_2N_2O–Si_3N_4 in Situ Composites. *Ceram. Int.* **2009**, *35*, 1851–1854.
21. Yuchen, P.; Shuqin, L.; Changqing, Y.; Jia lu, L. Thermal Shock Resistance of in Situ Formed Si_2N_2O–Si_3N_4 Composites by Gelcasting. *Ceram. Int.* **2009**, *35*, 3365–3369.
22. Singh, B. P.; Gaydardzhiev, S.; Ay, P. Stabilization of Aqueous Silicon Nitride Suspension with Dolapix A 88. *J. Dispersion Sci. Technol.* **2006**, *27*, 91–97.
23. Wagstaff, F. Crystallization and Melting Kinetics of Cristobalite. *J. Am. Ceram. Soc.* **1969**, *52*, 650–654.
24. Wagstaff, F. Crystallization Kinetics of Internally Nucleated Vitreous Silica. *J. Am. Ceram. Soc.* **1968**, *51*, 449–453.
25. Wang, L.Y.; Hon, M. H. The Effect of Cristobalite Seed on the Crystallization of Fused Silica Based Ceramic Core—a Kinetic Study. *Ceram. Int.* **1995**, *21*, 187–193.
26. Li, X.; Wu, P.; Zhu, D. The Effect of the Crystallization of Oxidation-Derived SiO_2 on the Properties of Porous Si_3N_4–SiO_2 Ceramics Synthesized by Oxidation. *Ceram. Int.* **2014**, *40*, 4897–4902.
27. Duan, R.; Roebben, G.; Vleugels, J.; Van der Biest, O. In Situ Formation of Si_2N_2O and TiN in Si3N4-Based Ceramic Composites. *Acta Mate.* **2005**, *53*, 2547–2554.

PART III
Thermal Engineering (TE)

CHAPTER 27

EXPERIMENTAL INVESTIGATION AND SIMULATION OF SPLIT INJECTION AT DIFFERENT INJECTION PRESSURES IN COMPRESSION IGNITION ENGINE FOR IMPROVING EMISSIONS

SUDHIR GIRISH AUTI[1,*], ASHOK J. KECHE[1], ARUN T. AUTEE[1], and HANUMANT M. DHARMADHIKARI[2]

[1]*Mechanical Engineering Department, Maharashtra Institute of Technology, Aurangabad, Maharashtra 431010, India*

[2]*Marathwada Institute of Technology, Aurangabad, Maharashtra 431010, India*

Corresponding author. E-mail: sudhirauti5@gmail.com

ABSTRACT

Emission control poses a challenge in the design of compression ignition (CI) engines due to stringent emission regulations, as the control over the emissions, that is, nitrogen oxide (NO_x) and particulate matter (PM) is very difficult in CI engines. This work is aimed at experimentally studying the effects of different split injection methods and dwell in between of the fuel injections on emissions from CI engine and combustion. These experiments are performed at different injection pressures. Various split injection methods at different injection pressure were employed by attaching the electronic fuel injection system to the engine. Moreover, computational fluid dynamics technique is used to explain the various experimental results. This study shows that 2-shot split injection method results in significant reduction in NO_x with an acceptable change in brake thermal efficiency (BTE), PM, and carbon monoxide (CO). This study also shows that injection pressure affects the emissions of NO_x and PM considerably.

27.1 INTRODUCTION

Compression ignition (CI) engines, also termed as diesel engines, are used to operate high-power vehicles due to enhanced fuel economy, thermal efficiency, and reliability. Hence, these engines are becoming preferred engines for many light-duty vehicles. Various literatures show that CI engines discharge lower amount of carbon monoxide (CO), carbon dioxide (CO_2), and hydrocarbons (HC) than gasoline engines. Conversely, the control over NO_x and particulate matter (PM) is more complicated in CI engines compared to gasoline engines,[1] whereas the regulations for pollutant emissions for CI engines have become more and more stringent, especially in last two decades. These regulations are taken into consideration for the development of modern engine technologies. Therefore, various emission-controlling methods were analyzed and implemented in engine to improve the emission control and to meet up the latest emission regulations as reported.[2]

Fuel injection system has been the primary focus in order to improve emissions and performance of diesel engines. Pierpont[3] showed that improvement in NO_x–soot trade-off could be achieved with high-pressure multiple-injection method. Various fuel injection methods on diesel engine were studied[4] to reduce emission level and improve the efficiency of the engine. Zheng and Kumar[5] showed that use of split injection method causes a reduction in NO_x and soot simultaneously by managing the correct injection timings and the dwell in between fuel injections.

Kouremenos et al.[6] showed a significant reduction in diesel NO_x by employing retarded injection timing. In the work of Ganesh et al.,[7] the effect of retarded injection timing on diesel engine was studied to reduce the NO_x and soot emission. However, this work also reported the increase in CO emissions. Brijesh et al.[8] observed an increase in NO_x with advanced injection timing. Husberg et al.[9] investigated the combustion characteristics in an optical engine with multiple advanced pilot injections and these experimental results were compared with those obtained from numerical calculations. Nehmer and Reitz[10] studied the effect of dwell in between injections on various emissions. Mobasheri et al.[11] observed the effect of dwell at injection pressure of 1000 bar on multiple-injection method. This study showed that lean mixture formed with higher dwell does not help the combustion for the next injection and reduced dwell leads to a region with rich fuel.

Various literatures reported above suggest that the balance between NO_x and PM can be enhanced with split injection methods and the dwell in between injections. Many of the research works were limited to a single injection pressure. Hence, this work was carried to evaluate the effects of different split injection methods on CI engine emissions and performance at various injection pressures. Simulation of engine operation was also performed to justify the results obtained from experimental work.

The experimental work was carried out to find the effects of different split injection methods for improving emissions and performance of CI engine. The detailed analysis of split injection methods at different injection pressures on various output characteristics such as NO_x, PM, CO, and brake thermal efficiency (BTE) was also carried out.

27.1.1 EXPERIMENTAL SETUP

Figure 27.1 shows the schematic diagram of experimental setup which is used in the present study. The setup consists of a single-cylinder, four-stroke, and variable compression ratio (VCR) CI engine coupled with a dynamometer (eddy current type). Essential instruments such as fuel line and in-cylinder pressure transducer, load cell, thermocouple, crank angle (θ) encoder, and so forth, were attached to achieve P–θ diagram which provides significant information about CI engine performance. Figure 27.2 shows the actual experiment setup. The main components of the test rig are VCR engine, dynamometer, fuel tank, high-pressure injection kit, injector driver, exhaust gas analyzer, and data acquisition system as shown in Figure 27.2. This setup comprised mainly four subsystems, that is, fuel control unit, air management unit, loading unit, and exhaust gas unit. An incorporated data acquisition system was used to obtain different parameters such as crank angle, temperature, pressure, load, and flow rates of air and fuel. Performance and combustion investigation was performed using "IC Engine Soft_8.5" software.

An electronic fuel injector unit was used for this comprehensive experimental investigation. The unit in Figure 27.3 consists of electric fuel pumps (low pressure and high pressure), a fuel filter, an electronic control unit, a rail pressure control valve, a solenoid injector, and a common rail. By the application of electronic fuel injection unit, it is capable to introduce the fuel up to three injections per cycle among an injection pressures varying from 220 to 1000 bar.

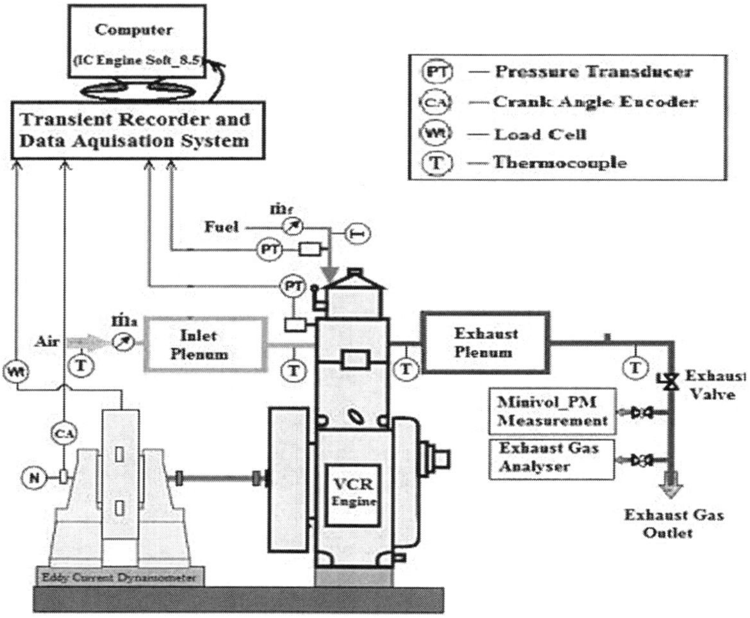

FIGURE 27.1 Schematic diagram of experimental setup.

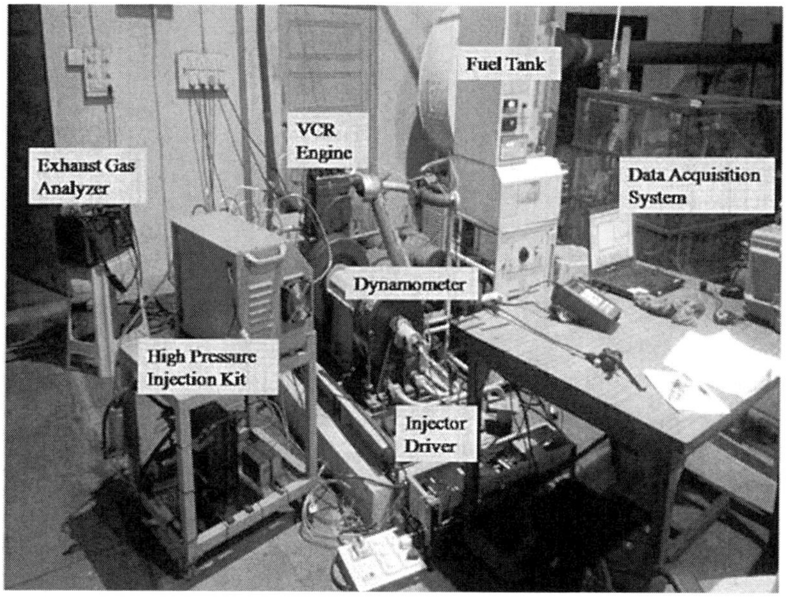

FIGURE 27.2 Actual experimental setup.

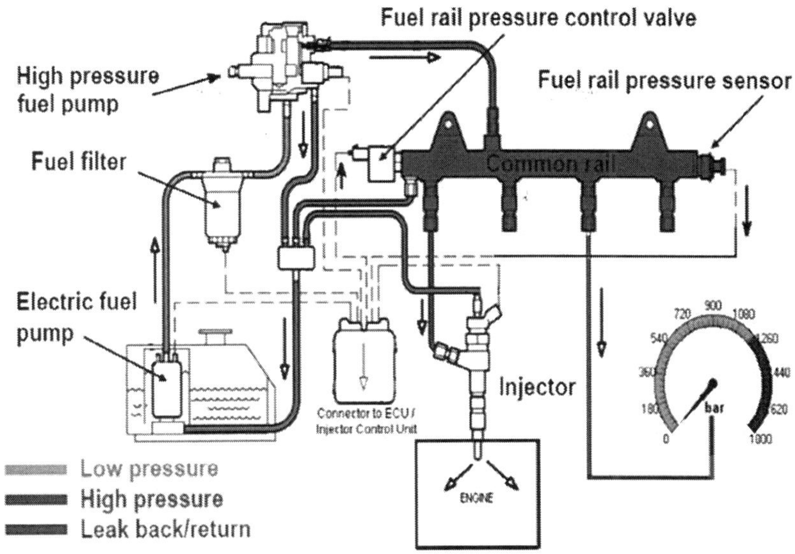

FIGURE 27.3 Electronic fuel injection unit.

The input parameters considered for the experimentations are the number of injections per cycle, start of injection (SOI), and the dwell in between fuel injections. The number of injections has been restricted to one to four injections per thermodynamic cycle. Identical mass of fuel in each injection was achieved by sharing the entire mass of fuel among the number of injections. It is clearly indicated in Figure 27.4 that entire fuel injected throughout the single injection method is shared uniformly into two injections for 2-shot split injection method. Based on the literature survey, 10 crank angle degree (CAD) and 15 CAD dwells between injections were selected for this investigation.[12] The experimental setup utilized in the present study was capable up to 450-bar injection pressure. Hence, experiments were conducted at 240, 340, and 440-bar injection pressures. The base SOI of the engine was set as 27°, that is, 27 CAD before top dead center (CAD bTDC) for at least one injection showed by Brijesh et al.[13]

The matrix of experimental runs was listed in Table 27.1. Experimental run 1 is at base operational conditions, that is, single-shot injection at 240, 340, and 440-bar injection pressures with 27 CAD bTDC SOI. Experimental runs 2 and 4 were completed through 2-shot split injection method with 10 and 15° dwell in between injections with first injection is happening at base SOI, that is, 27 CAD bTDC. Experimental runs 3 and 5 were completed through 3-shot split with 10 and 15° dwell in between injections and it was designed such a way so that second injection begins at 27 CAD bTDC.

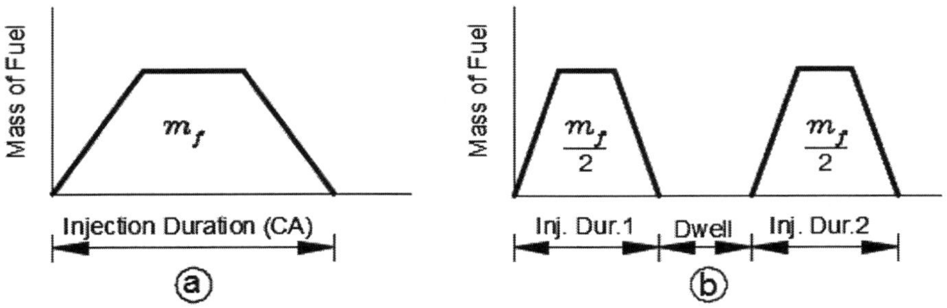

FIGURE 27.4 Fuel injection profile of (a) single and (b) 2-shot split injection method.

TABLE 27.1 Matrix of Experimental Runs.

Experiment runs	No. of injections	Injection pressure (bar)	Dwell between injection (CA)	Start of injection (CAD bTDC)			Method of injection
				Injection 1	Injection 2	Injection 3	
1	1	240, 340, and 440	–	27	–	–	Single (-27)
2	2		10	27	11	–	2-shot: 50(-27)-50
3	3		10	41	27	12	3-shot: 33-34(-27)-33
4	2		15	27	6	–	2-shot: 50(-27)-50
5	3		15	46	27	7	3-shot: 33-34(-27)-33

27.1.2 SIMULATION METHODOLOGY

The simulation study was carried to realize the chemical and physical aspects of the combustion process occurring in the cylinder. The commercial computational fluid dynamics (CFD) tool "Converge" was used for this study. For the simulation purpose, a 120° sector model was used to reduce the computational time. For modeling the turbulence,[14] a variant of k–ε renormalization group model was utilized. As suggested by Schmidt and Rutland,[15] the modified Kelvin–Helmholtz (KH) Rayleigh–Taylor (RT) breakup model was used in this study. The SAGE detailed chemistry model was used by Senecal et al.[16] for combustion. NO_x emissions were predicted using the extended Zeldovich NO_x model given by Hill and Smoot.[17]

Based on grid sensitivity study by Brijesh et al.,[18] dt=0.5 μs and $1.96 \times 1.96 \times 1.96$ mm^3 size of grids were deliberated as the time step and base grid size, respectively. Adaptive mesh refinement technique based on various gradients was used to further refine the grids during the simulation.

27.2 RESULTS AND DISCUSSION

This study was carried to understand the effects of different split injection methods and the dwell in the between fuel injections on combustion, performance, and emissions characteristics of CI engine. A detailed investigation of various outcomes such as NO_x, PM, CO, and BTE was carried out and reported in further sections. Combustion analysis was also studied experimentally to understand the physics involved various processes.

27.2.1 EFFECT OF SPLIT INJECTION METHODS ON PERFORMANCE AND EMISSIONS

The effect of different injection methods on NO_x and PM emissions is plotted in Figure 27.5. This figure shows that a substantial reduction in NO_x (between 32 and 40%) was achieved by using 2-shot and 3-shot injection methods compared to that of single injection method. It also shows that various split injection methods discharge higher amount of PM compared to that of single injection method. Similar results for NO_x and PM emission were also shown by Rajkumar et al.[19] They showed that as the dwell between injections increases, it reduces the NO_x and increases the PM. These observations are explained in further sections using appropriate combustion parameters.

The increase in fuel injection pressure from 240 to 440 bar leads to the increase in the values of peak pressure and temperature. Higher injection pressure reduces the size of fuel, and hence, more surface area of fuel droplets interacts with hot in-cylinder air. Efficient combustion at higher injection pressures leads to higher temperatures and hence higher NO_x. It is seen from Figure 27.5 that as the injection pressure increases, the NO_x also increases. Moreover, it is observed that as the injection pressure increases, PM emission decreases.

Figure 27.6 illustrated the effect of different injection methods on BTE and carbon monoxide (CO) emissions. BTE was reduced significantly for 3-shot [33-34(-27)-33] split injection method compared to that of single injection method. For 3-shot injection method, fuel is injected at around 41

CAD bTDC for the first injection. Some fraction of the fuel may be stuck in the squish and crevice region which reduces BTE.

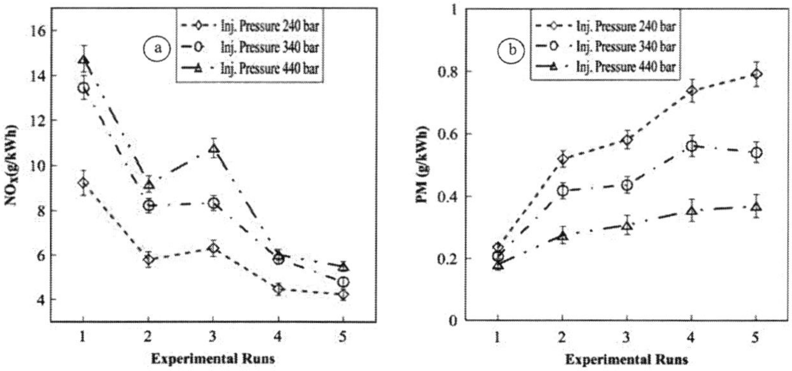

FIGURE 27.5 Effect of different injection methods on (a) NO_x and (b) particulate matter.

From Figure 27.6b, CO emission was observed higher for different split injection methods compared to that of single injection method. This increase in CO emission for 2-shot [50(-27)-50] split injection method found to be in a tolerable range (approximately 36%). When fuel is injected at 41 CAD bTDC tends to initiate vaporization and combustion too early. Therefore, there is more chance of wall wetting and losses in squish and crevice region, which reduces the quality of combustion and results in higher CO emissions. It indicates a drastic increase in CO with 3-shot [33-34(-27)-33] injection compared to other injection methods, hence not recommended.

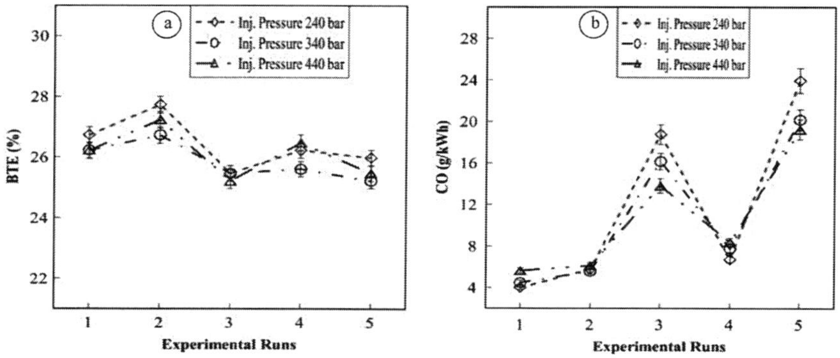

FIGURE 27.6 Effect of different injection methods on (a) brake thermal efficiency and (b) CO.

Further, a 39% reduction in NO_x with a tolerable variation in PM, CO, and BTE of 2-shot split injection method [50(-27)-50] with a dwell of 10 CAD compared to that of single injection method was observed and plotted in Figures 27.5 and 27.6.

27.2.2 EFFECT OF SPLIT INJECTION METHODS ON COMBUSTION CHARACTERISTICS

The effect of various split injection methods on NO_x and PM emissions has been described by mean gas temperature (MGT) and heat release rate (HRR) traces. The HRR traces and MGT traces for different injection methods were shown in Figure 27.7. It was observed that the magnitude of the combustion peak is higher for single injection method compared to 2-shot and 3-shot injection methods. Therefore, higher temperature and hence higher NO_x were observed for single injection method as shown in Figure 27.5a. In contrast, lower magnitude of the combustion peak for 2-shot and 3-shot injection methods compared to that of single injection method which caused educed NO_x emissions.

For producing PM, diffusion combustion phase plays a significant role. As inferred from Figure 27.7, heat release is highest for 2-shot method and least for single injection method during diffusion combustion. Higher PM is formed for 2-shot, while lower PM is formed for single injection method. Engine-out PM is also the function of the oxidation rate of PM. MGT traces were shown for different injection methods in Figure 27.7b.

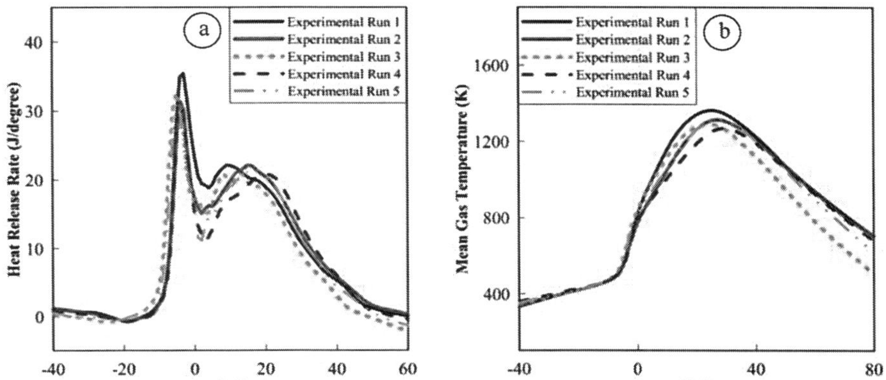

FIGURE 27.7 Different injection methods for (a) heat release rate traces and (b) mean gas temperature traces.

By the application of 2-shot injection method, the highest temperature in the consequent stage of the expansion process was successfully achieved. This implies to the increased rate of oxidation of PM compared to that of other injection methods, whereas a contrary trend is observed for single injection method. Combustion analysis was also performed for various injection methods with an injection pressure of 340 and 440 bar and similar results were obtained for HRR and MGT.

27.2.3 SIMULATION ANALYSIS

The simulation results (converge CFD code used for 120° sector model) was compared with the experimentally obtained results of base operating condition. The simulation result of pressure traces complements well with the experimental results. The relationship of pressure traces achieved from experiment and simulation (converge) results were shown in Figure 27.8. These results were close for a large range of crank angles (−60 to +60 CAD). Hence, this calibrated model was utilized for performing further simulation studies.

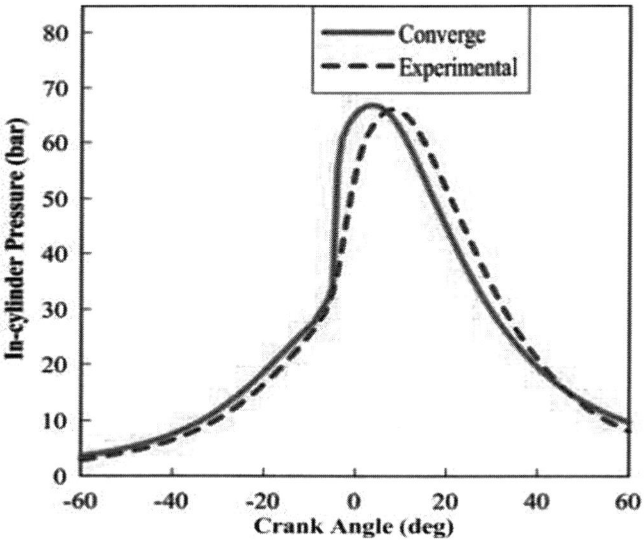

FIGURE 27.8 Comparison of pressure traces.

An extra injector was integrated into the simulation to attain 2-shot split injection. Same injection rate shape, injection duration, and fuel quantity were maintained for each injection. Simulation results indicate that NO_x emission is reduced by using 2-shot split injection method compared to that of single injection method, whereas PM and CO emissions are further increased. Furthermore, the developments of simulation results compare with the experimental results qualitatively. Therefore, contour plots achieved from simulations were utilized to comprehend the experimental findings.

Figure 27.9 indicates that the temperature contours at 1 CAD bTDC for (a) single injection method and (b) 2-shot split injection method with 10 CAD dwell. Mean average temperature was evaluated at 1 CAD bTDC and it measures 1393 and 1295 K for single injection and 2-shot split injection methods, respectively. This leads to reduced NO_x emission for 2-shot split injection method compared to that of single injection method.

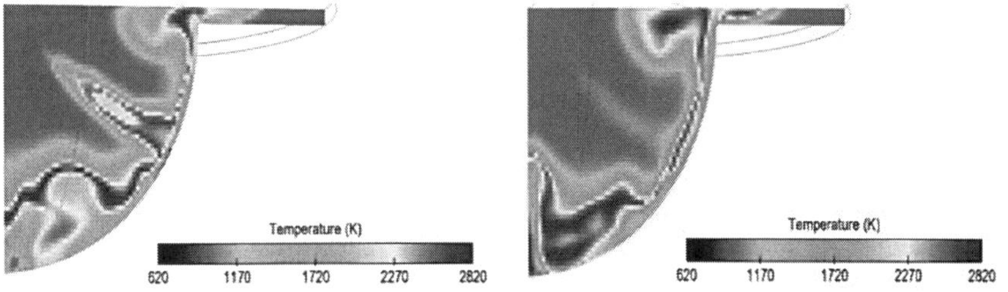

FIGURE 27.9 Temperature contours at 1 crank angle degree (CAD) before top dead center (bTDC) for (a) single injection method and (b) 2-shot split injection method with 10 CAD dwell.

Equivalence ratio contours at 7 CAD bTDC for (a) single injection method and (b) 2-shot split injection method with 10 CAD dwell were shown in Figure 27.10. The region of equivalence ratio is more for 2-shot injection, which also leads to higher PM and CO emissions for 2-shot split injection method compared to that of single injection method.

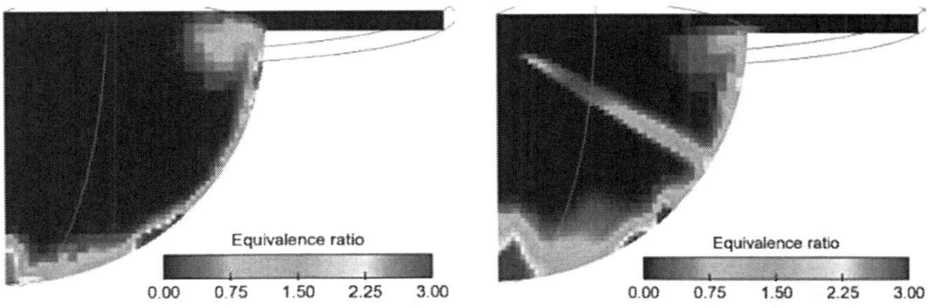

FIGURE 27.10 Equivalence ratio contours at 7 CAD bTDC for (a) single injection method and (b) 2-shot split injection method with 10 CAD dwell.

27.2.4 COMPARISON OF EXPERIMENTAL RESULTS

The quantitative comparison between single injection and 2-shot split injection methods at dwell of 10 CAD is shown in Table 27.2. A 39% decrease in NO_x without varying BTE was accomplished with 2-shot split injection method compared to that of single injection method; however, PM and CO emissions were found to be increased. Similar trends were obtained for injection pressure of 340 and 440 bar as mentioned in Table 27.2. Split injection method can be considered as a possible substitute for the exhaust gas recirculation (EGR) method.

TABLE 27.2 Quantitative Comparison Between Single Injection and 2-Shot Split Injection Methods.

Injection method	Injection pressure	Brake thermal efficiency (%)	NO_x (g/kWh)	Particulate matter (g/kWh)	CO (g/kWh)
Single	240	26.72	9.23	0.33	4.13
2-shot split with 10 dwell		27.73	5.79	0.51	5.63
% change (split compared to single)		3.77 (↑)	−37.66 (↓)	54.54 (↑)	36.31 (↑)
Single	340	26.22	13.46	0.27	4.47
2-shot split with 10 dwell		26.72	8.21	0.41	5.59
% change (split compared to single)		1.9 (↑)	−39.00 (↓)	51.85 (↑)	25.05 (↑)
Single	440	26.22	14.74	0.18	5.64
2-shot split with 10 dwell		26.23	9.17	0.27	6.13
% change (split compared to single)		0.038 (↑)	−37.78 (↓)	50.00 (↑)	8.68 (↑)

27.3 CONCLUSIONS

In the present work, the outcome of various split injection methods on CI engine with respect to performance, combustion, and emissions characteristics is studied. The main input parameters considered for the experimentations were the number of fuel injections per cycle and the dwell between fuel injections. A simulation study using converge CFD tool has also been performed to know the chemical and physical phenomenon of the process occurring in the cylinder. The conclusions drawn from this study are summarized below.

1. The 2-shot split injection method with a dwell of 10 CAD with an injection pressure of 240 bar offered 38% reduction in NO_x, 54% increase in PM, BTE improvement by 4%, and 36% increase in CO. For 340 bar injection pressure, 39% reduction in NO_x, 52% increase in PM, BTE improvement by 2%, and 25% increase in CO are observed. Similarly, for 440 bar injection pressure, 38% reduction in NO_x, 50% increase in PM, BTE vary a small amount, and 9% increase in CO was observed.
2. If the fuel injection pressure increases from 240 to 440 bar, the values of peak pressure and peak temperature increases. This causes better combustion in the engine. Hence, it is concluded that as the injection pressure increases, NO_x also increases.
3. The magnitude of the combustion peak is less for 2-shot and 3-shot injection methods as compared to that of single injection method resulting lower NO_x.
4. The oxidation rate of PM is more for 2-shot as compared to that of other injection methods, hence lower PM. The region of equivalence ratio is more for 2-shot injection. Hence, higher PM and CO emissions are produced from it.

With the application of split injection method, NO_x can be reduced. It is used as a substitute to EGR method or to minimize the quantity of EGR required to achieving advanced combustion modes in CI engine. It can also be used in heavy-duty engines along with EGR, to decrease the load on EGR and to attain different combustion modes in favor of large operating conditions.

ACKNOWLEDGMENT

The authors would like to gratefully acknowledge the support of IC Engines and Combustion Laboratory, Mechanical Engineering Department, Indian Institute of Technology Bombay, Mumbai 410076, India.

KEYWORDS

- **emissions**
- **compression ignition engine**
- **nitrogen oxide (NO_x)**
- **particulate matter (PM)**
- **split injection methods**
- **dwell**
- **injection pressure**
- **computational fluid dynamics (CFD)**
- **converge**

REFERENCES

1. Heywood, J. B. *Internal Combustion Engine Fundamentals*. McGraw-Hill, New York, 1988.
2. Brijesh, P.; Chowdhury, A.; Sreedhara, S. Effect of Ultra-Cooled EGR and Retarded Injection Timing on Low Temperature Combustion in CI Engines. *SAE International United States* **2013,** 2013-01-0321.
3. Pierpont, D. A.; Montgomery, D. T.; Reitz, R. D. Reducing Particulate and NO_x Using Multiple Injections and EGR in a D. I. Diesel. *SAE International United States* **1995,** 950217.
4. Asad, U.; Zheng, M.; Han, X.; Reader, G.; Wang, M. Fuel Injection Strategies to Improve Emissions and Efficiency of High Compression Ratio Diesel Engines. *SAE Int. J. Engines* **2008,** *1*, 1220–1233.
5. Zheng, M.; Kumar, R. Implementation of Multiple-Pulse Injection Strategies to Enhance the Homogeneity for Simultaneous Low NO_x and Soot Diesel Combustion. *Int. J. Thermal Sci.* **2009,** *48*, 1829–1841.
6. Kouremenos, D. A.; Hountalas, D. T.; Binder, K. B.; Raab, A., Schnabel, M. H. Using Advanced Injection Timing and EGR to Improve DI Engine Efficiency at Acceptable NO and Soot Levels. *SAE International United States* **2001,** 2001-01-0199.
7. Ganesh, V.; Deshpande, S.; Sreedhara, S. Numerical Investigation of Late Injection Strategy to Achieve Premixed Charge Compression Ignition Mode of Operation. *Int. J. Eng. Res.* **2016,** *17*, 469–478.
8. Brijesh, P.; Harshvardhan, A.; Sreedhara, S. A Study of Combustion and Emissions Characteristics of a Compression Ignition Engine Processes Using a Numerical Tool. *Int. J. Adv. Eng. Sci. Appl. Math.* **2014,** *6*, 17–30.
9. Husberg, T.; Denbratt, I.; Karlsson, A. Analysis of Advanced Multiple Injection Strategies in a Heavy-Duty Diesel Engine Using Optical Measurements and CFD-Simulations. *SAE International United States* **2008,** 2008-01-1328.

10. Nehmer, D. A.; Reitz, R. D. Measurement of the Effect of Injection Rate and Split Injections on Diesel Engine Soot and NO_x Emission. *SAE International United States* **1994,** 940668.
11. Mobasheri, R., Peng, Z., Mirsalim, S. Analysis the Effect of Advanced Injection Strategies on Engine Performance and Pollutant Emissions in a Heavy Duty DI-Diesel Engine by CFD Modeling. *Int. J. Heat Fluid Flow* **2012,** *33*, 59–69.
12. Brijesh, P.; Sreedhara, S. Experimental and Numerical Investigations of Effect of Split Injection Strategies and Dwell Between Injections on Combustion and Emissions Characteristics of a Diesel Engine. *Clean Technol. Environ. Policy* **2016,** *18*, 2325–2334.
13. Brijesh, P.; Chowdhury, A.; Sreedhara, S. Simultaneous Reduction of NO_x and Soot Using Ultra-Cooled EGR and Retarded Injection Timing in a Diesel Engine. *Int. J. Green Energy* **2014,** *12*, 347–358.
14. Chiodi, M. *An Innovative 3D-CFD-Approach Towards Virtual Development of Internal Combustion Engines*. Vieweg+Teubner Verlag: Braunschweig, 2010.
15. Schmidt, D. P.; Rutland, C. J. A New Droplet Collision Algorithm. *J. Compute. Phys.* **2000,** *164*, 62–80.
16. Senecal, P.; Pomraning, E.; Richards, K.; Briggs, T.; Choi, C.; Mcdavid, R.; Patterson, M. Multi-Dimensional Modeling of Direct-Injection Diesel Spray Liquid Length and Flame Lift-Off Length Using CFD and Parallel Detailed Chemistry. *SAE International United States* **2003,** 2003-01-1043.
17. Hill, S.; Smoot, L. Modelling of Nitrogen Oxides Formation and Destruction in Combustion Systems. *Prog. Energy Combust. Sci.* **2000,** *26*, 417–458.
18. Brijesh, P.; Abhishek, S.; Sreedhara, S. Numerical Investigation of Effect of Bowl Profiles on Performance and Emission Characteristics of a Diesel Engine. *SAE International United States* **2015,** 2015-01-0402.
19. Rajkumar, S.; Mehta, P.; Bakshi, S. Parametric Investigation for NO_x and Soot Emissions in Multiple-Injection CRDI Engine Using Phenomenological Model. *SAE International United States* **2011,** 2011-01-1810.

CHAPTER 28

EFFECTIVE MITIGATION OF NO_X EMISSIONS FROM DIESEL ENGINES WITH SPLIT INJECTIONS

R SINDHU, G. AMBA PRASAD RAO,* and K. MADHU MURTHY

Department of Mechanical Engineering, NIT Warangal, Warangal, Telangana 506004, India

**Corresponding author. E-mail: ambaprasadrao@gmail.com*

ABSTRACT

Good thermal efficiency and high fuel economy of diesel engines are actually welcome attributes from the standpoints of suppressing global warming and conserving energy sources. However, high emission of oxides of nitrogen (NO_x) and particulate matter associated with their unresolved trade-off is a major challenge that needs to be addressed by the researchers. Exhaust gas recirculation (EGR) and retardation of injection timings (ITs) are widely adopted in-cylinder techniques to lower NO emissions from diesel engines. Of late, a split/multiple injection strategy is being increasingly adopted for its potential to effectively address NO_x, soot, and piston work trade-offs. The present work deals with the development of a quasi-dimensional numerical model to predict engine performance and emissions under various operating parameters. EGR levels have been varied from 0 to 30%, IT has been varied from 16° before top dead center (bTDC) to TDC, and two stages of split injections (25/75 and 75/25) are employed and their effect on emissions is discussed. Split injection of 25/75 is observed to be superior in controlling NO emissions than increasing EGR levels and retarded ITs without a penalty on engine performance.

28.1 INTRODUCTION

Diesel engines inherently possess high thermal efficiency, good durability, and superior fuel economy compared to other engines making them the most desirable ones in heavy-, medium-, and light-duty vehicle applications.[1] However, high emission of oxides of nitrogen (NO_x) and particulate matter with their unresolved trade-off from diesel engines is a major challenge to be addressed by the researchers nowadays. Suitable after treatments, such as selective catalyst reduction, are effective in reducing diesel engine emissions but require extensive exhaust system modification. Moreover, conversion of NO_x to nitrogen at the exhaust of diesel engines using three-way catalytic converter is found to be less efficient due to excessive oxygen content present in the exhaust stream of compression ignition (CI) engines. Researchers are thus restoring to the less expensive and more efficient in-cylinder techniques, wherein various combustion modes are investigated to mitigate diesel engine emissions. Low-temperature combustion (LTC) is generally employed in modern diesel engines to simultaneously lower NO_x and soot emissions as it would restrain sudden release of fuel energy and promote air–fuel mixing.[2] LTC in systems with homogeneous charge compression ignition,[3] premixed charged compression ignition,[4] and so forth combustion modes is achieved due to intensive premixing of fuel and air and lean mixture combustion. In such systems, combustion phasing is largely decoupled from fuel injection timing (IT). Combustion is initiated mainly due to the chemical kinetics of gas mixture. Hence, these systems thus have the drawbacks of lack of control over ignition timing, limited operational speeds, and high CO and hydrocarbon (HC) emissions.[5]

In contrast, systems in which combustion phasing is largely dependent on IT result in enhanced diffusion phase due to hindered premixing process. In such systems, higher in-cylinder temperatures are realized and techniques used to lower in-cylinder temperatures are the focus of the present work. Downsizing of the engine,[6] retardation of ITs,[7] increased dilution rates using exhaust gas recirculation (EGR),[8] and so forth are some of the widely adopted techniques to achieve LTC conditions in the latter discussed systems. Moreover, diesel engine industry is now inclined toward the broader use of fuel injection systems with the advent of common rail direct injection, which provides the flexibility in choosing the desired fuel IT, number of injections, dwell period, injection pressure, and so forth, which increases the scope of improving engine performance and lowering engine-out emissions.[9]

In the present work, a crank-angle (CA)-resolved quasi-dimensional computational model is developed which is primarily based on derivations of the first law of thermodynamics and ideal gas equation. The performance and engine-out emissions are predicted and analyzed using the validated model with variation in IT, EGR, and two stages of split injections. The efficacies of the adopted techniques employed to lower emissions are discussed, drawing a comparison between conventional and split injection techniques.

28.2 METHODOLOGY

The present model deals with bowl-in-piston combustion chamber of a direct injection diesel engine. Analysis has been carried out on the so-called closed cycle, where all the valves are closed. The model takes into account different sub-models those govern/define/estimate main processes in the cylinder from intake valve close to exhaust valve open.

28.2.1 ENERGY CONSERVATION

Temporal variation of in-cylinder pressure and temperature is computed using the following numerical formulations at every CA step. Derivations from the first law of thermodynamics and perfect gas equations are used to arrive at the following equations

$$\frac{dP}{dt} = P\left[-\frac{\frac{dV}{dt}}{V} + \frac{\frac{dm}{dt}}{m} + \left(\frac{dT}{dt}\right)\frac{1}{T} - \frac{RT}{P}\left(\frac{\partial \rho}{\partial \phi}\right)\frac{d\phi}{dt} \right] \tag{28.1}$$

$$\frac{dT}{dt} = \frac{\left(\frac{\partial u}{\partial P}\right)\left[-\rho\frac{\frac{dV}{dt}}{V} + \rho\frac{\frac{dm}{dt}}{m} - \left(\frac{\partial \rho}{\partial \phi}\right)\frac{d\phi}{dt} - AA + \left(\frac{\partial u}{\partial \phi}\right)\frac{d\phi}{dt} \right]}{\left(\frac{\partial u}{\partial P}\right)\left(\frac{\partial \rho}{\partial T}\right) - \left(\frac{\partial \rho}{\partial P}\right)\left(\frac{\partial u}{\partial T}\right)} \tag{28.2}$$

where

$$AA = -RT\frac{\frac{dV}{dt}}{V} + \frac{1}{m}\left(\dot{Q} + \dot{m}_i h_i - \dot{m}_e h_e - \left(u\frac{dm}{dt} \right)_{cv} \right) \tag{28.3}$$

$$BB = 1 - \frac{P}{R}\left(\frac{\partial R}{\partial P}\right); CC = 1 + \frac{T}{R}\left(\frac{\partial R}{\partial T}\right) \tag{28.4}$$

28.2.2 ENGINE DYNAMICS

In the present work, to calculate the instantaneous stroke, volume, and rate of change of volume with respect to crankshaft angle rotation, derivations of slider crank mechanism are used.[10] The rate of change of volume can be written as (in time domain and crankshaft angle degrees converted to radians).

$$\frac{dV}{dt} = 3V_d N\left[\sin\frac{\pi N}{30}t + \frac{\sin\frac{\pi N}{30}t\cos\frac{\pi N}{30}t}{\left(r^2 - \sin^2\frac{\pi N}{30}t\right)^{\frac{1}{2}}}\right] \tag{28.5}$$

28.2.3 CLOSED CYCLE ANALYSIS

Closed cycle analysis corresponds to the modeling of compression, combustion, and expansion processes. In a CI engine, only air is compressed and till the start of fuel injection process, the term related to variation in equivalence ratio, φ, with respect to time becomes zero. When the fuel injection begins, the mixture composition varies. Temporal variation of pressure and temperature is then calculated using Equations 28.1 and 28.2, respectively. To solve the abovementioned equations, models for ignition delay, heat release rate, heat transfer, and chemical composition of species determining models are required which are discussed in the following sections.

28.2.4 IGNITION DELAY

Ignition delay is estimated once per engine cycle and is computed as the time between fuel injection (t_{inj}) and fuel ignition (t_{ign}). The model presented by Hardenberg and Hase[11] has been used in the present work, which points out the dependence of ID on properties of the fuel, mixture pressure, temperature, and equivalence ratio of the gas mixture before fuel injection.

$$ID(\deg) = \left(0.36 + 0.22S_p\right)\exp\left[E_a\left(\frac{1}{RT_a} - \frac{1}{17190}\right) + \left(\frac{21.2}{P_a - 12.4}\right)^{0.63}\right] \quad (28.6)$$

28.2.5 HEAT RELEASE RATE

The present model follows the work reported by Serrano et al.,[12] which involves four different Wiebe functions to predict heat release rates: one Wiebe function to reproduce pilot injection, two Wiebe functions to predict premixed and diffusion combustion phase, and finally a Wiebe function for late burning phase. The generic mathematical form of a Wiebe function to determine heat release rate (HRR) is given by

$$RoHR = \sum_{1}^{4}\left[\frac{C_i\left(m_i+1\right)}{d_{comb,i}}\left(\frac{\theta - SOC_i}{d_{comb,i}}\right)^{m_i} e^{\left[-C_i\left(\frac{\theta - SOC_i}{d_{comb,i}}\right)^{m_i+1}\right]} \cdot \beta_i\right] \quad (28.7)$$

where C_i is the parameter which indicates the completion of combustion; m_i controls the profile of ith phase of combustion; $d_{comb,i}$ is duration of ith combustion phase; θ is the instantaneous CA degree; SOC_i is the start of ith combustion phase; and β_i is the portion of fuel burnt in ith phase compared to the total amount of fuel injected.

28.2.6 HEAT TRANSFER MODEL

Modeling heat transfer losses between the gas mixture and the surrounding walls of the cylinder needs is quite a complex process and requires the use of empirical expressions. Expression generally used to calculate the forced convective heat transfer is[13]

$$Q_{loss} = A(\theta)_{cyl}\,\bar{h}_c\left(\bar{T}_{gas} - \bar{T}_{wall}\right) \quad (28.8)$$

In the present work, to model the forced convective heat transfer coefficient, Woschini's[13] model has been used. Hence, $\bar{h}_c$ in Equation 28.8 is given by

$$h_c = 3.26D^{\bar{m}-1}P^{\bar{m}}T_g^{\;0.75-1.62\bar{m}}w^{\bar{m}} \quad (28.9)$$

The burnt gas velocity, *w*, is expressed by Woschini[13] as

$$w = C_1 S_p + C_2 (T - T_m) = C_1 S_p + C_2 \frac{V_d T_{gr}}{P_r T_r} \left(P(\theta) - P_m \right) \tag{28.10}$$

28.2.7 CHEMISTRY OF COMBUSTION

The equilibrium composition of the gases produced by combustion of a general HC fuel with air is determined using this model. Ten combustion species are assumed to form and the general combustion equation can be written as

$$\begin{aligned} & m_a C_a H_b O_c + \frac{\left(b + \frac{a}{4} + \frac{c}{2} \right)}{\phi} \left(O_2 + 3.72 N_2 \right) \rightarrow \\ & x_1 H_2 O + x_2 H_2 + x_3 OH + x_4 H + x_5 N_2 \\ & + x_6 NO + x_7 CO_2 + x_8 CO + x_9 O_2 + x_{10} O \end{aligned} \tag{28.11}$$

28.2.8 NITRIC OXIDE FORMATION

A widely accepted NO formation kinetics scheme proposed by Lavioe[14] has been used in the present work. This model postulates NO formation during combustion process and rate of NO formation is given by the following expression

$$\frac{d[NO]}{dt} = \frac{2R_1 \left\{ 1 - \left([NO]/[NO]_e \right)^2 \right\}}{1 + \left([NO]/[NO]_e \right) R_1 / (R_2 + R_3)} \tag{28.12}$$

where $[NO]_e$ represents species concentration at equilibrium.

28.2.9 SOOT FORMATION

Net soot formed in diesel engines is generally the competition between the soot formed in early stages of combustion and soot oxidized in the later stages of combustion. A model proposed by Lipkea and Dejoode[15] is used to calculate the net soot formation rate. The models for formulating rates of soot formation and soot oxidation are given by

$$\frac{dm_{sf}}{dt} = A_{sf} m_{fb}^{0.8} P^{0.5} \exp\left(-E_{sf} / R_{mol} T\right) \quad (28.13)$$

$$\frac{dm_{sf}}{dt} = A_{sf} m_{fb}^{0.8} P^{0.5} \exp\left(-E_{sf} / R_{mol} T\right) \quad (28.14)$$

In Equations 28.13 and 28.14, subscripts *sf* and *so* denote soot formed and soot oxidized, respectively. Mass is expressed in kilograms, pressure in megapascal, and P_{O_2} is the partial pressure of oxygen. E_{sf} and E_{sc} in the equations are the corresponding activation energies and the values are taken as 82 × 106 and 120 × 106, respectively.

The evaluation of performance and emission characteristics from a compression ignition direct injection engine involves sequential determination of the following events: mass flow rate through intake and exhaust valves, chemical composition and thermodynamic properties of working fluid, effect of heat transfer, fuel injection rate, ignition delay, heat release rate, residual gases and nitric oxide, and soot formation. To calculate all the parameters involved, a computational model is developed using the mathematical relationships or models proposed in the present section.

28.3 VALIDATION OF THE NUMERICAL MODEL

A comparison has been made in this section between the values obtained from the theoretical model and experimental values. For this purpose, two sets of experimental data are considered. One set of data of in-cylinder pressure histories from a single-cylinder, water-cooled, direct injection diesel engine with 80 mm × 100 mm stroke and bore and with a compression ratio of 16.5:1 present in authors' laboratory and the other from the experiments carried out by Rakopoulos et al.[16] on Ricardo-Hydra engine have been used in the present work.

Figure 28.1 represents comparisons between predicted and experimental values of in-cylinder pressure histories diesel engine operating at 1500 rpm, at static IT of 23° before top dead center (bTDC) and at 25% load conditions.

Figure 28.2 shows a comparison between the predicted and experimental cylinder gas pressure traces obtained from[16] Figures 28.1 and a good agreement between the theoretical and experimental cylinder pressure curves, thus validating the model and its implementation.

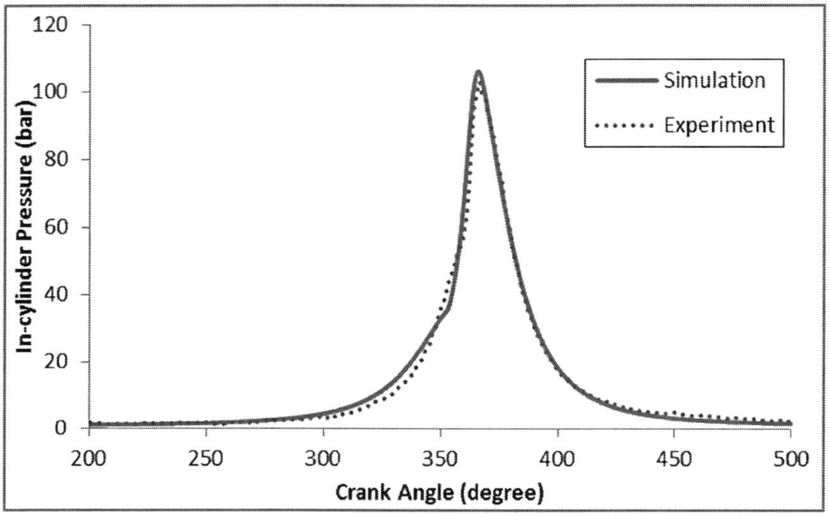

FIGURE 28.1 Comparison between predicted and experimental values for engine operating at 1500 rpm, at an injection timing (IT) of 23° before top dead center (bTDC) at 25% load.

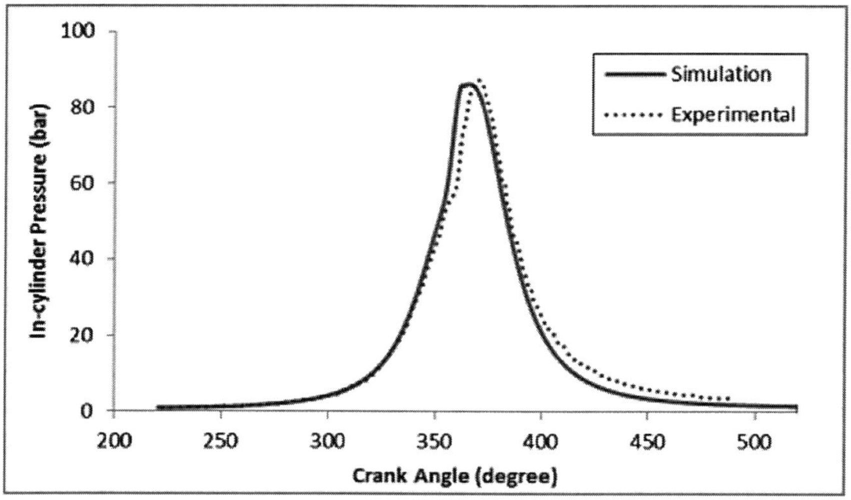

FIGURE 28.2 Predicted and experimental pressure (indicator) diagram versus crank angle (CA), at 2000 rpm, 75% of full load condition and 29° bTDC static IT.

28.4 RESULTS AND DISCUSSION

The effect of various in-cylinder techniques involving reduction of engine emissions (NO_x and soot) has been studied in the present section. A wide range of ITs, three levels of EGR, and two stages of split injections have been implemented and their effect on engine performance has been studied using in-cylinder pressure, temperature, heat release rates, and piston work plots as a function of CA.

28.4.1 STUDIES ON EFFECT OF INJECTION TIMING

Numerical experiments are carried out to study the effect of variation of fuel IT on engine performance and engine-out emissions (NO_x and soot) when the remaining operating conditions are kept unchanged. Figure 28.3 depicts NO–soot cumulative values on emission trade-off curve over a range of ITs from 20° bTDC to TDC.

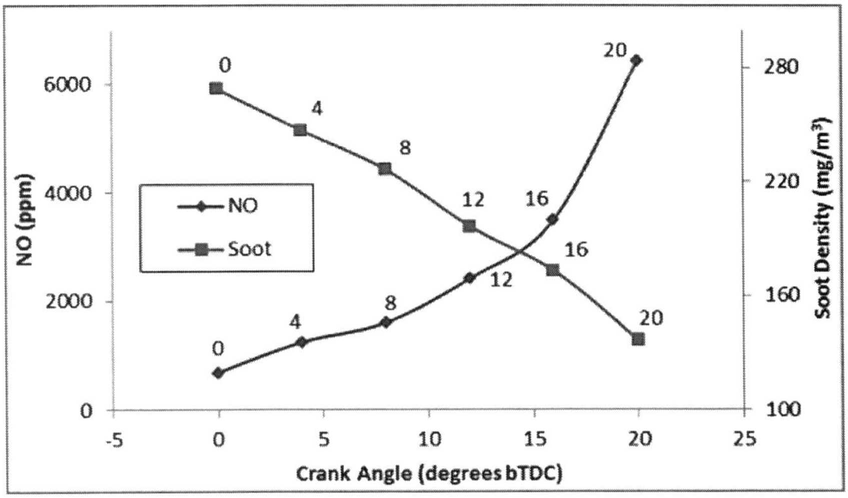

FIGURE 28.3 Comparison of predicted and NO–soot emissions over a range of ITs. Numbers in the figure indicate the ITs (bTDC).

Higher in-cylinder temperatures triggered by advanced ITs have been observed to generate more NO emissions and less soot emissions. This is due to the enhanced premixed phase of combustion increasing the resident time of the burnt charge in the combustion chamber. This increases the tendency of NO formation due to increased oxygen availability and high temperature, increasing net NO production. However, due to increase in in-cylinder temperatures, combustion rate is increased during premixed phase, ensuring complete combustion. Hence, the probability of fuel left unburnt in the combustion chamber decreases, reducing soot formation with consequent near oxidation reactions. Thus, NO can be reduced with retardation of IT; however, this leads to increase in soot emissions, increase in exhaust gas temperature, and increased specific fuel consumption rate.

28.4.2 STUDIES ON EFFECT OF EXHAUST GAS RECIRCULATION

It is well established that NO_x formation is highly dependent on in-cylinder temperatures. Thus, reduction in in-cylinder temperatures would reduce the probability of formation of NO_x. Thus, the widely adopted technique is to dilute the fresh charge by means of adding burned products, thereby reducing heat capacity of charge. This process, in turn, leads to lower in-cylinder temperatures and oxygen concentration available for combustion.

EGR lowers peak flame temperature of the combustible mixture due to the reduction in oxygen percentage in the cylinder. The kinetics of NO_x formation mainly depends upon the in-cylinder temperatures and oxygen content in the cylinder.[10] Figure 28.4 shows the main advantage of introducing EGR in reducing NO_x emissions. NO_x formation is reduced significantly with the increase in EGR percentage. It is seen that 20% EGR fraction led to a decrease of 55% reduction in EGR when the other operating parameters are kept constant and in-cylinder temperatures are above 1600 K.

However, diesel engine combustion is predominantly mixed controlled associated with fuel-rich pockets, which in the late part of combustion are struggling to find air, especially at high load operation. EGR worsens this situation by reducing oxygen availability and thus reducing the possibility to burn smoke free.

Although EGR implementation leads to a reduction in engine-out NO_x, it also has a profound effect on mixture-averaged specific heat and critical temperatures (before ignition and combustion). Deterioration in combustion quality and incomplete fuel combustion are observed with increase in EGR. All these effects lead to drop in thermal efficiency of the engine and reduction in loss of piston work. Hence, for the analysis, control, and design of diesel engine with EGR implementation, these effects are needed to be taken into consideration.

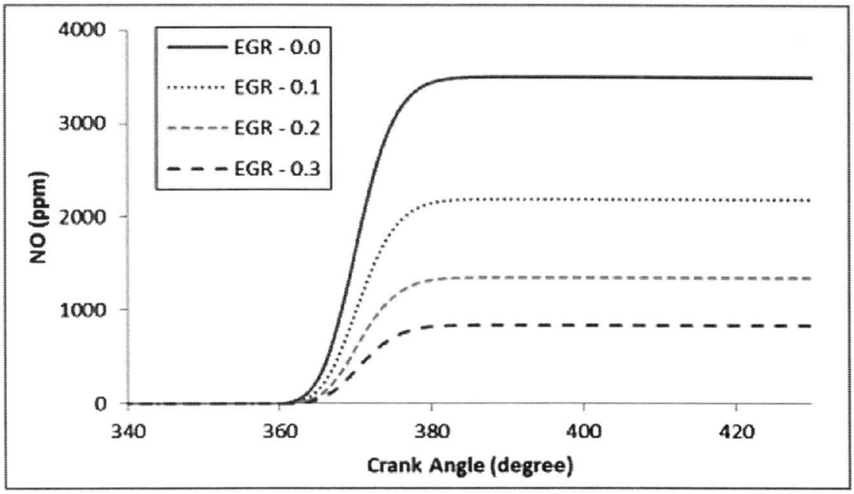

FIGURE 28.4 Predicted NO formation histories as a function of CA with increasing exhaust gas recirculation (EGR) fraction.

28.4.3 STUDIES ON EFFECT OF SPLIT INJECTION

In the present work, a proposed technique of split injection is employed, wherein a part of the total quantity of fuel (25–50%) is initially injected and the remaining fuel is injected with a certain injection pause (dwell period) over a single injection event. The schematic representation of the designed injection schemes is shown in Figure 28.5, which consists of one single injection (16° bTDC) and two split injections (25/75 and 50/50). IT of all the injection schemes is taken as 16° bTDC with the same amount of fuel injected in all cases (36.84 mg/mm^3). A split injection in the present work is represented as the amount of fuel injected in the first pulse/amount of fuel injected in the second pulse. The dwell period between the two pulses is kept at 8° of CA for both the split injection cases.

Figures 28.6 and 28.7 compare the HRR profiles of single and split injections and also depict injection profile of the considered split injections. From the figures, it can be apparently distinguished that as the amount of fuel injected in the first pulse increases, premixed phase of combustion increases and the HRR profile becomes similar to that of single injection.

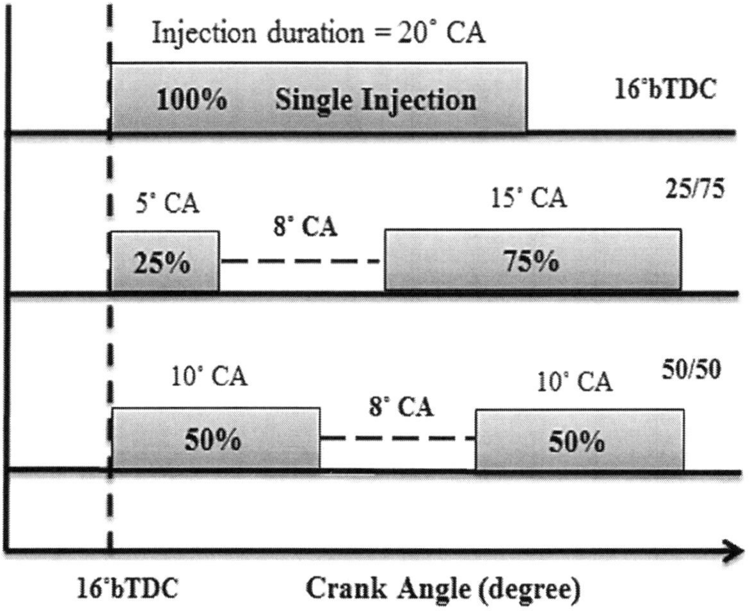

FIGURE 28.5 Schematic representation of designed injection strategies.

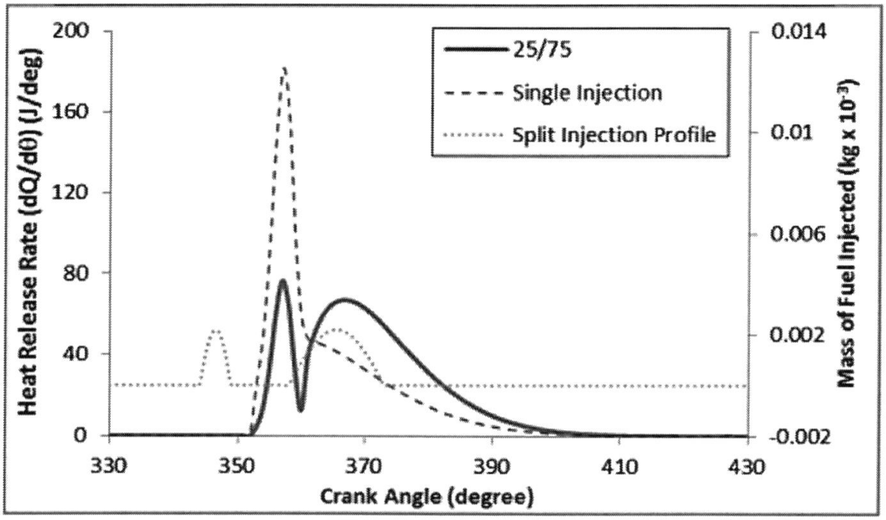

FIGURE 28.6 Heat release rate (HRR) profiles for single and split injection of 25/75. Injection pause between the two pulses is 8° CA.

In detail, as the amount of fuel injected in the first pulse decreases (25%), fuel participating in premixed phase of combustion also decreases (~25% of fuel). This further creates a conducive environment for the oncoming second injection, making the combustion predominantly mixed controlled. However, in the case of single injection, 100% of fuel is injected at a time increasing the amount of fuel burnt in premixed phase of combustion at early stages of combustion. In 25/75 case, a major portion of combustion is delayed by a period of dwell angle away from TDC as depicted in Figure 28.6a. This reduces the peak in-cylinder temperature in split injection cases due to the restrained premixed phase of combustion and off-phasing of a part of combustion away from TDC. This suppresses the favorable conditions of NO formation.

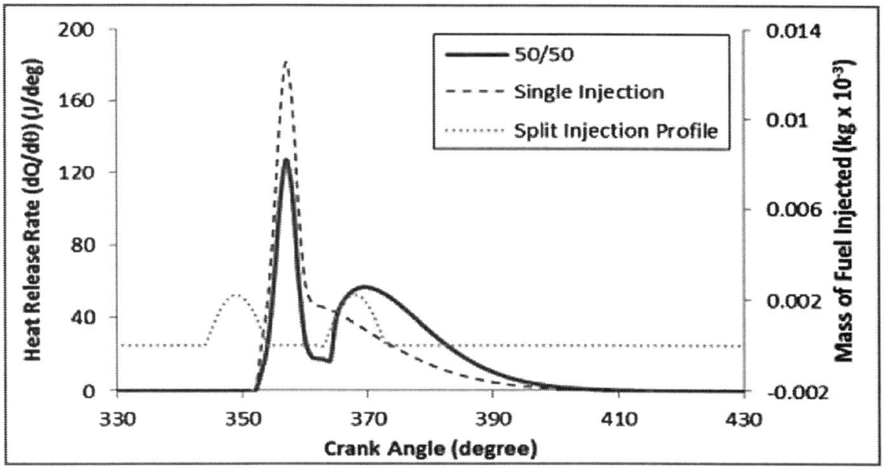

FIGURE 28.7 HRR profiles for single and split injections. Injection pause between the two pulses is 8° CA.

Figure 28.8 represents NO formation rates of single and split injections as a function of CA. From the figure, it can be derived that single injection has generated higher NO emissions followed by 50/50 and 25/75. This can be reasoned with an increased amount of fuel participating in premixed phase of combustion in single injection followed by 50/50 and 25/75. Moreover, with an increase in amount of fuel injected in the first pulse increases, resident time of burnt fuel increases, triggering favorable conditions for NO formation.

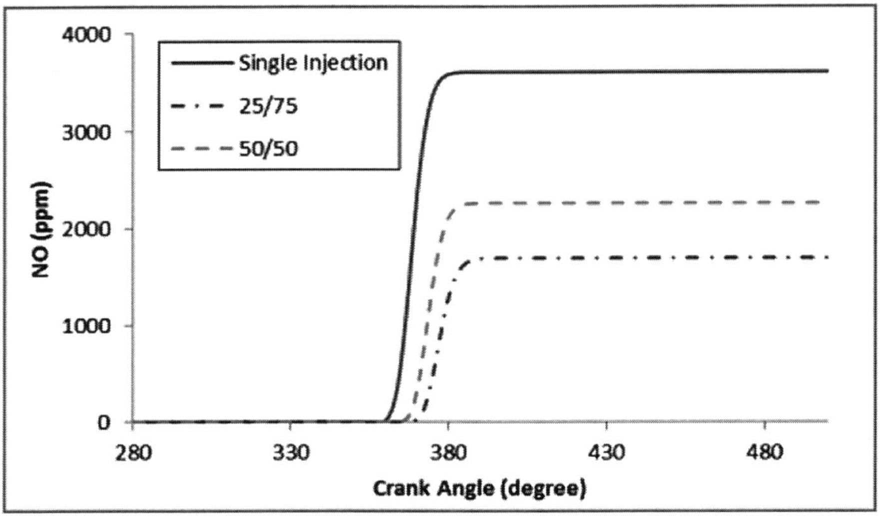

FIGURE 28.8 Predicted NO formation histories as a function of CA for different injection strategies.

28.4.4 COMPARISON OF NO REDUCTION TECHNIQUES

From the above discussion of employment of in-cylinder techniques for reduction of emissions, it can be observed that retarded IT of 8° bTDC, EGR of 20%, and split injection of 25/75 have resulted in similar levels of NO emissions. NO formation rates of IT of 8° bTDC, EGR of 20%, and split injection of 25/75 are shown in Figure 28.9. From the figure, it can be derived that NO formation in EGR of 20% case has started much earlier than IT of 8° bTDC and 25/75 cases. Hence, split injection of 25/75, wherein a major portion of combustion (75% of fuel) is retarded by a period of dwell angle, acts as a retarded combustion event.

Figure 28.10 depicts P–V (or indicator) diagram of IT of 8° bTDC, EGR of 20%, and split of 25/75. It can be observed that single injection conventional techniques have shown higher peak pressures when compared to 25/75 case. However, 25/75 case has shown an increase in pressure along the power stroke of the engine. Hence, split injection of 25/75 has resulted in lower NO emissions as that of conventional techniques, however, without much deterioration in engine piston work.

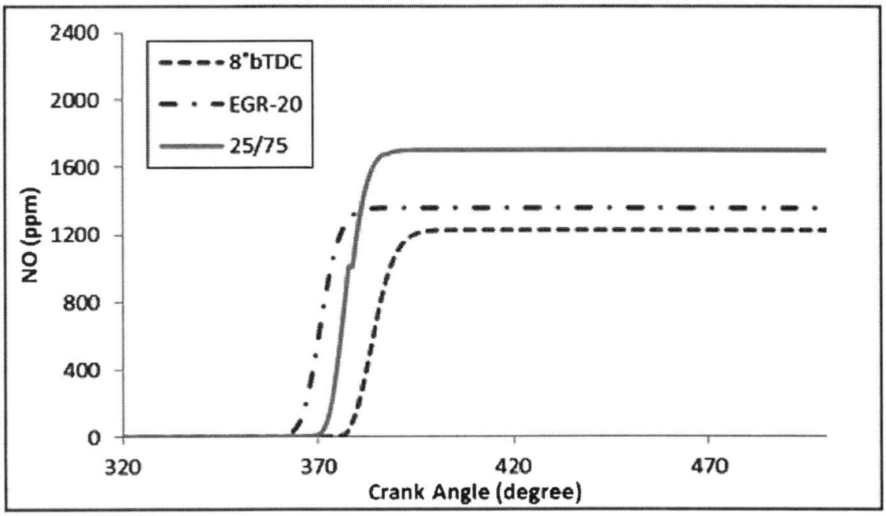

FIGURE 28.9 Comparison of NO formation for IT of 8° bTDC, EGR of 20%, and split of 25/75 cases.

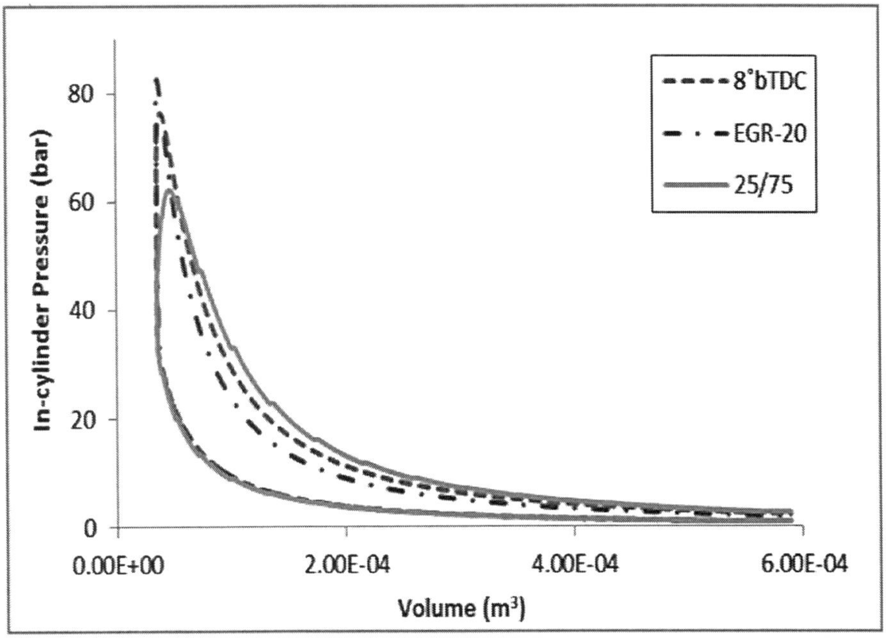

FIGURE 28.10 Comparison of P–V (or indicator) diagrams for IT of 8° bTDC, EGR of 20%, and 25/75 cases.

28.5 CONCLUSIONS

A diesel engine computational model has been developed in the present work using empirical, semiempirical, and physical laws that govern/predict/estimate the engine flow process. Variation of ITs, EGR, and two stages of split injections has been employed and their effect on engine-out emissions is discussed and the following conclusions are drawn:

1. The model is insensitive to engine geometry.
2. The model is able to establish the trade-off between NO_x and soot emissions with variation in fuel IT.
3. It is observed that EGR can be effectively used for reduction of NO_x emissions, however, with a loss in piston work of the engine.
4. Split injections are observed to restrain premixed phase of combustion when compared to single injection at the same IT.
5. Split injection with a smaller quantity of fuel injected in the first pulse (25/75) is observed to lower NO emissions.
6. Split of 25/75 is observed to be superior in controlling NO emissions compared to increasing EGR levels or retarding ITs.

KEYWORDS

- **diesel engine**
- **emissions**
- **NO_x-PM trade-off**
- **split injections**
- **dwell**

REFERENCES

1. Borman, G. L.; Ragland, K. W. *Combustion Engineering*. McGraw-Hill Science/Engineering/Math: New York, 1988.
2. Kook, S.; Bae, C.; Miles, P. C.; Choi, D.; Pickett, L. M. The Influence of Charge Dilution and Injection Timing on Low-Temperature Diesel Combustion and Emissions. Technical Report, SAE Technical Paper, 2005-01-3837, 2005. SAE International Transactions, Warrandale.

3. Helmantel, A.; Denbratt, I. HCCI Operation of a Passenger Car Common Rail DI Diesel Engine with Early Injection of Conventional Diesel Fuel. Technical Report, SAE Technical Paper, SAE 2004-01-0935, 2004. SAE International Transactions, Warrandale.
4. Neely G D, Sasaki S, and Leet J A, Experimental Investigation of PCCI-DI Combustion on Emissions in a Light-Duty Diesel Engine. Technical Report, SAE Technical Paper, 2004-01-0121, 2004. SAE International Transactions, Warrandale.
5. Gan, S.; Ng, H. K.; Pang, K. M. Homogeneous Charge Compression Ignition (HCCI) Combustion: Implementation and Effects on Pollutants in Direct Injection Diesel Engines. *Appl. Energy* **2011,** *88*(3), 559–567.
6. Laguitton, O.; Crua, C.; Cowell, T.; Heikal, M.; Gold, M. The Effect of Compression Ratio on Exhaust Emissions from a PCCI Diesel Engine. *Energy Convers. Manage.* **2007,** *48*(11), 2918–2924.
7. Kouremenos, D.; Hountalas, D.; Binder, K.; Raab, A.; Schnabel, M. Using Advanced Injection Timing and EGR to Improve DI Diesel Engine Efficiency at Acceptable NO and Soot Levels. Technical Report, SAE Technical Paper, 2001-01-0199, 2001. SAE International Transactions, Warrandale.
8. Zheng, M.; Reader, G. T.; Hawley, J. G. Diesel Engine Exhaust Gas Recirculation—A Review on Advanced and Novel Concepts. *Energy Convers. Manage.* **2004,** *45*(6), 883–900.
9. Huang, H.; Liu, Q.; Yang, R.; Zhu, T.; Zhao, R; Wang, Y. Investigation on the Effects of Pilot Injection on Low Temperature Combustion in High-Speed Diesel Engine Fueled with n-Butanol–Diesel Blends. *Energy Convers. Manage.* **2015,** *106*, 748–758.
10. Heywood, J. B.; et al. *Internal Combustion Engine Fundamentals.* Mcgraw-Hill: New York, 1988; Vol. 930.
11. Hardenberg, H.; Hase, F. An Empirical Formula for Computing the Pressure Rise Delay of a Fuel from its Cetane Number and from the Relevant Parameters of Direct-Injection Diesel Engines. SAE 790493, SAE International Transactions, Warrendale, 1999.
12. Serrano, J.; Arnau, F.; Dolz, V.; Tiseira, A.; Cervelló, C. A Model of Turbocharger Radial Turbines Appropriate to be Used in Zero- and One-Dimensional Gas Dynamics Codes for Internal Combustion Engines Modelling. *Energy Convers. Manage.* **2008,** *49*(12), 3729–3745.
13. Woschni, G. A Universally Applicable Equation for the Instantaneous Heat Transfer Coefficient in the Internal Combustion Engine. SAE 670931, SAE International Transactions, Warrandale, 1967.
14. Lavoie, G. A.; Heywood, J. B.; Keck, J. C. Experimental and Theoretical Study of Nitric Oxide Formation in Internal Combustion Engines. *Combus. Sci. Technol.* **1970,** *1*(4), 313–326.
15. Lipkea, W. H.; DeJoode, A. D. Direct Injection Diesel Engine Soot Modeling: Formulation and Results. SAE 940670, SAE International Transaction, Warrandale, 1994.
16. Rakopoulos, C.; Rakopoulos, D.; Giakoumis, E.; Kyritsis, D. Validation and Sensitivity Analysis of a Two Zone Diesel Engine Model for Combustion and Emissions Prediction. *Energy Convers. Manage.* **2004,** *45*(9), 1471–1495.

CHAPTER 29

COMBINED EFFECT OF CYLINDRICAL COMBUSTION CHAMBER SHAPE AND NOZZLE GEOMETRY ON THE PERFORMANCE AND EMISSION CHARACTERISTICS OF A COMPRESSION IGNITION ENGINE OPERATED ON PONGAMIA

MAHANTESH M. SHIVASHIMPI[1,*], S. A. ALUR[1,2],
S. N. TOPANNAVAR[1,3], and B. M. DODAMANI[1,4]

[1]*Mechanical Engineering Department, Hirasugar Institute of Technology Nidasoshi, Belagavi, Karnataka, India*

[2]*saalur.mech@hsit.ac.in*

[3]*sntopannavar.mech@hsit.ac.in*

[4]*bmdodamani.mech@hsit.ac.in*

[*]*Corresponding author. E-mail: shivashimpi@gmail.com*

ABSTRACT

The use of biodiesel in the diesel engine decreases the exhaust emissions but the performance of the engine is found to decrease. This is due improper mixing of air and inadequate turbulence. The performance of the engine can be improved by improving the mixing quality of biodiesel spray with air which can be achieved by modifying the combustion chamber shape and nozzle geometries. The present work investigates the combined effect of cylindrical combustion chamber (CCC) shape and five-hole nozzle geometry on the performance and emission characteristics of biodiesel operated diesel engine with baseline diesel fuel. Engine tests were carried out on a single-cylinder four-stroke direct injection (DI) diesel engines using various blends of Pongamia oil methyl esters (POME) with standard diesel as a fuel and compared with modified combined CCC shape and five-hole nozzle geometry. For comparison, the compression ratio of the engine is kept constant. The experimental results depict that brake thermal efficiency increases up to B60 blend of POME in both baseline diesel engine and modified diesel engine geometry. A drastic reduction in unburnt hydrocarbons and oxides of nitrogen (NO_x) emissions was observed with modified diesel engine geometry as compared to baseline diesel engine. The percentage of carbon monoxide (CO) emissions decreases more with increasing the percentage of POME in modified diesel engine geometry up to B40 blend, thereafter increase in carbon monoxide emissions as increases in blend in modified diesel engine geometry compared with baseline diesel engine. However, the percentage of carbon dioxide emission (CO_2) is more in modified diesel engine geometry as compare with baseline diesel engine for injection pressure 205 bars. The experiment is repeated with increase in injection pressure.

29.1 INTRODUCTION

The diesel and petrol fuels are vanishing spontaneously and the costs of the fossil fuels are increasing day by day. Hence, renewable fuels have been focused more on its capacity to replace the fossil fuels. However, due to environmental considerations, more attention is given to the use of biofuels. The main problem of vegetable oils of usage in internal combustion engine is with their higher viscosity, from 15 to 20 times higher than the regular diesel fuel. Many techniques such as blending with standard diesel fuel, micro-emulsification with methanol, thermal cracking, and conversion of vegetable oils into biodiesels by transesterification process are used to reduce the viscosity of the vegetable oils. All of the above, transesterification process is most commonly used for reducing the viscosity of oil. To improve the performance of diesel engine operated with biodiesel in comparing the conventional diesel engine with diesel fuel, there in need of altering the fuel properties, engine design parameters, and engine operating parameters. To enhance the performance and reduce the emissions, quick and improved air–biodiesel mixing is the essential condition. To improve the air and fuel mixing, quality of the biodiesel spray with air can be enhanced by the selection of appropriate nozzle injection pressure and suitable combustion chamber shape.

Jaichandar and Annamalai[1] investigated the performance and emission characteristics of various shapes of combustion chamber such as toroidal, shallow depth, and hemispherical combustion chamber (HCC) shapes by using 20% Pongamia oil methyl esters (POME) in standard diesel without changing the compression ratio (CR) of the engine. The result show that the toroidal combustion chamber shape shows higher brake thermal efficiency (BTE) compared to other two combustion chamber shapes and also observed that there is significant reduction of particulates, carbon monoxide (CO), and unburnt hydrocarbons (UBHCs) for toroidal combustion chamber shape compare to other two shapes but oxides of nitrogen (NO_x) in increased for toroidal combustion chamber shape. Jaichandar and Annamalai[2] conducted an experiment on diesel engine by using the toroidal reentrant and HCC shapes with varying the nozzle opening pressure for 20% POME in standard diesel fuel. The result revealed that the performance parameters such as BTE and specific fuel consumption (SFC) are improved for toroidal reentrant combustion chamber shape compared to baseline shape at high nozzle injection pressure and also observed reduction in emission characteristics but increase in oxides of nitrogen for toroidal reentrant combustion chamber shape at higher nozzle injection pressure as compared to baseline shape. Jaichandar et al.[3] show that the BTE is improved by 5.64% and the SFC is reduced by 4.6 but 11% of oxides of nitrogen has been increased for toroidal reentrant combustion chamber shape as compared to diesel engine operated with *ultra-low-sulfur diesel* due to appropriate air–fuel mixing and reduce injection timing. Rajashekhar et al.[4] investigated the combined effect of multichambered combustion chamber shape and nozzle injection pressure by using Jatropha as a biodiesel in diesel engine and results showed that the both performance and emission characteristics were optimum at 200-bar injection pressure. Karra and Kong[5] studies show that the nozzle size of 10-hole nozzle geometry gives lesser pressure drop and also performs better atomization, maximum air utilization, and reduction in oxides of nitrogen at full load condition of diesel engine. The experimental result by Saravanan et al.[6] reveals that by modification of combustion process, the oxides of nitrogen, and carbon monoxides emissions are highly reduced but increase in the BTE and UBHC emission with increase in the smoke density in biodiesel-fueled diesel engine. Lahane and Subramanian[7] suggested that varying the number of holes in nozzle injector reduces the oxides of nitrogen in biodiesel-fueled diesel engine and wall impingement minimizes the UBHC, carbon monoxide, and brake SFC (BSFC). The experimental work by Mobasheri and Peng[8] showed that by changing the combustion chamber shape, the emission characteristics are reduced but performance parameters remain the same. The experimental work by Balaji and Cheralathan[9] was carried out to observe the effect of nozzle injector pressure on DI diesel engine with neem methyl ester. The result showed that the performance characteristics such as BTE increases with decrease in BSFC and emission characteristics such as carbon monoxide, carbon dioxide, hydrocarbon, nitrogen oxides, and smoke intensity are reduced at optimum nozzle injection pressure for 240 bar. The experimental result by Khandal et al.[10] reveals that varying the injection timing, injection pressure, CR, and nozzle holes leads to enhanced performance parameters like BTE and meanwhile reduce the emission characteristics in biodiesel-fueled diesel engine; however, oxides of nitrogen emission increases as the number of holes increase in injector nozzle. Banapurmath et al.[11] studied the various combustion chamber shapes such as cylindrical, trapezoidal, and toroidal combustion chamber shapes in diesel engine. The results showed that toroidal combustion chamber shape showed higher performance parameters with reduced emission characteristics compared to remaining two combustion chamber shapes.

From the above summary of literature review, we found that there was minimum research work carried on combined effect of combustion chamber shape and nozzle geometries in the biodiesel-fueled diesel engine, hence analyzing the best and suitable combination of modified geometries that are studied in reference to performance and emission characteristics with consideration of cylindrical combustion shape and five-hole nozzle geometry in diesel engine fueled with POME.

29.2 PROPERTIES OF FUEL

All over India, the production of nonedible is increasing as the usage is increasing day by day because the availability of petroleum products are decreasing as the cost of production is more. Out of all nonedible oils, Jatropha and Pongamia have more importance in research activities. *Pongamia pinnata*, an essential plant having natural importance across the world and also one of the best potential biofuel crops, can grow anywhere in the waste lands. *P. pinnata* having higher potential exists everywhere in our country to bring the higher plantation growth. Basically, Pongamia seed contains 30–35% of the oil. POME has to be prepared usually by transesterification process. The properties of diesel, blends, and POME are as follows in Table 29.1.

TABLE 29.1 The Properties of Diesel, Blends, and Pongamia Oil Methyl Esters.

Properties	Blends					
	Diesel	**B20**	**B40**	**B60**	**B80**	**B100**
Density (kg/m^3)	830	820	842	851	868	894
Specific gravity	0.83	0.82	0.842	0.851	0.868	0.894
Flash point (°C)	63	83	89	110	118	180
Fire point (°C)	66	95	100	124	145	195
Calorific value (kJ/kg)	42,000	41,618	39,380	38,100	37,880	37,150

29.3 EXPERIMENTAL METHODOLOGY

The engine selected for experimentation was a single-cylinder, compression ignition, and direct injection (DI) engine. It was a naturally aspirated, four-stroke, water-cooled, and vertical engine. As shown in Figure 29.1 experimental setup, it is made to work under higher pressures and is widely used in agriculture and industrial sectors. The engine runs at a constant speed of 1500 rpm. The engine has overhead valve arrangement. The valves are controlled by push rods and a camshaft. The water required for the cooling of the engine is obtained by forced circulation of water from water pump through the water jacket. The specification of the engine is shown in Table 29.2.

FIGURE 29.1 Experimental test engine.

TABLE 29.2 Engine Specifications.

Engine parameter	Specifications
Engine	Kirloskar single-cylinder four-stroke direct injection (DI) diesel engine
Nozzle opening pressure	200–205 bar
Rated power	5 HP at 1500 rpm
Bore	80 mm
Stroke length	110 mm
Compression ratio	17.5:1
Displacement volume	660 cm^3
Arrangement of valves	Overhead
Combustion chamber	Open chamber (DI)
Cooling type	Water cooled
Loading	Mechanical-type loading dynamometer

The selection of biodiesel is usually based on the availability of biodiesel; the Pongamia methyl ester is used for the experimentation. The various blends such as B20, B40, B60, B80, and B100 are prepared in laboratory; we found the various fuel properties of prepared blends and standard diesel in laboratory using Abel's apparatus and Saybolt's viscometer. The experimentations are carried out with HCC shape and three-hole nozzle geometry for standard diesel, B20, B40, B60, B80, and B100 fuels with 0, 25, 50, 75, and 100% engine loading with help of mechanical loading type of dynamometer and emission readings are noted with help of gas analyzer. The piston shape is changed to cylindrical shape without altering the volume by filling the molten aluminum and machining process. The cylindrical combustion chamber (CCC) shape and five-hole nozzle geometry assembled into baseline engine setup. The same experimentation will be repeated with CCC shape and five-hole nozzle geometry as carried to HCC shape without changing the CR. Finally, the performance and emission characteristics of modified geometries were compared with baseline hemispherical piston geometry. The hemispherical and CCC shapes are shown in Figure 29.2. The three-hole injector nozzle and five-hole injector nozzle geometries are as shown in Figure 29.3.

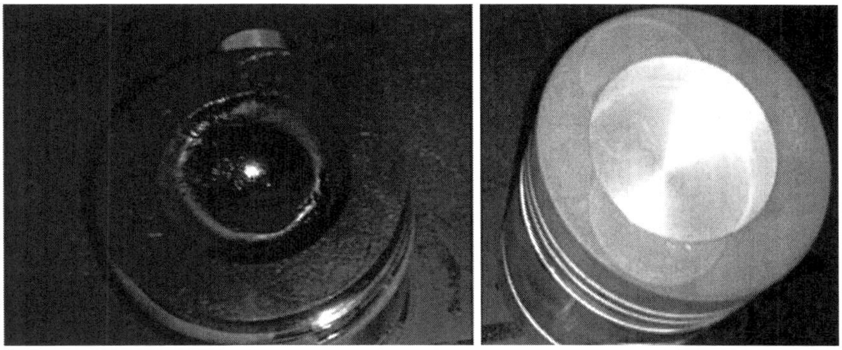

FIGURE 29.2 Hemispherical and cylindrical combustion chamber shapes.

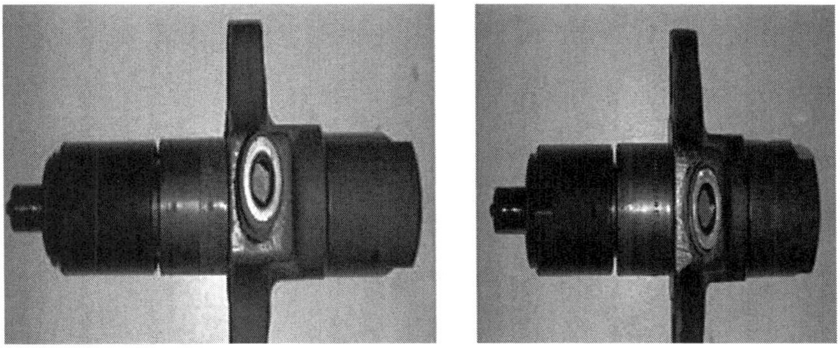

FIGURE 29.3 Three-hole and five-hole nozzle geometries.

29.4 RESULTS AND DISCUSSION

In this experiment, we investigated the comparison between the baseline geometry of HCC shape and three-hole nozzle geometry and modified geometry of CCC shape and five-hole nozzle geometry with reference to performance characteristics and emission characteristics without changing the CR at full load condition of diesel engine are as follows.

29.4.1 SPECIFIC FUEL CONSUMPTION

From Figures 29.4 and 29.5, we observed that the variation of SFC with brake power (BP) for baseline and modified engine geometries. The SFC increases from 0.26 to 0.29 kg/kW·h with increase in the blend for baseline geometry and for modified geometry, it varies from 0.28 to 0.35 kg/kW·h with increasing the blend. The SFC increases for biodiesel in both the shapes due to lesser calorific value. However, the SFC is more in modified geometry compared to baseline geometry due to poor mixing quality in modified shape.

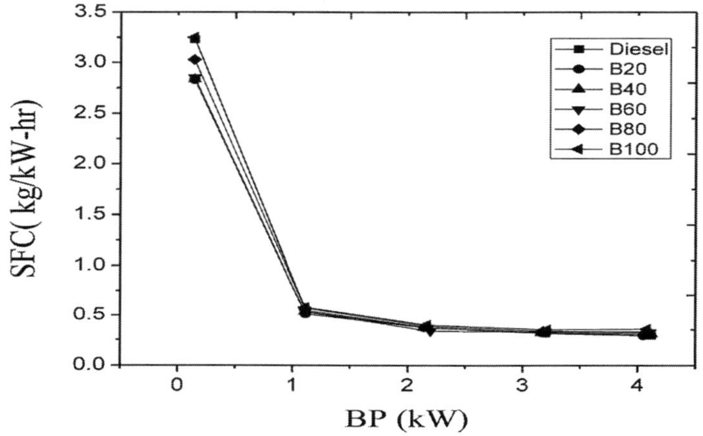

FIGURE 29.4 Variation of specific fuel consumption (SFC) with brake power (BP) for baseline geometry.

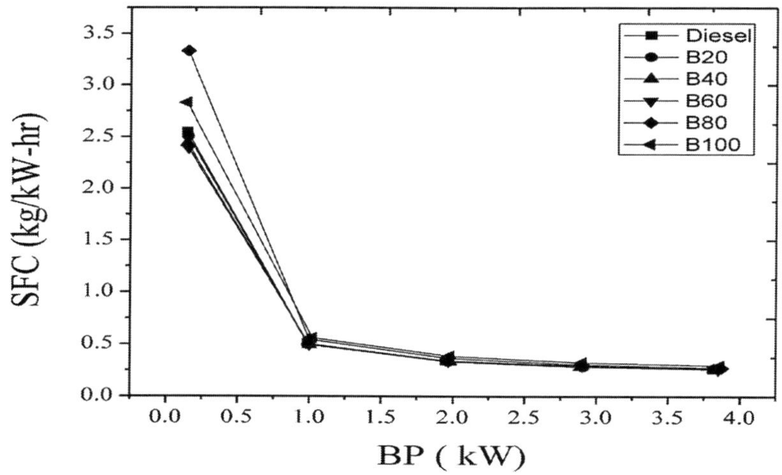

FIGURE 29.5 Variation of SFC with BP for modified geometry.

29.4.2 BRAKE THERMAL EFFICIENCY

From Figures 29.6 and 29.7, we observed that the variation of BTE with BP for baseline and modified engine geometries. The BTE increases from 32.79 to 34.69% with increase in the blend for baseline geometry, and it increases from 29.58 to 29.83% for modified engine geometry; however, the decrease in BTE is observed for modified geometry for pure biodiesel (B100). This is due to improved combustion characteristics in baseline geometry.

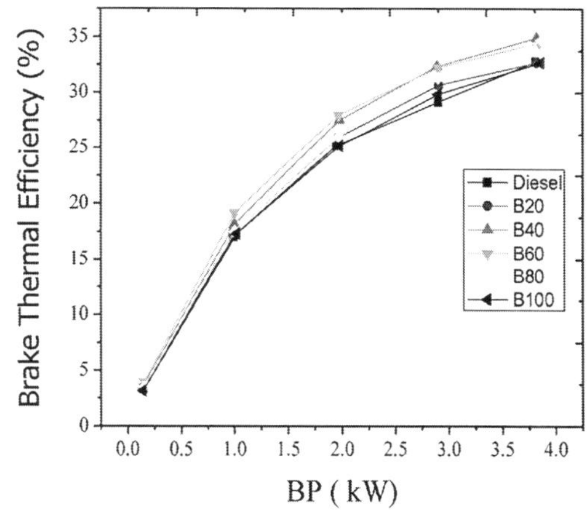

FIGURE 29.6 Variation of BTE with BP for baseline geometry.

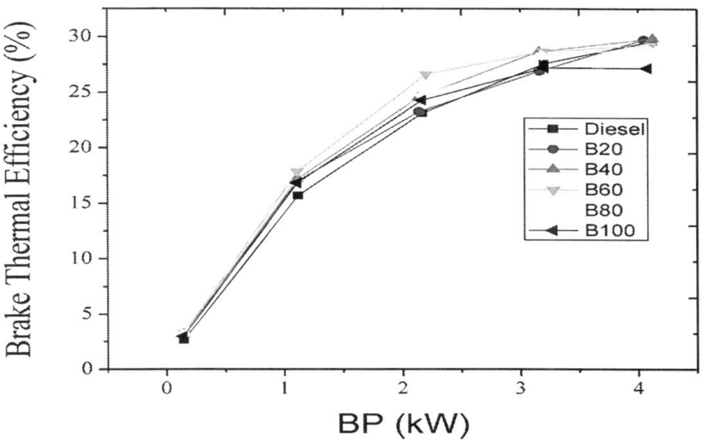

FIGURE 29.7 Variation of BTE with BP for modified geometry.

29.4.3 CARBON MONOXIDE

From Figures 29.8 and 29.9, we observed that the variation of carbon monoxide with BP for baseline and modified engine geometries. The percentage of CO emission decreases from 0.225 to 0.009 for baseline geometry, for modified geometry, it varies from 0.743 to 0.672 but the percentage of CO emission decreases up to B40 blend and thereafter it increases with increasing the blend for baseline geometry, whereas the percentage of CO emission increases up to B80 blend and thereafter it decreases with the B100 blend for modified geometry. However, the percentage of CO is more in baseline geometry compared to modified geometry at 50% load condition due to lower temperature in modified geometry at full load condition.

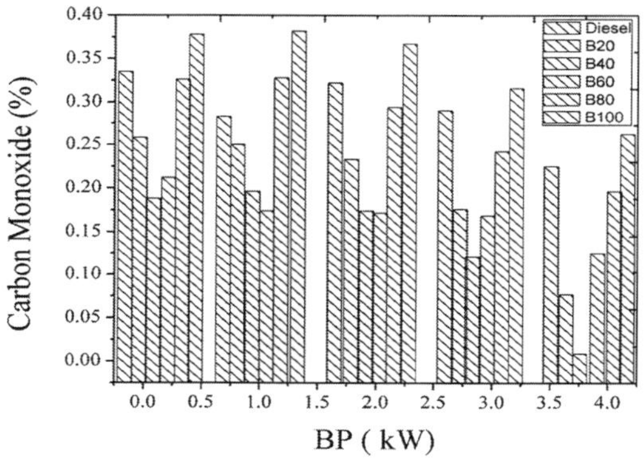

FIGURE 29.8 Variation of CO with BP for baseline geometry.

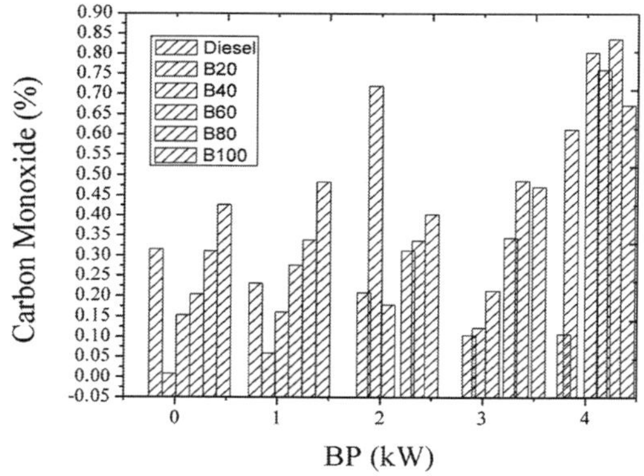

FIGURE 29.9 Variation of CO with BP for modified geometry.

29.4.4 CARBON DIOXIDE

From Figures 29.10 and 29.11, we observed that the variation of carbon dioxide with BP for baseline and modified engine geometries. The percentage of CO_2 varies from 7.55 to 6.03% in baseline geometry, but for modified geometry, it varies from 7.33 to 10.83%; however, the percentage of CO2 is more in modified geometry compared to baseline geometry due to improved combustion in HCC shape.

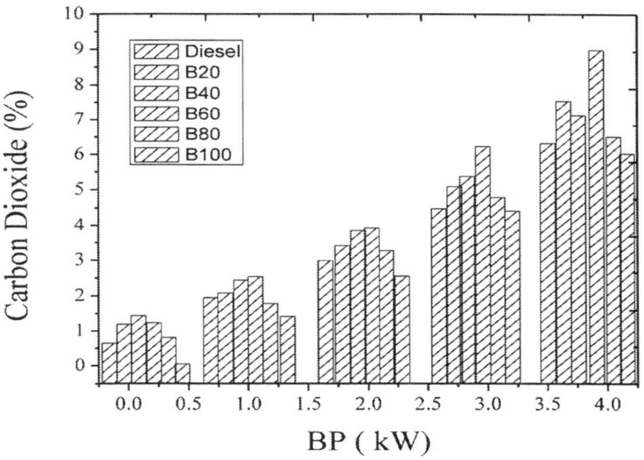

FIGURE 29.10 Variation of CO_2 with BP for baseline geometry.

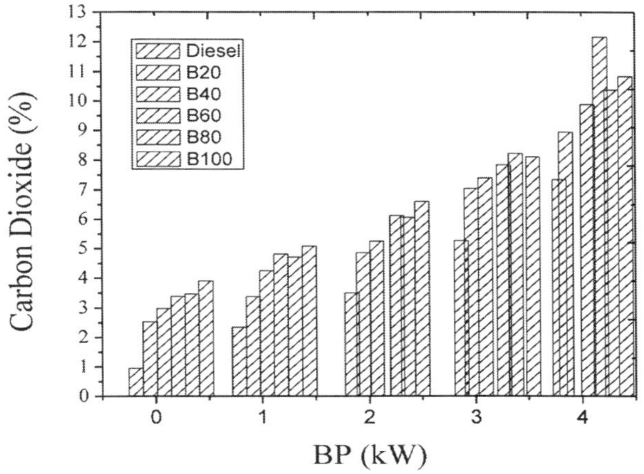

FIGURE 29.11 Variation of CO_2 with BP for modified geometry.

29.4.5 UNBURNT HYDROCARBONS

From Figures 29.12 and 29.13, we observed that the variation of UBHCs with BP for baseline and modified engine geometries. The UBHC emission decreases as load increases in both geometries due to the high content of oxygen in Pongamia biodiesel and better air swirl but comparatively UBHC emission reduces up to 50% load condition and thereafter increases in modified geometry and also lower UBHC are observed for modified shape than the baseline geometry.

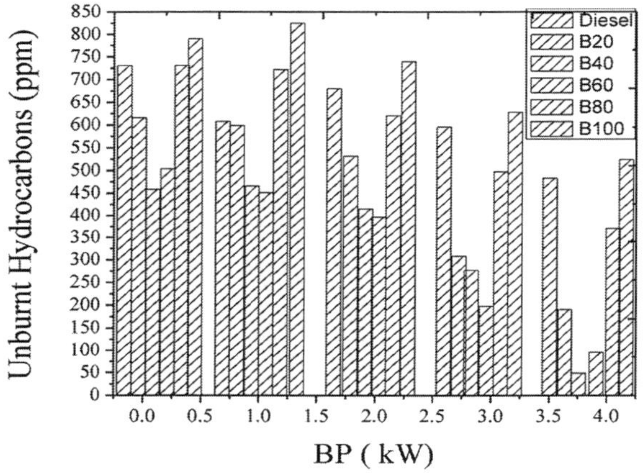

FIGURE 29.12 Variation of unburnt hydrocarbons (UHC) with BP for baseline geometry.

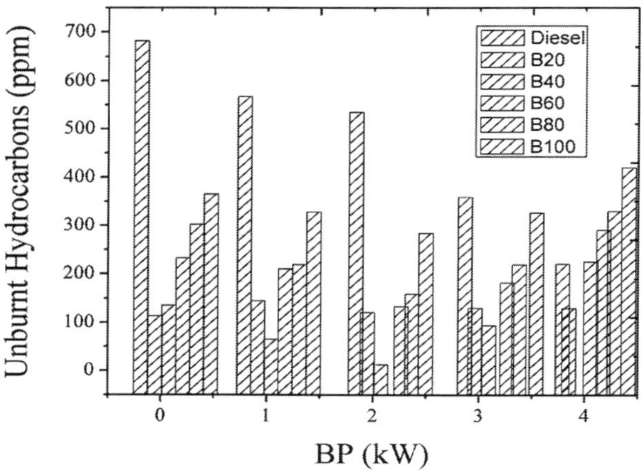

FIGURE 29.13 Variation of UHC with BP for modified geometry.

29.4.6 OXIDES OF NITROGEN

From Figures 29.14 and 29.15, we observed that the variation of oxides of nitrogen with BP for baseline and modified engine geometries. The better mixing of fuel and air and faster burning process in both geometries NO_x level increases in both the shapes. The NO_x emission decreases from 2103 to 1914 ppm as blend increases for baseline geometry, and it varies from 1028 to 892 ppm as increase in the blend for modified shape. However, the 45% of NO_x emission has reduced in modified geometry comparatively baseline geometry due to low temperature attained in modified geometry.

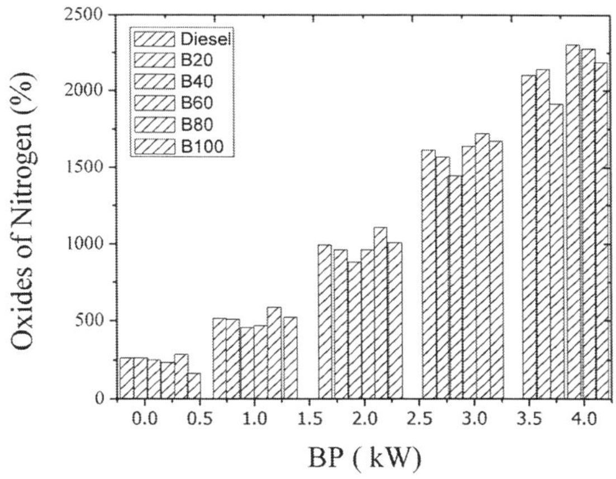

FIGURE 29.14 Variation of NO_x with BP for baseline geometry.

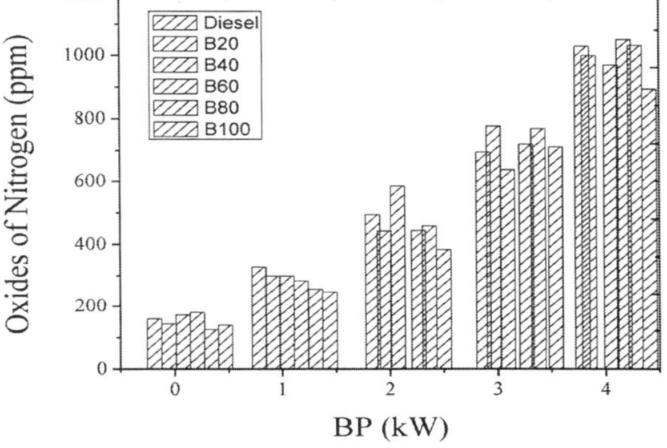

FIGURE 29.15 Variation of NO_x with BP for modified geometry.

29.5 CONCLUSIONS

The SFC has shown reduction with increase in BP for both geometries but it decreases in the case of baseline geometry due to improved combustion. BTE has improved with increase in BP for both geometries and it is observed that among all blends, B40 for baseline geometry and B40 for modified geometry give more efficiency. The BTE increases as blends increase in both geometries observed at all load. The modified geometry gives lower emissions as compared to baseline geometry especially there is a 45% of the reduction in the NO_x for modified geometry as compared to the baseline geometry. The modified geometry utilized more oxygen than that of baseline geometry due to complete combustion.

ACKNOWLEDGMENT

Our sincere thanks to the VGST and Karnataka Science and Technology Promoting Society (KSTePS) Government of Karnataka in Collaboration with DST Government of India for their continuous support and help extended to research equipment.

KEYWORDS

- **combustion chamber shapes**
- **emission**
- **Pongamia**
- **biodiesel**
- **diesel engine**

REFERENCES

1. Jaichandar, S.; Annamalai, K. Effects of Open Combustion Chamber Geometries on the Performance of Pongamia Biodiesel in a DI Diesel Engine. *Fuel* **2012,** *98*, 272–279.
2. Jaichandar, S.; Annamalai, K. Combined Impact of Injection Pressure and Combustion Chamber Geometry on the Performance of a Biodiesel Fueled Diesel Engine. *Energy* **2013,** *55*, 330–339.

3. Jaichandar, S.; Senthil Kumar, P.; Annamalai, K. Combined Effect of Injection Timing and Combustion Chamber Geometry on the Performance of a Biodiesel Fueled Diesel Engine. *Energy* **2012,** *47*(1), 388–394.
4. Rajashekhar, C. R.; Chandrashekar, T. K., Umashankar, C.; Kumar, R. H. Studies on Effects of Combustion Chamber Geometry and Injection Pressure on Biodiesel Combustion. *Trans. Can. Soc. Mech. Eng.* **2012,** *36*(4), 429–438.
5. Karra, P. K.; Kong, S.-C. Experimental Study on Effects of Nozzle Hole Geometry on Achieving Low Diesel Engine Emissions. *J. Eng. Gas Turbines Power* **2010,** *132*(2), 22802.
6. Saravanan, S., Nagarajan, G.; Sampath, S. Combined Effect of Injection Timing, EGR and Injection Pressure in Reducing the NO_x Emission of a Biodiesel Blend. *Int. J. Sustain. Energy* **2014,** *33*(2), 386–399.
7. Lahane, S.; Subramanian, K. A. Impact of Nozzle Holes Configuration on Fuel Spray, Wall Impingement and NO_x Emission of a Diesel Engine for Biodiesel–Diesel Blend (B20). *Appl. Therm. Eng.* **2014,** *64*(1), 307–314.
8. Mobasheri, R.; Peng, Z. CFD Investigation of the Effects of Re-Entrant Combustion Chamber Geometry in a HSDI Diesel Engine. *World Academy Sci. Eng. Technol. Int. J. Mech. Aerosp. Ind. Mechatronic Manuf. Eng.* **2013,** *7*(4), 770–780.
9. Balaji, G.; Cheralathan, M. Experimental Investigation of Varying the Fuel Injection Pressure in a Direct Injection Diesel Engine Fuelled with Methyl Ester of Neem Oil. *Int. J. Ambient Energy* **2017,** *38*(4), 356–364. DOI: 10.1080/01430750.2015.1111846.
10. Khandal, S. V.; Banapurmath, N. R.; Gaitonde, V. N.; Hosmath, R. S. Effect of Number of Injector Nozzle Holes on the Performance, Emission and Combustion Characteristics of Honge Oil Biodiesel (HOME) Operated DI Compression Ignition Engine. *J. Pet. Environ. Eng.* **2015,** *6*(215), 1–18. DOI: 10.4172/2157-7463.1000215.
11. Banapurmath, N. R.; Chavan, A. S.; Bansode, S. B.; Patil, S.; Naveen, G.; Tonannavar, S.; Keerthi Kumar, N.; Tandale, M. S. Effect of Combustion Chamber Shapes on the Performance of Mahua and Neem Biodiesel Operated Diesel Engines. *J. Pet. Environ. Eng.* **2015,** *6*(230), 1–7. DOI: 10.4172/2157-7463.1000230.

CHAPTER 30

AN EXPERIMENTAL STUDY ON THE ROLE OF DIESEL ADDITIVE AND BIODIESEL ON PERFORMANCE EMISSION TRADE-OFF CHARACTERISTICS OF A DI CI ENGINE

G. RAVI KIRAN SASTRY[1,2], JIBITESH KUMAR PANDA[2,*], RABISANKAR DEBNATH[2,3], and RAM NARESH RAI[2,4]

[1]*grksastry1@rediffmail.com*

[2]*National Institute of Technology, Agartala, Tripura, India*

[3]*rabi101991@gmail.com*

[4]*nareshray@yahoo.co.in*

[*]*Corresponding author. E-mail: jibiteshpanda90@gmail.com*

ABSTRACT

In this work, ethanol was blended with palm methyl ester (PME) and conventional diesel fuel. The effect of ethanol concentration and PME on diesel and both the performance and emissions were investigated. Both low and high concentrations of ethanol and biodiesel were studied. The results showed that ethanol, biodiesel, and diesel blends lead to a positive effect on the reduction in the nitrogen oxide. Brake thermal efficiency (BThE) was better in different blends (methyl ester of palm oil, ethanol, and diesel) as compared to pure diesel. Moreover, a comparative trade-off analysis was done among BThE, brake specific fuel consumption, and NO_x to reflect the performance and emission characteristics at a time.

30.1 INTRODUCTION

With uninterrupted rise in population, energy requirements have also increased which has resulted in the extensive use of the fossil fuel resources. Due to high rate of fossil fuels depletion, these are going to be exhausted. Therefore, the need of hour is to look beyond conventional sources of fuel and find alternate sources of fuel. Few detectives have already found biodiesel from some of these oils. The possibility of annual palm oil production in India is very high.[5–7] Alcohols have been widely used in compression ignition engines as alternative fuels. Although alcohols are cheaper than standard diesel fuel, there are challenges with respect to utilization of alcohols in diesel engines and blending these fuels with diesel. This research was focused on the engine performance, emission, and trade-off characteristics of PME and diesel additive like ethanol, when used in a diesel engine.[1–5,8,10]

Nomenclature	
PME	Palm methyl esters
D100	Pure diesel
BL 1	20% PME + 2% ethanol + 78% diesel
BL 2	20% PME + 4% ethanol + 76% diesel
BL 3	20% PME + 6% ethanol + 74% diesel
BL 4	20% PME + 80% diesel
BSFC	Brake specific fuel consumption
BThE	Brake thermal efficiency
NO_x	Nitrogen oxide

30.2 EXPERIMENTAL PROCEDURE AND SPECIFICATIONS

The engine was used for investigational work containing single-cylinder, four-stroke, conventional fossil fuel engine, which is associated to eddy current type dynamometer (Fig. 30.1). Specifications of the engine are shown in Table 30.1.

A gas analyzer (TESTO-350) is used for the measuring of nitric oxide (NO_x). NO_x was in form of ppm which was fitted at the exhaust end; by this preparation, we can grow the emission physiognomies. The experiments are led by diesel (D100) and blends (BL) 1, 2, 3, and 4 at different load circumstances on the engine from 0 to 100% in suitable steps at a compression ratio (CR) 17:5. The performance investigation is carried out on a direct injection (DI) diesel engine using D100, the biodiesel, ethanol, and their blends. The load range taken is from no load to full load. Table 30.2 shows the basic fuel properties of biodiesel used in this study in addition to ethanol and diesel.

TABLE 30.1 Engine Specification.

Make and type	Kirloskar, single-cylinder, four-stroke diesel
Engine type	Vertical compression ignition engine
Stroke length	110 mm
Swept volume	661 cc
Compression ratio	17.6
Power	3.5 kW
Rated speed	1500 rpm
Bore size	87.5 mm
Dynamometer	
Make	Power mag
Type	Eddy current
Load measurement method	Strain gauge
Maximum load	12 kg
Cooling	Water

TABLE 30.2 Properties of Biodiesel, Diesel, and Ethanol.

Properties	Diesel	PME	Ethanol
Density at 150°C (g/cc)	0.815	0.880	0.789
Viscosity at 400°C (cSt)	2.46	4.43	1.22
Flash point (°C)	55	175	13
Cetane number	52	49	41.4

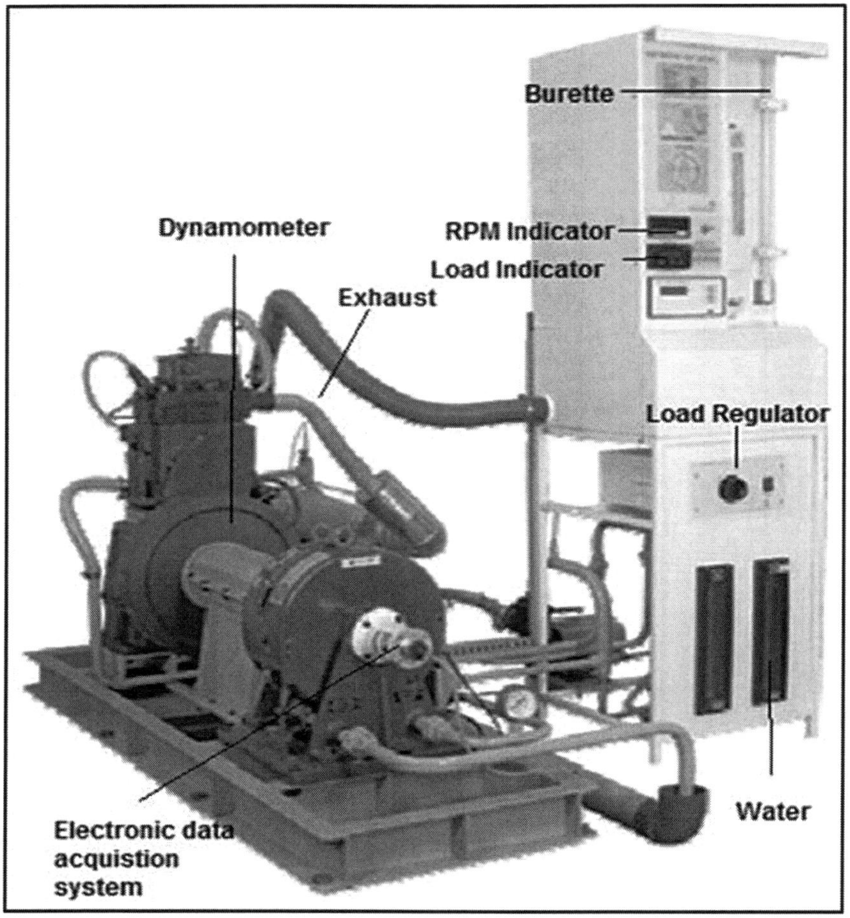

FIGURE 30.1 Experimental setup.

30.3 RESULTS AND DISCUSSION

30.3.1 ENGINE PERFORMANCE

30.3.1.1 BRAKE SPECIFIC FUEL CONSUMPTION

The brake specific fuel consumption (BSFC) values for different blends at different loads are shown in Figure 30.2. The BSFC values of all fuels decrease with increase in the load. BSFC for ethanol and biodiesel blends

(BL1, BL2, BL3, and BL4) was always lower than mineral diesel due to the presence of diesel additives that help for proper combustion.[9]

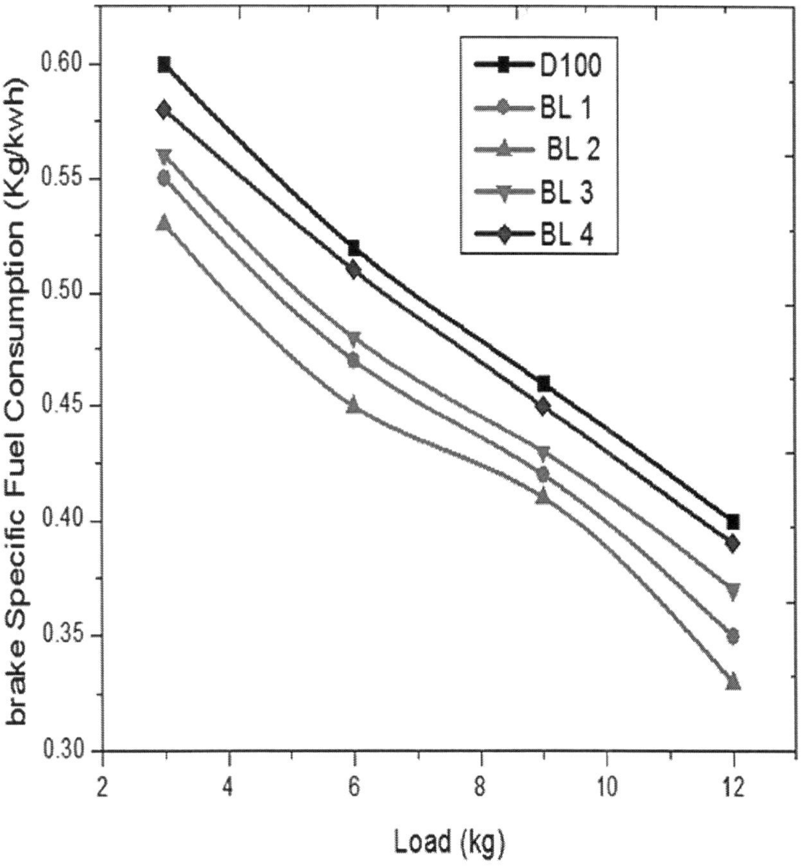

FIGURE 30.2 Brake specific fuel consumption (BSFC) as a function of load for different fuel blends.

30.3.1.2 BRAKE THERMAL EFFICIENCY

The brake thermal efficiency (BThE) of DI diesel engine found for different blends is shown in Figure 30.3 as a role of load for C: R (17:5). The maximum BThEs were BL2 and BL1, respectively. This could be attributed to the presence of increased amount of oxygen and presence of ethanol in all blends, which might have caused better burning as compared to D100.[4]

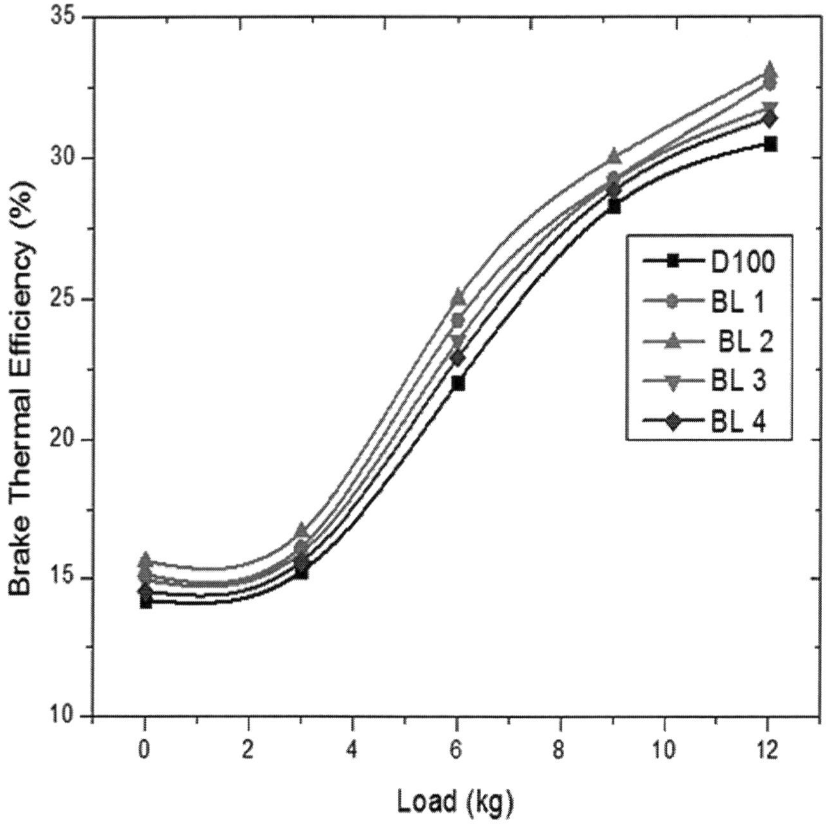

FIGURE 30.3 Brake thermal efficiency (BthE) as a function of load for different fuel blends.

30.3.2 ENGINE EMISSION

30.3.2.1 OXIDES OF NITROGEN EMISSION

NO_x emissions increase in all cases as the load increases due to higher combustion temperature, as seen in Figure 30.4. In the case of biodiesel–diesel–ethanol blends, ethanol addition reduces NO_x emissions as compared to diesel fuel, which might be because of the cooling effect of ethanol leading to lower cylinder combustion temperatures where NO_x is formed, in spite of the previously noted higher exhaust temperature at low load, but does not contribute to NO_x production there because of lower exhaust temperatures than in-cylinder temperatures.[9–11]

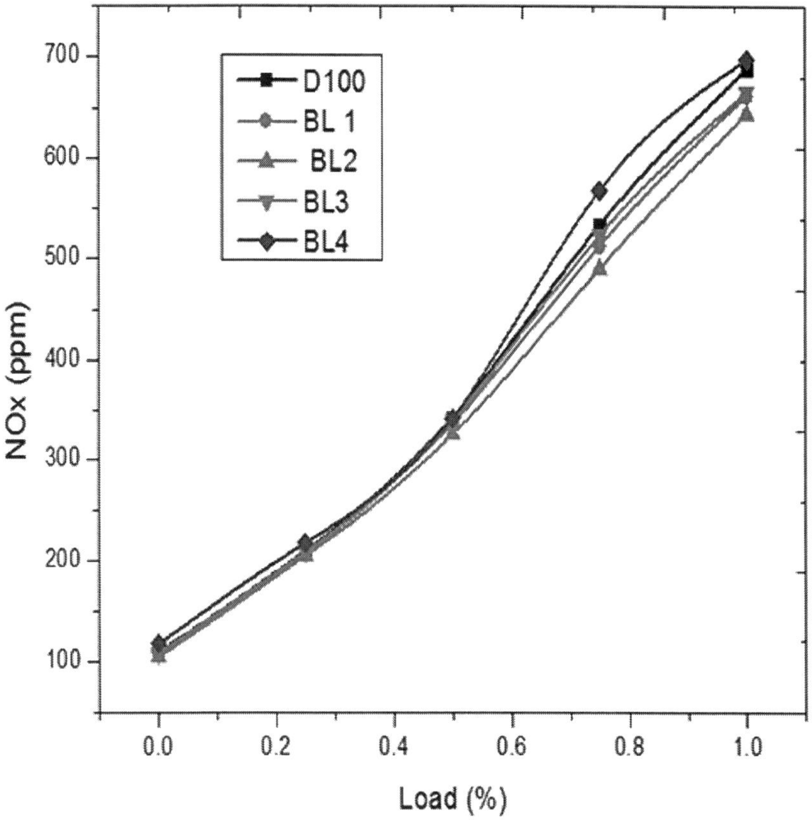

FIGURE 30.4 Oxides of nitrogen (NO_x) as a function of load for different fuel blends.

30.3.3 TRADE-OFF ANALYSIS

The present study deals with a complete trade-off analysis carried on full-load levels involving NO_x, BSFC, and BThE diesel equivalent, which has been summarized in Figure 30.5. Thus, this study deals with a scope to explore the best possible fuel combination at full-load conditions, which will simultaneously reduce NO_x at optimum fuel consumption and BThE.

Figure 30.5 shows the trade-off graph at full-load condition It is evident from the graph, compared to D100 and all other blends, the BL2 yielded enhancement in BThE with a reduction in BSFC and NO_x emissions.

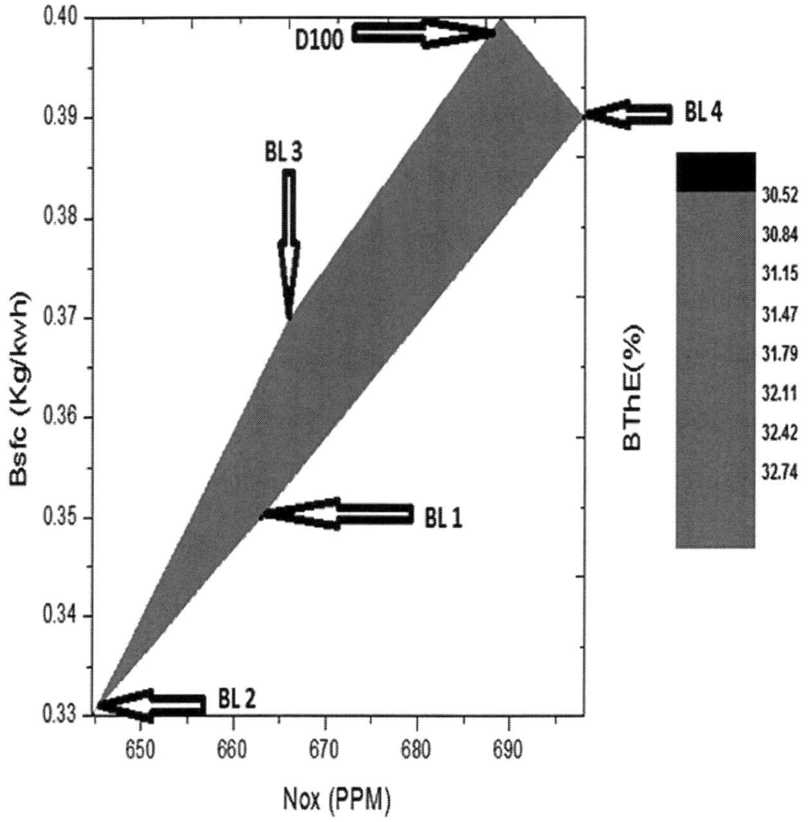

FIGURE 30.5 Trade-off between NO_x-BSFC with reference to BThE at 100% load condition.

30.4 CONCLUSIONS

The goal of the current experimental investigation was to study the performances and emissions of palm biodiesel and ethanol blends with D100. The successive deductions are drawn from the investigational results:

- Palm biodiesel and ethanol appear to have a latent to be used as a supernumerary fuel in DI engines without any modification.
- The BSFC (Fig. 30.3) all blends are better than D100. BSFC (Fig. 30.2) is lower for all blends of PME, diesel additive, and diesel than D100 at all load conditions.

- Key exhaust contaminant such as NO_x (Fig. 30.4) is in summary with the use of methyl easter and the additive.
- In trade-off study, it has been observed that at full-load condition, BL2 is the optimum blend for our uses.
- From experiment, it is concluded that BL2 can substitute the D100 for DI diesel engine to achieve better performance and less emission.

ACKNOWLEDGMENT

The authors acknowledge the research to the Department of ME and PE, National Institute of Technology, Agartala.

KEYWORDS

- **palm oil**
- **ethanol**
- **emissions**
- **biodiesel blends performance**
- **trade-off study**

REFERENCES

1. Aydın, F.; Öğüt, H. Effects of Using Ethanol-Biodiesel-Diesel Fuel in Single Cylinder Diesel Engine to Engine Performance and Emissions. *Renew. Energy* **2017,** *103*, 688–694.
2. Alviso, D.; et al. Development of a Diesel-Biodiesel-Ethanol Combined Chemical Scheme and Analysis of Reactions Pathways. *Fuel* **2017,** *191*, 411–426.
3. Bakar, R. A., Semin; Ismail, A. R. Fuel Injection Pressure Effect on Performance of Direct Injection Diesel Engines. *Am. J. Appl. Sci.* **2008,** *5*(3), 197–202.
4. Celýkten, I. An Experimental Investigation of the Effect of the Injection Pressure on Engine Performance and Exhaust Emission in Indirect Injection Diesel Engines. *Appl. Therm. Eng.* **2003,** *23*, 2051–2060.
5. Hu, N.; et al. Volatile Organic Compound Emissions from an Engine Fueled with an Ethanol-Biodiesel-Diesel Blend. *J. Energy Inst.* **2017,** *90*(1), 101–109. https://doi.org/10.1016/j.joei.2015.10.003.
6. Ksastry, G. R. Bio-Diesel Bio-Degradable Alternative Fuel for Diesel Engines. Readworthy Publications Pvt. Ltd. ISBN 13: 978-81-89973-50-6, ISBN 10: 81-89973-50-9.

7. Murugesan, A.; Umarani, C.; Subramanian, R.; Nedunchezhian, N. Bio-Diesel as an Alternative Fuel for Diesel Engines—a Review. *Renewable Sustainable Energy Rev.* **2009,** *13*, 653–662.
8. Sathiyagnanam, A. P.; Saravanan, C. G.; Dhandapani, S. *Effect of Thermal-Barrier Coating plus Fuel Additive for Reducing Emission from DI Diesel Engine*. Proceedings of the World Congress on Engineering, London, U. K., June 30–July 2, 2010, Vol. 2 WCE 2010.
9. Torres, J. J.; et al. Ultrafiltration Polymeric Membranes for the Purification of Biodiesel from Ethanol. *J. Cleaner Prod.* **2017,** *141*, 641–647.
10. Venu, H.; Madhavan, V. Influence of Diethyl Ether (DEE) Addition in Ethanol-Biodiesel-Diesel (EBD) and Methanol-Biodiesel-Diesel (MBD) Blends in a Diesel Engine. *Fuel* **2017,** *189*, 377–390.
11. Verma, P.; Dwivedi, G.; Sharma, M. P. Comprehensive Analysis on Potential Factors of Ethanol in Karanja Biodiesel Production and Its Kinetic Studies. *Fuel* **2017,** *188*, 586–594.

CHAPTER 31

A NOVEL EXPERIMENTAL STUDY OF THE PERFORMANCE AND EMISSION CHARACTERISTICS ON A DIRECT INJECTION CI ENGINE UNDER DIFFERENT BIODIESEL AND DIESEL ADDITIVE FUEL BLENDS

G. RAVI KIRAN SASTRY[1,*], JIBITESH KUMAR PANDA[1,2], RABISANKAR DEBNATH[1,3], and RAM NARESH RAI[1,4]

[1]*National Institute of Technology, Agartala, Tripura, India*

[2]*jibiteshpanda90@gmail.com*

[3]*rabi101991@gmail.com*

[4]*nareshray@yahoo.co.in*

[*]*Corresponding author. E-mail: grksastry1@rediffmail.com*

ABSTRACT

The direct injection diesel engines used for commercial and transport applications are main causes of disaster of fossil fuel reduction and environmental degradation due to emissions. In order to address these difficulties, in this study, mahua methyl ester (MME) and ethanol, an alternative methyl ester, have been considered to be used as a substitute fuel in light of greener emissions related to pure diesel. The engine performance parameters studied were brake specific fuel consumption, brake thermal efficiency, and oxides of nitrogen (NO_x) by using diesel fuel alone and the different blend fuels. The performance of engine with blend fuel was found to be better than the pure diesel. The unburned hydrocarbon emission was reduced significantly and the NO_x emission decreased at different load condition.

31.1 INTRODUCTION

The rising anxieties due to environmental pollution caused by the fossil fuels and the realization that they are non-renewable have led to search for more environment-friendly and alternative fuels.[4] In between various options explored for diesel fuel, methyl ester has been reported to be one of the robust candidates for drops in environmental pollutions.[2,9] Numerous countries including India have already initiated substituting the pure diesel by a certain amount of biofuel. Since India is not independent in edible oil manufacturing, some nonedible oil seeds obtainable in the country are required to be selected for biodiesel production. With plenty of woodland and plant-based nonedible oils existing in our country such as karanja, Jatropha, mahua, sal, neem, and rubber, much effort has not been made to use esters of these nonedible oils as excessive for pure diesel except Jatropha and karanja.[3] Rare detectives have previously obtained biodiesel from some of these oils. Mahua methyl ester (MME) has an estimated annual manufacturing of possibly 181,000 t in India. Alcohols have been widely used in compression ignition engines as substitute fuels. Although alcohols are economically preferred than pure diesel fuel, there are challenges with respect to utilization of alcohols in diesel engines and blending these fuels with diesel. This investigation was fixated on the engine performance and emission characteristics of esterified vegetable oil and diesel additive like ethanol, when used in a diesel engine.[2,7,9]

Nomenclature	
MME	Mahua methyl esters
D100, B100	Pure diesel, pure biodiesel
BL1	20% MME + 2% ethanol + 78% diesel
BL2	20% MME + 4% ethanol + 76% diesel
BL3	20% MME + 6% ethanol + 74% diesel
BL4	20% MME + 80% diesel
BSFC	Brake specific fuel consumption
BThE	Brake thermal efficiency
NO_x	Oxides of nitrogen
UHC	Unburnt hydrocarbon

31.2 EXPERIMENTAL PROCEDURE AND SPECIFICATIONS

The engine used for investigational work contains single-cylinder, four-stroke, conventional fossil fuel engine, which is associated to eddy current type dynamometer (Fig. 31.1). Specifications of the engine are shown in Table 31.1.

TABLE 31.1 Engine Specification.

Make and type	Kirloskar, single-cylinder, four-stroke diesel
Engine type	Vertical compression ignition engine
Stroke length	110 mm
Swept volume	661 cc
Compression ratio	17.6
Power	3.5 kW
Rated speed	1500 rpm
Bore size	87.5 mm
Dynamometer	
Make	Power mag
Type	Eddy current
Load measurement method	Strain gauge
Maximum load	12 kg
Cooling	Water

A five-gas analyzer was used for the measurement of unburned hydrocarbon (UHC) and oxides of nitrogen (NO_x). UHC and NO_x were measured as parts per million which was fitted at the exhaust; by this preparation, we can get the emission characteristics. The experiments are led by using diesel (D100) and blends (BL; 1, 2, 3, 4) at different load conditions on the engine from 0 to 100% in suitable steps at a compression ratio (CR) 17:5. The performance test was carried out on a single-cylinder direct injection (DI) diesel engine using pure diesel, the methyl ester of mahua oil, ethanol, and their blends with diesel. The load range taken is from no load to full load. Table 31.2 shows the basic fuel properties of biodiesel used in this study in addition to ethanol and diesel.

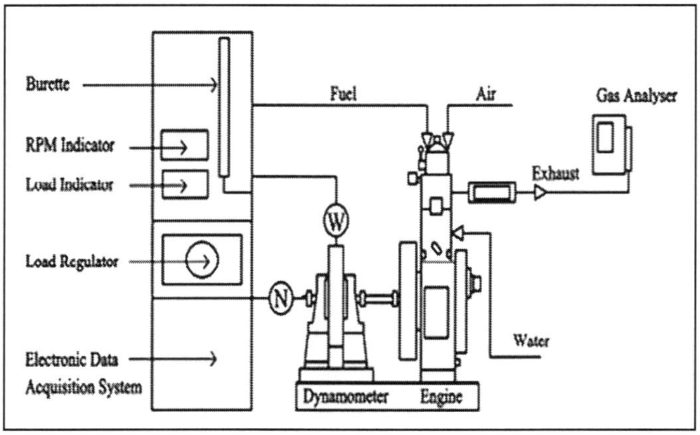

FIGURE 31.1 Schematic diagram of experimental setup.

TABLE 31.2 Properties of Biodiesel, Diesel, and Ethanol.

Properties	Diesel	MME	Ethanol
Density at 150°C (g/cc)	0.815	0.916	0.790
Viscosity at 400°C (cSt)	2.95	5.8	1.1
Flash point (°C)	70	129	12.77
Cetane number	46	48	6

31.2.1 FUEL

After transesterification, the color of crude mahua oil changed from yellow to reddish yellow, and on an average, 80% of recovery of methyl ester was possible. The various fuel properties of mahua oil and mahua methyl ester were determined. The characteristics of methyl ester are close to D100, and therefore, biodiesel can replace the D100, if needed.

31.3 RESULTS AND DISCUSSION

31.3.1 ENGINE PERFORMANCE

31.3.1.1 BRAKE SPECIFIC FUEL CONSUMPTION

The brake specific fuel consumption (BSFC) values for different blends at different loads are shown in Figure 31.2. The BSFC values of all fuels decrease with increase in load. BSFC for ethanol and biodiesel blends (BL1, BL2, BL3, and BL4) was always lower than mineral diesel due to the presence of diesel additives that help in proper combustion.[4]

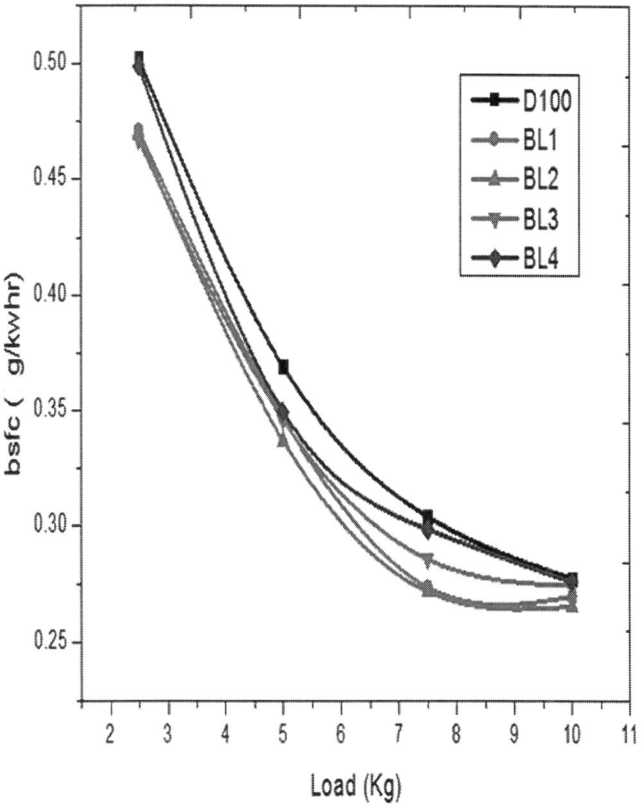

FIGURE 31.2 Brake specific fuel consumption as a function of load for different fuel blends.

31.3.1.2 BRAKE THERMAL EFFICIENCY

The brake thermal efficiency (BThE) of DI diesel engine found for different blends is shown in Figure 31.3 as a function of load for compression ratio of 17:5. The maximum BThEs were BL2 and BL1, respectively. This could be attributed to the presence of increased amount of oxygen and presence of ethanol in all blends, which might have resulted in its improved combustion as compared to pure diesel.[5,6,8]

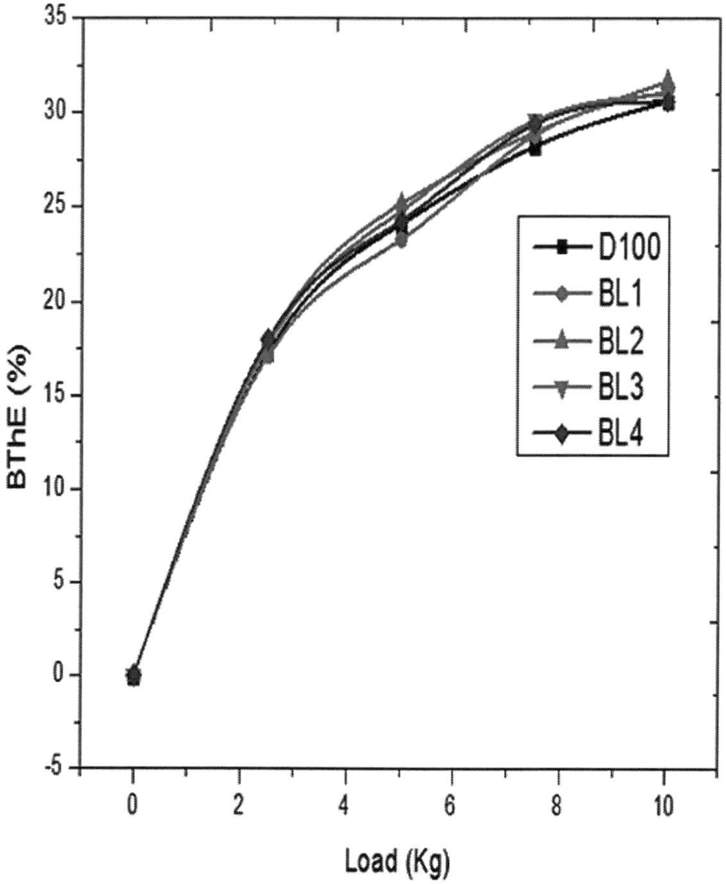

FIGURE 31.3 Brake thermal efficiency as a function of load for different fuel blends.

31.3.2 ENGINE EMISSION

31.3.2 1 UNBURNED HYDROCARBON EMISSIONS

UHC emissions are generated due to incomplete combustion. As the load increases, more complete combustion is achieved and less UHC is generated as seen in Figure 31.4.[3,6,8] Ethanol blended biodiesel–diesel fuels indicate a reduction of UHC emissions at low concentrations and an increase at high concentrations.

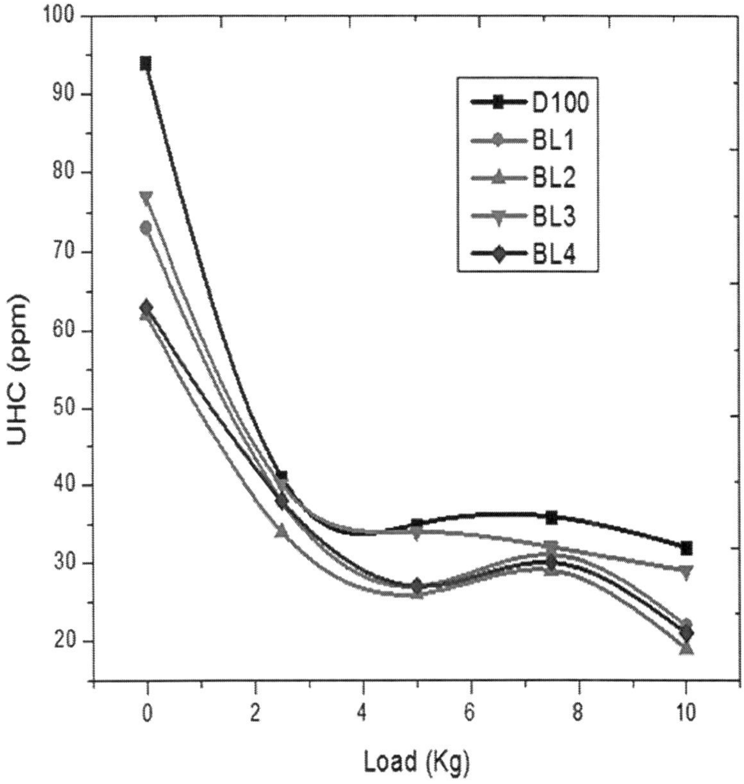

FIGURE 31.4 Unburned hydrocarbon as a function of load for different fuel blends.

31.3.2 2 OXIDES OF NITROGEN EMISSION

NO_x emissions increase in all cases as the load increases due to higher combustion temperature, as seen in Figure 31.5. In the case of biodiesel–diesel–ethanol blends, ethanol addition showed reduced NO_x emissions as compared to diesel fuel, which might be because of the cooling effect of ethanol leading to lower cylinder combustion temperatures where NO_x is formed, in spite of the previously noted higher exhaust temperature at low load, but does not contribute to NO_x production there because of lower exhaust temperatures than in-cylinder temperatures.[1,3,5,6,8]

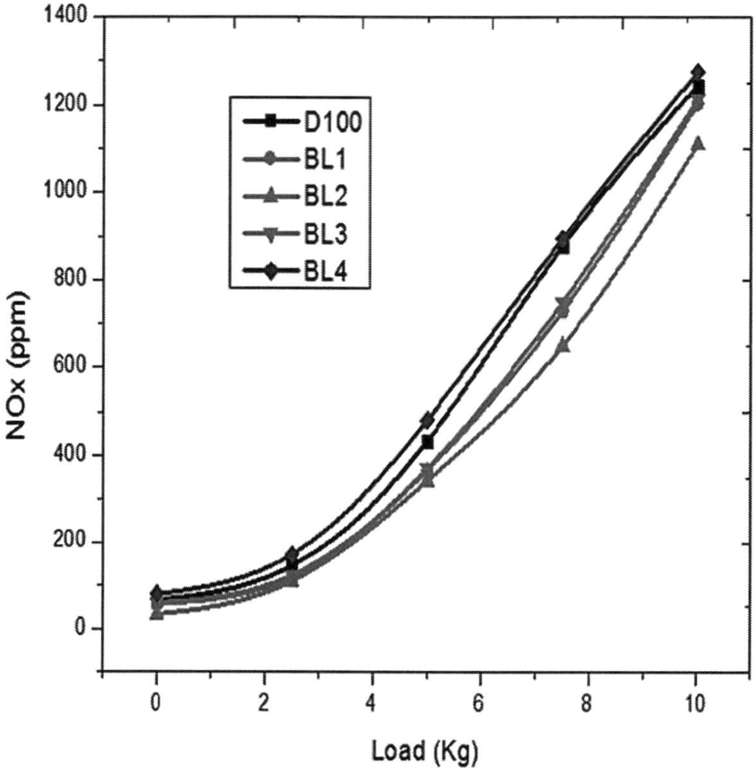

FIGURE 31.5 Oxides of nitrogen as a function of load for different fuel blends.

31.4 CONCLUSIONS

The goal of the current experimental investigation was to study the performances and emissions of mahua methyl ester and ethanol blends with D100. The successive deductions are drawn from the investigational results:

- Mahua biodiesel and ethanol seem to have a potential to be used as substitute fuel in DI diesel engines without any change.
- The BThE (Fig. 31.3) all blends are better than pure diesel. BSFC (Fig. 31.2) is lower for all blends of MME, diesel additive, and diesel than D100 at all load conditions.
- Most of the major exhaust pollutants such as HC (Fig. 31.4) and NO_x (Fig. 31.5) are reduced with the use of biodiesel and the additive.

- From experiment, it is concluded that BL2 might substitute the D100 for diesel engine for getting improved performance and less emission.

ACKNOWLEDGMENT

The authors acknowledge the support of the Department of ME and PE, National Institute of Technology, Agartala for this research.

KEYWORDS

- **Mahua oil**
- **ethanol**
- **emissions**
- **biodiesel blends**
- **performance**

REFERENCES

1. Amini, Z.; et al. Biodiesel Production by Lipase-Catalyzed Transesterification of *Ocimum Basilicum* L. (Sweet Basil) Seed Oil. *Energy Convers. Manage.* **2017,** *132*, 82–90.
2. Basha, S. A.; Raja Gopal, K.; Jebaraj, S. A Review on Biodiesel Production, Combustion, Emissions and Performance. *Renew. Sustain. Energy Rev.* **2009,** *13*(6), 1628–1634.
3. Boloy, R. A. M.; et al. Thermoeconomic Analysis of Hydrogen Incorporation in a Biodiesel Plant. *Appl. Therm. Eng.* **2017,** *113*, 519–528.
4. da Silva, M. A. V.; et al. Comparative Study of NO_x Emissions of Biodiesel-Diesel Blends from Soybean, Palm and Waste Frying Oils Using Methyl and Ethyl Transesterification Routes. *Fuel* **2017,** *194*, 144–156.
5. Hoseini, S. S.; et al. The Effect of Combustion Management on Diesel Engine Emissions Fueled with Biodiesel-Diesel Blends. *Renew. Sustain. Energy Rev.* **2017,** *73*, 307–331.
6. Kakati, J.; Gogoi, T. K.; Pakshirajan, K. Production of Biodiesel from Amari (*Amoora wallichii* King) Tree Seeds Using Optimum Process Parameters and its Characterization. *Energy Convers. Manage.* **2017,** *135*, 281–290.
7. Patil, P. D.; et al. Biodiesel Fuel Production from Algal Lipids Using Supercritical Methyl Acetate (Glycerin-Free) Technology. *Fuel* **2017,** *195*, 201–207.
8. Sundus, F.; Fazal, M. A.; Masjuki, H. H. Tribology with Biodiesel: A Study on Enhancing Biodiesel Stability and its Fuel Properties. *Renew. Sustain. Energy Rev.* **2017,** *70*, 399–412.

9. Syed, A.; et al. Experimental Investigations on DI (Direct Injection) Diesel Engine Operated on Dual Fuel Mode with Hydrogen and Mahua Oil Methyl Ester (MOME) as Injected Fuels and Effects of Injection Opening Pressure. *Appl. Therm. Eng.* **2017,** *114*, 118–129.
10. Venu, H.; Madhavan, V. Influence of Diethyl Ether (DEE) Addition in Ethanol-Biodiesel-Diesel (EBD) and Methanol-Biodiesel-Diesel (MBD) Blends in a Diesel Engine. *Fuel* **2017,** *189*, 377–390.

CHAPTER 32

PRODUCTION AND PURIFICATION OF BIOGAS FROM BIODEGRADABLE WASTE

P. SAI CHAITANYA[1,*], T. V. S. SIVA[2,3], and K. SIMHADRI[1]

[1]*GMR Institute of Technology, Rajam, Srikakulam District, Andhra Pradesh, India*

[2]*Sasi Institute of Technology and Engineering Tadepalligudem, West Godavari District, Andhra Pradesh, India*

[3]*tvenkatasatyasiva@gmail.com*

[*]*Corresponding author. E-mail: nandu.343@gmail.com*

ABSTRACT

Energy demand in rural areas of developing countries has driven researchers toward harnessing energy sources available to them. One of those recognized sources available in rural areas is waste organic materials. This present work examines the continuous process method of biogas production from cattle wastes at various conditions under anaerobic digestion. Biogas technology enables rural dwellers to obtain cheap, high-grade fuel as well as organic fertilizer through the resources available to them locally.[14,16] A biogas program helps to solve environmental problems arising from the disposal of organic waste materials. Treatment of sewage would improve the quality of drinking water supply. Biogas technology helps to conserve foreign exchange for energy importing countries. This work mainly focuses on the production and purification of biogas from biodegradable waste, in this connection, the cow dung was used as a raw material and it was digested in the designed anaerobic digester. The biogas and fertilizer were produced and raw biogas was purified by using designed sections such as steel wool column, CO_2 scrubber, and silica gel column. The test results show that the biogas was produced with respect to the number of days at ambient conditions. The methanogenic bacterium is more active at moderate temperatures. Methane composition is higher after purification when compared with raw biogas. The digested effluent sludge is a high-quality organic fertilizer, which is richer than inorganic manure, and it fixes well in the soil.

32.1 INTRODUCTION

The demand of the world's fossil fuels and their price increase due to the global population rise, and then the use of alternate energy resources is the most important source for the future. These alternate resources can be waste from various sources, that is, organic waste,[12,22] cow dung,[1,8,19] food waste,[9] cattle slurry,[3,5,21,24] garden waste,[10] pig slurry,[2] agricultural waste, municipal waste, industrial waste,[26] poultry litter,[18,21] and so forth. The present paper deals with "the production and purification of biogas from cow dung." The biogas production takes place in an anaerobic digester. The by-product of anaerobic digestion (AD) of organic materials is commonly referred to as "biogas" because of the biological nature of gas production. Biogas technology refers to the production of a combustible gas and a value added fertilizer by the anaerobic fermentation of organic materials under certain controlled conditions of temperature, pH, C/N ratio,[13] and so forth.

By Al-Rousan and Zyadin,[4] a thermophilic biodigester unit was built subsurface with 22 m^3 capacity (15 m^3 manure tanks plus 7 m^3 biogas holder) in a relatively small scale dairy farm. The daily feed was about 500 L of cow slurry (150–200-L cow manure) and the remaining were production liquids. The retention time was approximately 25–30 days and the seasonal temperature measured was about 18–20°C. The unit was thermally insulated; therefore, the temperature fluctuation was slightly around +/−3°C. The daily biogas production was estimated at 7 m^3 equivalent to 4 kg of liquefied petroleum gas (LPG) or 11 (12.5 kg) LPG cylinders per month worth US$140. The methane percentage was 60% in relatively warm temperature (18–20°C) and approximately 56% in colder temperature.

32.2 MATERIALS AND METHODS

32.2.1 MATERIALS AND CHEMICALS

Chemicals: Sodium hydroxide, potassium dichromate, and ferrous ammonium sulfate. These chemicals are used for calculation of pH and chemical oxygen demand (COD) of the material undergoing AD in the digester.

Materials: Silica gel and steel wool for purifying the obtained gas.

32.2.2 EXPERIMENTAL PROCEDURE

Specifications:

Anaerobic digester volume: 20 L
Steel wool bottle volume: 500 mL
Water scrubbing unit volume: 5 L
Moisture removal unit volume: 3 L
Gas storage: 2 L

Figure 32.1 shows the schematic diagram of the production and purification of Biogas setup. The AD process[6,17,26] takes place in three main phases: hydrolysis, acid, and methane, with each phase characterized by the main activity of a certain group of bacteria. During the AD process, the bacteria decompose the organic matter in order to produce the energy necessary to their metabolism, of which methane is a by-product. Our digester consists of 12 L of mixture (water and cow dung in the ratio of 1:1). For creating the anaerobic condition, we can purge nitrogen gas or we can maintain it as a closed system. Then, anaerobic process takes place in three phases (Fig. 32.2):

1. *Phase 1: Hydrolysis:* This is the first phase of the process. Here, complex molecules of proteins, cellulose, lipids, and other complex organics in the vegetable waste are solubilized into glucose, amino acids, and fatty acids. This stage is also known as the polymer breakdown stage 2.
2. *Phase 2: Acid phase:* In this phase, facultative acid-forming bacteria convert the solubilized organic matter to organic acids. The principal acids produced are acetic acid, propionic acid, butyric acid, and ethanol. The optimum temperature of this phase is 30°C.

3. *Phase 3: Methane phase:* The third phase results in the production of methane by methanogenic bacteria. They convert the acids produced in the second phase into methane and carbon dioxide. The AD process is controlled effectively by this group of bacteria; they are very sensitive to pH, substrate composition, and temperature. If the pH level in the digester drops below 6.0, methane formation ceases and there is a more acid accumulation bringing the digestion process to a halt.

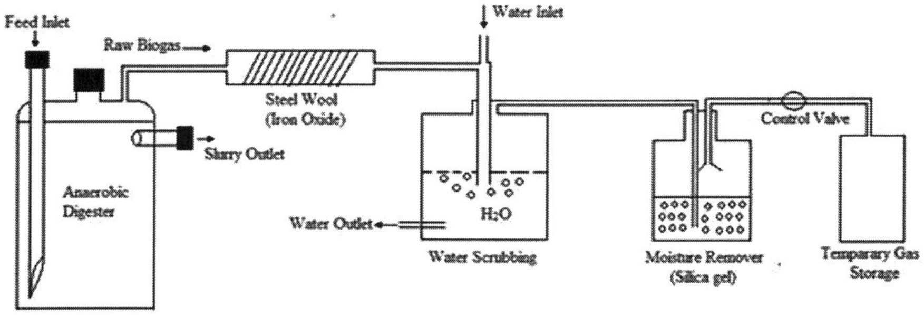

FIGURE 32.1 Schematic diagram of the production and purification of biogas setup.

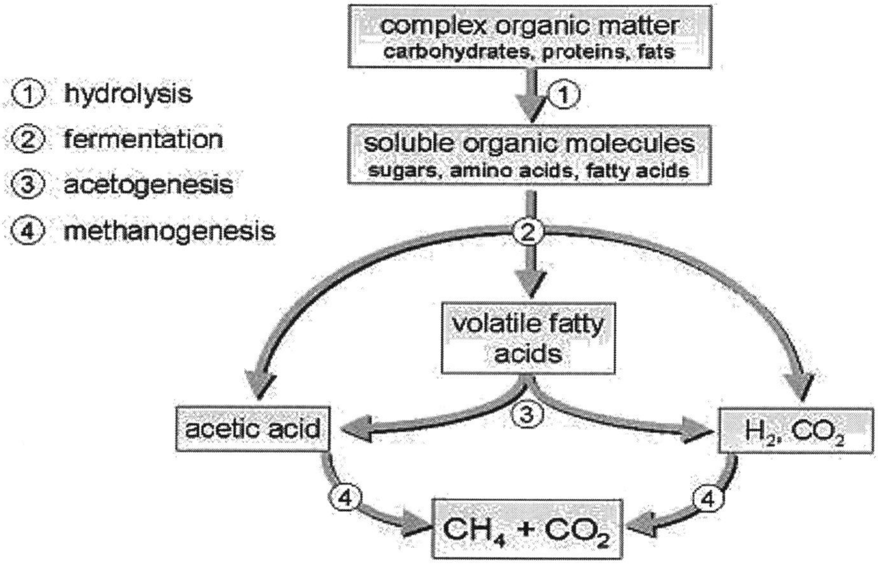

FIGURE 32.2 Anaerobic process phases.

32.2.3 PRODUCTS OF ANAEROBIC DIGESTION

The AD process creates biogas, fiber, and liquor. In order for a scheme to be financially viable, uses for all the products need to be developed and balanced. The balance of the different products from AD is shown in the figure below.

1. *Biogas:* This is made of mostly methane (approximately 60%) and carbon dioxide with traces of other gases such as ammonia and hydrogen sulfide. The gas is considered to be an environmental-friendly energy source for a number of reasons. The first is that the carbon dioxide is from an organic source with a short carbon cycle. It does not contribute to increasing carbon dioxide concentrations. The amount of gas produced is dependent on the quality of the feedstock. The biogas produced through the AD process needs to be cleaned as soon as possible after generation for two main reasons; hydrogen sulfide produced is foul smelling and corrosive as it reacts with moisture in the biogas to form sulfurous acid during combustion; this could damage the engines for power generation. Carbon dioxide, on the other hand, occupies space without providing any extra benefit, though the removal of hydrogen sulfide is of a greater concern. Methods of removing the gases are outlined below.
2. *Digestate:* The digestate produced can be further separated into the liquor and fiber, the quality of which is dependent on the quality of the feedstock.
3. *Fiber:* This is an organic material composed largely of lignin. It is bulky and contains a small amount of plant nutrients. It can be used as a soil conditioner.
4. *Liquor:* This is often used as a fertilizer. It is rich in nutrients and as a result of its high water content; it has additional irrigation benefits. The level of potentially toxic waste is low for the feedstock as it essentially is a vegetable waste. The digestate would pose little or no problems in terms of handling and utilizing them. There are two main types of AD process: the mesophilic digestion[2,20,25] and thermophilic digestion[9,11].

32.2.4 FERMENTED SLURRY AS FERTILIZER

The slurry which comes out of a biogas unit constitutes good quality manure free form weed seeds, foul smell, and pathogens. It contains a full range of plant nutrients in the digested slurry as is indicated in following Table 32.1.

TABLE 32.1 Comparison of Plant Nutrient Content in Digested Slurry (DS) and Farmyard Manure (FYM).

Plant nutrient	Digested slurry (%)	Farmyard manure (%)
Nitrogen (N)	1.5–2.0	0.5–1.0
Phosphorus (P_2OS)	1.0	0.5–0.8
Potash (K_2O)	1.0	0.5–0.8

During the digestion process, gaseous nitrogen (N) is converted to ammonia (NH_3).[23] In this water soluble form, the nitrogen is available to the plant as a nutrient. A particularly nutrient-rich fertilizer is obtained if not only dung, but also urine is digested.

It may be noted that 2.5 kg of fresh dung gives 1 kg of air-dry biogas manure. A mixture of solid and liquid fermented material gives the best yields. Fermented slurry with a lower C/N ratio has better fertilizing characteristics; compared with fresh manure, increase in yield of 5–15% is possible. Particularly good harvests are obtained from the combined use of compost and fermentation slurry. Approximately 70% of the total solids in the slurry can be expected to come out in the sludge; fermentation does not change the form and quantity of these elements. Depending upon the input material used, the sludge contains elements like nitrogen, phosphorus, potassium and also several trace elements like boron, calcium, copper, iron, magnesium, zinc, and so forth. Table 32.2 shows the fertilizer value of biogas manure.

TABLE 32.2 Fertilizer Value of Biogas Manure.

N, P, K in 1000 kg (dry) biogas manure	Equivalent chemical fertilizer
17 kg N_2	37 kg urea
15 kg P_2O_5	94 kg superphosphate
10 kg K_2O	17 kg muriate of potash

If fermentation slurry is to be stored before spreading in the field, it should be covered with soil in layers. This reduces evaporative losses. As already mentioned, the sludge can be used as organic fertilizer. Generally, it is applied either as it comes out from the digester or after dilution with irrigation water.

Furthermore, some manorial value of the sludge may be lost by allowing it to dry. In an alternate method, sludge is led by a channel to a filter bed with an opening at the opposite end of the slope in the bottom. A compact layer of green or dry leaves is made in the filter bed. Water from the sludge filters down and flows out of the opening into a pit. This water can be reused for preparing fresh slurry. The semi-solid residue left on top of the bed has the consistency of dung and can be transported and stored in a pit for use when required.

32.3 EXPERIMENTAL SETUP DESCRIPTION

AD is a collection of processes by which microorganisms break down biodegradable materials in the absence of oxygen. The process is used for industrial or domestic purposes to manage waste and/or to produce fuels. AD is widely used as a source of renewable energy. The process produces biogas, consisting of methane, carbon dioxide, and traces of other "contaminant" gases. This biogas can be used directly as fuel, in combined heat and power gas engines or upgraded to natural gas-quality biomethane. The nutrient-rich digestate is also produced and can be used as fertilizer. The accurate pH is maintained, so that there will be only methanogenic bacteria which results in the formation of biogas. Experimental setup was shown in Figure 32.3.

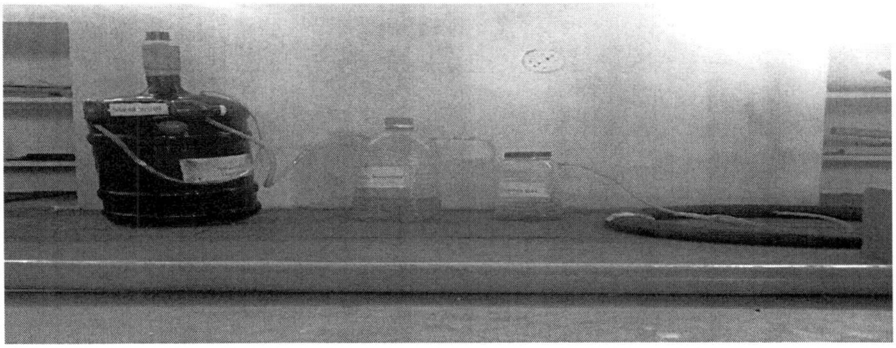

FIGURE 32.3 Experimental setup.

The production of biogas depends upon the following factors:

- Temperature: 30–60°C[1,4,7,9,13,23] is the best range
- pH value: 7.4, best results are obtained in the range of 7.0–7.2, though 6.5–7.6 range [1,4,10,15,19,21,23,26] is considered satisfactory. It is calculated using, generally pH of our feed was less than 7 so by adding 94 mL NaOH, we obtained pH as 7.4, and it is calculated periodically and said to be constant so that we can form methanogenic bacteria.

- Influent solid concentration: Optimum gas production is obtained with 1:1 slurry of cow dung and water.
- Toxicity: The slurry should not contain any toxic materials as they affect anaerobic bacteria.
- COD: 1560 ppm. It is calculated by the dichromate reflux method by using ferrous ammonium sulfate and potassium dichromate solution. Blend sample if necessary and pipet 50.00 mL into a 500-mL refluxing flask. For samples with a COD of >900 mg O_2/L, use a smaller portion diluted to 50.00 mL. Add 1 g $HgSO_4$, several glass beads, and very slowly add 5.0 mL sulfuric acid reagent, with mixing to dissolve $HgSO_4$. Cool while mixing to avoid possible loss of volatile materials. Add 25.00 mL 0.04167 M $K_2Cr_2O_7$ solution and mix. Attach the flask to the condenser and turn on cooling water. Add remaining sulfuric acid reagent (70 mL) through open end of condenser. Continue swirling and mixing while adding sulfuric acid reagent. Cover the open end of the condenser with a small beaker to prevent foreign material from entering refluxing mixture, and continue reflux for 2 h. Cool and wash down condenser with distilled water. Disconnect reflux condenser and dilute the mixture to about twice its volume with distilled water. Cool to room temperature and titrate excess $K_2Cr_2O_7$ with FAS, using 0.10–0.15 mL (two to three drops) ferroin indicator. Although the quantity of ferroin indicator is not critical, use the same volume for all titrations. Take as the end point of the titration the first sharp color change from blue-green to reddish brown that persists for 1 min or longer. Duplicate determinations should agree within 5% of their average. Samples with suspended solids or components that are slow to oxidize may require additional determinations. The blue-green color may reappear. In the same manner, reflux and titrate a blank containing the reagents and a volume of distilled water equal to that of the sample.

32.3.1 REMOVAL OF H_2S USING STEEL WOOL

The gas coming out of the digester enters the bottle containing steel wool and removes H_2S gas. The presence of H_2S causes corrosion to the metal parts. The effect of H_2S on non-ferrous metals in components such as pressure regulators, gas meters, valves, and mountings is more serious. The combustion product SO_2 combines with water vapor and badly corrodes burners, gas lamps, and engines. Running engines with H_2S containing gas can reduce service time to the first general overhaul by 10–15%. Even small traces of H_2S can damage plants.

32.3.2 CO_2 SCRUBBING

After the hydrogen sulfide was removed by the steel wool, the raw biogas passes into the water scrubbing unit for further purification. When carbon dioxide is dissolved in water, carbonic acid is formed. It is a weak acid. Complete separation of CO_2 can improve the heating value of the gas by about 80% on a volume basis and nearly 225% on weight basis. Methane burns faster than biogas and hence will yield a higher specific output and thermal efficiency compared to biogas as engine fuel. Moreover, the absence of CO_2 in the gas will enable an additional volume of air to be inducted into the engine cylinder, thereby improving the volumetric efficiency of the engine. It, therefore, seems very attractive to find means of separating biogas into methane and CO_2, which can serve the dual purpose of improving the quality of combustion as well as obtaining CO_2 as a commercial by-product.

32.3.3 REMOVAL OF MOISTURE CONTENT USING SILICA GEL

The gas leaving the scrubbing unit is purified and has some water vapors. Water vapors are the leading corrosion risk factor. We use silica gel for absorbing that moisture content present in the gas.

32.3.4 STORAGE

The storage of biogas is not easy. Its critical temperature and pressures are −82.5°C and 47.5 bar, respectively. Most commonly used biogas storage options are on propane or butane tanks and commercial gas cylinders up to 200 bar. Depending on the application of biogas the storage facilities vary.

32.3.5 TESTING

When tested, if we observe a clear blue flame, then it is an indication of high-quality biogas. Composition of gas can be calculated using gas chromatography.

32.4 GAS ANALYSIS, PURIFICATION, AND PRODUCTION

32.4.1 GAS ANALYSIS

The gas obtained is stored in a glass sampler and by using a gas chromatography technique we can find the composition of that gas. As the glass sampler is high in cost and very rarely available, we have chosen an alternative method for finding the composition of methane in the sample taken.

- 30 mL of 5% NaOH is taken in a rubber cork bottle and 20 mL of biogas is injected into it.
- It is shaken for 5 min and left for 10 min.
- The CO_2 present in biogas is absorbed by NaOH and forms sodium bicarbonate.
- Methane present in biogas is separated with the help of syringe.

32.4.2 PURIFICATION

Scrubbing is an operation that removes H_2S from raw biogas; as a result the hydrogen sulfide is converted into black iron sulfide by the steel wool. The steel wool used before and after scrubbing is shown in Figure 32.4.

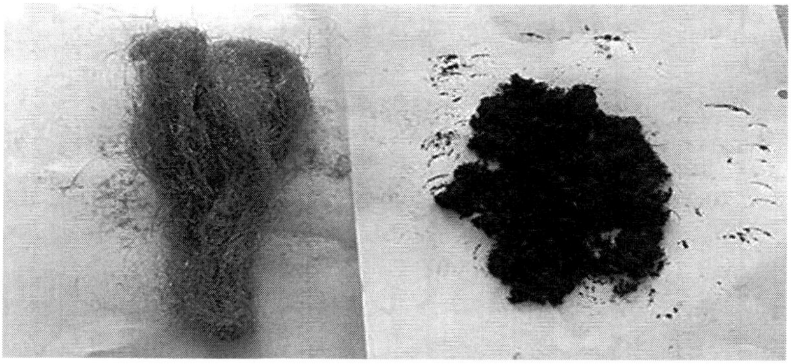

FIGURE 32.4 Purification using steel wool.

The color of the silica gel was changed from blue to pink after absorbing the moisture from the purified biogas. The silica gel used before and after absorption is shown in Figure 32.5.

FIGURE 32.5 Purification using silica gel.

32.4.3 PRODUCTION

It is clear that cow dung is an effective feedstock for AD and could significantly enhance the cumulative biogas production. Gas production from cow dung and methane production from cow dung after purification is shown in Figures 32.6 and 32.7.

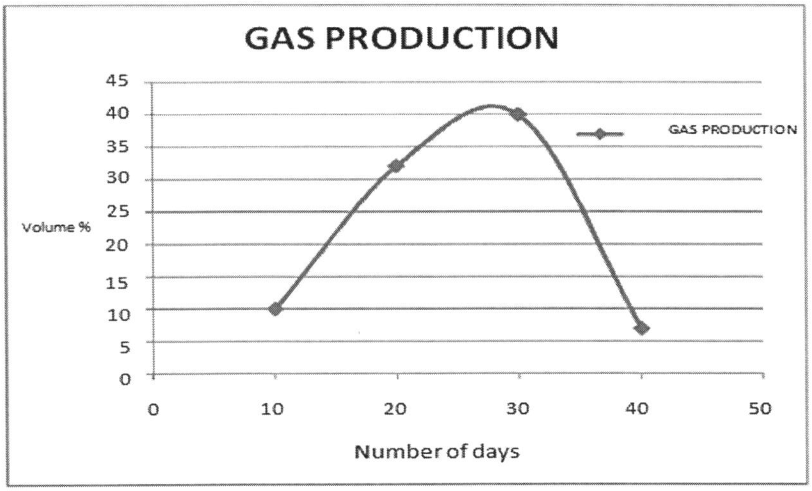

FIGURE 32.6 Gas production with respect to time (number of days).

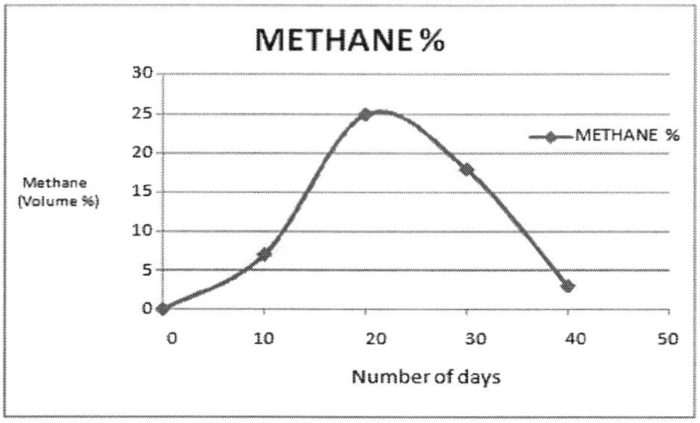

FIGURE 32.7 Methane production with respect to time.

32.5 A CASE STUDY ON PRODUCTION OF BIOGAS FROM DIFFERENT FEEDSTOCKS

The biogas production was different for different substrates because the bacteria responsible for the breakdown of substrate were different. While amylolytic bacteria are good for cow dung, proteolytic bacteria are best for poultry manures. What are good for the farmers, based on this study, are poultry manures, but the supply of these substrates can be a problem at times.

Cow dung users can have a continuous supply of substrate from animals on a daily basis. This is one reason that the use of cow dung can be recommended for the long-term use. For large farms, where there is a continuous process of rearing poultry, biogas production by this method would be the best as continuous organic loading of reactors. Figure 32.8 shows the gas production from different materials in different time slots.

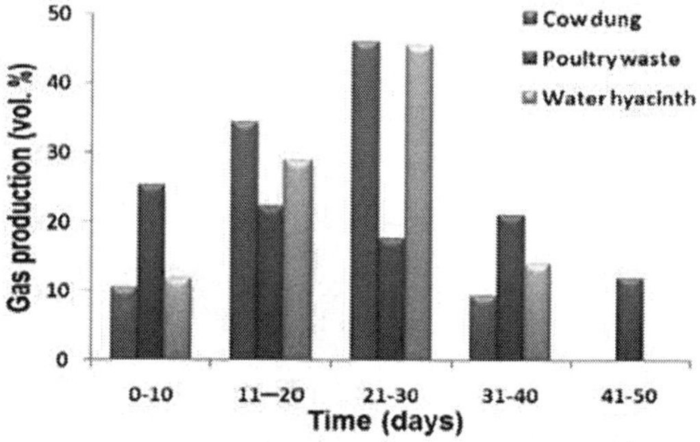

FIGURE 32.8 Gas production (vol.%) from different materials in different time slots.

32.6 CONCLUSION

- In the literature study, we estimated that the traces of impurities are present in biogas. Removal of these impurities is essential to use this gas as fuel for various applications.
- It is proved that the concentration of methane available in the purified biogas was higher than that of raw biogas.
- The approximate methane concentration of purified and raw biogas was 40 and 25%, respectively.
- Biogas technology can be a viable development option for developing countries for energy production and substitution if properly managed and marketed.
- Biogas as a fuel, meeting the requirements of the compositional standards should,

  a) Provide safe operation of the engine whether stationary or automotive and associative equipment.
  b) Protect the fuel system from the detrimental effects of corrosion, poisoning, and liquid or solid deposition and
  c) Do not emit any pollutants or the greenhouse gases after combustion, beyond prescribed limit.

This study investigated the effectiveness of cow dung for biogas production and presented the performance characteristics of the AD in batch and semicontinuous operations. Under above cited operations, cow dung quality is improved by 47% VS reduction[19,22] and approximately 48.5% COD reduction[1,6,7,23] with biogas yield of 0.15 L biogas/kg VS added. Despite large variations in pollutants concentrations, an improved performance of AD of the biodegradable fraction of cow dung was achieved. The results showed that cow dung might be one of feedstocks for the efficient biogas production, fertilizer, and waste treatment.

KEYWORDS

- **biogas**
- **anaerobic digestion**
- **biodegradable waste**

REFERENCES

1. Abubakar, B. S. U. I.; Ismail, N. Anaerobic Digestion Of Cow Dung For Biogas Production. *ARPN J. Eng. Appl. Sci.* **2012,** *7*, 169–172.
2. Adebayo, A. O.; Jekayinfa, S. O.; Linke, B. Effect Of Co-Digesting Pig Slurry with Maize Stalk on Biogas Production at Mesophilic Temperature. *J. Multidiscip. Eng. Sci. Technol.* **2015,** *2*, 2295–2300.
3. Adebayo, A. O.; Jekayinfa, S. O.; Linke, B. Effect of Co-Digestion on Anaerobic Digestion of Cattle Slurry with Maize Cob at Mesophilic Temperature. *J. Energy Technol. Policy* **2013,** *3*(7), 47–54.
4. Al-Rousan, A.; Zyadin, A. A Technical Experiment on Biogas Production from Small-Scale Dairy Farm. *J. Sustainable Bioenergy Syst.* **2014,** *4*, 10–18.
5. Budiyono, B.; Widiasa, I. N.; Seno, J.; Sunarso, S. Influence of Inoculum Content on Performance of Anaerobic Reactors for Treating Cattle Manure using Rumen Fluid Inoculum. *Int. J. Eng. Technol.* **2009,** *1*, 109–116.
6. Demirer, G. N.; Chen, S. Anaerobic Digestion of Dairy Manure in a Hybrid Reactor with Biogas Recirculation. *World J. Microbiol. Biotechnol.* **2005,** *21*, 1509–1514.
7. Fantozzi, F.; Buratti, C. Anaerobic Digestion of Mechanically Treated OFMSW: Experimental Data on Biogas/Methane Production and Residues Characterization. *Bioresour. Technol.* **2011,** *102*, 8885–8892.
8. Forhad Ibne Al Imam, Md.; Khan, M. Z. H.; Sarkar, M. A. R.; Ali, S. M. Original Article Development of Biogas Processing from Cow Dung, Poultry Waste, and Water Hyacinth. *Int. J. Nat. Appl. Sci.* **2013**.
9. Forster-Carneiro, T.; Pérez, M.; Romero, L. I. Influence of Total Solid and Inoculum Contents on Performance of Anaerobic Reactors Treating Food Waste. *Bioresour. Technol.* **2008,** *99*(15), 6994–7002.
10. Gupta, P.; Singh, R. S.; Sachan, A.; Vidyarthi, A. S.; Gupta, A. Study on Biogas Production by Anaerobic Digestion of Garden-Waste. *Fuel* **2011,** *95*, 495–498.
11. Hansen, K. H.; Angelidaki, I.; Ahring, B. K. Improving Thermophilic Anaerobic Digestion Of Swine Manure. *Wat. Res.* **1999,** *33*, 1805–1810.
12. Hansen, T. L.; Jansen, J. C.; Davidsson, S.; Christensen, T. H. Effects of Pre-Treatment Technologies on Quantity and Quality of Source-Sorted Municipal Organic Waste for Biogas Recovery. *Waste Manage.* **2006,** *27*, 398–405.
13. Keener, H. M.; Wicks, M. H.; Michel, F. C.; Eckinci, K. New Technologies and Approaches for Composting Broiler Manure. First International Poultry Meat Congress, Antalya, Turkey, May 2011.
14. Khalid, A.; Naz, S. Isolation and Characterization of Microbial Community in Biogas Production from Different Commercially Active Fermentors in Different Regions of Gujranwala. *Int. J. Water Resour. Environ. Sci.* **2013,** *2*, 28–33.
15. Magdum, S. S.; Minde, G. P.; Adhyapak, U. S.; Kalyanraman, V. Competence Evaluation of Mycodiesel Production by Oleaginous Fungal Strains: Mucor Circinelloides and Gliocladium Roseum. *Int. J. Energy Environ.* **2015,** *6*, 377–382.
16. Mshandete, A. M.; Parawira, W. Biogas Technology Research in Selected Sub-Saharan African Countries – a Review. *Afr. J. Biotechnol.* **2009,** *8*, 116–125.
17. Ngumah, C. C.; Ogbulie, J. N.; Orji, J. C.; Amdi, E. S. Biogas Potential of Organic Waste in Nigeria. *J. Urban Environ. Eng.* **2013,** *7*, 110–116.

18. Nnabuchi, M. N.; Akubuko, F. O; Augustine, C.; Ugwu, G. Z. Assessment of the Effect of Co-Digestion of Chicken Dropping and Cow Dung on Biogas Generation. *Global J. Sci. Front. Res. Phys. Space Sci.* **2012,** *12*(7), 21–26.
19. Okoye, B. O.; Igbokwe, P. K.; Ude, C. N. Comparative Study of Biogas Production from Cow Dung and Brewer's Spent Grain. *Int. J. Res. Adv. Eng. Technol.* **2016,** *2*, 19–21.
20. Rabiu, A.; Yaakub, H.; Liang, J. B.; Samsudin, A. A. Increasing Biogas Production of Rumen Fluid Using Cattle Manure Collected at Different Time as a Substrate. *IOSR J. Agric. Vet. Sci.* **2014,** *7*, 44–47.
21. Rao, A. G.; Gandu, B.; Sandhya, K.; Kranti, K.; Ahuja, S.; Swamy, Y. V. Decentralized Application of Anaerobic Digesters in Small Poultry Farms: Performance Analysis of High Rate Self Mixed Anaerobic Digester and Conventional Fixed Dome Anaerobic Digester. Bioresour. Technol. **2013,** *144*, 121–127.
22. Rico, C.; Diego, R.; Valcarce, A.; Rico, J. L. Biogas Production from Various Typical Organic Wastes Generated in the Region of Cantabria (Spain): Methane Yields and Co-digestion Tests. *Smart Grid and Renew. Energy* **2014,** *5*, 128–136.
23. Strik, D. P. B. T. B.; Domnanovich, A. M.; Holuba, P. A pH-Based Control of Ammonia in Biogas During Anaerobic Digestion of Artificial Pig Manure and Maize Silage. *Process Biochem.* **2006,** *41*, 1235–1238.
24. Sunarso, S.; Seno, J.; Widiasa, I N.; Budiyono, B. The Effect of Feed to Inoculums Ratio on Biogas Production Rate from Cattle Manure Using Rumen Fluid as Inoculums. *Int. J. Waste Resour.* **2012,** *2*, 1–4.
25. Syed, W. S.; Nadeem, M.; Ikram-ul-Haq, Khan, F. A. Production of Biogas by Mesophilic Bacteria Isolated from Manure. *South Asian J. Life Sci.* **2013,** *1*, 12–18.
26. Ye Chen, Jay J. Cheng, Kurt S. Creamer, Inhibition of Anaerobic Digestion Process: A Review. *Bioresour. Technol.* **2007,** *99*, 4044–4064.

CHAPTER 33

EXPERIMENTAL AND STATISTICAL INVESTIGATION OF DENSITY AND VISCOSITY OF BIODIESEL AND DIESEL BLENDS

SAYYED SIRAJ[1*] and H. M. DHARMADHIKARI[2]

[1] *Maharashtra Institute of Technology, Aurangabad (MS), India*

[2] *Marathwada Institute of Technology, Aurangabad (MS), India*

[*]*Corresponding author. E-mail: lucky.sartaj@gmail.com*

ABSTRACT

In this research, commercially available fossil diesel fuel blended with the two different nonedible biodiesels, namely karanja biodiesel and the neem biodiesel. The blends were prepared on the basis of volume which starts from B0 (neat fossil diesel), B10, B20, B30 ... up to B100 (neat biodiesel) in the step of 10. The physical properties of the fossil diesel and biodiesel blends such as density and viscosities were measured as per the ASTM standards test methods. The regression analysis was used for predicting the density and viscosities for the blends of diesel and biodiesel. The new correlations between density and viscosities for diesel and biodiesel blends were developed by using linear, quadratic, and cubic regression analysis and compared. Therefore, according to the results, as the concentration of the fuel blends increases, the density and viscosity also increase. Finally, ±2.49% uncertainty was present in the estimation of density and viscosities from the regression-generated correlations.

33.1 INTRODUCTION

The reservoirs of fossil fuels are depleting all over the world and the energy crises are one of the important issues at the global level. Fossil fuels have been utilized everywhere in the world for various applications such as industrial, transportation sectors, locomotives and marine, and so forth. The emission produced by the fossil fuels after burning is dangerous for the environment in general and human beings in particular.[1] It increases the air pollution and the global warming. The variety of crude oil extracted from the various plant sources is one of the potential sources available for the applications of fossil diesel is the biodiesel.[2] In the transesterification process, the long chain of triglycerides and the alcohol in the presence of catalyst (basically KOH, NaOH) forms ethyl esters and glycerol. If the fatty acid content in the oil is more than 2.5%, the process is known as esterification followed by transesterification.[3] The biodiesel can be utilized in four ways: direct use or blending, microemulsion, thermal cracking or heating or pyrolysis, and transesterification. The most common method is the transesterification of biodiesel. The biodiesel is produced by the alkali-catalyzed process in most of the time.[4] With high free fatty acid (FFA) in the feedstock will react undesirably with the alkali catalyst, thereby forming soap. About 2.5 wt.% FFA is acceptable in the alkali-catalyzed process. If FFA percentage is more than 2.5%, then a pre-step will be required called esterification, followed by transesterification. Figure 33.1 shows the biodiesel production by catalyzed transesterification process.[5]

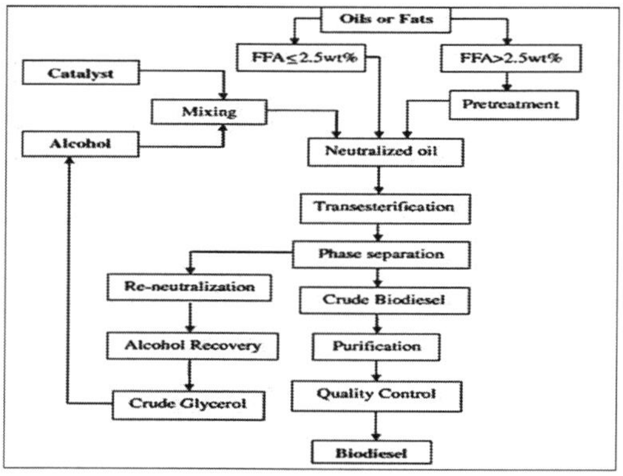

FIGURE 33.1 Alkali-catalyzed transesterification process.

Source: Adapted from ref 5.

33.1.1 FEEDSTOCK OF BIODIESEL IN INDIA

In developing countries like India, the biofuels are being given serious considerations as prospective sources of energy in the future. As per the demand of fossil fuels and the crises of fossil fuels, the fossil fuels will require an alternate solution. The alternative fuels, that is, biodiesels are clean and produced from edible and nonedible sources. The biodiesel blend (B20) is suitable for compression ignition engines with little or no modifications. Biodiesel is simple to use, biodegradable, nontoxic, and essentially free of sulfur and aromatics. The biodiesels are same as that of fossil diesel and do not require any special infrastructure for storage purpose. The products such as carbon monoxide, particulate matters, and unburned hydrocarbons, which come from engine emissions after combustion process, can be reduced when biodiesel is used as a fuel in the conventional diesel engines. When biodiesel is blended with fossil diesel, it improves ignition ability of the engine because biodiesel has higher cetane number.[6] In India, there is a vast requirement of fossil fuel, though the edible oil is not suitable for biodiesel production. In India, plants such as Jatropha, karanja, neem, and mahua contain approximately 30% or more oil contents in their seeds, fruits, kernels, and nuts.

33.1.2 KARANJA (*Pongamia pinnata*)

For a long time, plants like karanja have been utilized as curative agents for different illnesses. In our day-to-day life, we see different kinds of herbal products that are extensively prepared from fruits or seeds of karanja in the market. The karanja, also called as Pongamia, is a genus having only one species, that is, *Pongamia pinnata*. This belongs to family Leguminosae and subfamily Papilionaceae. It is medium-sized glabrous, perennial tree that grows in the littoral regions of Southeastern Asia and Australia. *P. pinnata* is preferred species for controlling soil erosion and bounding sand dunes because of a dense network of lateral roots. Root, bark, leaves, flower, and seeds for this plant also have medicinal properties and traditionally used as medicinal plants.[7,20] All parts of the plant have been used as crude drug for the treatment of tumors, piles, skin diseases, wounds, and ulcers. In the traditional systems of medicines, such as Ayurveda and *Unani*, the *P. pinnata* plant is used. The Pongamia seed contains about 40% oil, which can be converted to biodiesel by transesterification method. Pongamia biofuel requires no engine modifications when blended with diesel in proportions as high as 20%.[21]

33.1.3 NEEM (*Melia azadirachta*)

Neem (*Melia azadirachta*) is of Meliaceae family. It is one of the two species in the genus *Azadirachta* and is native to India and Burma, growing in tropical and semitropical regions. Neem is a fast-growing tree and can reach up to a height of 15–20 merrily to 35–40 m. It bears an ovoid fruit, 2 × 1 cm^2 and each seed contains one kernel. The seed kernels, which weigh 0.2 g, constitute some 50–60%. The fruit yield per tree is 37–55 kg. Neem oil can be used as soaps, medicinal, and insecticide.[8,22] Major fatty acid compositions of oil are palmitic acid 19.4%, stearic acid 21.2%, oleic acid 42.1%, linoleic acid 14.9%, and arachidic acid 1.4%. Neem oil is unusual in containing nonlipid associates often loosely termed as "bitters" and organic sulfur compounds that impart a pungent, disagreeable odor.[9,23] The fatty acid composition of karanja and neem oil has been reported in Table 33.1.

TABLE 33.1 Fatty Acid Composition of Various Biodiesels.

Formal name of free fatty acid (FFA)	Ratio	Chemical formula	FFA composition of karanja oils (%)	FFA composition of neem oils (%)
Caprylic	8:0	$C_8H_{16}O_2$	–	–
Capric	10:0	$C_{10}H_{20}O_2$	–	–
Lauric	12:0	$C_{12}H_{24}O_2$	–	–
Myristic	14:0	$C_{14}H_{28}O_2$	–	0.2–0.26
Palmitic	16:0	$C_{16}H_{32}O_2$	3.7–7.9	13.6–16.2
Palmitoleic	16:1	$C_{16}H_{30}O_2$	–	–
Margaric	17:0	$C_{17}H_{34}O_2$	–	–
Stearic	18:0	$C_{18}H_{36}O_2$	2.4–8.9	14.4–24.1
Oleic	18:1	$C_{18}H_{34}O_2$	44.5–71.3	49.1–61.9
Linoleic	18:2	$C_{18}H_{32}O_2$	10.8–18.3	2.3–15.8
Linolenic	18:3	$C_{18}H_{30}O_2$	–	–
Arachidic	20:0	$C_{20}H_{40}O_2$	–	0.8–3.4
Gondoic	20:1	$C_{20}H_{38}O_2$	–	–
Behenic	22:0	$C_{22}H_{44}O_2$	–	–
Erucic	22:1	$C_{22}H_{42}O_2$	–	–
Lignoceric	24:0	$C_{24}H_{48}O_2$	1.1–3.5	–

Note: X:Y ratio shows the number of carbon atoms and the double bonds in a chain.

Source: Adapted from ref 10.

33.2 LITERATURE REVIEW

33.2.1 DENSITY

The term density for liquid or a material is defined as the ratio of mass per unit volume. The specific gravity is nothing but the ratio of density of a substance to the density of reference substance (usually water). Many researchers prefer this dimensionless term.[11] The density of fossil fuel (diesel) and biodiesel can be measured by two different test methods, first by ASTM standard D941 and second by European standard of EN ISO 3675 and EN ISO 12185 test methods. [19] Density can be measured by the instruments such as hydrometer, pycnometer, and anton-Paar density meter at room temperature 25°C.[12] From the different research papers, it has been observed that the average density 5–8% is higher than the corresponding fossil diesel value. Due to unsaturation and decrease in chain length of methyl esters, the density of fuel increases. This high density of fuel can affect the fuel consumption as the fuel introduced into the combustion chamber is determined volumetrically.[13,24] As we know, the biodiesels fuels have a higher density than the fossil diesel, which means that volumetrically operating fuel pumps will inject greater mass of biodiesel than the fossil diesel fuel. That is why the flow is controlled by volume, the expected peak power reduction for engines using is only 5–7% less than the fossil diesel because more (g/mL) would flow and vaporize more efficiently given a set throttle (volume). The biodiesel has an average density, that is, crude oil density, 12% higher than the fossil diesel. Biodiesel's higher specific gravity and density relative to fossil diesel means that on road, biodiesel blends are normally made by splash blending the biodiesel fuel on top of the conventional diesel fuel or fossil fuel. It should be noted that biodiesel produces more than three times the energy as the same amount of fossil fuel. Tat and Gerpen measured experimentally the specific gravity of 20, 50, 75, and 100% soybean biodiesel as a function of temperature in the temperature range of crystallization temperature to 100°C using the standard hydrometer method. The results indicate that the biodiesel and its blends demonstrate temperature-dependent behavior. They developed first-degree linear regression equation as given in Equation 33.1.

$$SG = a + bT \tag{33.1}$$

where SG is the specific gravity of blended biodiesel, T is the temperature in °C, and a and b are constants that depend on the percentage of biodiesel and diesel blends. Tate et al. also reported similar linear equation for three different biodiesels, namely canola oil biodiesel, soybean oil biodiesel, and fish oil biodiesel. This equation was developed based on the data obtained which was available in the range of 20–300°C. Tate et al. had also suggested that Equation 33.2 can be used to determine the specific gravity of different biodiesel blends at a standard temperature. In this equation, the specific gravity of the blend has been considered to be proportional to mass fractions of the constituents.

$$SG_{blend} = \Sigma\, SG_i M_i \tag{33.2}$$

where SG_{blend} is the specific gravity of the blend, SG_i is the specific gravity of component i, and M_i is the mass fraction of component i. Alptekin and Canakci carried out the experimental tests on different biodiesels made of soybean oil, waste palm oil, sunflower oil, corn oil, canola oil, and cottonseed. They suggested first-degree empirical equation, which can relate the density of biodiesel blend with the percentage of biodiesel used shown in Equation 33.3.

$$D = Ax + B \tag{33.3}$$

where D is the density (g/cm^3), A and B are constants which vary with the type of the biodiesel, and x is the biodiesel fraction.

33.2.2 KINEMATIC VISCOSITY

The viscosity is the measurement of internal fluid friction or resistance of oil to flow, which tends to oppose any dynamic change in the fluid motion. The viscosity can be measured by two different ways: kinematic viscosity and dynamic viscosity.[14] The fossil diesel and biodiesel kinematic viscosity can be measured by two different test methods: first by ASTM standard D445 and second by European standard of EN ISO 3104 and EN ISO 3105 test methods. The viscosity ranges given as per ASTM D445 and EN ISO 3104/05 standards are 3.5–5.0 and 1.9–6.0 mm^2/s.[19] Many of the researchers have used Redwood viscometer, Setavis kinematic viscometer, and Cannon-Fenske viscometer of tube size 75, 100 in the viscometer bathtub to measure the viscosity. Normally, the kinematic viscosity can be estimated in Equation 33.4.[15,25]

$$\begin{aligned}\text{Kinematic viscosity} &= \text{Calibration constant (mm}^2/\text{s}^2) \\ &\quad \times \text{efflux time of flow (s) in mm}^2/\text{s}\end{aligned} \tag{33.4}$$

The crude vegetable oils have higher viscosity than the fossil diesel though crude oils are not suitable to use as a fuel in compression ignition engines, at least not without prior heating (viscosity decreases exponentially with increase in temperature) and only for relatively small blending ratios. The crude oil has higher viscosity than biodiesel and biodiesel has higher viscosity than fossil diesel; the crude oil has viscosity 10–17% higher than the biodiesels. Several structural features influence the kinematic viscosities of fatty acid methyl esters FAME, such as chain length, degree of unsaturation, double-bond orientation, and type of ester head group. Factors such as longer chain length and larger ester head group result in the increases in kinematic viscosity.[16] Increasing the degree of unsaturation results in a decrease in kinematic viscosity and as the temperature of oil is increased, its viscosity decreases and it is, therefore, able to flow more readily. Double-bond orientation also impacts kinematic viscosity.[17,26] Viscosity is the most important property of lubricating oil as it affects the wear rate of engine components. Relatively higher viscosity of biodiesel helps in plugging the clearance between piston rings and cylinder liner effectively, thus reducing blow by losses and fuel dilution of lubricating oil.[18] The literature shows that the viscosity of biodiesel can be estimated from well-known mixing laws such as the Grun–Nissan and Katti–Chaudhari laws, which were originally proposed by Arrhenius. The law has given in mathematical form in Equation 33.5.

$$\ln \eta_{max} = x_1 \ln \eta_1 + x_2 \ln \eta_2 \tag{33.5}$$

where η_{max} is the kinematic viscosity (mm^2/s) of the mixture, η_1 and η_2 are the kinematic viscosities of components 1 and 2, and x_1 and x_2 are the mass or volume fractions of components 1 and 2. Alptekin and Canakci investigated the variation of viscosity as a function of different percentages of blends of biodiesel. The test was conducted at 40°C for a wide variety of biodiesels. They developed second-degree empirical equation to calculate the viscosity of blended biodiesel taking the fraction of biodiesel in the mixture as the main parameters in Equation 33.6.

$$\eta_{blend} = Ax^2 + Bx + C \tag{33.6}$$

where η is the kinematic viscosity of biodiesel (mm^2/s), A, B, and C are coefficients and x is the biodiesel fraction. Riazi and Al-Qtaibi developed Equation 33.7.

$$\frac{1}{\mu} = A + \frac{B}{I} \tag{33.7}$$

where μ is the dynamic viscosity (cP), A and B are constants specific to each component, and I is the refractive index. A modified equation was proposed by Tat and Gerpen to determine the viscosity of the biodiesel at different temperatures in Equation 33.8.

$$\ln(\eta) = A + \frac{B}{T} + \frac{C}{T^2} \tag{33.8}$$

where A, B, and C are constants, T is the temperature in K, and η is the kinematic viscosity (mm²/s). Finally, from Equation 33.8, Pegg et al. developed Equation 33.9 to calculate the dynamic viscosity of B100 as a function of temperature in the temperature range of 277–573 K.

$$\ln(\eta) = -2.4343 + \frac{216.66}{T} + \frac{293523}{T^2} \tag{33.9}$$

where η is the dynamic viscosity (mPa) and T is the temperature (K). By considering the carbon number, Krisnangkura et al. proposed different equations, that is, Equations 33.10 and 33.11 to determine the viscosity of biodiesels with long or short carbon structure at different temperatures.

$$\ln(\eta c_{12} - c_{18}) = -2.177 - 0.202z + \frac{403.66}{T} + \frac{109.772}{T}z \tag{33.10}$$

$$\ln(\eta c_{6} - c_{12}) = -2.915 - 0.158z + \frac{492.12}{T} + \frac{103.35}{T}z \tag{33.11}$$

where $\eta c_{12} - c_{18}$ and $\eta c_{6} - c_{12}$ are the kinematic viscosities of biodiesels with a number of carbon atoms varying from 12 to 18 and 6 to 12, respectively (mm²/s), T is the temperature in K, and z is the carbon number. In Equations 33.10 and 33.11, the number of carbon atoms is the required a priori which limits the use of equations.

33.3 EXPERIMENTATION

The oils are mainly of two types, edible oils and the nonedible oils. The karanja oil and neem oil come in nonedible oil category. Normally, the properties of biodiesel are measured by the ASTM D6751 standard and for fossil diesel by ASTM D975 standard. The physical properties density and viscosity are measured by the ASTM test methods. The density of karanja and neem biodiesel and diesel blends is measured by taking the weights of blends measured by the weighing balance system and the volume of the blends by specific gravity bottle. The density estimated as a derived quantity. The kinematic viscosity at 40°C is measured by the Cannon-Fenske tube of size 100 mm^2/s^2 with constant of 0.0293 in a viscometer bathtub and efflux time taken by the racer digital stopwatch. Table 33.2 provides the experimental values for density and kinematic viscosity for karanja oil methyl esters (KOME) and neem oil methyl esters (NOME).

TABLE 33.2 Experimental Values of Density and Kinematic Viscosity for KOME and NOME Blends.

Blends	(ρ) for KOME in kg/m^3	(ρ) for NOME in kg/m^3	(ν) for KOME in mm^2/s	(ν) for NOME in mm^2/s
B0	807	807	3.809	3.809
B10	810	846	5.303	4.688
B20	814	853	5.655	6.153
B30	820	861	6.065	8.673
B40	827	876	7.032	11.574
B50	835	884	9.376	14.181
B60	848	893	10.343	19.426
B70	851	901	11.486	27.806
B80	867	909	12.892	32.960
B90	875	915	13.830	48.785
B100	878	930	14.357	61.325

The density for fossil diesel is 807 kg/m^3 and for KOME and NOME is about 878 kg/m^3 and 930 kg/m^3, which is approximately 7–16% higher than the fossil diesel. The kinematic viscosity for fossil diesel is 3.809 mm^2/s and for KOME and NOME is about 14.357 and 61.325 mm^2/s, which is approximately 2.76 times for KOME and 15.10 times for NOME in an absolute manner. Figure 33.2 shows the information of density and kinematic viscosity for karanja and neem biodiesel and diesel blends. In Figure 33.1, the diesel values are considered as a base value and from that density and kinematic viscosity for karanja and neem biodiesel and diesel blends are estimated in an absolute manner.

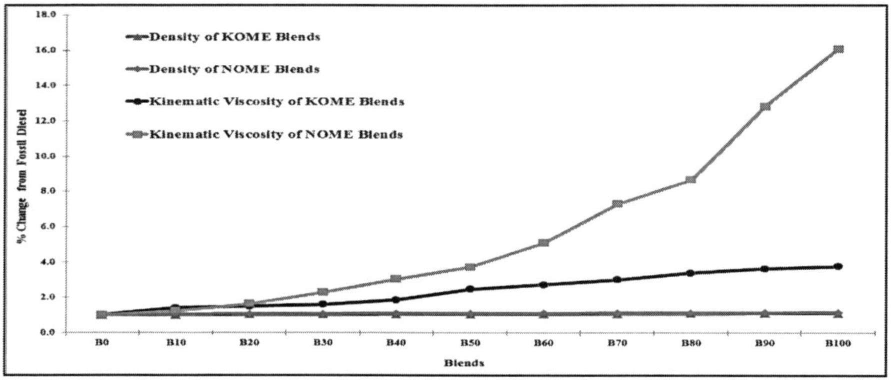

FIGURE 33.2 Blends versus % change from fossil diesel.

33.4 MODELING

This study has used the regression analysis. In brief, the regression analysis is the statistical technique that identifies the relationship between two or more quantitative variables: a dependent variable, whose value is to be predicted, and an independent or explanatory variable (or variables), which is known. The goal of regression analysis is to determine the values of parameters for a function that cause the function to best fit a set of data observations that you provide. Most of the researchers had worked on the characterization of biodiesel and they had predicted some formulations to estimate the values for physical properties from the correlations. The generalized equation is given below to estimate the density from the kinematic viscosity or vice versa.

Density is a function of kinematic viscosity,

$$\rho_{\mathrm{Est.}} = \mathrm{f}\,(\vartheta) \text{ or } \vartheta_{\mathrm{Est.}} = \mathrm{f}\,(\rho)$$

The general equation to predict the density of KOME and NOME and diesel blends,

$$\rho_{(Est.)} = K_{(Bxx)} \pm K_{1\ (Bxx)} x\, \upsilon \pm K_{2\ (Bxx)} x \upsilon^2 \pm K_{3\ (Bxx)} x\, \upsilon^3$$

where $\rho_{(Est.)}$ is the estimated density value in kg/m^3; *Bxx* stands for biodiesel blends, that is, B0, B10, B20, and so forth; and K, K_1, K_2, K_3 are the constant values given by the regression equations.

33.4.1 ESTIMATION OF DENSITY FOR KOME

To estimate the density from the kinematic viscosity, the linear, quadratic, and cubic equations are given below:

$$\rho_{Est.} = 776.07 + 6.94208 \times \vartheta$$

$$\rho_{Est.} = 792.3 + 2.807 \times \vartheta - 0.2237 \times \vartheta^2$$

$$\rho_{Est.} = 787.1 + 4.840 \times \vartheta - 0.163 \times \vartheta^2 + 0.00872 \times \vartheta^3$$

The constants for KOME are given by the regression equations, which are listed in Table 33.3.

TABLE 33.3 Constants for Different Equations.

Density	Constant K	Constant $K_1 \times v$	Constant $K_2 \times v^2$	Constant $K_3 \times v^3$
Linear	+776.07	+6.9420	0.0000	0.0000
Quadratic	+792.3	+2.8070	+0.2237	0.0000
Cubic	+787.10	+4.8400	−0.0163	+0.00872

33.4.2 ESTIMATION OF DENSITY FOR NOME

To estimate the density from the kinematic viscosity, the linear, quadratic, and cubic equations are given below:

$$\rho_{Est.} = 844.296 + 1.61977 \times \vartheta$$

$$\rho_{Est.} = 822.1 + 4.192 \times \vartheta - 0.04195 \times \vartheta^2$$

$$\rho_{Est.} = 797.6 + 8.932 \times \vartheta - 0.2370 \times \vartheta^2 + 0.002072 \times \vartheta^3$$

The constants for NOME are given by the regression equations, which are listed in Table 33.4.

TABLE 33.4 Constants for Different Equations.

Density	Constant K	Constant $K_1 \times v$	Constant $K_2 \times v^2$	Constant$K_3 \times v^3$
Linear	+844.29	+1.6197	0.0000	0.0000
Quadratic	+822.10	+4.1920	−0.04195	0.0000
Cubic	+797.60	+8.9320	−0.2370	+0.00207

33.5 RESULTS AND DISCUSSION

The density and kinematic viscosity of karanja oil methyl esters for different blends are given in Table 33.5, and simultaneously, the estimated values for density by linear, quadratic, and cubic equations are also given in Table 33.5.

TABLE 33.5 KOME Density Values from Different Equations.

Blends	Exp. (ρ)	Exp. (ν)	Est. (ρ) by linear	Est. (ρ) by quadratic	Est. (ρ) by cubic
B0	807	3.809	802.51	806.24	805.78
B10	810	5.303	812.88	813.48	813.61
B20	814	5.655	815.33	815.33	815.53
B30	820	6.065	818.17	817.55	817.80
B40	827	7.032	824.89	823.10	823.36
B50	835	9.376	841.16	838.28	838.23
B60	848	10.343	847.87	845.26	845.06
B70	851	11.486	855.81	854.05	853.76
B80	867	12.892	865.57	865.67	865.47
B90	875	13.830	872.08	873.91	873.99
B100	878	14.357	875.74	878.71	879.03

Figure 33.3 shows the correlation between the experimental and estimated density of karanja oil methyl esters and its blends. The experimental density is the actual measured value of the biodiesel blends and the estimated density value can be calculated from the formula which is derived from the statistical software. The density estimated from linear, quadratic, and cubic regression equations is correlated with the experimental density of karanja oil methyl esters blended with fossil diesel. From Figure 33.3, it is concluded that the points are nearer to the experimental density line, which means that there is a strong correlation between them. Table 33.6 contents the experimental values of density and kinematic viscosity of neem oil methyl esters and it also has the information of the estimated value of density by linear, quadratic, and cubic equations.

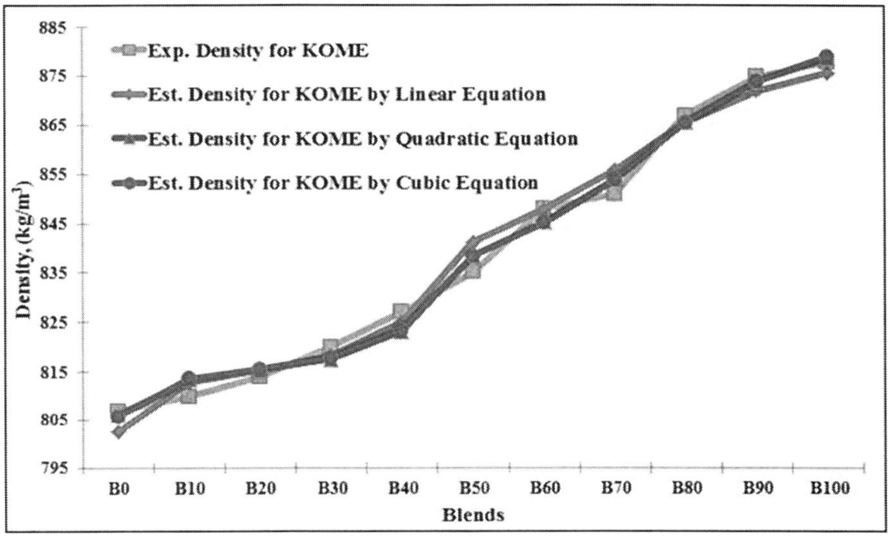

FIGURE 33.3 KOME blends versus experimental and estimated density.

TABLE 33.6 NOME Density Values from Different Equations.

Blends	Experimental (ρ)	Experimental (ν)	Estimated (ρ) by linear	Estimated (ρ) by quadratic	Estimated (ρ) by cubic
B0	807	3.809	850.47	837.46	828.30
B10	846	4.688	851.89	840.83	834.48
B20	853	6.153	854.26	846.31	844.07
B30	861	8.673	858.34	855.30	858.59
B40	876	11.574	863.04	865.00	872.44
B50	884	14.181	867.27	873.11	882.51
B60	893	19.426	875.76	887.70	896.87
B70	901	27.806	889.34	906.23	907.27
B80	909	32.960	897.68	914.70	908.72
B90	915	48.785	923.32	926.77	909.87
B100	930	61.325	943.63	921.41	931.92

Figure 33.4 shows the correlation between the experimental and estimated density of neem oil methyl esters and its blends.

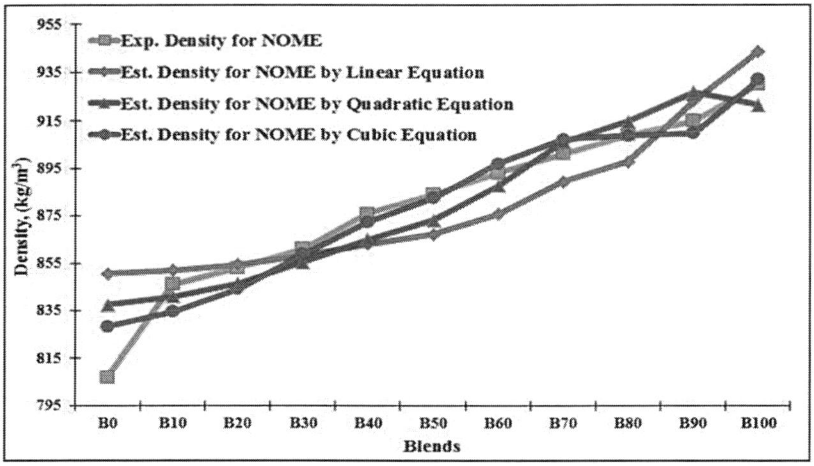

FIGURE 33.4 NOME blends versus experimental and estimated density.

The experimental density is the actual measured value of the biodiesel blends and the estimated density value can be calculated from the formula which is derived from the statistical software. The density estimated from linear, quadratic, and cubic regression equations is correlated with the experimental density of neem oil methyl esters blended with fossil diesel. From Figure 33.4, it is concluded that the points are away from the experimental density line, which means that there is no strong correlation between them.

33.5.1 UNCERTAINTY ANALYSIS

Three runs of the test were performed under the identical conditions to check the repeatability of all the results. In general, the repeatability of the results was found to be within 2%. Each reading of the basic quantities measured is the average of three values. From the uncertainty analysis, we found that the error ±2.49% is present in the estimated value given by Table 33.7.

TABLE 33.7 Uncertainty Calculations.

Uncertainty parameters	Error
Density	1.006611
Kinematic viscosity	5.193489
Total	±2.49%

33.6 CONCLUSIONS

From the modeling, a new method of density or kinematic viscosity estimation for nonedible oil biodiesel is presented. The aim of this work is to show that it is possible to estimate the density from the viscosity or vice versa. It is established that the density of biodiesel and diesel blends has the predominant effect on the kinematic viscosity and vice versa. From the result, it is observed that the linear regression has less prediction accuracy than the quadratic regression equation and quadratic has some time less and same accuracy than the cubic regression equation. For KOME and NOME, cubic regression equations have the accuracy of 99.0 and 94.0%, respectively. The estimation has $\pm 2.49\%$ uncertainty.

KEYWORDS

- **Karanja**
- **neem**
- **blends**
- **biodiesel**
- **density**
- **viscosity**
- **modeling**
- **regression**

REFERENCES

1. Agarwal, A. K.; Rajamanoharan, K. Experimental Investigations of Performance and Emissions of Karanja Oil and its Blends in a Single Cylinder Agricultural Diesel Engine. *Appl. Energy* **2009,** *86*(1), 106–112.
2. Alptekin, E.; Canakci, M. Determination of the Density and the Viscosities of Biodiesel–Diesel Fuel Blend. *Renew. Energy* **2008,** *33*(12), 2623–2630.
3. Anand, K.; Sharma, R. P.; Mehta, P. S. A Comprehensive Approach for Estimating Thermo-physical Properties of Biodiesel Fuels. *App. Therm. Eng.* **2011,** *31*(2), 235–242.
4. Bello, E. I.; Otu, F. Effects of Blending on the Properties of Biodiesel Fuels. *J. Emerg. Trends Eng. Appl. Sci.* **2012,** *3*(3), 556–562.

5. Bobade, S. N.; Khyade, V. B. Detail Study on the Properties of *Pongamia pinnata* (Karanja) for the Production of Biofuel. *Res. J. Chem. Sci.* **2012.** ISSN: 2231–606X.
6. Demirbas, A. Progress and Recent Trends in Biodiesel Fuels. *Energy Convers. Manage.* **2009,** *50*(1), 14–34.
7. Food Safety and Standards Authority of India (FSSAI). Manual of Methods of Analysis of Foods (Oil and Fats), *Ministry of Health and Family Welfare*, Government of India, New Delhi, 2012.
8. Giakoumis, E. G. A Statistical Investigation of Biodiesel Physical and Chemical Properties, and Their Correlation with the Degree of Unsaturation. *Renew. Energy* **2013,** *50*, 858–878.
9. Gitte, B. M.; Siraj, S.; Dharmadhikari, H. M. Performance and Emission Characteristics of Diesel Engine Fuelled with Biodiesel and its Blends: a Review. *Int. J. Eng. Res. Technol.* **2013,** *2*(10) (ESRSA Publications), 3235–3243.
10. Gupta, A. K.; Gupta, A. K. Biodiesel Production from Karanja Oil. *J. Sci. Ind. Res.* **2004,** *63*(1), 39–47.
11. Gupta, A.; Sharma, S. K.; Toor, A. P. An Empirical Correlation in Predicting the Viscosity of Refined Vegetable Oils. *Indian J. Chem. Technol.* **2007,** *14*(6), 642.
12. Hossain, A. K.; Davies, P. A. Plant Oils as Fuels for Compression Ignition Engines: a Technical Review and Life-Cycle Analysis. *Renew. Energy* **2010,** *35*(1), 1–13.
13. Huang, D.; Zhou, H.; Lin, L. Biodiesel: An Alternative to Conventional Fuel. *Energy Procedia*, **2012,** *16*, 1874–1885.
14. Kesari, V.; Rangan, L. Development of Pongamia Pinnata as an Alternative Biofuel Crop-Current Status and Scope of Plantations in India. *J. Crop. Sci. Biotechnol.* **2010,** *13*(3), 127–137.
15. Leung, D. Y.; Wu, X.; Leung, M. K. H. A Review on Biodiesel Production Using Catalyzed Transesterification. *Appl. Energy* **2010,** *87*(4), 1083–1095.
16. Lin, L.; Cunshan, Z.; Vittayapadung, S.; Xiangqian, S.; Mingdong, D. Opportunities and Challenges for Biodiesel Fuel. *Appl. Energy* **2011,** *88*(4), 1020–1031.
17. Mathur, Y. B.; Poonia, M. P.; Jethoo, A. S. Economics, Formulation Techniques and Properties of Biodiesel: a Review. *Univer. J. Environ. Res. Technol.* **2011,** *1*(2), 124–134.
18. Moser, B. R. Biodiesel Production, Properties, and Feedstocks. In *Biofuels*; Springer: New York, 2011; pp 285–347.
19. National Renewable Energy Laboratory (NREL) Innovation for Our Energy Future (2009), Biodiesel Handling and Use Guide Fourth Edition. *Revised January 2009; NREL/TP – 540–4672.*
20. Padhi, S. K.; Singh, R. K. Non – Edible Oils as the Potential Source for the Production of Biodiesel in India: A Review. *J. Chem. Phar. Res. J. Chem.* **2010,** *2*(5), 599–608.
21. Rao, P. V. Effect of Properties of Karanja methyl Ester on Combustion and NO_x Emissions of a Diesel Engine. *J. Pet. Technol. Altern. Fuels*, **2011,** *2*(5), 63–75.
22. Sangwan, S.; Rao, D. V.; Sharma, R. A. A Review on *Pongamia pinnata* (L.) Pierre: A Great Versatile Leguminous Plant. *Nat. Sci.* **2010,** *8*(11), 130–139.
23. Sayyed Siraj, R.; Uttarwar, L.; Pagey, S.; Suryawanshi, R. Effect of Acid and Iodine Value of Karanja Oil Methyl Ester (KOME) and its Statistical Correlation with Gross Calorific Value. *Int. J. Res. Eng. Technol.* **2013,** *2*(11), 680–685.
24. Sivaramakrishnan, K.; Ravikumar, P. Determination of Higher Heating Value of Biodiesels. *Int. J. Eng. Sci. Technol. (IJEST)*, **2011,** *3*(11), 7981–7987.

25. Sivaramakrishnan, K.; Ravikumar, P. Determination of Cetane Number of Biodiesel and its Influence on Physical Properties. *ARPN J. Eng. Appl. Sci.* **2012,** *7*(2), 205–211.
26. Tesfa, B.; Mishra, R.; Gu, F.; Powles, N. Prediction Models for Density and Viscosity of Biodiesel and Their Effects on Fuel Supply System in CI Engines. *Renew. Energy* **2010,** *35*(12), 2752–2760.

CHAPTER 34

COMPARISON OF RSM, ANN, FACTORIAL DoE, AND FUZZY LOGIC NETWORK FOR DIESEL ENGINE PERFORMANCE PARAMETER AND EMISSION ANALYSIS

PRAMOD K. TIWARI[1], ATUL G. LODHEKAR[1,*], and SUHAS C. KONGRE[2,3]

[1]*Mechanical Engineering Department, Rajiv Gandhi Institute of Technology, Mumbai, Maharashtra, India*

[2]*Mechanical Engineering Department ASP Pipri, Wardha, Maharashtra, India*

[3]*Suhas_kongre@rediffmail.com*

[*]*Corresponding author. E-mail: atullondhekar@gmail.com*

ABSTRACT

Nowadays, due to the high demand for petroleum product, people are looking for another alternative fuel, in which biodiesel is on top. Researchers perform the experiment with various combinations of fuels. To have maximum information in less number of experiments, various predication models are present. In this chapter, response surface methodology (RSM), artificial neural network (ANN), factorial design of experiment, and fuzzy inference system (FIS) method have been compared for modeling and prediction of engine parameter and emission characteristics. For initial experimental data, the experiment is performed on single-cylinder, four-stroke compression ignition engine with biodiesel–ethanol–diesel as fuel. For modeling, four factors are considered, that is, biodiesel (0–20%), ethanol (0–20%), speed (1500 rpm), and load (0–5 kg). RSM and factorial method is performed by using Design Expert and Minitab statistical software, respectively, and MATLAB is used in developing ANN and FLN model. In ANN, three factors are taken as input and output response and it is generated by using transfer function. In current work, feed forward with back propagation neural network model is used. In FIS, multi-input multi-output fuzzy models are developed, trained, and validated with experimental data. The developed RSM and FIS describe the process with high accuracy. Optimal condition obtained by RSM is better than factorial design. However, ANN shows advantage over RSM on basis of absolute deviation and coefficient of determination (R^2).

34.1 INTRODUCTION

Fossil fuels are widely used as transportation and machinery energy source due to its high heating power, availability, and quality combustion characteristics, but its reserve is depleting day by day. The declining reserves of fossil fuels and the growing environmental concerns have stimulated scientists and industries to search for alternative fuels for petrol and diesel engines.[4,12,14]

Biodiesel, the promising alternative fuel to fossil fuel, is produced from renewable biological resources. It is nontoxic, biodegradable, environmentally friendly, has high flash point, and can be blended with diesel. Among biodiesel, the edible and nonedible category is present. In these, nonedible biodiesel, namely neem, karanja, rubber, and Jatropha, had more attraction.[3,5,11,17] For selection of biodiesel, the various literatures have been studied.

Agarwal and Dhar[1] carried out experimental study for performance, emission, and combustion characteristics of compression ignition (CI) engine with karanja biodiesel. They found that 20% of karanja biodiesel blend can be utilized in an unmodified direct injection (DI) CI engine. Gangil et al.[8] experimentally evaluated performance and emission characteristics of diesel engine for blends (B20, B40, B60, B80, and B100) of karanja biodiesel and commercial diesel. The results obtained indicated the better fuel properties and engine performance at B40. Senthil and Silambarasan[16] experimentally studied combustion analysis of Jatropha methyl esters and Pongamia methyl esters with the addition of ethanol as fuel in a diesel engine. The fuel blends [Pongamia–ethanol (50–50) and Jatropha–ethanol (50–50)] are used in the DI CI engine. Heat release rate (HRR) and cumulative HRR are maximum for Pongamia–ethanol (50–50) when compared with the neat diesel fuel. Therefore, karanja biodiesel is selected. Owing to its increase in performance and emission characteristics such as brake specific fuel consumption

(BSFC), NO_x emission, it is required to blend with the additive. For selection of additives, some literatures are reviewed.

Atmanlı et al.[2] focused on finding diesel fuel, n-butanol, and cotton oil optimum blend ratios for diesel engine applications by using response surface methodology (RSM). The experimental test fuels were prepared by choosing seven different concentrations. The optimum component concentration was determined as 65.5 vol.% diesel, 23.1 vol.% n-butanol, and 11.4 vol.% cotton oil. Khoobbakht et al.[10] investigated the effect of operating factors of engine load and speed as well as blended levels of biodiesel and ethanol in diesel fuel on the emission characteristics of DI diesel engine. The experiment design was done based on central composite rotatable design of RSM. To study various factors and interaction effect, the numbers of experiments go on increasing. Therefore, to reduce trails of experiment, number of method are available, namely design of experiment techniques, soft computing technique includes artificial neural network (ANN), fuzzy logic system, genetic algorithm, support vector machine (SVM), and so forth.

Sakthivel[15] investigated the potential of fuzzy inference system (FIS) to predict the performance, combustion, and exhaust emissions of the CI engine at different injection timings (21°, 24°, 27° before top dead center) using fish oil biodiesel. They confirmed the applicability of the developed FIS models with high degree of accuracy and minimum time demand that can effectively replace the costly and time-consuming real-life experiments and trails. Fayyazbakhsh and Pirouzfar[7] investigated the effect of ethanol–diesel blends with different types of additives through experimental modeling and optimization method (factorial design). Hosoz et al.[9] investigated the applicability of adaptive network-based FIS (ANFIS) approach for modeling the performance parameters and exhaust emissions of a diesel engine employing various fuels. The ANFIS model yielded a good statistical performance.

Elfghi[6] presented a comparative study and combined application between RSM and ANN based on DOE strategy in the modeling and prediction of the research octane number (RON). They found that ANN methodology had a very obvious advantage over RSM for both data fitting and estimation capabilities. Moon et al.[13] developed artificial intelligence (AI)-based thermal control logics and compared their performances for identifying potentials as an advanced thermal control method in buildings. They developed three AI-based control logics, namely fuzzy-based control, ANFIS-based, and ANN-based control. In conclusion, they found that adaptive AI-based control method has potential to maintain interior air temperature more comfortably.

The present work focused on the comparison of modeling techniques which are RSM, factorial design, ANN, and FIS. For initial data, the experiment is conducted on single-cylinder, four-stroke CI engine at constant speed of 1500 rpm. The three factors of input, that is, biodiesel, ethanol, and the load, are considered. The response considered is representing performance and emission characteristics, that is, brake power (BP) (kW), BSFC (kg/kW·h), hydrocarbon (HC) (ppm) emission, CO (%) emission, NO_x (ppm) emission, and CO_2 (%) emission.

34.2 EXPERIMENTAL TECHNIQUES AND METHODOLOGY

34.2.1 ENGINE AND EXPERIMENTAL SETUP

The experiment is conducted on single-cylinder, four-stroke, DI, and air-cooled compression diesel engine. The engine test rig consists of diesel engine, eddy current dynamometer, exhaust gas analyzer, and fuel tank. The engine specification is shown in Table 34.1.

TABLE 34.1 Engine Specification.

Trademark	Kirloskar
Speed	1500 rpm
Bore × stroke	87.5 mm × 110 mm
Compression ratio	17.5:1
Method of cooling	Air cooled with radial fan

First, experiments are conducted with diesel only and readings recorded. Then test is carried with karanja oil as biodiesel blended with diesel. The ratio of 10% karanja biodiesel and 90% diesel referring as K10 and similarly K15, K18, and K20 blend test are conducted. In this, the speed of the engine is kept constant at 1500 rpm and the load is varied in the range of 0–5 kg. Last, test conducted with karanja biodiesel, ethanol, and diesel blend as fuel. In this, K10+E10 referring to 10% of karanja oil, 10% of ethanol, and 80% diesel as blended fuel. Among all possible combination, random test was conducted with K10+E10, K15+E15, K18+E18, and K20+E20. To avoid the effect of environmental condition, the engine is initially run for few minutes (for warming) until the temperature of cooling water and exhaust gases reached to stable condition.

34.2.2 FACTORIAL DESIGN

Factorial is a good statistical technique for designing experiments; creating models, aiming effects of multiple variables, and finding optimum conditions for propose responses.[7] For designing model, D-optimal approach of factorial design with statistical approach was employed to reduce the number of experiment for various blended fuels. One of advantage over the statistical method is that it enables all possible main effect and interaction effect of all variables at different levels. The response could be written as a function of variables by arranging multiple regressions using the least squares method to fit the equation below:

$$Y = a_0 + a_1X_1 + a_2X_2 + a_3X_3 + a_{12}X_1X_2 + \ldots \quad (34.1)$$

where a_0 is the general coefficient, a_i is the coefficient of linear effects, and a_{ij} represents the coefficient of quadratic effects. X_1, X_2, X_3, and X_4 denote the model variables. The combinations of variables (e.g., X_1X_2) represent an interaction between the exclusive variables in that term.[7]

In this study, D-optimal type of factorial design has been used in developing the model. The modeling is done by "Minitab 17.0" statistical software. The input parameters, namely biodiesel (%), ethanol (%), and load (kg) are categorized in three levels and one replicate. Therefore, total 27 experiment data sets are taken as input and response were BP (kW), BSFC (kg/kW·h), HC (ppm), CO (%), NO_x (ppm), and CO_2 (%) selected. Optimization had been done by using response optimizer function available in Minitab software.

34.2.3 RESPONSE SURFACE METHOD

Three experimental factors are included in the present study, namely biodiesel (10–20%), ethanol (0–20%), and load (0–5 kg). For response, two performance parameters, that is, BSFC and BP, and four emission parameters, that is, HC, CO, NOX, and CO_2 are considered. Each factors reading were taken in three levels, so that it can be fitted in model of central composite design of RSM. The total number of experiment reading in central composite design can be calculated from the method given by Khoobbakht et al.

Factorial experiment points allow clear estimates of all main effects and two-factor interactions. The axial points allow the estimation of the pure quadratic effects of variables. On considering three levels of each factor, the total of 15 runs of the experiment were required to develop model for predication and optimization. The minimum and maximum range of experiment variables were considered with their values in coded and uncoded form is presented in Table 34.2. A multiple regression analysis and the quadratic model were carried out to evaluate different coefficients and equations to be utilized for prediction of the response parameters. The first-order and second-order equation used by RSM is referred by Atmanlı. Moreover, analysis of variance (ANOVA) is performed on experimental data. The conclusion was based on results of "Fisher's test" (F-test) and "Student's t-test" to evaluate the statistical significance of model and regression coefficients, respectively.

TABLE 34.2 Experiment Variables and Their Levels for Central Composite Design.

Variable		Variable		Level
	Code	−1	0	+1
Biodiesel	A (%)	10	15	20
Ethanol	B (%)	0	10	20
Load	C (kg)	0	2	4

The predication and optimization was carried out by using "Design Expert 7.0.0" software, where each response is transformed to a dimensionless desirability value (D) and it ranges between D = 0, which points that the response is completely unacceptable, and D = 1, which points that the response is more desirable. The optimization module of software uses gradient descent method to calculate the optimum factor levels. This method needs one initial starting point and defined restriction criterions for screening on the response surfaces determined by using the calculated mathematical equations.[2] The optimization criterions defined in Design Expert and the restrictions are given in Table 34.3.

TABLE 34.3 Optimization Criterion.

Response	Criteria	Minimum value	Maximum value
Hydrocarbon (HC) (ppm)	Minimum	71	164
CO (%)	Minimum	0.082	0.201
NO_x (ppm)	Minimum	140	613
CO_2 (%)	Minimum	1.44	3.52
Brake power (BP) (kW)	Maximum	0	1.19
Brake specific fuel consumption (BSFC) (kg/kW·h)	Maximum	0	1.179

34.2.4 ARTIFICIAL NEURAL NETWORK

ANN, similar to the human neural structure and its learning process, has been increasingly applied to analyze diesel engine performance and emission characteristics. ANN utilizes connectivity and transfer functions between input and output neurons for calculating the optimal response. Different from mathematical models such as regression models or proportional integral derivative controllers, ANN models have adaptability through a self-tuning process, so can decide accurately without external expert intervention for retuning model parameters.[13]

Various studies proved that, as engine performance and emission parameters, the ANN-based predicted model have advantage over the mathematical model. ANN model can take multiple input variables to predict response variables. The available experimental data is divided into three parts, training (70%), validating (15%), and testing data (15%) set. Using "MATLAB 2016a" and its neural network toolbox, ANN model was developed to calculate response variables. The model structure is shown in Figure 34.1. The model consists of an input layer, a hidden layer, and an output layer. A training goal mean squared error (MSE) was set to 0.0 with a maximum 20 epochs, 0.75 learning rate, and 0.9 momentum based on literature. The Levenberg–Marquardt algorithm was used as a training algorithm due to its training speed and accuracy.[6,13,18]

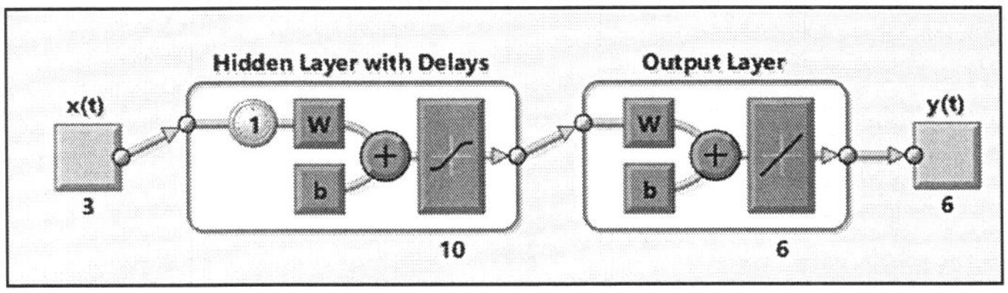

FIGURE 34.1 Structure of neural network layers.

34.2.5 FUZZY INFERENCE SYSTEM

Fuzzy logic, developed by Zadeh, deals with the degree of truth or falsity of phenomena. Fuzzy logic include three basic steps: (1) fuzzification (input stage) for drawing input values and converting them to fuzzy type through transfer function such as triangular, trapezoidal, Gaussian distribution curve, sigmoidal functions, and so forth; (2) inferencing (processing stage) for making inferred results based on previously fuzzified values and linguistic if–then rule statements; and (3) defuzzification (output stage) for summing the result by linguistic rules and then converting it to specific output values.[13] From previous studies, fuzzy logic has difficulties in optimal fuzzy rules and membership function. To overcome this difficulty, a new approach integrating fuzzy logic and neural network has been introduced. Neuro-adaptive learning techniques are implemented for tuning membership function parameters, to permit the FIS to track given input/output data. The ANN and ANFIS tuning process of membership function parameters are similar, since both employ training data sets consisting of input–output data and training algorithm.[9,13]

A Sugeno-type ANFIS model was developed using the neuro-fuzzy logic toolbox in "MATLAB 2016a." The structure is shown in Figure 34.2. In total, 48 experiment data sets are divided into 32 training data, 8 checking data, and 8 testing data. Many parameters for the training model followed the recommended default values in the neuro-fuzzy logic toolbox. The ANFIS structure is shown in Figure 34. 2. The neuro-fuzzy algorithm should be trained using a proper set of training data so that the outputs can be estimated based on the input–output data. Therefore, the data was trained to identify the parameters of Sugeno-type FIS based on the hybrid algorithm combining the least square method and the back propagation gradient descent method.[9] After training model, the same validated and tested for another set experimental data. The criterions used for measuring the network performance were the correlation coefficient (R), mean relative error, and root MSE. Definitions of criterions were given by Sakthivel.

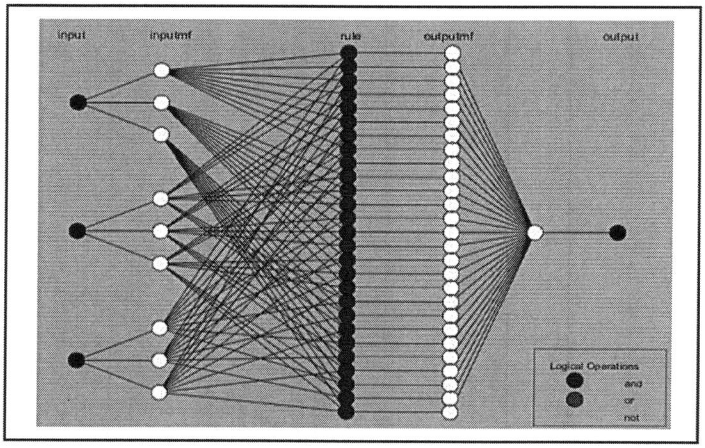

FIGURE 34.2 Structure of fuzzy inference system (FIS).

34.3 RESULTS AND DISCUSSIONS

In this study, performance and emission characteristics of the diesel engine are predicated by using RSM, factorial design, ANN and ANFIS, and optimized by using RSM and factorial design. The different models are presented in the section.

34.3.1 RESPONSE SURFACE METHODOLOGY

The response variables results are presented in Table 34.4 based on the criteria discussed in methodology. The main and interaction effect of different input parameters on response variable can be studied through counterplot. For BP, Figure 34.3 represents counterplot, ethanol, and load factor is significant in linear model based on "F-value" and *p*-value in ANOVA. In interaction, the ethanol and load interaction having *p*-value less than 0.001, so it is significant for fitting. Similarly, for BSFC, Figure 34.4 represents counterplot. In BSFC, among linear and quadratic term, only load factor is significant.

TABLE 34.4 Precision Index of Different Response by Response Surface Methodology.

Response	R^2 (%)	Adjusted R^2 (%)	Predicated R^2 (%)
BP	99.56	99.23	97.71
BSFC	99.95	99.85	99.73
CO	99.46	99.05	97.78
CO_2	95.71	93.33	88.32
HC	99.55	98.56	96.51
NO_x	99.79	99.67	99.31

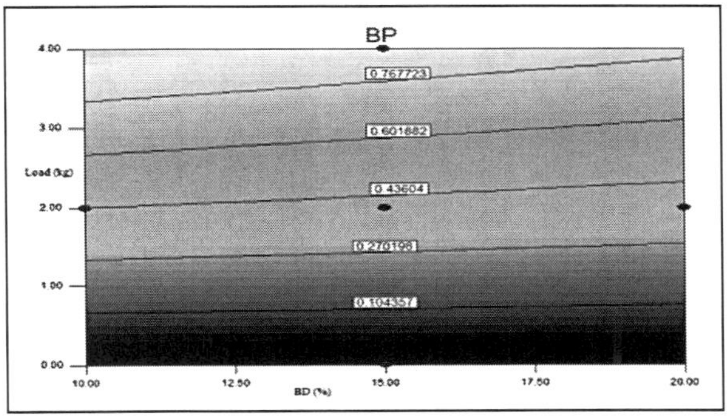

FIGURE 34.3 The counterplot of brake power (BP) versus biodiesel (BD) and load.

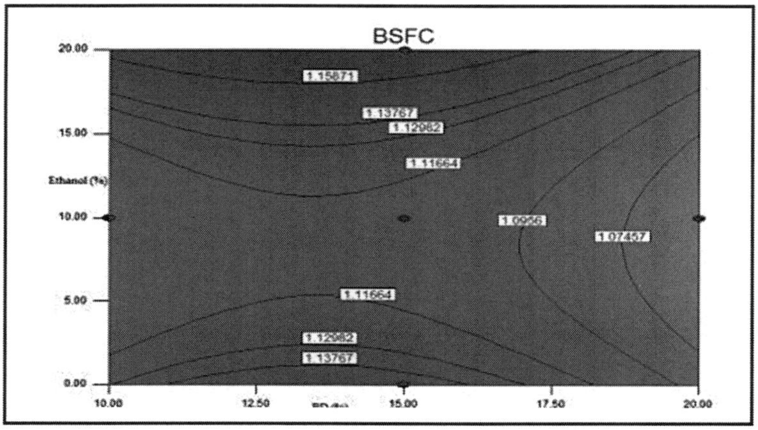

FIGURE 34.4 The counterplot of brake specific fuel consumption versus BD and ethanol.

In emission characteristics, the four response variables are considered, that is, HC, CO, NO_x, and CO_2. The counterplot of CO emission is shown in Figure 34.5. In this, ethanol is significant in the linear and quadratic term. The counterplot of CO_2 emission is shown in Figure 34.6. Among linear, interaction, and quadratic term, ethanol and model are significant. In HC emission, counterplot shown in Figure 34.7, ethanol and load in the linear term, biodiesel–load and ethanol–load in interaction term, and ethanol in the quadratic term are significant. Similarly, in NO_x emission, all linear terms are significant. The counterplot is shown in Figure 34.8.

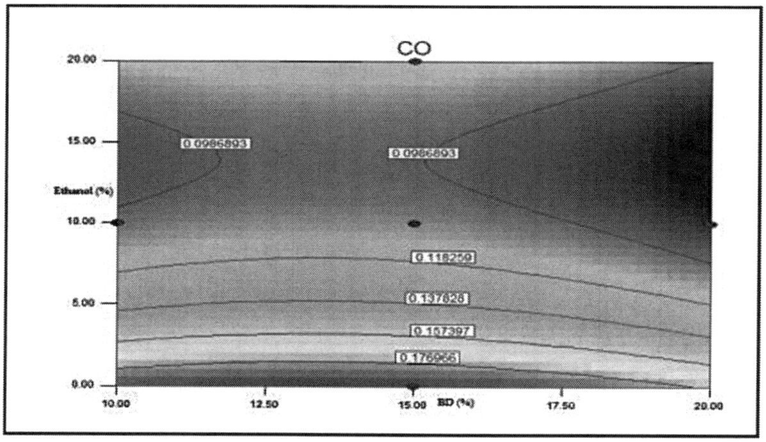

FIGURE 34.5 The counterplot of CO emission versus BD and ethanol.

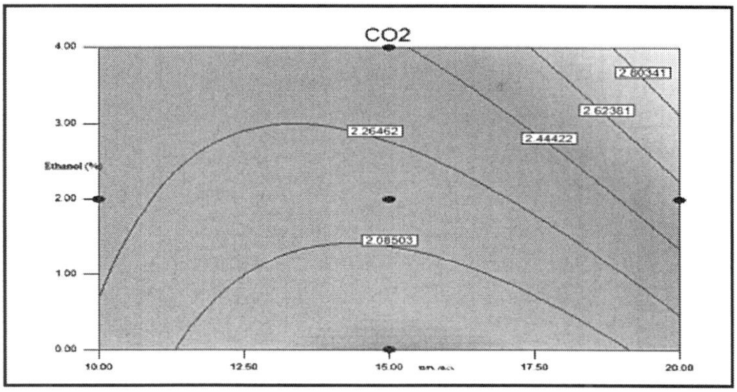

FIGURE 34.6 The counterplot of CO_2 emission versus BD and ethanol.

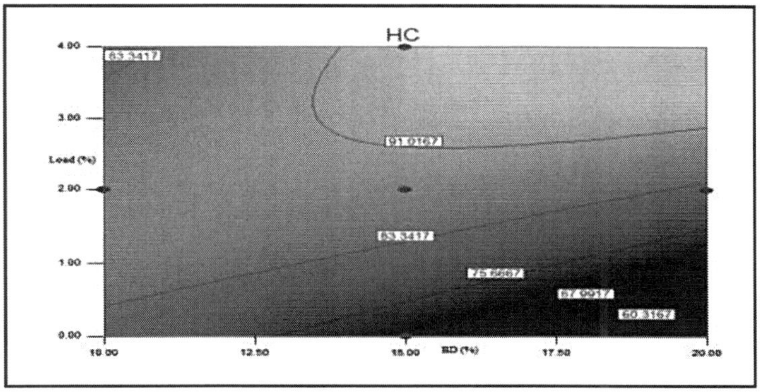

FIGURE 34.7 The counterplot of hydrocarbon (HC) emission versus BD and load.

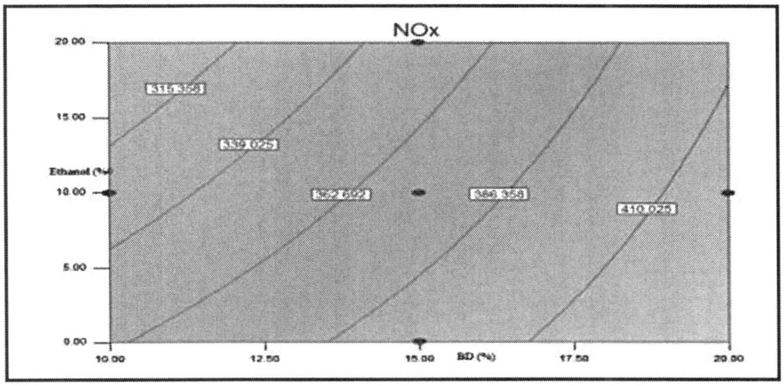

FIGURE 34.8 The counterplot of NO_x emission versus BD and ethanol.

34.3.2 FACTORIAL DESIGN

The result based on the coefficient of determination (R^2), adjusted R^2, and predicated R^2 is presented in Table 34.5. The main and interaction effect of input parameters with response variables were much more similar to RSM.

TABLE 34.5 Precision Index of Different Response by Factorial Design.

Response	R^2 (%)	Adjusted R^2 (%)	Predicated R^2 (%)
BP	96.34	95.87	94.60
BSFC	16.35	5.44	Large residual
CO	64.24	59.58	52.66
CO_2	87.99	86.43	82.76
HC	42.13	34.67	20.06
NO_x	99.6	99.55	99.44

The counterplot of BP and HC emission is shown in Figures 34.9 and 34.10, respectively.

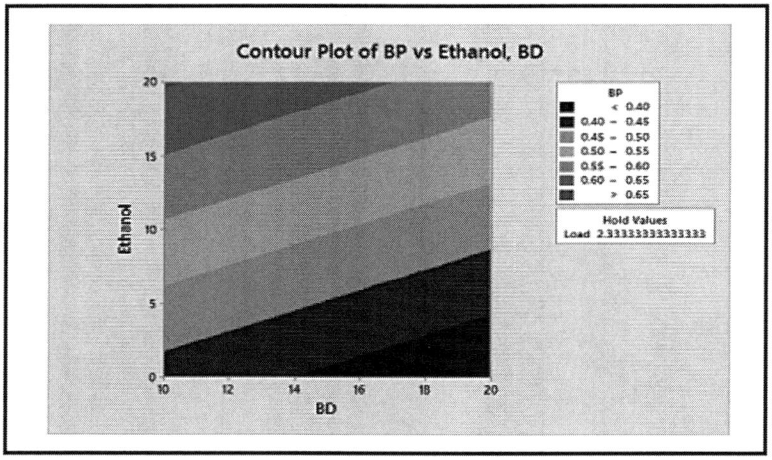

FIGURE 34.9 The counterplot of BP versus ethanol and BD.

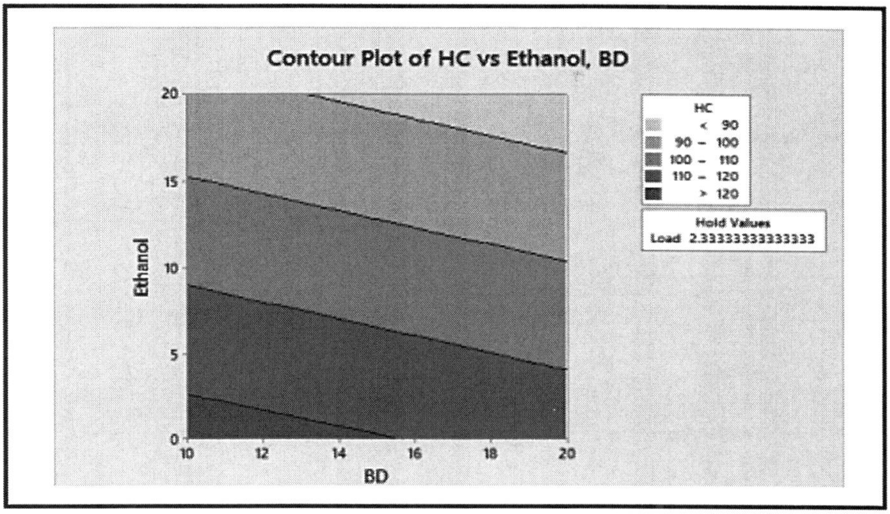

FIGURE 34.10 The counterplot of HC emission versus ethanol and BD.

34.3.3 ARTIFICAL NEURAL NETWORK

ANN performed with feed-forward-back propagation network-type results, in MSE and coefficient of correlation (R), is shown in Table 34.6. The log–sigmoid transfer function is used for modeling. The model developed with 10 neurons. The response plot against the input variables is shown in Figure 34.11 for three phases, that is, training, validation, and testing. Figure 34.12 shows error histogram for same phases. On retraining neural network, MSE and coefficient of correlation (R) improve.

TABLE 34.6 Mean Squared Error and R Results.

Phase	Mean squared error	R
Training	15.63	0.9997
Validation	101.37	0.9978
Testing	468.707	0.9901

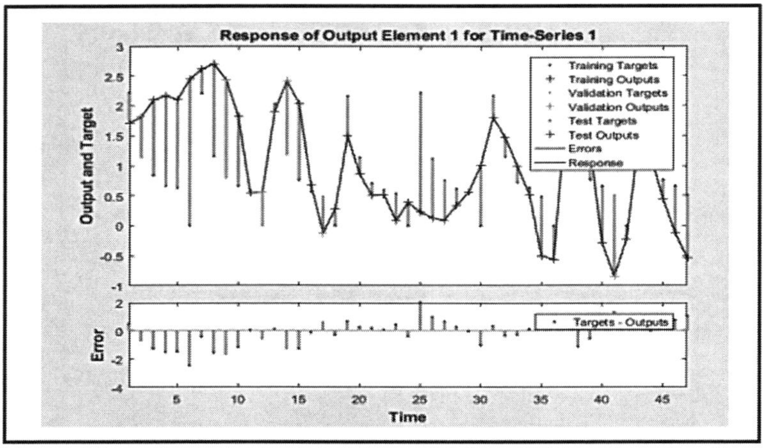

FIGURE 34.11 Response plot.

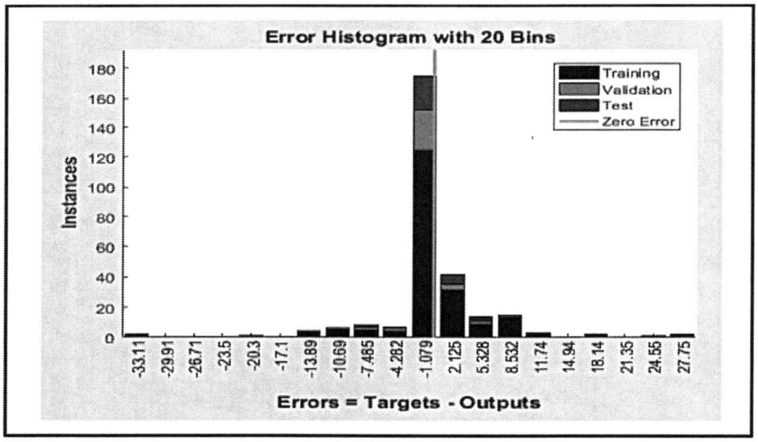

FIGURE 34.12 Error histogram.

34.3.4 FUZZY INFERENCE SYSTEM

ANFIS performed with gaussmf membership function and hybrid optimization method, consisting of regression and back propagation method, is used and result for training, checking, and testing is shown in Table 34.7. FIS output for BP on training, checking, and testing data is shown in Figures 34.13 and 34.14. Moreover, surface plot of response BP against biodiesel and load is shown in Figure 34.15.

TABLE 34.7 Average Testing Error.

Response	Training	Checking	Testing
BP	0.0199	0.228	0.154
BSFC	0.4058	1.72	0.7145
HC	2.805	26.014	13.454
NO_x	8.35	20.469	23.185
CO_2	0.0477	0.6172	0.315
CO	0.0026	0.0328	0.173

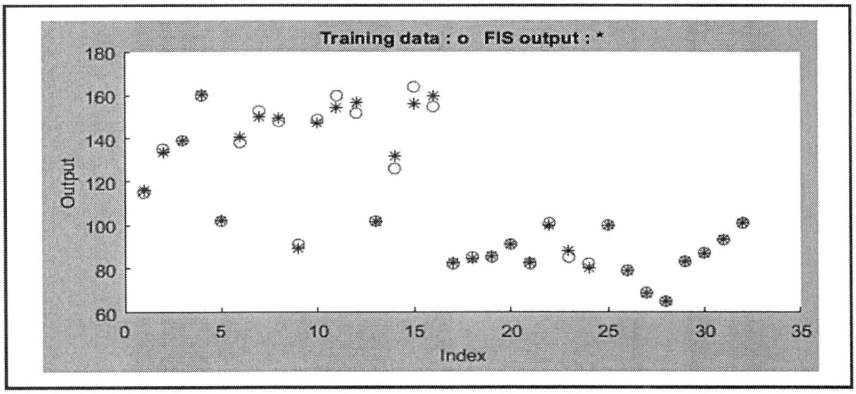

FIGURE 34.13 FIS output during training.

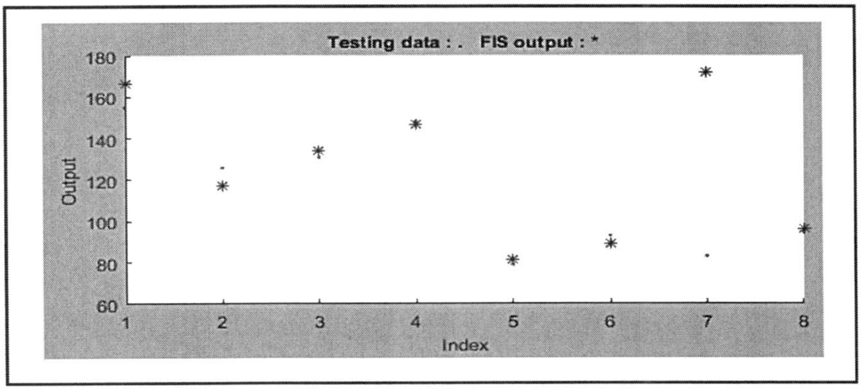

FIGURE 34.14 FIS output during testing.

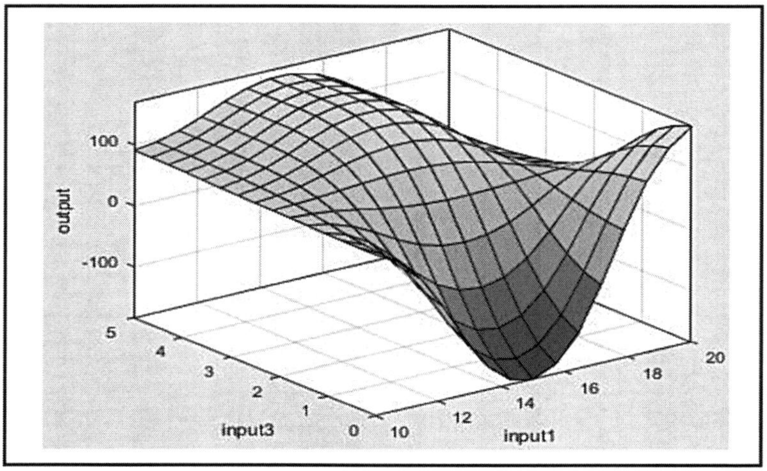

FIGURE 34.15 Surface plot for BP versus BD and load.

34.3.5 OPTIMIZATION

Optimization can be done by criteria defined in methodology. RSM gives optimal level as 10% biodiesel, 19.8% ethanol, and 2.86 kg load. At optimal level, value of BP, BSFC, HC, CO, NO_x, and CO_2 are 0.862 kW, 1.172 kg/kW·h, 95.23 ppm, 0.11%, 358.15 ppm, and 1.896%, respectively. The factorial design gives optimal level of 10% biodiesel, 20% ethanol, and 3.18 kg load. At optimal level, value of BP, BSFC, HC, CO, NO_x, and CO_2 are 0.838 kW, 0.640 kg/kW·h, 95.9 ppm, 0.091%, 405.87 ppm, and 1.703%, respectively.

34.3.6 COMPARISON OF METHODS

Based on predication and optimization ability, RSM gives more precise results as compare to other methods which are contradiction with the studies of Elfghi and Moon et al. due to factors such as experiment design, experiment data involving number of parameter and replication, software used, and so forth. Moreover, RSM requires less set of experimental data as compared to ANN and ANFIS. For optimization, RSM results confirmed with postexperiment at optimal level, showing good agreement with predication.

34.4 CONCLUSION

The modeling techniques and developed model, namely ANN, RSM, factorial design, and ANFIS, are helpful in design and optimization with effective performance and reduced emission in lesser run of experiment. The optimal level obtained as 10% biodiesel, 19.8% ethanol, and 2.86 kg load is confirmed with the experiment. In comparison, RSM gives precise result and having ability of predication and optimization. Therefore, this study will be useful in selection of modeling technique among ANN, RSM, factorial design, and ANFIS.

KEYWORDS

- **artificial neural network**
- **response surface methodology**
- **adaptive network-based fuzzy inference system**
- **brake power**
- **brake-specific fuel consumption**

REFERENCES

1. Agarwal, A.; Dhar, A. Experimental Investigations of Performance, Emission and Combustion Characteristics of Karanja Oil Blends Fuelled DICI Engine. *Renew. Energy* **2013,** *52*, 283–291.
2. Atmanlı, A.; Yüksel, B.; Ileri, E.; Karaoglan, A. D. RSM Based Optimization of Diesel–n-Butanol –Cotton Oil Ternary Blend Ratios to Improve Engine Performance and Exhaust Emission Characteristics. *Energy Convers. Manage.* **2015,** *90*, 383–394.
3. Bajpai, S.; Das, L. Experimental Investigations of an IC Engine Operating with Alkyl Esters of Jatropha, Karanja and Castor Seed Oil. 4th Inter. Conf. on Adv. in Energy Res. *Energy Procedia* **2013,** *54*(14), 701–717.
4. Bhuiya, M.; Rasul, M.; Khan, M. M. K.; Ashwath, N.; Azad, A. K.; Hazrat, M. A. Prospects of 2nd Gen. BD as a Sustain Fuel–Part 2: Properties, Perf. and Emission Charact.... *Renew. Sustain. Energy Rev.* **2016,** *55*,1129-1146.
5. Dhar, A.; Agarwal, A. Performance, Emissions and Comb. Charac. of Karanja BD in a Transportation Engine. *Fuel* **2014,** *119*, 70–80.
6. Elfghi, F. M. A Hybrid Statistical Approach for Modeling and Optimization of RON: A Comparative Study and Combined Application of RSM and ANN based on Design of Experiment. *Chem. Eng. Res. Des.* **2016,** *113*, 264–272. ISSN: 0263-8762.

7. Fayyazbakhsh, A.; Pirouzfar, V. Investigating the Influence of Additives-Fuel on Diesel Engine Performance and Emissions: Analytical Modeling and Experimental Validation. *Fuel* **2015,** *171*, 0016–2361.
8. Gangil, S.; Singh, R.; Bhavate, P.; Bhagat, D.; Modhera, B. Evaluation of Engine Performance and Emission with Methyl Ester of Karanja Oil. *Perspect. Sci.* **2016,** *8*, 2213–0209.
9. Hosoz, M.; Ertunc, H. M.; Karabektas, M.; Ergen, G. ANFIS Modelling of the Performance and Emissions of a Diesel Engine Using Diesel Fuel and Biodiesel Blends. *Appl. Therm. Eng.* **2013,** *60*, 24–32.
10. Khoobbakht, G.; Najafi, G.; Karimi, M. Optimization of Operating Factors and Blended Levels of Diesel, Biodiesel and Ethanol Fuels to Minimize Exhaust Emissions of Diesel Engine Using Response Surface Methodology. *Appl. Ther. Eng.* **2015,** *12*, 143.
11. Liaquat, A. M.; Masjuki, H.; Kalam, M.; Rizwanul Fattah, I.; Hazrat, M.; Varman, M.; Mofijur, M.; Shahabuddin, M. Effect of Coconut BD Blended Fuels on Engine Performance and Emission Characteristics. 5th BSME International Conference on Thermal Engineering. *Procedia Eng.* **2013,** *56*, 583–590.
12. Masjuki H. H.; Kalam, M. A. An Overview of Biofuel as a Renew. Energy Source: Develop. and Challenges. 5th BSME Inter. Conference on Thermal Engg. *Procedia Eng.* **2013,** *56*, 39–53.
13. Moon, J. W.; Jung, S. K.; Kim, Y.; Han, S.-H. Comparative Study of Artificial Intelligence-Based Building Thermal Control Methods e Application of Fuzzy, Adaptive Neuro-Fuzzy Inference System, and Artificial Neural Network. *Appl. Therm. Eng.* **2011,** *31*, 2422–2429.
14. Patel, C.; Agarwal, A.; Tiwari, N.; Lee, S.; Lee, C.; Park, S. Comb., Noise, Vibrations and Spray Charac. for Karanja BD Fuelled Engine. *App. Ther. Eng.* **2016,** *16*, S1359–4311.
15. Sakthivel, G. Prediction of CI Engine Performance, Emission and Combustion Characteristics Using Fish Oil as a Biodiesel at Different Injection Timing Using Fuzzy Logic. *Fuel* **2016,** *183*, 214–229.
16. Senthil, R.; Silambarasan, R. Combustion Analysis of Jatropha Methyl Esters and Pongamia Methyl Esters with the Addition of Ethanol as Fuel in a Diesel Engine. *Int. J. Ambient Energy* **2014,** 962090.
17. Takase, M.; Zhao, T.; Zhang, M.; Chen, Y.; Liu, H.; Yang, L.; Wu, X. An Expatiate Review of Neem, Jatropha, Rubber and Karanja as Multipurpose Non-Edible BD Resources and Comparison of Their Fuel, Engine and Emission Properties. *Renew. Sustain. Energy Rev.* **2015,** *43*, 495–520.
18. Tasdemir, S.; Saritas, I.; Ciniviz, M.; Allahverdi, N. Artificial Neural Network and Fuzzy Expert System Comparison for Prediction of Performance and Emission Parameters on a Gasoline Engine. *Expert Syst. Appl.* **2011,** *38*, 13912–13923.
19. 19. Xiaoyan, S. H. I.; Yunbo, Y. U.; HE, H.; Shuai, S; Dong, H.; LI, R. Combination of BD-Ethanol-Diesel Fuel Blend and SCR Catalyst Assembly to Reduce Emissions from a Heavy-Duty Diesel Engine. *J. Env. Sci.* **2008,** *20*, 177–182.

CHAPTER 35

MODELING OF DEGREE OF HYBRIDIZATION OF FUEL CELL-ULTRACAPACITOR FOR HYBRID SPORT-UTILITY VEHICLE

VENKATA KOTESWARA RAO K.[1,*], G. NAGA SRINIVASULU[1,2], and VENKATESWARLU VELISALA[1,3]

[1]*Department of Mechanical Engineering, NIT Warangal, Warangal, Telangana, India*

[2]*gnsnitw@gmail.com*

[3]*2venkee@gmail.com*

[*]*Corresponding author. E-mail: kvkr79@gmail.com*

ABSTRACT

A sport-utility vehicle is demonstrated in Advanced Vehicle Simulator (ADVISOR) with a power device fuel cell (FC)/ultracapacitor hybrid electric drivetrain utilizing approved part models. The net power developed from the FC stack and ultracapacitor, electric traction drive, vehicle mass is fixed. High power density quality of ultracapacitor is a major preferred standpoint for FC hybrid power train, and that can improve the mileage and acceleration of the vehicle. The simulations are carried out in two different drive cycles: city driving cycle (urban dynamometer driving schedule; UDDS) and highway driving cycle (highway fuel economy test). From the simulation results, it is observed that the fuel economy and regeneration efficiency increase for an urban driving cycle (UDDS). Simulation results with respect to fuel economy show that the degree of hybridization is beneficial and it varies for different drive cycles. The simulation plots demonstrate that ultracapacitor basically helps the energy component stack to take care of the vehicle power requirement and to accomplish a superior execution and a higher mileage.

35.1 INTRODUCTION

Hybridization of energy component control source with other two secondary energy sources, for example, batteries and ultracapacitors will assist the fuel cell (FC) for a longer time. In hybrid power source, the FC power module provides the main power consistently during the acceleration phase, while other secondary power source gives supplementary power increasing speed and peak load operation and captures the regeneration braking energy during vehicle deceleration. Hence, the stress on the FC power module and cost will be diminished. The transient performance of the power train and the energy storage efficiency will be improved. Ultracapacitor has the nature of more power density and moderately less energy density. It can permit many years of cycle life and overall increased performance of the batteries. The capacitance of ultracapacitor is more than conventional capacitors, permitting enough energy requirements for increasing acceleration performance.[1,2] The energy storage capacity of ultracapacitors is 75–150 Wh. The fuel economy benefit with the ultracapacitors is 10–15% higher than with the equivalent weight of batteries due to the higher efficiency of the ultracapacitors and more efficient engine operation.[3]

Two hybrid power sources such as FC–battery hybrid source and FC–ultracapacitor hybrid source are designed and simulated using Advanced Vehicle Simulator (ADVISOR) to achieve a better performance and higher fuel economy.[4] Large sport-utility vehicle (SUV) is modeled with FC power module as main power source and battery as a secondary energy source for hybrid electric power train in ADVISOR by utilizing individual approved part models and achieved improved energy efficiency by varying degrees of hybridization for different drive cycles.[5] A prediction type power management approach for FC/battery plug-in hybrid vehicles with the aim of developing overall system efficiency during its operation is adopted to achieve optimal hybridization value in this chapter. The main objective of the suggested strategy is that if the absolute amount of energy demand to complete a certain drive cycle can be reliably desired, then the energy stored in the energy storage device can be consumed in an optimal aspect that allows the FC power device to operate in its most efficient regime.[6] Toyota has outlined a FC electric vehicle in view of F/B hybrid utilizing a nickel–metal hydride (Ni–MH) battery pack as an auxiliary energy source, and Honda has made another model depending on F/UC hybrid utilizing ultracapacitor cells as power barriers. It is expected that FC/battery/ultracapacitor (F/B/C) hybrid can result in outstanding system performance and energy efficiency.[7] A control standard is set for using proton exchange membrane FC energy as fundamental power source and supercapacitors as the secondary power source for hybrid electric vehicle applications. The technique depends on DC interface voltage control, and FC component is basically working in steady-state conditions in order to limit the mechanical stresses of FC power module and to ensure a decent synchronization between fuel stream and FC current,[8] assurance of the hybridization degree as per drivability necessities, the examination of the energy flows, and the calculation of the ideal hydrogen utilization. The outcomes demonstrate that hybridization permits a noteworthy change in the hydrogen economy through the recouped energy from braking. The entire study is performed with a detailed model of the fuel cell hybrid system in the ADVISOR.[9]

This chapter illlustrates an ADVISOR model depending on approved individual segment models that are exhibited to research the effect of hybridization to enhance the mileage of a large SUV in various drive cycles. The targets of this examination are to comprehend the effective cooperation of FC energy units and ultracapacitors, and to decide an ideal framework.

35.2 VEHICLE DESCRIPTION

The SUV selected for this simulation study depends on all-wheel drivetrain of Chevrolet 2000 model that was changed over to an FC hybrid power vehicle. For this demonstration, the existed external shape of the vehicle remains unchanged, and the existed engine power train is supplanted with FC/ultracapacitor arrangement for hybrid electric power train. The key technical specifications for this vehicle are mentioned in Table 35.1.

TABLE 35.1 Vehicle Parameters.

Coefficient of drag	0.45
Vehicle frontal area (m^2)	3.17
Rolling resistance coefficient	0.008
Vehicle mass (kg)	2700

35.3 COMPONENT MODELS

35.3.1 ELECTRIC DRIVETRAIN

Vehicle all-wheel traction drive comprises a pair of induction motors of each 83 kW AC power to deliver the tractive power of 166 kW. This power rating is sufficient for acceleration and gradeability performances coordinating to the conventional vehicle (5.3 l V8 motor 210 kW). The engines were integrated with sun and planet gear arrangement, replacing the traditional four-speed self-shifting transmission to achieve a maximum speed of 80 mph. Inverter and motor models depend on the approved ADVISOR models.[10]

35.3.2 FUEL CELL (FC) MODEL

The FC stack module operates at 1.7 bar at maximum power level using a double-rotating screw-type compressor and it depends on estimations from 110-cell 20 kW gross power modules.[11] A validated model of hybrid power source vehicle data in[12,13] using ADVISOR.

35.3.3 ULTRACAPACITOR MODEL

The simple ultracapacitor resistor–capacitor (RC) model is included a resistor R, which gives the ultracapacitor's ohmic deficiency, additionally called as equivalent series resistor (ESR) with capacitor C, which recreates the capacitance of ultracapacitor during the state of charge (SOC) variation impacts. The specifications of ultracapacitor are mentioned in Table 35.2.

TABLE 35.2 Ultracapacitor Specifications.

Ultracapacitor parameters	Value
Capacitance	2700±20% (F)
Internal resistance (de)	0.001±25% (Ω)
Leakage current	0.006 (A), 72 h, 25°C
Operating temperature	−40 to 65°C
Rated current	100 (A)
Voltage	2.5 (V)
Volume	0.6 (I)
Weight	0.725 (kg)

Figure 35.1 shows the simple RC circuit model. The current remains same during charging and discharging which is 30 A. The capacitance, rated voltage, capacitance, and ESR are 470 F, 2.5 V, and 2.5 mΩ, respectively.[14]

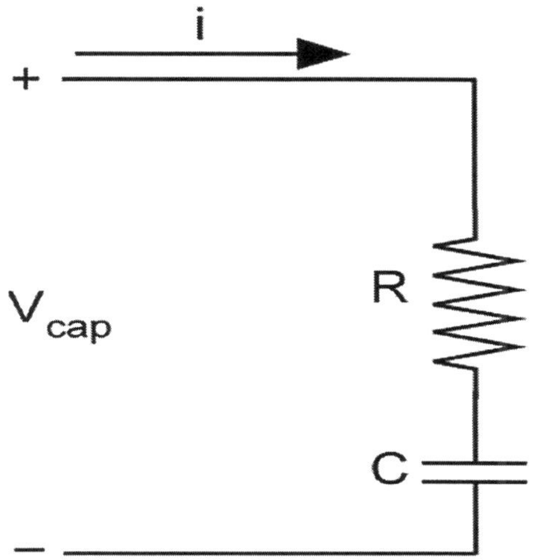

FIGURE 35.1 Ultracapacitor resistor–capacitor model.

35.3.4 VEHICLE ADVISOR MODEL

Table 35.1 represents the electric drivetrain, which is fixed, and FC power module (variable size) and ultracapacitor packs are included for the fuel cell hybrid electric vehicle in ADVISOR model. Varying vehicle setups using FC energy component segment sizes from zero (a pure ultracapacitor-powered vehicle) to a pure FC power source (without ultracapacitor) are chosen to examine the range of hybrid ratio with same vehicle mass, and hence, performance. For every setup, the desired power is regulated through drivetrain control and extra auxiliary loads. For this present vehicle model, a supply of around 166 kW and auxiliary loads (steering control, power brakes, 12 V loads) are around 1500 W. For the abovementioned power demands, around 170 kW of power-delivered FC stack and ultracapacitors are required. The potential to deliver supply 170 kW of energy for the high-voltage drive system of the vehicle assures the vehicle performance.

Figure 35.2 demonstrates some case study results with respect to time for the highway fuel economy test (HWFET) driving cycle for hybrid case. There is no variation in ultracapacitor SOC during the drive cycle to give repeated SOC-corrected hydrogen economy results. FC energy unit requires a base power during control system operation. The ADVISOR model is utilized to predict the mileage and hybrid power train proportions for various drive cycles.

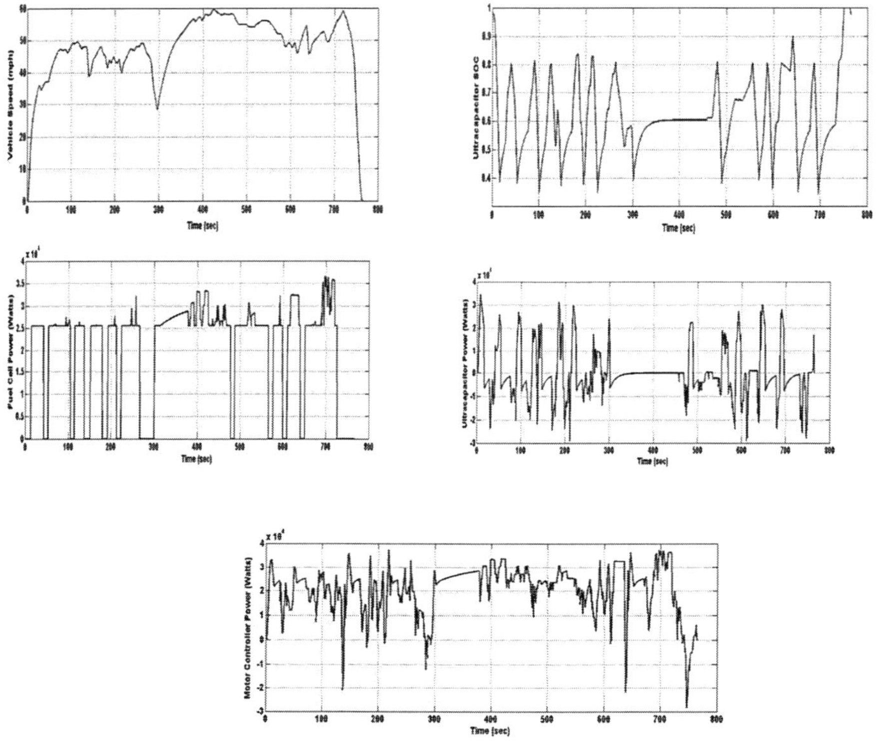

FIGURE 35.2 ADVISOR results for highway driving cycle.

35.4 HYBRIDIZATION RESULTS

A step incremental power of 10 kW for FC module and ultracapacitor module are considered for this model. The lower limit of FC power is set so as to assure that the vehicle maintains a steady speed of 103 kph (65 mph) on straight roads. A power demand of 30 kW supplied by the FC stack decides the minimum tractive power designs at 40 kW as total FC power. The SUV model vehicle can cover the power range from this minimum power level to the most extreme net power of 170 kW for the only FC source. The rest of the power not provided by the FC unit decides the volume of ultracapacitor required for a dual-power framework.

The level of hybrid ratio is determined by the volume of gross FC stack power in a hybrid component to gross stack power for the only FC power module design (225 kW). This component is likewise near the proportion of FC power to hybrid FC power in addition to ultracapacitor power (170 kW). Table 35.3 records the scope of module sizes used to provide roughly steady accessible power.

TABLE 35.3 Power Source Size Ratios.

Hybrid ratio	Fuel cell (FC) gross power	FC net power	ESS power (UC)
0.00	0	0	170
0.18	40	30	140
0.22	50	38	132
0.25	60	43	127
0.29	70	50	120
0.34	80	59	111
0.39	90	67	103
0.44	100	75	95
0.5	110	85	85
0.55	120	94	76
0.6	140	103	67
0.66	150	113	57
0.72	160	123	47
0.78	170	134	36
0.85	180	145	25
0.95	195	163	12
1	225	170	0

Hybridization empowers the size of FC to be diminished by using an energy storage system, that is, during peak loads, it permits the FC and the system operates more competently. When the power request is low, the FC provides the desired power. The usage of stored energy permits speedy beginning-up of the power device and favors the recognition of braking energy.

35.5 HYBRID FUEL ECONOMY RESULTS

Fuel economy plots shown in Figures 35.3–35.6 rely on the drive cycle dynamics. The primary increment is because of the expansion in FC stack size and productivity. The consistent power request is constantly over the FC least power criteria. The efficiency rises to some degree with energy component measure, then remains moderately consistent or diminishes before rising and dropping off once more. The FC power device estimate keeps on expanding and the ultracapacitor limit diminishes.

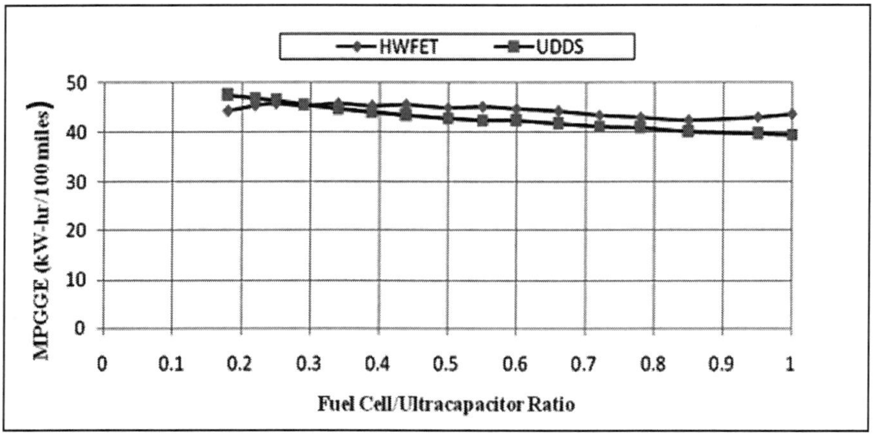

FIGURE 35.3 Fuel economy variation with hybrid ratio.

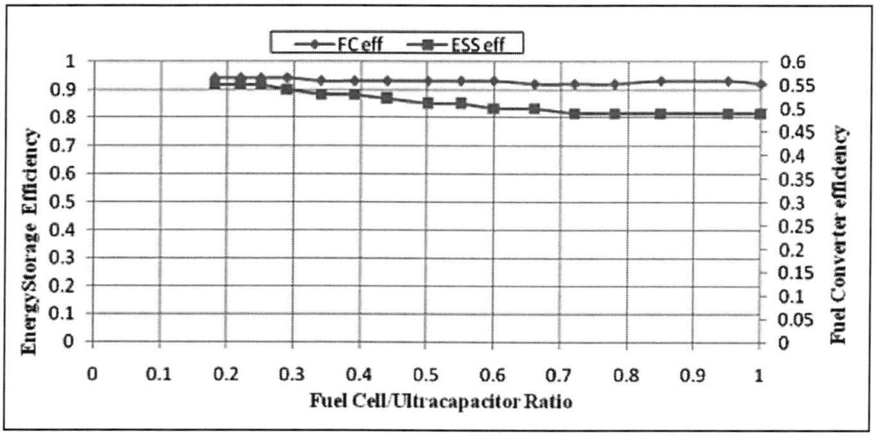

FIGURE 35.4 Highway fuel economy test component efficiency variation with hybrid ratio.

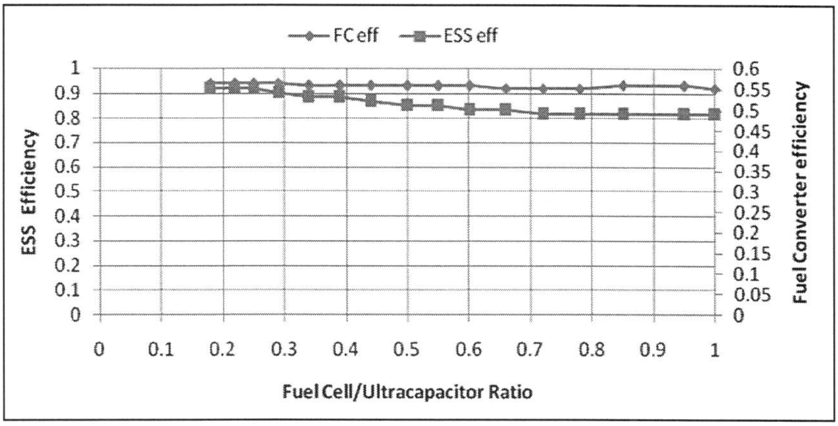

FIGURE 35.5 Urban dynamometer driving schedule component efficiency variation with hybrid ratio.

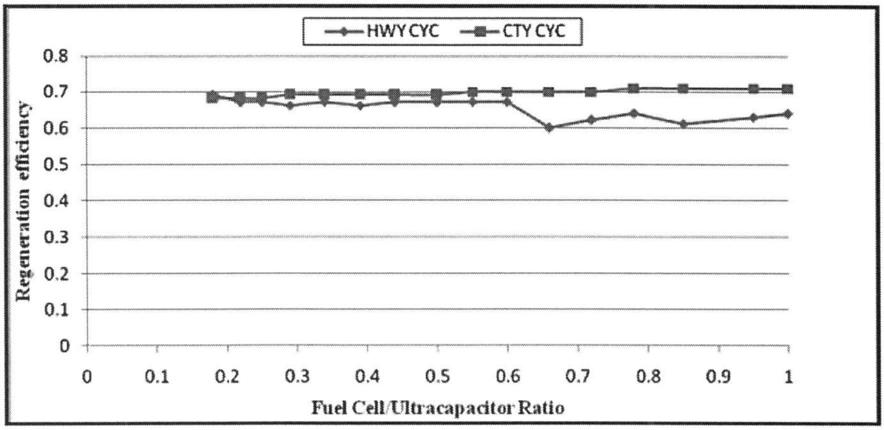

FIGURE 35.6 Regeneration efficiency variation with hybrid ratio.

A substantial dominant part of the energy conversion is achieved at the base energy component control level. The majority of highway hydrogen economy is accomplished at the peak fuel efficiency. With an increase in the FC stack size, the minimum power level as well as the hybrid ratio increase. The drop in hydrogen economy for the higher hybrid ratio is due to lower ability to trap regeneration braking energy as the ultracapacitor capacity is diminished.

35.6 CONCLUSIONS

The fixed constraints enforced on the current results are:

- Total output power from FC and ultracapacitor
- Electric traction drivetrain
- Vehicle performance (as a result of above)
- Total vehicle mass

FC energy unit may be profitable by diminishing its size to maintain a strategic extreme execution at minimum load or on/off operation because of least power request. For the same hybrid ratio (0.60), FC–ultracapacitor power train fuel economy is higher by 35.46% when compared with FC battery in urban dynamometer driving schedule (UDDS).

Ultracapacitor has the nature of peak power density allowing sufficient energy storage for acceleration power requirements and fuel economy. The benefit of ultracapacitor is that it can withstand over a broad range of SOC, hence increased durability. Its higher specific power characteristic to provide peak power surge is beneficial for hybrid power applications. The large storage capacitance of the ultracapacitor is utilized to meet the power shortfall during start-up and transient operation.

By increasing the modules of ultracapacitor, there will be an increase in the vehicle fuel economy, whereas increasing the modules of battery reduces the fuel economy because of incremental mass of battery and small specific power with respect to an increase in hybrid ratio.

KEYWORDS

- **ultracapacitor**
- **hybridization**
- **Powertrain**
- **drive cycle**

REFERENCES

1. Ultracapacitors Challenge the Battery. [Online], 2004, http://www.worldandi.com (accessed Jan 12, 2017).

2. Burke, A. Ultracapacitors: Why, How, and Where is the Technology. *J. Power Sources* **2000,** *91*, 37–50.
3. Burke, A. Batteries and Ultracapacitors for Electric, Hybrid, and Fuel Cell Vehicles. *Proc. IEEE* **2007,** *95*(4), 806–820.
4. Gao, W. Performance Comparison of a Fuel Cell-Battery Hybrid Powertrain and a Fuel Cell-Ultracapacitor Hybrid Powertrain. *IEEE Trans. Vehicular Technol.* **2005,** *54*(3), 846–855.
5. Atwood, P.; Gurski, S.; Nelson, D. J.; Wipke, K. B. Degree of Hybridization Modeling of a Fuel Cell Hybrid Electric Sport Utility Vehicle, SAE Paper 2001–01–0236.
6. Bubna, P.; Brunner, D.; Advani, S. G.; Prasad, A. K. Prediction-Based Optimal Power Management in a Fuel Cell/Battery Plug-In Hybrid Vehicle. *J. Power Sources* **2010,** *195*, 6699–6708.
7. Thounthong, P.; Sethakul, P.; Rael, S.; Davat, B. Performance Evaluation of Fuel Cell/Battery/Supercapacitor Hybrid Power Source for Vehicle Applications, *IEEE Ind. Appl. Soc. Annu. Meet. IAS* **2009,** *2009*, 1–8.
8. Thounthong,P.; Raëlb, S.; Davat, B. Control Strategy of Fuel Cell/Supercapacitors Hybrid Power Sources for Electric Vehicle. *J. Power Sources* **2006,** *158*, 806–814.
9. Feroldi, D.; Serra, M.; Riera, J. Design and Analysis of Fuel-Cell Hybrid Systems Oriented to Automotive Applications. *IEEE Trans. Vehicular Technol.* **2009,** *58*(9), 4720–4729.
10. Senger, R. D.; Merkle, M. A.; Nelson, D. J. Validation of ADVISOR as a Simulation Tool for a Series Hybrid Electric Vehicle. SAE Paper 981133, Technology for Electric and Hybrid Vehicles, 1998, SP-1331; pp 95–115.
11. Fuchs, M.; Barbir, F.; Husar, A.; Neutzler, J.; Nelson, D. J.; Ogburn, M. J.; Bryan, P. Performance of an Automotive Fuel Cell Stack. VA, SAE Paper 2000–01–1529, 5.
12. Ogburn, M. J.; Nelson, D. J.; Wipke, K.; Markel, T. *Modeling and Validation of a Fuel Cell Hybrid Vehicle*, Proceedings of the 2000 Future Car Congress, Arlington, VA, April 2–6, SAE Paper 2000–01–1566, 13.
13. Ogburn, M. J.; Nelson, D. J. Systems Integration and Performance Issues in a Fuel Cell Hybrid Electric Vehicle. SAE Paper 2000–01–0376, Fuel Cell Power for Transportation 2000, SP-1505, 2000, 125–137.
14. [Online]. http://www.maxwell.com/pdf/uc/datasheets/PC2500.pdf (accessed Jan 25, 2017).

CHAPTER 36

COMPUTATIONAL ANALYSIS OF CERAMIC REGENERATIVE STORAGE AIR HEATER FOR A HYPERSONIC WIND TUNNEL

S. NAGENDRA BABU[1,*], SANDIP CHATTOPADHYAY[1,2], D. R. YADAV[1], and G. AMBA PRASADA RAO[3]

[1]*Defence Research and Development Laboratory (DRDL), Hyderabad, Telangana, India*

[2]*sandip.cpdy@gmail.com*

[3]*Department of Mechanical Engineering, National Institute of Technology, Warangal, Telangana, India*

[*]*Corresponding author. E-mail: nagendra.drdo@gmail.com*

ABSTRACT

For testing objects under hypersonic regimes, provision of high pressure and high temperature, clean air is essential in a hypersonic wind tunnel (HWT) facility. For simulation of such conditions, ceramic brick-based storage heaters are the preferred choice. To operate the HWT for a Mach number of 6.5, air is to be heated to a temperature of 750 K. For this purpose, a regenerative storage air heater (RSAH) is selected and the design is carried out through an iterative process. Sizing of the heater is done based on the thermal duty requirements. In order to evaluate the performance of RSAH computational analysis using ANSYS software is carried out for heating, blowdown, and reheating cycles. Scale-down approach is followed with symmetric 3D computational fluid dynamics geometry consisting of last circumferential layer, insulation layers, and steel casing of heater. Moreover, enumerates challenges in arriving at the geometry for the analysis. Bed profile temperatures for heating, blowdown, and reheating are presented. The size of the heater is observed to be adequate for an exit air temperature during the 40-s blowdown period. The study revealed that the insulation layers to prevent the heat loss while safeguarding the shell are adequate. The analysis also helped in achieving the heating time of 7 h and reheating time of 3 h.

36.1 INTRODUCTION

Hypersonic wind tunnel (HWT) facilities normally operate as pressure vacuum driven enclosed tunnels. In these facilities, air is stored at high pressure, regulated to required stagnation pressure, and is expanded to the vacuum conditions to get the required velocity in the test section.[1] High-pressure air is preheated in order to avoid liquefaction of air in test section due to large expansion ratios. Preheating conditions depend on the pressure ratio and Mach number required in the test section. Ceramic cored brick regenerative storage air heater (RSAH) is the best choice for this type of heating to generate hot and clean air.[2] RSAH contains a stack of ceramic mass with multiple holes aligned from top to bottom. This mass of ceramic is referred as the core of the heater sitting on a grate at the bottom.[3] The core is surrounded by layers of refractory bricks acting as insulation to protect the shell from experiencing high temperature and also to minimize the heat loss from the core to the environment. RSAH is operated in three cycles namely heating cycle (charging cycle), blowdown cycle (discharging cycle), and reheating cycle (recharge cycle), respectively.[4] In heating cycle, the heater bed is heated to the required temperature by blowing hot combustion gases through the holes in the core of the bed. The combustion gases are produced by burning LPG in the burner placed on top of the heater. Heating is stopped once the core reaches the required temperature. In blowdown cycle, the cold air at required pressure and mass flow rate is sent from the bottom of hotbed to top. During this process, the air gets heated to required temperature and the ceramic bed of the heater loses some of its heat content. In reheating cycle, the bed is reheated from the existing condition to required temperature as performed in heating cycle.

The proposed HWT shall work at Mach number of 6.5 and requires the RSAH to operate at a temperature of 750 K and deliver air at a maximum flow rate of 150 kg/s. The design is carried out based on the thermal duty requirements. While arriving at the size, based on the thermal duty, thermal inertia factors shall be used. As the configuration along with the ceramic bricks and insulation is evolved for the first time, practical thermal inertia factor of this configuration is not available. Hence, a thermal inertia factor of seven is used and the size of the heater core is arrived at. The size of the heater core is about 2.5 m in diameter and 5 m in length. The core has around 18,500 holes for the airflow and heat transfer as shown in Figure 36.1. Computational study of the full configuration is essential to get the accurate performance of the heater before proceeding with the realization. To overcome this challenge, a scaled-down geometry that represents the full-scale heater performance is evolved and the transient conjugate heat transfer analysis is performed. Method is validated by analyzing the results in radial planes.

36.2 OBJECTIVE

The following are the objectives of the computational fluid dynamics analysis:

- Develop the scaled-down geometry architecture to simulate the equivalent performance of full-scale heater.
- Carry out the conjugate transient thermal analysis for heating cycle, blowdown cycle, and reheating cycle to understand the conditions of air and core at the end of the respective cycles.
- Arrive at the heating and reheating cycle times based on the temperature profile of the bed.

36.3 COMPUTATIONAL ANALYSIS: GEOMETRY AND MESH

ANSYS™ suite[5,6] of tools is used for carrying out the simulation. A 3D computational analysis of asymmetric geometry is performed. The following subsections provide the details of geometry, mesh, and solver used for the analysis.

36.3.1 GEOMETRY

The geometry of the RSAH is modeled with insulation layers and outer shell. The RSAH bricks are arranged in the circular manner to form a core diameter and it contains circumferential rings of bricks altogether having

18,500 holes. Computational fluid dynamics (CFD) simulation of full-size heater with end to end geometry (of the order ~5.5 m) and these large number of holes result in a large number of mesh elements. Hence, the symmetric scaled geometry is chosen for the analysis. In the heater, numbers of bricks are arranged in circumferential rings with an inter-circumferential ring gap as shown Figure 36.1. Owing to this arrangement, core bricks in the last circumferential ring are connected to the insulation layers and will be affected by the heat loss to the environment through the insulation layers. Therefore, it is decided to carry out the simulation of the heater with brick in last ring, insulation layers, and casing as the geometry. Geometry consists of 96 holes and arc outer diameter of the core corresponds to 2.4 m. After the simulation, results are analyzed specifically to confirm this behavior.

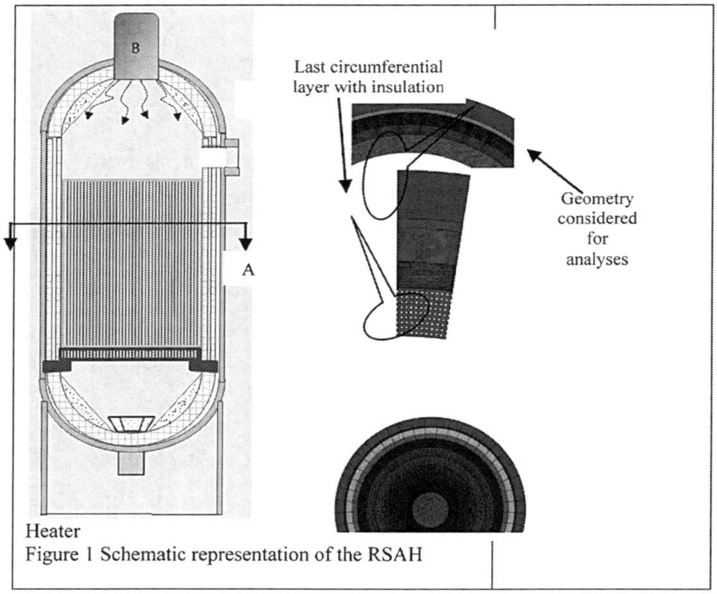

FIGURE 36.1 Schematic representation of the regenerative storage air heater.

36.3.2 MESH FOR REGENERATIVE STORAGE AIR HEATER

Considering the complexity of the flow, a hexahedral mesh to capture the flow is considered. In order to resolve the mesh near the wall, a very fine mesh is generated at the wall. Meshing for the flow domain is performed in integrated computer-aided engineering and manufacturing (ICEM)-CFD.

ANSYS ICEM-CFD blocking techniques provide an optimal platform to generate hexahedral meshes[6] with minimal effort and maximum control of the mesh quality. The block-wise mesh generated resulting in a mesh of approximately 13-million cells.

36.4 SOLVER AND CASE DEFINITIONS

Flow analysis has been carried out with ANSYS Fluent solver.[7]

36.4.1 SOLVER SETTINGS

Pressure-based coupled solver[8] along with k–ε turbulence model[9–11] with scalable wall functions (when Y+~12)[12] has been used to solve the flow and turbulence equations. Flow and turbulence have been discretized with second-order upwind[12] with cold face liner < 1 for all the heating, blowdown, and reheating cycles. Temperature-dependent material properties are considered for all the cases of analysis.

36.4.1.1 CASE 1 INITIAL HEATING CYCLE

The initial heating cycle is simulated to heat the cored brick from room temperature (300 K) to high temperature (~800 K). The following are the input conditions used for the simulation.

- Brick, insulation, and shell are initialized with room temperature.
- Top of the core is designated as inlet and flue gas enters the bricks top end with 800 K, and exits through bottom.
- Flue gas inlet is specified with given mass flow rate of 0.010378 kg/s and the pressure is calculated accordingly.
- Flue gas outlet is set to atmospheric conditions.
- Simulation is carried for 7 h.

36.4.1.2 CASE 2 BLOWDOWN CYCLE

Blowdown cycle is simulated to heat the air by passing through cored brick which is heated during the initial heating. The following are the input conditions used for the simulation

- Temperature profile of the heater core, insulation, and shell after heating for 7 h is taken as the input temperature for the simulation.
- Air enters the bricks at the bottom end with 300 K temperature and exits through the top.
- Air inlet is specified with given mass flow rate of 0.791715 kg/s at temperature 300 K and the pressure is imposed as 100 bar.
- Outlet is given outflow condition with extrapolated from the interior, while for the solver to start atmospheric conditions are given.
- Runtime of the simulation is 40 s.

36.4.1.3 CASE 3 REHEATING CYCLE

Reheating cycle is simulated to reheat the core so that core can be used for the next blowdown. The following are the input conditions used for the simulation.

- Temperature profile of the heater core, insulation, and shell after the 40-s blowdown cycle is taken as the input temperature for the simulation.
- Flue gas enters the bricks from the top end with 800 K temperature and exits through the bottom.
- Flue gas inlet is specified with given mass flow rate of 0.010378 kg/s at temperature 800 K and the pressure is atmospheric.
- Outlet is initially set to atmospheric conditions.
- Simulation runtime is 3 h.

36.5 RESULTS AND DISCUSSIONS

36.5.1 CASE 1 HEATING CYCLE RESULTS

After 7 h of heating, the temperature distribution in solid and fluid zones is extracted and the results are shown as contours in Figure 36.2. As the flue gas flows from top to bottom of the heater through the holes, it loses its energy to the core and insulation layers. Because of this phenomenon, the temperature of the flue gas drops by the time it reaches the bottom of the core.

Figure 36.3 shows the raise in temperature of the core at top and bottom planes during 7 h of heating. As per the simulation, top of the brick reaches almost at the steady state by the end of 7 h. The temperature at the bottom of the core starts increasing from 4 h onwards. It is observed that flue gas losses its complete energy while leaving the bottom of the core till first 3 h of heating, after 3 h, as the temperature of the core at the bottom increases, flue gas comes out at a higher temperature.

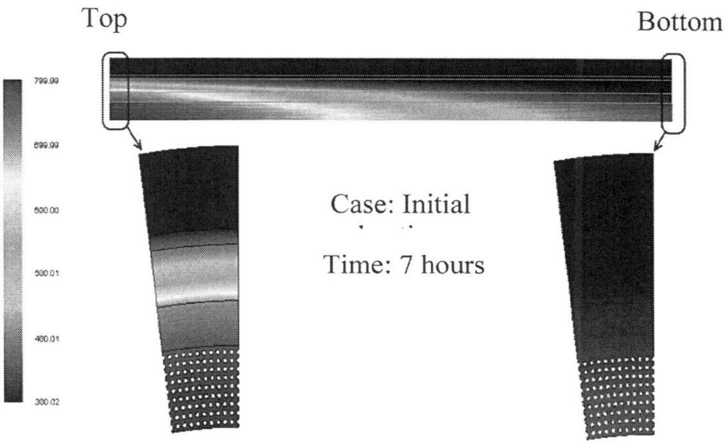

FIGURE 36.2 Temperature contours across all solid and fluid zones at the inlet and outlet.

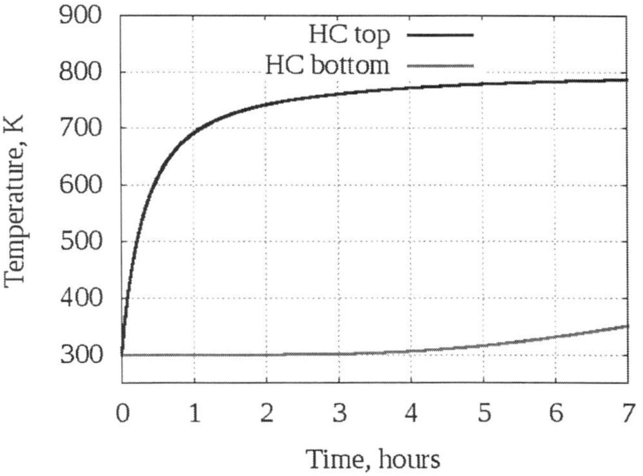

FIGURE 36.3 Surface average brick temperature history.

Figure 36.4 shows the temperature of the insulation layers during 7 h of heating. Raise in temperature of the shell is very small from the simulation results which confirm the adequacy of insulation thickness. Figure 36.5 shows the distribution of heat flux from one layer to the other layer. The simulation shows that the layers of insulation are effective in preventing the heat transfer from the core to the shell. From the Figure 36.5, it is clear that the heat flux from cold face liner to cerablanket and from cerablanket to shell is negligible.

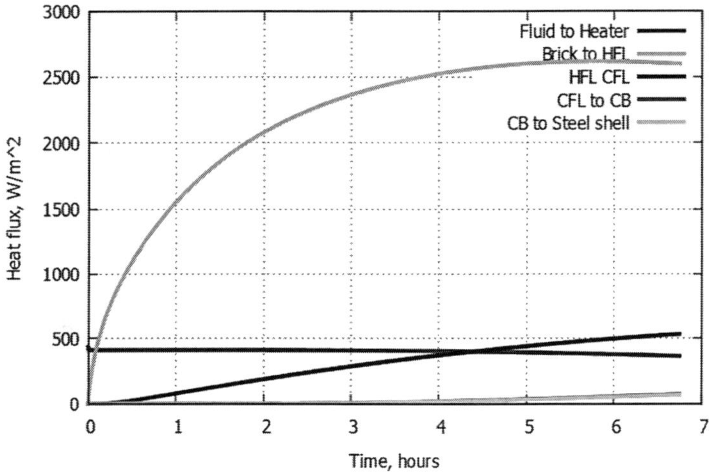

FIGURE 36.4 Temperature history of insulation and shell.

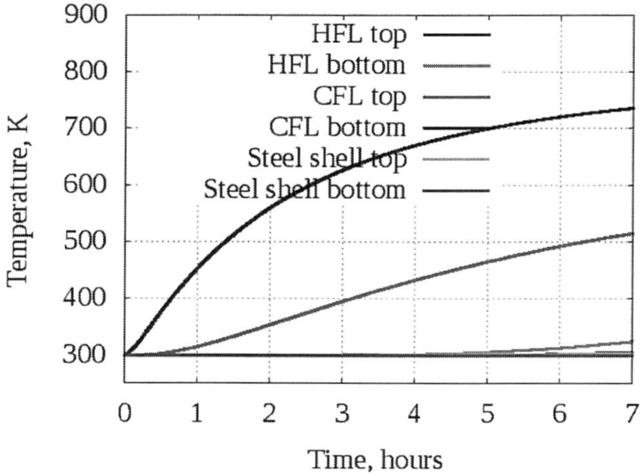

FIGURE 36.5 Average heat flux history.

In order to understand the distribution of the temperature from the core to the shell, the following profiles are extracted. Five radial lines from the core to the shell are selected as shown in Figure 36.6 and the temperature along these radial lines is extracted. Figure 36.7 shows the temperature on radial line 1 at the top and 1.1 m from top planes. Oscillations shown in the temperature in Figure 36.7 up to a radial distance of 0.15 m represents the transition from fluid to solid and solid to fluid zones as the lines considered for extraction of the data passes through the holes.

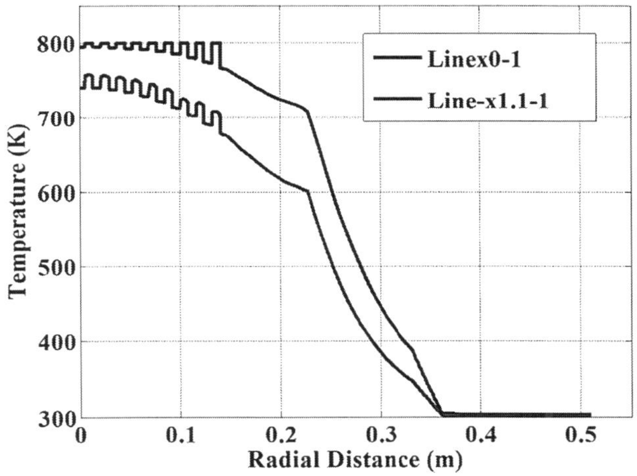

FIGURE 36.6 Radial temperature profiles at the end of 7 h.

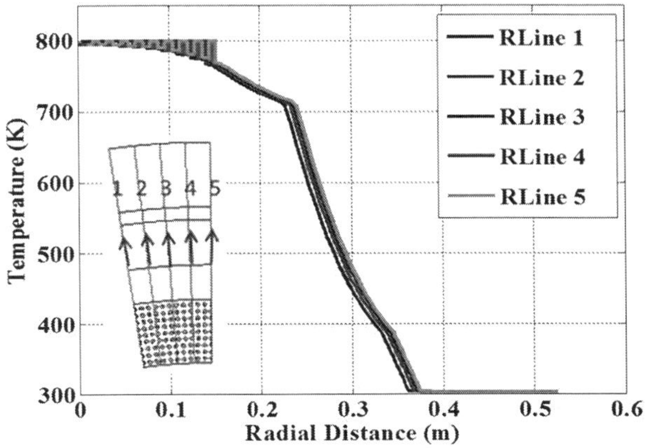

FIGURE 36.7 Radial temperature profile along two radial line at the end of 7 h (1.1 m from top).

Figure 36.8 shows the distribution of temperature along the length of the bed at five different radial locations. From these plots, it is clear that the temperature of the bed is uniform and almost constant up to a radial distance of 35 mm, which covers three holes from the inner edge of the brick. The effect of loss of heat on the insulation layers is not seen up to a distance of 35 mm. Based on this observation, it can be concluded that the size of the core of the brick considered for simulation represents the behavior of the whole core.

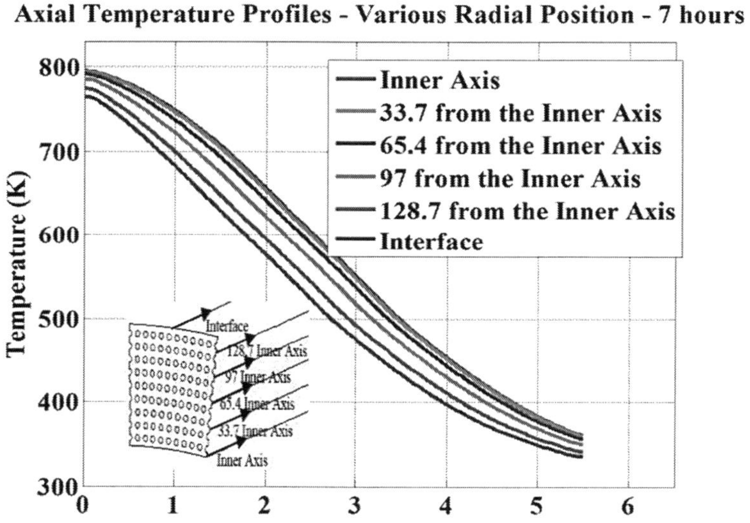

FIGURE 36.8 Axial temperature profiles in heater core at various radial positions (7 h).

Figure 36.9 shows the temperature profile of the heater core at three different time intervals. In order to validate the approach and the CFD method, results obtained during heating are compared with the test results available in the literature.[13] Based on the experimental data available for the flue gas temperature during the experiment, data of the experiment and analysis are brought on to the common scales and compared. The results are plotted in Figure 36.9. This shows a good agreement between the experimental data and the analysis results. Based on Figure 36.9, it is concluded that the geometry, mesh, and solver model selected for solving the problem represents the exact physics of the problem. Therefore, the same geometry, mesh, and solver models are used for carrying out blow-down and reheat analysis.

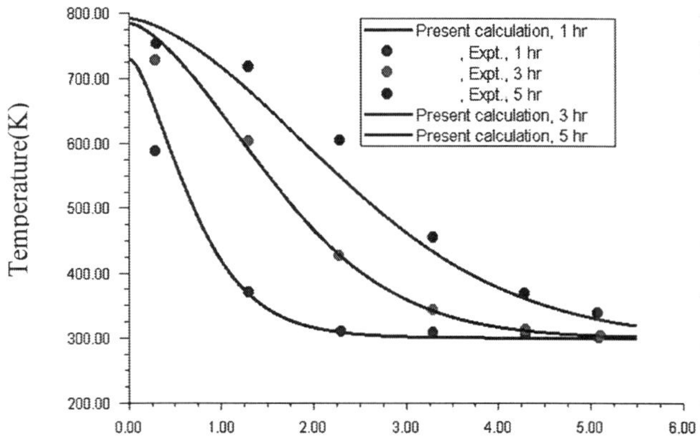

FIGURE 36.9 Axial temperature profiles compared with experimental data.

36.5.2 CASE 2 BLOWDOWN CYCLE RESULTS

Blowdown simulation is done for the duration of 40 s. After the simulation, the temperature distribution in solid and fluid zones is extracted. Figure 36.10 shows the temperature of the air coming out of the heater during the blowdown duration. Overall, 45 K drop in temperature for the air is found for the 40 s of blowdown.

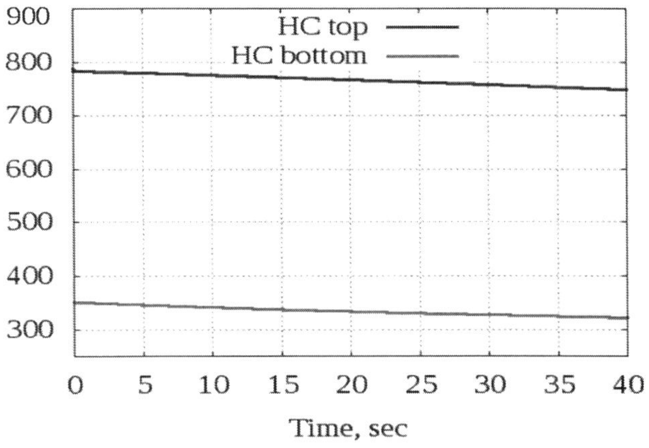

FIGURE 36.10 Air temperature history during blowdown.

Figure 36.11 shows drop in core top and bottom temperatures. From the plot, it is seen that the drop in core top temperature is about 35 K. No drop in temperature of the insulation layers and shell is found during 40 s of blowdown.

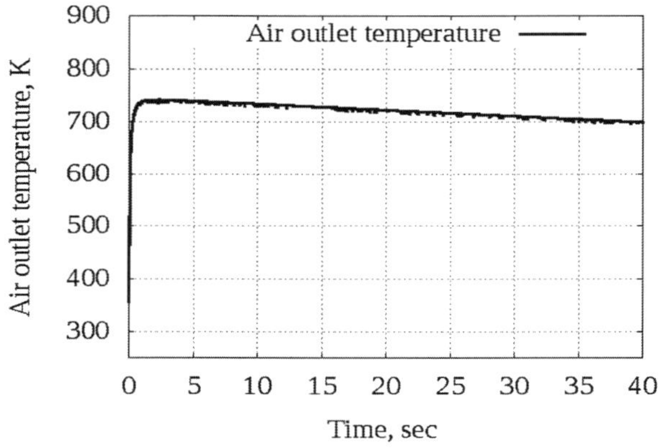

FIGURE 36.11 Core top and bottom temperature history during blowdown.

Temperature is extracted from the five radial lines from the core to the shell at the end of the 40 s blowdown. Figure 36.12 shows the radial distribution of the temperature at different planes along the length of the heater during the blowdown. From these plots, it is seen that there is some raise in the temperature at a radial distance of 0.15 m, which indicates the beginning of insulation layers.

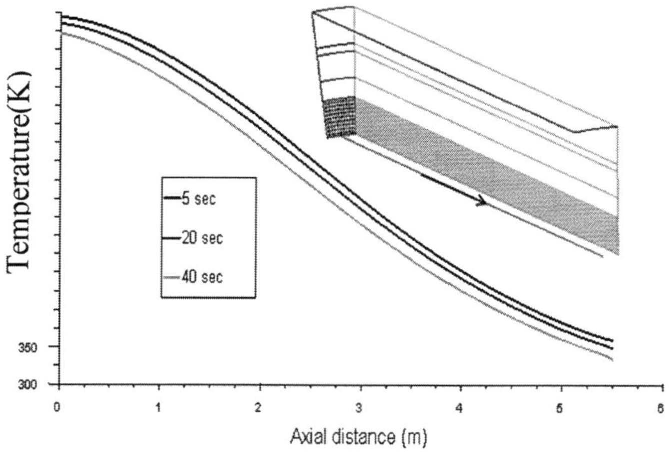

FIGURE 36.12 Radial temperature profiles at the end of 40 s (top of the heater plane).

From this profile of the temperature in these graphs, it can be understood that the insulation layers are not losing much of their energy during the blowdown. Figure 36.13 gives the details of the heater core temperature along the length of the heater at different time intervals; it can be inferred that there is about 35 K average temperature drop in the core of the heater during the blowdown.

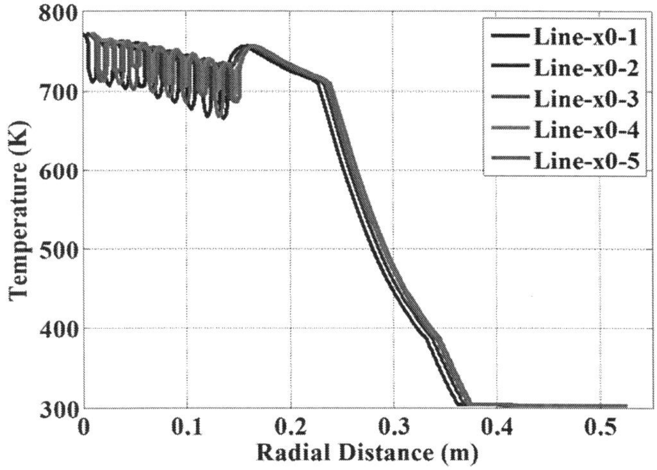

FIGURE 36.13 Axial temperature profiles in heater core at different time intervals.

36.5.3 CASE 3 REHEATING CYCLE RESULTS

In post-blowdown, the heater core is reheated to make it ready for the next blowdown. In order to understand the time required for bringing the heater core to the required temperature, one more heating simulation is carried out. It is found that 3 h of reheating brought the core back to the condition which is suitable for next blowdown.

Figure 36.14 shows the top and bottom temperature profile of the core during reheating. Top of the core reached a temperature of about 780 K within an hour during the reheating. However, further heating is required to ensure that the rest of the bed reaches the required temperature profile as that of the initial heating cycle.

Figure 36.15 shows the temperature of the core at the end of heating, blowdown, and reheat cycles. Based on this, it is clear that the 3 h of reheating is more than sufficient to bring the heater to the initial heating conditions.

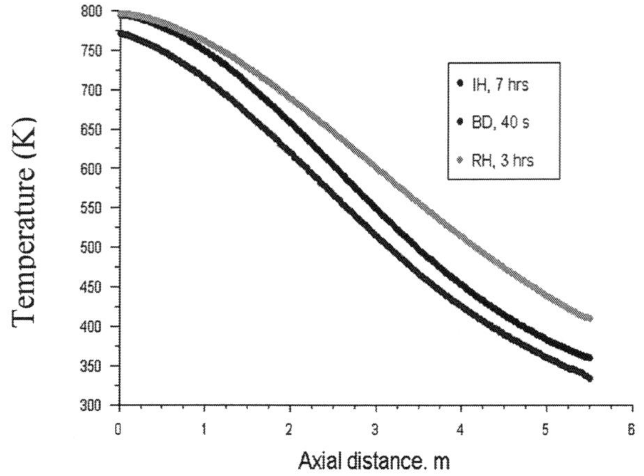

FIGURE 36.14 Heater core temperature profiles during reheating.

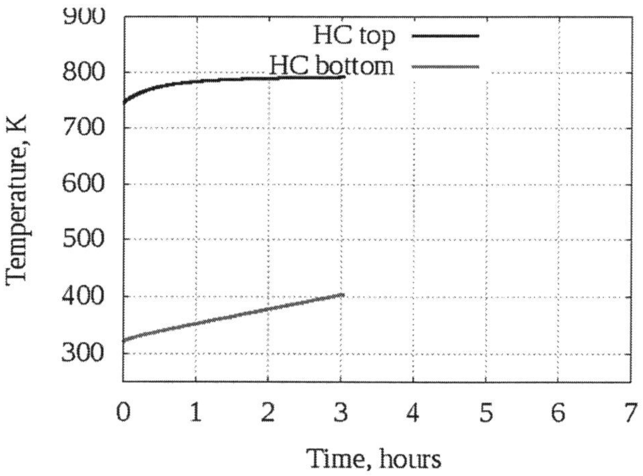

FIGURE 36.15 Axial core temperature profile, at the end of the initial heating, blowdown, and rheating cycles.

36.6 CONCLUSIONS

- Transient conjugate CFD approach is developed for the analysis of the flow through the heater core.
- Geometry selected for analysis proved to represent the physics of the full-scale heater. Precise block structured, mesh for heat transfer analysis has been generated to achieve wall y+ lesser than 20.

Unsteady conjugate heat transfer analysis using μRANS model has been successfully executed.

- Simulation has been successfully carried out for heating, blowdown, and reheating cycles.
- Simulation results show that initial heating of 7 h is required to bring the core to the required condition of the blowdown.
- Air temperature during the blowdown is observed to reduce by 45 K which is acceptable. Reheating of the heater bed is done in 3 h.

ACKNOWLEDGMENTS

The authors would like to thank authorities of DRDL for encouragement and Mr. M. S. R. Chandra Murthy, Scientist, DRDL, Hyderabad and Mr. P. K. Sinha, Scientist, DRDL, Hyderabad for their valuable suggestions during the progress of this work.

KEYWORDS

- **heating cycle**
- **blowdown cycle**
- **ceramic bricks**
- **hypersonic wind tunnel**
- **Mach 6.5**
- **reheating cycle**
- **core**

REFERENCES

1. Chu, P. H.; Marks berry, C. L.; Saari, D. P. High Temperature Storage Heater Technology for Hypersonic Wind Tunnels and Propulsion Test Facilities. *AIAA/CIRA 13th International Space Planes and Hypersonic Systems and Technologies Conference,* Capua, Italy, May 16–20, 2005.
2. Saari, D. P.; Jauch, C. E.; Chu, P. H. Clean Air Regenerative Storage Heater Technology for Propulsion Test Facilities. *16th AIAA/DLR/DGLR International Space Planes and Hypersonic Systems and Technologies Conference,* 2009, AIAA 2009–7377.

3. Lee, Y. J.; Kang, S. H.; Yang, S. S. Korea Aerospace Research Institute, Daejeon, 305–333, Korea Development of the Scramjet engine Test Facility in Korea Aerospace Research Institute, *27th AIAA Aerodynamic Measurement Technology and Ground Testing Conference,* Illinois AIAA: Chicago, June 28–July 1, 2010, 2010–4792.
4. Steinle, F. W. Jr.; Laster, M. L. High-Temperature, Continuous-Flow Heater Technology. *14th AIAA/AHI Space Planes and Hypersonic Systems and Technologies Conference,* Canberra, Australia, November 6–9, AIAA 2006–8045.
5. ANSYS Geometry and ANSYS ICEM CFD 15.0, ANSYS Inc, 2014.
6. Silieti, M.; Kassab, A. J.; Divo, E. Film Cooling Effectiveness: Comparison of Adiabatic and Conjugate Heat Transfer CFD Models. *Int. J. Therm. Sci.* **2009,** *48,* 2237–2248.
7. ANSYS FLUENT 15.0, ANSYS Inc, 2014
8. Thomas, E. D.; Board, D. G.; Sean, D. B. Evaluation of CFD Simulations of Film Cooling Performance on a Turbine Vane Including Conjugate Heat Transfer Effects. *Int. J. Heat and Fluid Flow* **2014,** *50,* 249–286.
9. Ebrahimi, H. B.; Ryder, R. C., Jr. CFD Simulations for AEDC Combustion Air Heater (CAH) for APTU Facility, *14th AIAA/AHI Space Planes and Hypersonic Systems and Technologies Conference,* 2006, AIAA 2006–8046.
10. Mohammad, D. A.; Masoud, R.; Ammar, A. A. CFD Modeling of a Radiant Tube Heater. *Int. Commun. Heat Mass Transfer* **2014,** *39,* 432–438.
11. Esmaeil, P.; Masoud, A. Using CFD to Study Combustion and Steam Flow Distribution Effects on Reheater Tubes Operation. *J. Fluids Eng. ASME* **2011,** *133,* 051303.
12. Jong, C. J.; Dong, G. K. CFD Analysis of Thermally Stratified Flow and Conjugate Heat Transfer in a PWR Pressurizer Surgeline. *J. Pressure Vessel Technol. ASME* **2010,** *132,* 021301.
13. Chandy, N.; Joseph, S. S.; Pandian, S. Thermal Analysis of a Cored Brick Heater, CP-34, NCWT-02.

CHAPTER 37

EFFECTS OF VELOCITY SLIP AND TEMPERATURE JUMP ON HEAT TRANSFER PHENOMENA OF SPHERICAL PARTICLES IN NEWTONIAN FLUIDS

RAHUL RAMDAS RAMTEKE[1,2] and NANDA KISHORE[1,*]

[1]*Department of Chemical Engineering, Indian Institute of Technology, Guwahati 781039, Assam, India*

[2]*r.ramteke@iitg.ernet.in*

[*]*Corresponding author. E-mail: nkishore@iitg.ernet.in*

ABSTRACT

Velocity slip along the solid surface can arise in the several applications such as aerosols, flow through porous materials, suspension, capillary flows, polymer flow through extruders, flow along smooth solid surfaces, and so forth. Thus, in this chapter, the effects of velocity slip and temperature jump on force convective heat transfer from spherical particles to Newtonian liquids have been numerically investigated. The governing dimensionless continuity, momentum and energy equations are solved using semi-implicit marker and cell algorithm implemented on a staggered grid arrangement in spherical coordinates. The present numerical results obtained in the range of dimensionless slip parameter ($0.01 \leq \lambda_v \leq 100$), temperature jump ($0.01 \leq \lambda_T \leq 10$), Prandtl number ($1 \leq Pr \leq 100$) at Reynolds number, $Re = 20$. The isotherm contours are presented for better understanding of heat transfer phenomena around spherical particles. Furthermore, the effects of dimensionless parameters on the local Nusselt and average Nusselt numbers are thoroughly discussed. The effect of slip and temperature jump on heat transfer affect in opposite manner, that is, large slip on the solid surface increases convection heat transfer along the solid surface, while large temperature jump decreases the heat transfer due to reducing the magnitude of the temperature gradient in the fluids.

37.1 INTRODUCTION

The well-known boundary condition to solve the energy equation is either constant wall temperature or constant wall flux. These boundary conditions are violated in many practical applications. Experimental observation shows that the temperature of a fluid at a surface may not be equal to the temperature of surface. This difference is known as temperature jump and it can be determined by the heat flux at the fluid–solid interface. Based on the analogy of slip flow, Poisson suggested that the temperature jump can be described as:

$$T_s - T_w = \mathcal{L}_T \left(\frac{\partial T}{\partial n} \right) \tag{37.1}$$

where $\mathcal{L}_T$ is temperature jump length. Smoluchowski[1] experimentally confirmed this hypothesis and proposed an equation for temperature jump length as:

$$\mathcal{L}_T = \left(\frac{2-\sigma_T}{\sigma_T} \right) \left(\frac{2\gamma}{\gamma+1} \right) \frac{Kn}{Pr} \tag{37.2}$$

where σ_T, γ, Pr, and Kn are thermal accommodation coefficient, specific heat ratio, Prandtl number, and Knudsen number, respectively. A general temperature jump condition for rarefied multicomponent gas flows is also described by Zade et al.[2] Gokcen and MacCormack[3] proposed a new temperature jump condition to simplify the Smoluchowski jump condition, and to yield the correct heat transfer in free molecular flow in the limiting case of very large Knudsen numbers. Some researchers investigated the effect of slip velocity and temperature jump on the heat transfer from a sphere in a low Reynolds number flow.[4–6] Taylor[5] analytically investigated the effects of velocity slip and temperature jump on the heat transfer from a sphere to low Reynolds number slip flow at small and large values of Peclet number. Mimaki et al.[7] presents the experimental results on the conductive heat transfer from a sphere to pure gases and their mixtures and also derived its analytical solution which is in good agreement with the experimental results. They obtained the following correlation for Nusselt number based on their analytical results which is in good agreement with experimental data in range of Knudsen number $0.016 \leq Kn \leq 0.8$ for various mixtures of hydrogen, helium, and nitrogen gases.

$$Nu = \frac{2}{1 + \frac{15}{2} \alpha_{mix}^{-1} Kn} \tag{37.3}$$

Finally, the Nusselt number predicted by Equation 37.3 also well agrees with the predictions obtained using the expression recently derived by Feng and Michaelides[8] in the limit $Pe \rightarrow 0$. Kavanau and Drake[9] experimentally and analytically studied the overall heat transfer from sphere to rarefied gas in subsonic flow at intermediate Reynolds number ($1.7 \leq Re \leq 124$) in the presence of temperature jump boundary condition. Vasudeviah and Balamurugan[10] obtained an analytical expression for the mean Nusselt number at $Re < 1$ and Pr of 0(1). They considered slip velocity boundary condition in their calculations and neglected the temperature jump. Most of the literatures of heat transfer in slip flow past solid spherical particle are available for $Re < 1$, rotating sphere, particle motion under rarefied condition, microfluidic, and nanofluidic systems.[11–17] There is very less number of literature on momentum and heat transfer study for viscous fluid flow past a spherical particle at intermediate Reynolds number with linear velocity slip and temperature jump boundary condition. Therefore, the aim of present work is to numerically analyze the heat transfer phenomena of the spherical particles in Newtonian fluids with linear velocity slip and temperature jump boundary condition at the solid–fluid interface.

37.2 PROBLEM DESCRIPTION AND MATHEMATICAL FORMULATION

The relative motion between a heated sphere and an infinite expanse of viscous fluid with linear velocity slip and temperature jump at the solid–fluid interface is considered as shown in Figure 37.1. The temperature of the heated sphere is T_w, whereas the temperature of the surrounding cold fluid is T_0. The relative velocity between the sphere and the fluid is U_0. Because of this temperature difference and the relative velocity, the convective heat transfer takes place from the heated sphere to the surrounding Newtonian fluid. Further assumed that the sphere is maintained at constant temperature T_w and there is no temperature gradient within the sphere, that is, the Biot number is negligible. The Newtonian fluid considered herein is assumed to be incompressible and its thermophysical properties are independent of the temperature difference $T_w - T_o$, that is, the temperature gradient between the solid sphere and the outer fluid is in the moderate range. Further the Navier's linear velocity slip model is used. The scaling of this linear velocity slip model yields the relation $\tau_{r\theta} = \lambda_v \times v_\theta$, where λ_v is a dimensionless slip parameter defined as $\lambda_v = (\beta R/\mu)$; here, β is a scalar positive coefficient function of the slip. Furthermore, the first-order temperature jump boundary condition is applied at the solid–fluid interface, that is, $T_f - T_w = \lambda_T \left(\frac{\partial T}{\partial r} \right)$, where λ_T is a dimensionless temperature jump parameter defined as $\lambda_T = \left(\frac{2 - \sigma_T}{\sigma_T} \right) \left(\frac{2\gamma}{1+\gamma} \right) \frac{Kn}{Pr}$. The continuity, momentum, and energy equations describe the transfer characteristics of this problem and are presented in their dimensionless form as follows:

- Equation of continuity

$$\nabla \cdot V = 0 \tag{37.4}$$

- Equation of motion

$$\frac{DV}{Dt} = -\nabla P + \frac{\nabla \cdot \tau}{Re} \tag{37.5}$$

- Equation of energy

$$\frac{DT}{Dt} = \frac{\nabla^2 T}{Pe} \tag{37.6}$$

The scaling parameters used to nondimensionalize Equations 37.4–37.6 are U_o, R, ρU_o^2, (R/U_o), and $(T_w - T_o)$ for velocity components, radial distance, pressure, time, and temperature difference, respectively. Similarly, the dimensionless boundary conditions required to solve this heat transfer problem are as follows:

- Surface of the sphere ($r = 1$):

$$v_r = 0;\ \tau_{r\theta} = \lambda_v \times v_\theta;\ T_f - T_w = \lambda_T \left(\frac{\partial T}{\partial r} \right) \tag{37.7}$$

- Outer free stream boundary ($r = R_\infty$):

$$v_r = -\cos\theta;\ v_\theta = sin\theta;\ T = 0 \tag{37.8}$$

- Central axis ($\theta = 0, \pi$):

$$v_\theta = 0; \frac{\partial v_r}{\partial \theta} = 0; \frac{\partial T}{\partial \theta} = 0 \tag{37.9}$$

In order to obtain fully converged temperature field in the entire computational domain, the governing Equations 37.4–37.6 along with the boundary conditions Equations 37.7–37.9 are numerically solved using a computational fluid dynamics-based segregated approach. In other words, the momentum equations are first solved to obtain the fully converged velocity and pressure fields; and then using the obtained velocity field as input, the energy Equation 37.6 is solved to obtain fully converged steady temperature field over entire domain.

Once the temperature field is known, the surface Nusselt number and average Nusselt number can be evaluated by using following Equations 37.10 and 37.11:

$$Nu_\theta = \frac{h(2\mathrm{R})}{k} = -2\left[\frac{\partial T}{\partial r}\right]_{r=1} \tag{37.10}$$

$$Nu_{avg} = \frac{1}{2}\int_0^\pi Nu_\theta \sin\theta d\theta \tag{37.11}$$

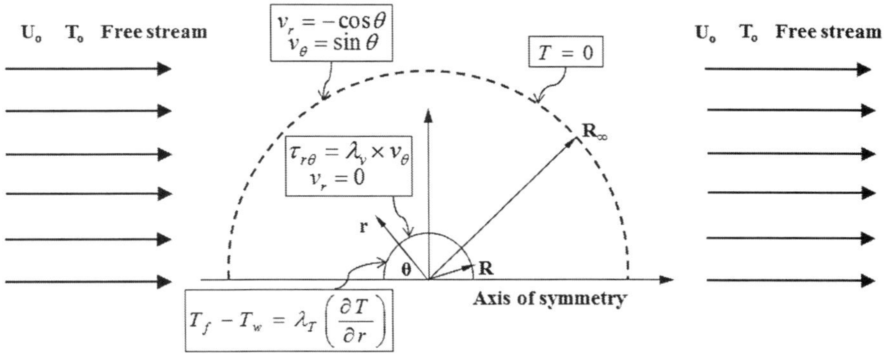

FIGURE 37.1 Schematic of flow domain and boundary conditions.

37.3 NUMERICAL METHODOLOGY

The governing momentum equations along with aforementioned boundary conditions are solved using a semi-implicit marker and cell (SMAC) method. This algorithm is a result of modification of the explicit marker and cell algorithm developed for free surface flows. Primarily, modifications are to incorporate semi-implicit formulation and accordingly modifying the pressure corrector equation. The SMAC approach is implemented using a finite difference method in staggered grid arrangement of the spherical coordinates. The temporal and pressure terms of momentum equations are discretized using a first-order forward differencing scheme, whereas the convective and diffusive terms are discretized using quadratic upstream interpolation for convective kinematics and second-order central differencing schemes, respectively. Finally, a false-transient time stepping method is adopted to obtain the steady velocity and pressure fields by solving the momentum equations by solving Equations 37.4 and 37.5 using SMAC algorithm. Then this fully converged steady velocity field is used as input to solve the energy Equation 37.6 using a similar time stepping approach to obtain fully converged steady temperature field in the entire computational domain. The fully converged temperature field is then used to evaluate the near surface kinematics such as the isotherm contours, surface, and average Nusselt numbers as described in previous section for various combinations of pertinent dimensionless parameters.

37.4 RESULTS AND DISCUSSION

The following ranges of dimensionless parameters are considered to discuss the heat transfer phenomena of single spheres in Newtonian fluids with velocity slip and temperature jump boundary condition at the solid–fluid interface: $Re=20$; $Pr=1, 10, 50, 100$; $\lambda_v=0.01, 0.1, 1, 5, 10, 50, 100$; and $\lambda_T=0.01, 0.1, 1, 2, 5, 10$.

37.4.1 ISOTHERM CONTOURS

Figure 37.2 shows the isotherm contours around a spherical particle for $\lambda_v=0.01$ in Newtonian fluid with $Pr=1$ (upper half) and $Pr=100$ (lower half) at $Re=20$ for different values of temperature jump parameter. At the small value of Prandtl number $Pr = 1$, the isotherm contours are uniformly carried away in the direction of flow for all values of temperature jump parameter λ_T. However, depending on value of Prandtl number isotherm, contours may be carried in the flow direction due to convection effects. Thus, at $Pr = 100$ (lower half) large amount of isotherm contours are carried away in the flow direction and thermal boundary layer becomes thinner. As the temperature jump parameter increases from 0.01 to 2, (Fig. 37.2d) isotherm contours shrinks around the sphere because it decreases the magnitude of the temperature gradient in fluid and as the result heat transfer decreases with the increasing temperature jump parameter. Similarly, Figure 37.3 represents the effect of temperature jump parameter on isotherm contours at slip parameter equals 100. As the slip parameter increases from 0.01 to 100, magnitude of slip reduces at surface of spheres. Therefore, by increasing the temperature jump parameter, there is reduction in heat transfer from sphere to fluid because the effect of slip parameter and temperature jump on heat transfer affect in opposite manner, that is, large slip on solid surface increases the convective heat transfer along the surface while large temperature jump decreases the convective heat transfer due to reduction in the magnitude of the temperature gradient in the fluid.

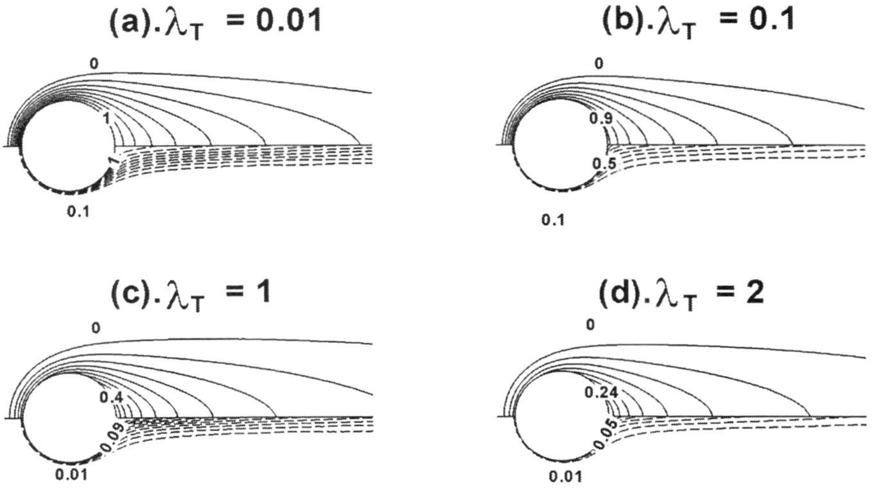

FIGURE 37.2 Isotherm contours around a single sphere at $\lambda_v = 0.01$ when Pr = 1 (upper half) and $Pr = 100$ (lower half).

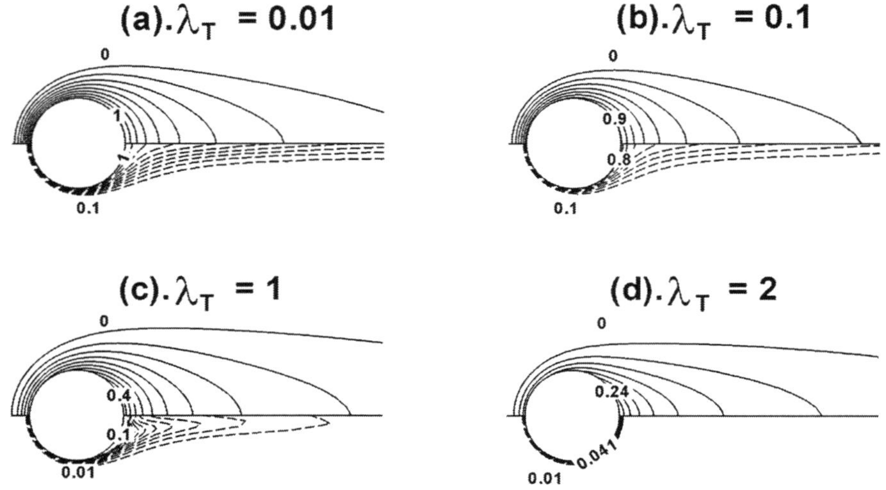

FIGURE 37.3 Isotherm contours around a single sphere at $\lambda_v = 100$ when Pr = 1 (upper half) and $Pr = 100$ (lower half).

37.4.2 SURFACE NUSSELT NUMBER

Figure 37.4 shows the effect of temperature jump parameter λ_T on the surface Nusselt around a spherical particle at different values of slip parameter λ_v in Newtonian fluid for $Pr = 1$ at $Re = 20$. At $Pr = 1$, as the value of slip parameter increases, the value of Nusselt number at front stagnant point decreases for the range of temperature jump parameter $0.01 \leq \lambda_T \leq 1$, whereas surface Nusselt number remains almost unchanged for other values of λ_T. Further, the overall heat transfer rate decreases as the value of slip parameter increases due to reduction in the magnitude of slip at the fluid–solid interface. For all value of temperature jump parameter and slip parameter as one traverse from the front stagnant point to the rear end, the surface Nusselt number decreases; however, for $\lambda_T > 1$ as the value of slip parameter increases, there is less change in the magnitude of Nusselt number along the surface of the sphere. This is due to the more reduction in the magnitude of the temperature gradient in the fluids. As the value of Prandtl number increase to 100 (Fig. 37.5), the magnitude of the surface Nusselt number around the surface of spherical particle increases due to the increasing the contribution of convective heat transfer. Therefore, the rate of heat transfer increases with increasing Prandtl number. Other trends are similar as observed in case of Prandtl number = 1 (Fig. 37.4) for all values of slip parameter and temperature jump parameter.

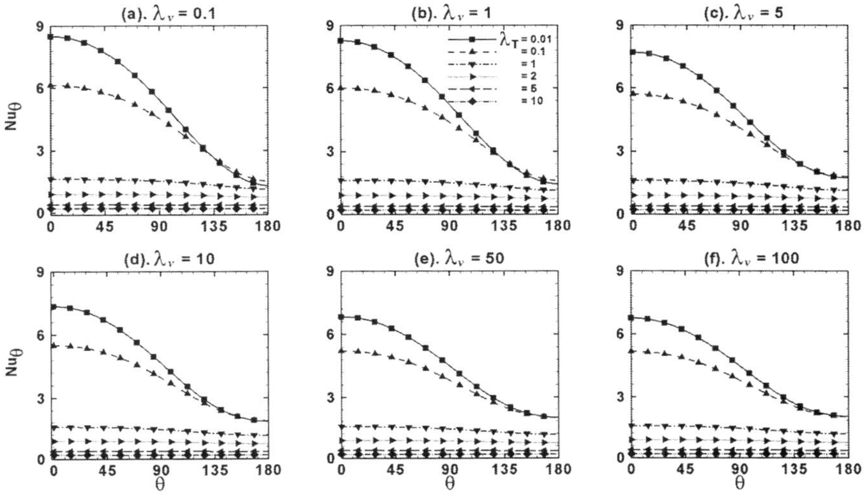

FIGURE 37.4 Surface Nu number around a single sphere at $Pr = 1$ for different values of slip and temperature jump parameters.

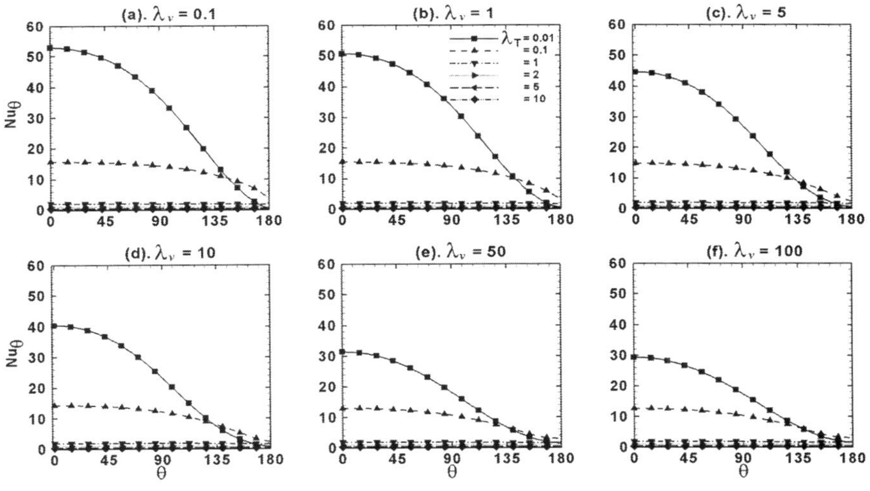

FIGURE 37.5 Surface Nu number around a single sphere at $Pr = 100$ for different values of slip and temperature jump parameters.

37.4.3 AVERAGE NUSSELT NUMBER

Figure 37.6 shows the variation of the average Nusselt numbers for Newtonian fluid past a single slip sphere for different slip and temperature jump parameter at $Re=20$. The average Nusselt number increases with increasing Prandtl number at all values of slip parameter and temperature jump parameter. At the small value of Pr, the effect of the slip parameter and temperature jump parameter on the average Nusselt number is insignificant. Further, the average Nusselt number decreases with the increasing the slip parameter. This is due to convection near the particle surface decreases as particle surface becomes approached to the no-slip boundary condition, that is, $\lambda_v \approx 1000$ (no-slip). Furthermore, the average Nusselt number decreases with the increasing the temperature jump parameter because of reduction in the magnitude of the temperature gradient in the fluid. In summary, heat transfer rate increases with decreasing slip parameter and/or temperature jump parameter.

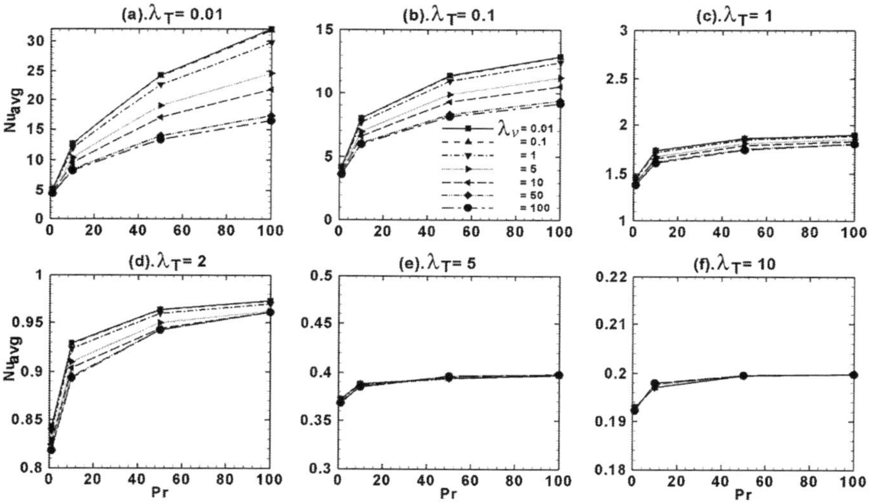

FIGURE 37.6 Average Nu number of a sphere in Newtonian fluid for different value of slip and temperature jump parameters.

37.5 CONCLUSION

The isotherm contours and the local and average Nusselt numbers of single spheres in Newtonian fluids with velocity slip and the temperature jump boundary condition at the fluid–solid interface are numerically analyzed. Isotherm contours shrink around spherical particle with the increasing slip parameter and temperature jump parameter. The effect of slip and temperature jump on heat transfer affect in opposite manner, that is, large slip on the solid surface increases the convective heat transfer along the surface while large temperature jump decreases heat transfer due to the reduced temperature gradient at the surface. For all value of temperature jump parameter and slip parameter as one traverse from the front stagnant point to the rear end, the surface Nusselt number decreases; however for $\lambda_T > 1$ as the value of slip parameter increases, there is much less change in the value of Nusselt number along the surface of the particle. This is due to the more reduction in the magnitude of the temperature gradient in fluids. Regardless the values of the slip parameter and temperature jump parameter, the surface Nusselt number increases with the Prandtl number. The average Nusselt number decreases with the increasing slip parameter as well as temperature jump parameter.

KEYWORDS

- **velocity slip**
- **temperature jump**
- **spheres**
- **Newtonian fluids**
- **heat transfer phenomena**

REFERENCES

1. Smoluchowski, M. Ueber Warmeleitung in Verdunnten Gasen. *Annalen der Physik und Chemie* **1898,** *64*, 101–130.
2. Zade, A. Q.; Renksizbulut, M.; Friedman, J. Boundary Conditions for Multi-Component Slip-Flows Based on the Kinetic Theory of Gases. *Proceedings of the Sixth International ASME Conference on Nanochannels, Microchannels and Minichannels, ICNMM 2008*, Darmstadt, Germany, June 23–25, 2008.
3. Gokcen, T.; MacCormack, R. W. Nonequilibrium Effects for Hypersonic Transitional Flows Using Continuum Approach. AIAA Paper No. 1987–1115, 1987, Presented at the 27th Aerospace Sciences Meeting, Reno, NV, January 1989.
4. Mohajer, B.; Aliakbar, V.; Shams M.; Moshfegh, A. Heat Transfer Analysis of a Micro-spherical Particle in the Slip Flow Regime by Considering Variable Properties. *Heat Transfer Eng.* **2015,** *36,* 596–610.
5. Taylor, T. D. Heat Transfer from Single Spheres in a Low Reynolds Number Slip Flow. *Phys. Fluids* **1963,** *6,* 7–12.

6. Strom, H.; Sasic, S. Heat Transfer Effects on Particle Motion Under Rarefied Conditions. *Int. J. Heat Fluid Flow* **2013,** *43*, 277–284.
7. Mimaki, H.; Endo, Y.; Takashima, Y. Heat Transfer from a Sphere to Rarefied Gas Mixture. *Int. J. Heat Mass Transfer* **1966,** *9*, 1435–1448.
8. Feng, Z. G.; Michaelides, E. E. Heat and Mass Transfer Coefficients of Viscous Spheres. *Int. J. Heat Mass Transfer* **2001,** *44,* 4445–4454.
9. Kavanau, L. L.; Drake, R. M. Heat Transfer from Spheres to a Rarefied Gas in Subsonic Flow. California University: Berkeley, 1953.
10. Vasudeviah, M.; Balamurugan, K. Heat Transfer in a Slip-Flow Past a Sphere. *Fluid Dyn. Res.* **1998,** *22,* 281–296.
11. Vajjha, S. R.; Das, D. K. A Review and Analysis on Influence of Temperature and Concentration of Nanofluids on Thermophysical Properties, Heat Transfer and Pumping Power. *Int. J. Heat Mass Transfer* **2012,** *55,* 4063–4078.
12. Martin, M. J.; Boyd, I. D. Momentum and Heat Transfer in a Laminar Boundary Layer with Slip Flow. *J. Thermophy. Heat Transfer* **2006,** *20,* 1–6.
13. Bao, F. B.; Lin, J. Z. Burnett Simulation of Gas Flow and Heat Transfer in Micro Poiseuille Flow. *Int. J. Heat Mass Transfer* **2008,** *51,* 4139–4144.
14. Dongari, N.; Agrawal, A. Analytical Solution of Gaseous Slip Flow in Long Microchannels. *Int. J. Heat Mass Transfer* **2007,** *50,* 3411–3421.
15. Rached, J.; Daher, N. Numerical Prediction of Slip Flow and Heat Transfer in Micro-Channels. *Proceedings of the 6th Annual Engineering Students Conference*, American University of Beirut: Beirut, Lebanon, 2007; pp 109–114.
16. Kakac, S.; Vasiliev, L. L.; Bayazitoglu, Y.; Yener, Y. *Microscale Heat Transfer-Fundamentals and Applications;* Springer: The Netherlands, 2005.
17. Minkowycz, W. J.; Sparrow, E. M.; Abraham, J. P. *Nanoparticle Heat Transfer and Fluid Flow;* Series: *Computational & Physical Processes in Mechanics & Thermal Science*, ISBN: 978-1-4398-6192-9; CRC Press, New York, 2013.

CHAPTER 38

NATURAL CONVECTION AROUND A HORIZONTAL CYLINDER POSITIONED ALONG A CIRCLE INSIDE A SQUARE ENCLOSURE

SATYENDRA KUMAR PANKAJ[1,*], HARISHCHANDRA THAKUR[1,2], and PRAVEEN CHOUDHARY[1,3]

[1]*Department of Mechanical Engineering, School of Engineering, Gautam Buddha University, Greater Noida, Uttar Pradesh 201312, India*

[2]*harish@gbu.ac.in*

[3]*chsmech@gmail.com*

[*]*Corresponding author. E-mail: satymech92@gmail.com*

ABSTRACT

This chapter investigates natural convection around a hot horizontal cylinder placed eccentrically in a cold square enclosure. To attend eccentricity in all possible direction, the cylinder position is changed along a circle inside the enclosure around its center. The effects of changing position on the phenomena of natural convection are shown by corresponding flow and thermal field developments and finally in terms of Nusselt number. Effect of various convectional strengths has also been included in the analysis by varying Rayleigh number (10^4–10^6). The Nusselt number value is found to be the maximum when the cylinder is positioned on the vertical axis below from the center.

38.1 INTRODUCTION

Natural convection is a phenomenon in which heat transfer takes place due to the density difference from a hot body to cold body and has a vital role in the performance of the design of the thermal manufacturing application. Due to engineering and industrial importance, a lot of research has been carried out on the enclosure.

Asan[1] found the natural convection effect in between two isothermal concentric square ducts with the parameters Rayleigh number up to 10^6 and dimension ratios L^* 1/5, 3/10, 3/5. Numerical simulation was applied to reach optimum result under the effect of Prandtl and Rayleigh number.

Kumar De and Dalal[2] analyzed the issue of characteristic convection around a square, level, warmed chamber put inside a walled in area in the scope of $10^3 \leq Ra \leq 10^6$. Impacts of the fenced in area geometry were surveyed utilizing three distinctive angle proportions by putting the square chamber at various statures from the base. The whole problem was divided in two cases: (1) prismatic cylinder is assumed to be at constant temperature and (2) cylinder is heated under uniform heat flux condition. The outcomes of different parameters such as geometric aspect ratio, condition of heated prismatic cylinder at various locations were studied and under constant wall temperature condition results were satisfactory. The geometry of cylinder did not affect the results.

Ghaddar[3] investigated the geometrical aftereffects of common convection from a consistently warmed level chamber put in an expansive air-filled rectangular walled in area. He watched that stream and warm conduct relied upon warmth fluxes forced on the internal chamber inside the isothermal walled in area.

Cesini et al.[4] experimentally and numerically investigated the heat transfer effect in terms of Nusselt number. The results were found approximately similar to the experimental values. A rectangular domain was taken for the numerical study with cold vertical walls and bottom insulated wall. Finite volume method was adopted for numerical simulation. For the experimental analysis, various temperature differences ($\Delta=0$, 0.35, 1.3, 6.3 and 7.8 K), aspect ratio W^* (2.1, 3.6, 4.3), and Rayleigh number (1.3×10^3 to 7.5×10^4) play a major role in finding the valuable results.

Shu and Zhu[5] concentrated on the change in the flow-field, and isotherms between a cold square enclosure and heated inward round cylinder regarding the span of the inward roundabout chamber.

Moukalled and Acharya[6] considered three diverse viewpoint proportions (r/L) of the chamber sweep r to the walled in area stature L in the scope of $10^4 \leq Ra \leq 10^7$. They took the cold square enclosure and a heated annulus cylinder placed in it. Due to the temperature difference, thermal boundary layer and streamlines are formed to depict the natural convection effect in terms of Nusselt number. As the Rayleigh number raises from 10^4 to 10^7, the convection regime becomes dominant and heat transfer phenomenon is observed to be maximum for the $Ra = 10^7$ at aspect ratio 0.3.

Shu et al.[7] conducted numerical examinations on natural convection between an external square walled in area and an internal roundabout chamber as per the unconventionality and rakish position of the internal roundabout chamber at a Rayleigh number of 3×10^5. Natural convection between square wall and circular chambers for $Ra=3 \times 10^5$ and also a predefined angle proportion of (r/L) were methodically investigated. It was found that the worldwide course, stream division, and the top space between the square external walled in area and the roundabout inward chamber had noteworthy consequences for the crest slant.

Kim et al.[8] as of late explored natural convection prompted by a temperature contrast between a cool external square cylinder and a hot internal cylinder for various Rayleigh numbers in the range of 10^3–10^6. The cylinder moves horizontally or diagonally with parameters δ and €. δ is the horizontal shift and € is the diagonal shift which influences the heat transfer phenomenon directly. Shifting of the cylinder in horizontal and diagonal direction provides the optimum value of a Nusselt number and heat transfer.

38.2 PROBLEM STATEMENT AND GOVERNING EQUATIONS

Figure 38.1 shows various positions of the inner cylinder in the computational domain in the given problem. It requires the cold square enclosure of length L and a hot circular cylinder of radius r placed inside it. The position of circular cylinder varies eccentrically at distance of $0.25L$ from the center along a circle of radius R. Various positions are assigned at angles 0°, 45°, 90°, 225°, and 270° from the horizontal axis of the cross section. The heat transfer effect is studied at various Rayleigh number ranging 10^4–10^6. The isotherms and velocity plots are drawn at these constraints. Figure 38.2 shows the mesh generation in the square enclosure, which plays a major role in the computational domain. The outside surface is considered as cold body, that is, T_C and the circular cylinder is considered as hot body, that is, T_H.

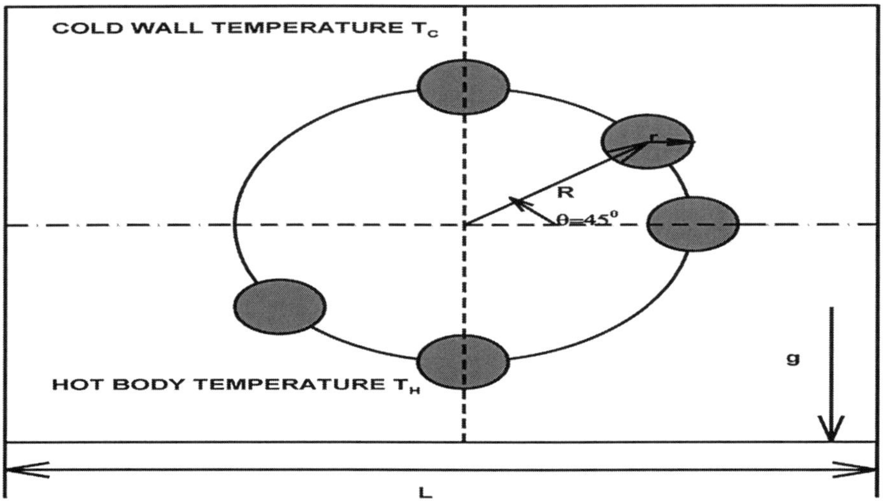

FIGURE 38.1 Computational domain.

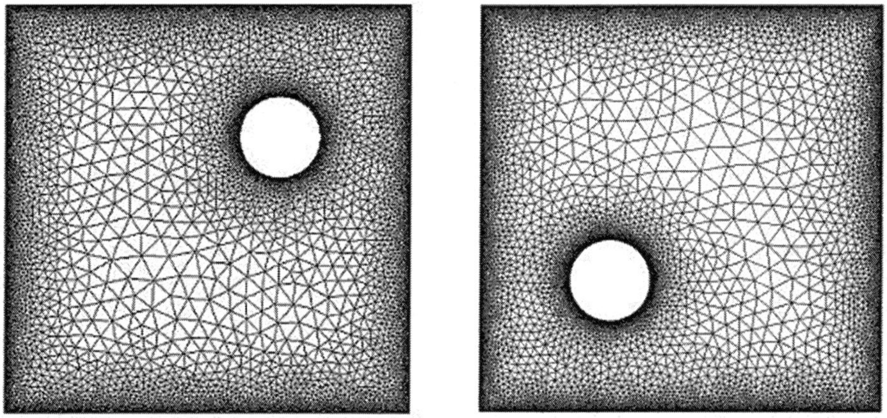

FIGURE 38.2 Mesh generation.

For the present problem, the stream is thought to be steady, laminar, incompressible, and two-dimensional. Viscous dissemination is ignored and thermophysical properties are thought to be steady, aside from thickness in the lightness term in Y-momentum equation, that is, the Boussinesq estimation is used. By applying these assumptions, the governing equations in nondimensional form are depicted as

$$\frac{\partial U}{\partial X} + \frac{\partial V}{\partial Y} = 0 \tag{38.1}$$

$$U\frac{\partial U}{\partial X} + V\frac{\partial U}{\partial Y} = -\frac{\partial P}{\partial X} + \frac{\partial^2 U}{\partial X^2} + \frac{\partial^2 U}{\partial Y^2} \tag{38.2}$$

$$U\frac{\partial V}{\partial X} + V\frac{\partial V}{\partial Y} = -\frac{\partial P}{\partial Y} + Gr\cdot\theta + \frac{\partial^2 V}{\partial X^2} + \frac{\partial^2 V}{\partial Y^2} \tag{38.3}$$

$$U\frac{\partial \theta}{\partial X} + V\frac{\partial \theta}{\partial Y} = \frac{1}{Pr}\left(\frac{\partial^2 \theta}{\partial X^2} + \frac{\partial^2 \theta}{\partial Y^2}\right) \tag{38.4}$$

Dimensionless variables are defined by

$$Pr = \frac{\nu}{\alpha},\ Gr = \frac{g\beta L^3 (T_H - T_L)}{\nu^2},\ \theta = \frac{T - T_L}{T_H - T_L} \tag{38.5}$$

Boundary conditions related to the above governing equations are given by:
On hot cylinder walls, $U=0$, $V=0$ and $\theta=1$ and on enclosure walls,

$$U = 0,\ V = 0 \text{ and } \theta = 0. \tag{38.6}$$

Table 38.1 shows the validation of the results from the Moukalled and Acharya which is almost equal to required value. Table 38.2 shows the variation of Nusselt number along the cold enclosure (cold body), and Table 38.3 depicts the variation of Nusselt number on the hot cylinder.

TABLE 38.1 Validation.

Ra	Surface-averaged Nusselt number at cold cylinder wall		
	Present study	**Moukalled and Acharya[6]**	**Difference percentage**
10^4	2.144	2.071	3.524
10^5	3.8999	3.825	1.934
10^6	6.1455	6.107	0.600

TABLE 38.2 Averaged Nusselt Number for Enclosure (Cold Body).

Ra	R=0.25L				
	θ=0°	θ=45°	θ=90°	θ=225°	θ=270°
10^4	2.760	2.570	2.536	2.658	2.757
10^5	3.756	3.522	3.681	3.860	4.173
10^6	6.190	6.434	6.072	6.392	6.878

TABLE 38.3 Average Nusselt Numbers for Circular Cylinder (Hot Body).

Ra	R=0.25L				
	θ=0°	θ=45°	θ=90°	θ=225°	θ=270°
10^4	16.616	14.669	15.066	15.542	15.987
10^5	23.152	21.062	21.658	23.718	21.452
10^6	38.108	38.036	36.090	40.200	42.938

38.3 RESULT AND DISCUSSIONS

Figure 38.3a shows at $\theta=0°$ vortices are formed at the right side of the hot cylinder. At $\theta=45°$ and $\theta=90°$, two symmetric vortex are formed along the vertical line which is almost same for all Rayleigh number, and at $\theta=225°$, six vortexes are formed around the hot cylinder due to the conduction effect. In Figure 38.3b, at $\theta=225°$, two vortexes are formed at both regions left and right side of the cylinder. At $\theta=270°$, nearly symmetric two counter rotating vortices are formed; two at the left and one at the right of the cylinder and the third is below the cylinder. In Figure 38.3c, for $Ra=10^6$, isotherms tend to become steep due to higher convective regions.

Figure 38.4 depicts the average Nusselt number variation along the cold wall; as the Rayleigh number increases, the Nusselt number increases and becomes maximum at $\theta=270°$. For $Ra=10^4$ and 10^6 at $\theta=90°$, it is minimum and for $Ra=10^5$ at $\theta=225°$, it goes maximum due to the vortices formation. Figure 38.5 depicts the behavior of average Nusselt number along the hot circular cylinder again at $\theta=270°$; it becomes maximum.

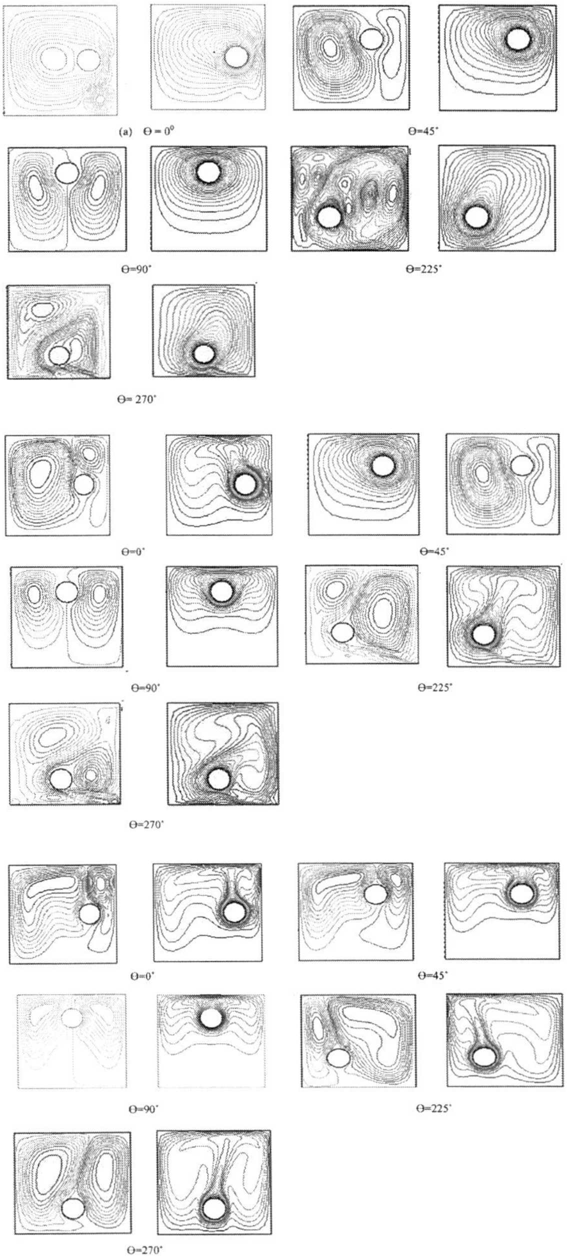

FIGURE 38.3 (a) Streamlines and isotherms at different angle positions for $Ra=10^4$, (b) streamlines and isotherms at different angle positions for $Ra=10^5$, (c) streamlines and isotherms at different angle positions for $Ra=10^6$.

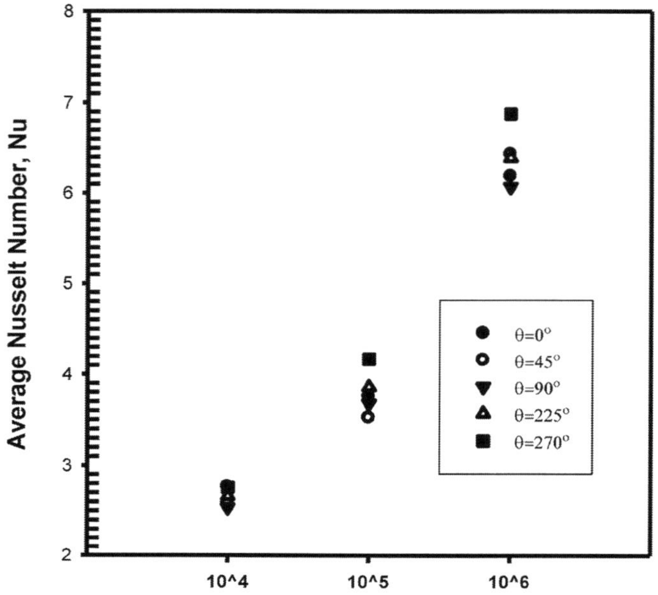

FIGURE 38.4 Average Nusselt number plot for cold wall.

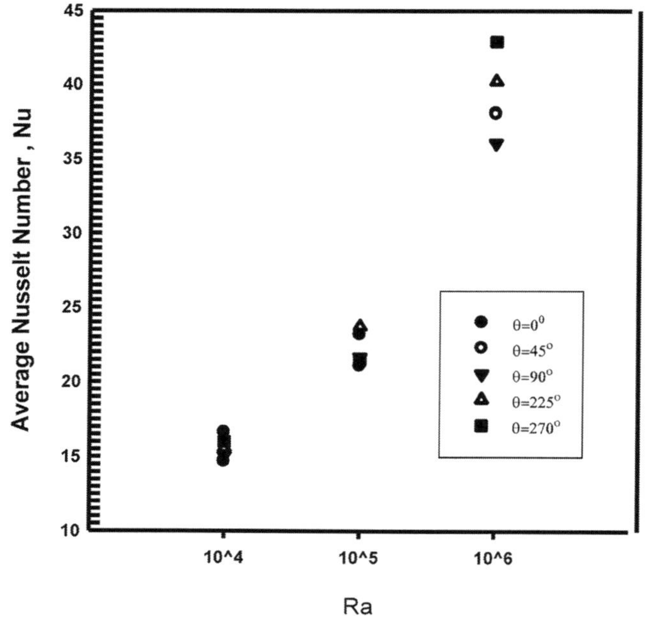

FIGURE 38.5 Average Nusselt number plot for hot circular cylinder.

38.4 CONCLUSION

Numerical reproduction of natural convection along warmed cylinder inside a frosty square nook is performed. A parametric review is directed to see the impacts of the Rayleigh number, cylinder position, and chamber introduction on the marvels of warmth exchange.

1. As the Rayleigh number increases, the value of Nusselt number also increases.
2. The estimation of Nusselt number gets to be distinctly most extreme for $Ra=10^6$ at $\theta=270°$ and least for $Ra=10^4$.
3. Along the vertical axis at $\theta=270°$, heat transfer rate is maximum.

ACKNOWLEDGMENT

Numerical analysis was performed on ANSYS Fluent 15. We are thankful to our guide Dr. H. C. Thakur.

NOMENCLATURE

g—Acceleration due to gravity, m/s^2
Gr—Grashof number
K—Thermal conductivity, W/m·K
L—Length of square enclosure, m
Nu—Surface averaged Nusselt number
Pr—Prandtl number
P—Pressure, Pa
P—Dimensionless pressure
u, *v*—Velocity components in x- and y-direction, m/s
h—Convective heat transfer coefficient
T_L—Lower temperature of cold wall
T_H—Higher temperature of inside body, °C
U, *V*—Dimensionless velocity components in X- and Y-directions
x, *y*—Rectangular coordinates
X, *Y*—Dimensionless rectangular coordinates
R—Distance of single cylinder from center
r—Radius of the hot circular cylinder

GREEK SYMBOLS

α—Thermal diffusivity of the fluid
β—Coefficient of volumetric expansion of the fluid
ν—Kinematic viscosity of the fluid
θ—Dimensionless temperature

KEYWORDS

- **natural convection**
- **square enclosure**
- **cylinder inside square enclosure**
- **Rayleigh number**

REFERENCES

1. Asan, H. Natural Convection in an Annulus Between Two Isothermal Concentric Square Ducts. *Int. Commun. Heat Mass Transfer* **2000,** *27,* 367–376.
2. De, A. K.; Dalal, A. A Numerical Study of Natural Convection Around a Square, Horizontal, a Heated Cylinder Placed in an Enclosure. *Int. J. Heat Mass Transfer* **2006,** *49,* 4608–4623.
3. Ghaddar, N. K. Natural Convection Heat Transfer Between a Uniformly Heated Cylindrical Element and its Rectangular Enclosure. *Int. J. Heat Mass Transfer* **1992,** *35,* 2327–2334.
4. Cesini, G.; Paroncini, M.; Cortella, G.; Manzan, M. Natural Convection from a Horizontal Cylinder in a Rectangular Cavity. *Int. J. Heat Mass Transfer* **1999,** *42,* 1801–1811.
5. Shu, C.; Zhu, Y. D. Efficient Computation of Natural Convection in a Concentric Annulus Between an Outer Square Cylinder and an Inner Circular Cylinder. *Int. J. Numer. Methods Fluids* **2002,** *38,* 429–445.
6. Moukalled, F.; Acharya, S. Natural Convection Between Concentric Horizontal Circular and Square Cylinders. *J. Thermophys. Heat Transfer* **1996,** *10,* 524–531.
7. Shu, C.; Xue, H.; Zhu, Y. D. Numerical Study of Natural Convection in an Eccentric Annulus Between a Square Outer Cylinder and a Circular Inner Cylinder Using DQ Method. *Int. J. Heat Mass Transfer* **2000,** *44,* 3321–3333.
8. Kim, B. S.; Lee, D. S.; Ha, M. Y.; Yoon, H. S. A Numerical Study of Natural Convection in a Square Enclosure with a Circular Cylinder at Different Vertical Locations. *Int. J. Heat Mass Transfer* **2008,** *51,* 1888–1906.

CHAPTER 39

NUMERICAL INVESTIGATION OF MIXED CONVECTION IN A LID-DRIVEN SQUARE CAVITY WITH AND WITHOUT FIN

ASHISH SAXENA[1], NG YIN KWEE EDDIE[1], JAYA KRISHNA DEVANURI[2], JALAIAH NANDANAVANAM[3,*], and PARAMESHWARAN RAJAGOPALAN[3]

[1]*School of Mechanical and Aerospace Engineering, Nanyang Technological University, Singapore*

[2]*Department of Mechanical Engineering, NIT Warangal, Telangana, India*

[3]*Department of Mechanical Engineering, BITS, Pilani, Hyderabad Campus, Hyderabad, Telangana, India*

[*]*Corresponding author. E-mail: jalaiah@hyderabad.bits-pilani.ac.in*

ABSTRACT

Mixed convection in a lid-driven square cavity with an adiabatic fin placed at the middle of the left and bottom wall is studied. A two-dimensional (2D), steady, and laminar flow with stable vertical temperature gradient conditions is simulated. Vorticity-stream function formulation is employed to analyze the energy transport and the resulting governing equations are discretized using finite difference method. Grid convergence index method is used to confirm the reliability of results obtained from the adopted grid. Top wall of the cavity moves from left to right and right to left direction alternatively, while the other three walls are stationary. An isothermal hot and cold temperature boundary condition is applied, respectively, at the top and bottom walls, while an adiabatic boundary condition is applied to the vertical sidewalls of the cavity and to the fin. Taking the fin width as 2.5% of the cavity's side, its length is varied from 5 to 20%. Cavity's thermo-hydraulics is investigated at Richardson number (Ri) equal to 1. Results are presented in terms of streamlines, isotherms, average, and local Nusselt numbers.

39.1 INTRODUCTION

Studying the thermo-hydraulics of a cavity/enclosure has become an attractive numerical research domain since a long time. Lid-driven cavity (LDC) is considered as a suitable model for validating one's own numerical codes and to simulate fluid flow and heat transfer in various applications such as glass production,[1] water flow in lakes, reservoirs, and so forth.[2] Prasad and Koseff[3] studied the mixed convection heat transfer in an insulated LDC of a rectangular cross section for a wide range of Richardson and Reynolds numbers. Reviewing the published experimental and numerical results, Shankar and Deshpande[4] concluded that two-dimensional (2D) direct numerical simulations are quite promising in studying the LDC problems over complex and costly experimental trials.

Khanafer[5] and Yapici and Obut[6] studied the effect of modified heated wall on cavity's flow and heat transfer results. The effect of a heat generating or isothermal square/cylindrical block inside the cavity at centric and/or eccentric positions was also reported.[7,8] This trend in geometry modification was continued further, to study the effect of an inclined thin fin of arbitrary length,[9] conductive triangular fins,[10] and finned enclosure[11] on the resting isothermal walls of a cavity.

Using the finite volume method, Shi and Khodadadi[12] investigated the role of an isothermal cold fin of varied length, anchored on one of the three isothermal cold resting walls of the cavity with negligible natural convection, while the isothermal hot top wall was moving from left to right. They observed that the presence of a fin slows the flow near the wall to which it is attached and thus lowers the heat transfer capacity, more so for a longer fin. These authors, in their other work[13] studied the effect of a cold fin on an isothermal hot vertical wall of a differentially heated cavity induced only by natural convection. Substantial modification in the flow field and heat transfer was observed due to the blockage effect and extra heating, which in turn depends on the fin's position and length and the Rayleigh number. Later, Ben-Nakhi and Chamkha[14] extended the work of Shi and Khodadadi[13] by placing an inclined hot fin on the hot vertical wall. Apart from the Rayleigh number, the fin angle and length were found to have significant effects on the average Nusselt number of the heated wall including the fin.

Darzi et al.[15] investigated the effect of fin(s), their height and Richardson number on the mixed convection heat transfer of a differentially heated LDC. Heat transfer enhancement was observed for the cavity with fewer fins and high Richardson number in comparison to the cavity with no fins. Xu et al.[16] numerically simulated the natural convection flows of a differently heated rectangular cavity with an adiabatic fin on each sidewall. They observed that the flow near the finned wall changes from a steady to periodic unsteady flow, at a critical Rayleigh number that is sensitive to the fin length. Later, Xu and Saha[17] studied the transition to an unsteady flow due to an adiabatic fin on the sidewall of a square cavity.

Based on the literature review presented and to the best of authors' knowledge, the results for a differentially heated LDC with an adiabatic fin at Ri = 1 has not been reported so far. Hence, a numerical investigation is carried out to present the thermo-hydraulics of a cavity with a thin adiabatic fin of varied sizes anchored to either left or bottom wall.

39.2 NUMERICAL MODELING

A 2D square cavity filled with incompressible fluid (air, $Pr = 0.70$) and top wall moving in +x- and −x-direction is considered in the present study as shown in Figure 39.1. An adiabatic thin rectangular fin of constant thickness (2.5% of the cavity's side) with three different lengths (5, 10 and 20% of the cavity's side) is placed at the mid-position of either left or bottom wall. It is assumed that the flow is 2D, steady, laminar, and incompressible. The fluid's thermophysical properties are assumed to be constant, except for the buoyancy term in the momentum equation. The buoyancy term is written on the basis of the Boussinesq approximation. It is further assumed that the radiation heat transfer, viscous dissipation, and pressure work are negligible. Further, the Navier–Stokes and energy equations are written in the dimensionless form of vorticity-stream function, Grashof number, Reynolds number, and Prandtl number as given below:

$$\psi_{xx} + \psi_{xx} = -\omega \tag{39.1}$$

$$u\omega_x + v\omega_y = \frac{1}{Re}\left(\omega_{xx} + \omega_{yy}\right) + \frac{Gr}{Re^2}\theta_x \tag{39.2}$$

$$u\theta_x + v\theta_y = \frac{1}{RePr}\left(\theta_{xx} + \theta_{yy}\right) \tag{39.3}$$

where $u = \psi_y, + v = \psi_x$ (39.4)

$$\omega = v_x - u_y \tag{39.5}$$

$$\theta = \frac{T - T_C}{T_H - T_C} \tag{39.6}$$

39.2.1 BOUNDARY CONDITIONS

Left wall: $u = v = 0, \theta_y = 0$
Right wall: $u = v = 0, \theta_y = 0$
Top wall: $u = \pm 1, v = 0, \theta = 1$
Bottom wall: $u = v = 0, \theta = 0$
Fin walls: $u = v = 0, \theta_y = 0$

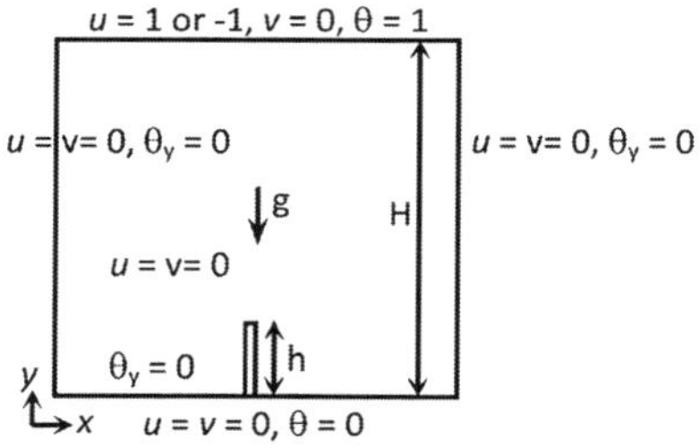

FIGURE 39.1 Schematic diagram of the cavity with a fin on bottom wall.

Using the finite difference method (FDM), the convection and diffusion terms are discretized with the second-order central difference scheme. The governing equations are solved iteratively using the Gauss–Seidel method, till the convergence criterion of 10^{-6} is achieved for ψ, ω, and θ. The current numerical scheme's grid convergence is examined, in terms of grid convergence index (GCI)[18,19] for mixed convection (Ri = 1) in an air-filled square cavity with a moving heated lid at the top and stationary cold wall at the bottom. Taking a grid refinement factor (r) of 2, the cavity is discretized with three different square grids, namely, A: 50×50, B: 100×100, and C: 200×200. Treating the Nusselt number and vorticity as solution parameters for GCI calculations, the values are drawn, respectively, on the top moving lid (namely, minimum, maximum, and average) and at the vertical midline of the cavity (namely, minimum, maximum, and midpoint). Denoting the GCI between two successive grids, say A and B as GCI_{AB}, it is defined as, $GCI^{AB} = \frac{1.25\varepsilon^{AB}}{r_{AB}^{p}-1}$. Here, r_{AB} is the grid refinement factor of the two grids, that is, A and B, and is evaluated as the ratio of their representative cell size (m). Therefore, $r_{AB} = m_A/m_B$ where $m = \left(\frac{\sum_{i=1}^{N} Area\,of\,the\,cell_i}{N}\right)^{1/2}$ and N is the total of number of cells present in a grid. ε^{AB} is an approximate relative error of a solution parameter (ϕ), and expressed mathematically as $|(\phi_B - \phi_A)/\phi_B|$. Lastly, p is the order of convergence, and expressed mathematically as $p = \ln\left|\frac{\phi_A - \phi_B}{\phi_B - \phi_C}\right| / \ln(r)$

Table 39.1 summarizes the results of GCI calculations. It is observed that the refinement of grid from coarser to finer yields a drop in the GCI ($GCI^{BC} < GCI^{AB}$), indicating a tendency for grid independency. Further, the solution parameters, namely vorticity and Nu are extrapolated using the Richardson extrapolation method[18] and included the same in Table 39.1. It is found that the extrapolated solution parameters (ϕ_{ext}) are in good agreement with the results of the finest grid $\phi_{ext} = \phi_C + \left(\frac{\phi_C - \phi_B}{r^p - 1}\right)$. The average Nusselt number derived from the finest grid (i.e., 200×200) is compared with the results of refs 20–22 and listed in Table 39.2. Good agreement in results validates the present numerical scheme and thus facilitated to investigate the effect of adiabatic fin in a square cavity in mixed convection regime.

TABLE 39.1 Summary of the Grid Convergence Index Calculations.

	Vorticity (ω) at the cavity's mid-vertical line			**Nusselt number (Nu) at the cavity's top wall**		
	ω_{max}	ω_{min}	$\omega_{cavity\ mid}$	Nu_{max}	Nu_{min}	Nu_{avg}
ϕ_A	0.6889	−8.5878	0.4836	5.1704	0.0878	1.3935
ϕ_B	0.6955	−8.5353	0.4593	5.8265	0.0661	1.4050
ϕ_c	0.6997	−8.5422	0.4518	6.2462	0.0592	1.4066
ϕ_{ext}	0.7079	−8.5411	0.4483	6.9917	0.0560	1.4069
p	0.6104	2.9287	1.6812	0.9125	1.4566	2.8540
GCI^{AB}	0.0223	0.0012	0.0299	0.1563	0.2443	0.0017
GCI^{BC}	0.0145	0.0002	0.0095	0.0784	0.1017	0.0002

TABLE 39.2 Average Nusselt Number for Validation.

Ri (Re=100 and $Gr=10^4$)	Nu_{avg} (on the top wall)			
	Iwatsu et al.[20]	Oztop et al.[21]	Tiwari and Das[22]	Present study
1.00	1.34	1.33	1.47	1.40

39.3 RESULTS AND DISCUSSION

Results are presented here for the cavity with an adiabatic fin of varied size ($h/H=0$, 0.05, 0.1, and 0.2) anchored on to either left or bottom wall. For the fin positioned on the left wall, the top wall's direction of movement is alternately changed. Figure 39.2 shows the streamline contours of the cavity with a fin on the left wall. Figure 39.3 presents the local Nusselt number distribution along the bottom and top walls of the cavity with a fin on the left wall, while the top wall moves from left to right.

39.3.1 WITHOUT A FIN

Driven by the induced buoyancy and inertia forces under mixed convection conditions, the cavity experiences two vortices, whose shape, size, and orientation are strongly affected by the direction of the top wall's motion. Rotating in opposite direction to each other, these two vortices come in contact near the middle of the cavity as shown in Figure 39.2, justifying to refer them henceforth as top vortex and bottom vortex of the cavity. Due to the close proximity of inertia forces offered by the top wall, the top vortex grows a little bigger than the bottom vortex. Interestingly, these two vortices appear in as a mirror image when the top wall's direction is altered as shown in Figure 39.2.

The local Nusselt number distribution on bottom and top walls of the cavity without a fin ($h/H=0$) is shown in Figure 39.3. Affected by the inertia forces, the thermal gradients at the leading edge of the top wall (refer isotherms shown in Fig. 39.4) are very strong; hence, the local Nusselt number is very high. The gradual growth in velocity/temperature profile along the wall length results in the corresponding decline in the local Nusselt number. Except for the minor bias/effect caused by the direction of rotation of the bottom vortex, the thermal gradients on the bottom wall are almost same along its length and so the Nusselt number. Interestingly, the average Nusselt number on these two walls (Table 39.3) is almost equal, indicating that the net heat exchange due to the convection by these two walls is same.

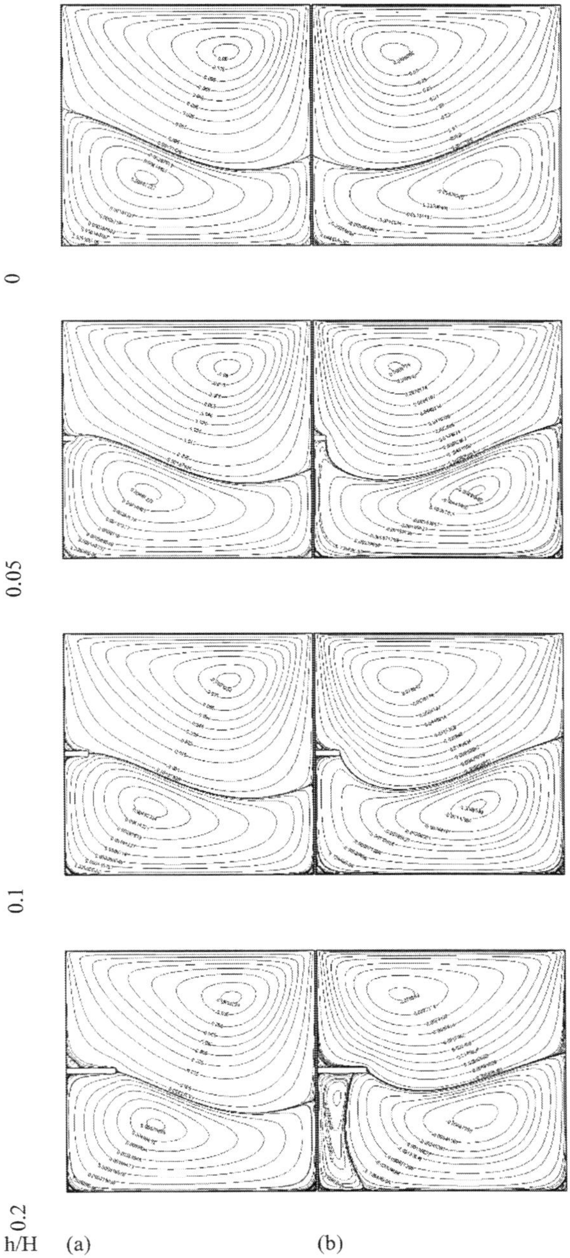

FIGURE 39.2 Streamline contours of the cavity with a fin on the left wall, as the top wall moves from (a) left to right and (b) right to left.

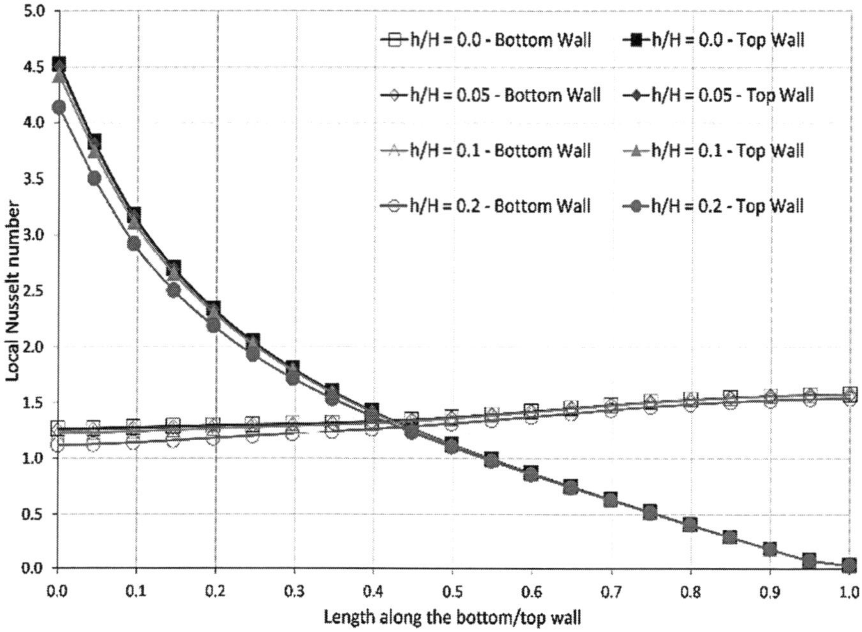

FIGURE 39.3 Local Nusselt number variation along the bottom and top walls of the cavity with a fin on the left wall.

39.3.2 WITH A FIN ON THE LEFT WALL

Figure 39.2 shows the streamline contours of the cavity with varied fin sizes on the left wall as the top wall moves toward either right or left. For $h/H = 0.05$ and 0.1, the fin has hardly shown an effect on the vortex dynamics, as it lies exactly at the place where the top and bottom vortices see each other. The same is reflected in the local and average Nusselt number values of the bottom and top walls (Fig. 39.3 and Table 39.3).

TABLE 39.3 Average Nusselt Number Due to the Fin on Left Wall.

Fin size (h/H)	Top wall moves from left to right		Top wall moves from right to left	
	Bottom wall	Top wall	Bottom wall	Top wall
0.0	1.402	1.407	1.402	1.405
0.05	1.399	1.402	1.391	1.390
0.1	1.387	1.388	1.355	1.352
0.2	1.329	1.329	1.234	1.233

As the fin size is increased to $h/H=0.2$, the cavity's main flow experiences a deviation in its original vortex flow dynamics. This effect is visible more in the case of top wall moving from right to left, wherein a small vortex forms right below the fin. This small vortex, comparable to the fin length, is secondary in nature and rotates in the opposite direction of the bottom vortex. With the presence of multiple vortices in the cavity, the resistance for heat exchange by convection increases, hence a drop in the average Nusselt number for the top and bottom walls (Table 39.3). It should be noted that the heat exchange by convection is high for the cavity with no fin, and it decreases as the fin length increases.

39.3.3 WITH A FIN ON THE BOTTOM WALL

Acknowledging the role of fin's size and position on the cavity's flow dynamics, a fin of different sizes is now placed at the center of the bottom wall to evaluate the convective heat exchange. Since the fin lies at the center of the bottom wall, that is, a symmetric place to the left and right walls, the direction for top wall motion is considered only one, that is, from left to right.

With the introduction of a fin on the bottom wall, the primary vortex adjacent to the bottom wall starts realigning its shape in tune with the fin length, while rotating in the same direction as shown in Figure 39.4. Owing to the corner effect caused by the fin presence, two vortices, almost equal and very small in size are found next to each side of the fin. With the increment in fin size, the bottom vortex gradually manifests a bottleneck shape near the fin, all due to the simultaneous restriction caused by the fin and top vortex, while the vortex continues to have a full circulation from left wall to the right wall. When the fin size grows further and come closer to the top vortex, the bottom vortex split into two smaller vortices, as shown in Figure 39.4 for the case, $h/H=0.2$. Interestingly, the newly formed two smaller vortices appearing on each side of the fin continue to rotate in the same direction as that of the bygone bottom vortex. To avoid coercion while in rotation, these two smaller vortices confluence at the tip of the fin with the top vortex, thus causing the latter to change its shape accordingly.

Figure 39.5 shows the local Nusselt number distribution along the top and bottom walls of the cavity with an adiabatic fin on the bottom wall. Notably, the local Nusselt number distribution on the top wall remains same regardless of the fin size. This can be attributed to the fact that the thermal gradients close to the top wall, as shown in Figure 39.4 through isotherm

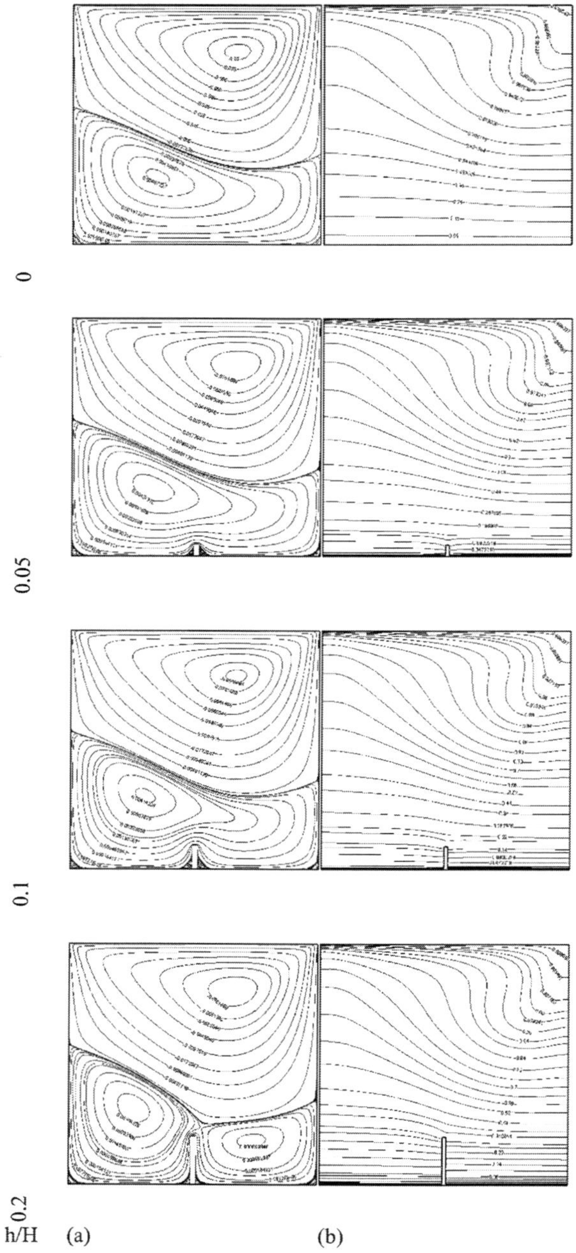

FIGURE 39.4 (a) Streamline and (b) isotherm contours of the cavity with a fin on bottom wall as the top wall moves from left to right.

contours, are not much affected by the adiabatic fin at bottom wall. The little variation in the average Nusselt number values of top wall, as given in Table 39.4, supports this argument. It should be also noted that the local Nusselt number distribution is qualitatively almost same as that of the cavity with fin on the left wall shown in Figure 39.3.

Barring the sudden dip due to the adiabatic fin presence, the local Nusselt number profile on the bottom wall follows almost the same as shown in Figure 39.3. A close observation at Figure 39.5 for the local Nusselt number values of the bottom wall on left side of the fin reveals that they are low for a longer fin and high for a shorter fin. This trend goes exactly opposite on the right side of the fin, that is, local Nusselt numbers are low for a shorter fin and high for a longer fin. Therefore, for any given fin size, the local *Nu* values found to be high on one side and low on the other side of fin. This leads to offer almost the same average Nusselt number for the fin sizes studied and listed in the Table 39.4.

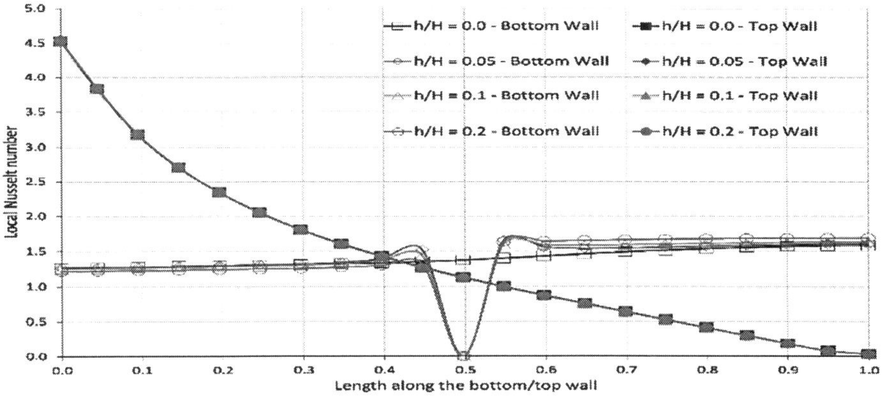

FIGURE 39.5 Local Nusselt number variation on bottom and top walls of the cavity with a fin on the bottom wall.

TABLE 39.4 Average Nusselt Number Due to the Fin on Bottom Wall.

Fin size (h/H)	Top wall moves from left to right	
	Bottom wall	Top wall
0.0	1.402	1.407
0.05	1.439	1.403
0.1	1.435	1.401
0.2	1.436	1.404

39.4 CONCLUSIONS

Effect of adiabatic fin on mixed convection is investigated using FDM for a differentially heated and lid-driven square cavity. A fin with sizes (h/H=0.05, 0.1, and 0.2) are anchored to either the left or bottom wall of the cavity. The top wall's direction of motion is investigated for the fin placed on the left wall. Following are the conclusions noted from the present work.

39.4.1 FIN ON THE LEFT WALL

- For a given fin size, the average Nusselt number is almost the same on the top and bottom walls but is lesser for the top wall moving from right to left.
- As the fin size increases, the average Nusselt number decreases for both the top and bottom walls of the cavity.
- When the fin size (h/H) is equal to 0.2, a secondary vortex forms right below the fin and to the left of the bottom vortex, for the top wall moving from right to left.

39.4.2 FIN ON THE BOTTOM WALL

- For the fin size, the average Nusselt number is almost constant respectively for the top and bottom walls. Owing to the rigorous vortex dynamics in the proximity of bottom wall, the average Nusselt numbers are slightly higher for it.
- When the bottom vortex experiences an increased restriction for circulation, predominantly due to the fin's presence, the former splits into two and both rotate in the same direction.

NOMENCLATURE

g—Acceleration due to gravity (m/s^2)
h—Fin length (m)
H—Height/side of the square cavity (m)
Nu—Local Nusselt number, $-\theta_n$
Nu_{avg}—Average Nusselt number of the wall, $\frac{\int Nu\,dx}{\int dx}$ or $\frac{\int Nu\,dy}{\int dy}$
Pr—Prandtl number, ν/α

Re—Reynolds number, UH/v
Ri—Richardson number, Gr/Re^2
Gr—Grashof number, $g\beta(T_H - T_C)\cdot H^3/v^2$
T—Temperature (K)
U—Velocity of the side wall (m/s)
u—Dimensionless velocity in the x-direction
v—Dimensionless velocity in the y-direction
x—Dimensionless distance along the x-coordinate
y—Dimensionless distance along the y-coordinate

GREEK SYMBOLS

α—Thermal diffusivity (m^2/s)
β—Thermal expansion coefficient of the fluid (K^{-1})
ν—Kinematic viscosity (m^2/s)
θ—Dimensionless temperature, $(T - T_C)/(T_H - T_C)$
ω—Dimensionless vorticity, $v_x - u_y$
ψ—Dimensionless stream function, $\int u\partial y$ or $-\int v\partial x$

SUBSCRIPTS

C—Cold wall
H—Hot wall
n—Partial derivative in normal direction, $\frac{\partial}{\partial n}$
x—Partial derivative in x-direction, $\frac{\partial}{\partial x}$
y—Partial derivative in y-direction, $\frac{\partial}{\partial y}$

KEYWORDS

- **lid-driven cavity**
- **mixed convection**
- **finite difference method**
- **adiabatic fin**

REFERENCES

1. Pilkington, L. A. B. The Float Glass Process. *Proc. Royal Society London A.* **1969,** *314,* 1–25.
2. Imberger, J.; Hamblin, P. F. Dynamics of Lakes, Reservoirs, and Cooling Ponds. *Annu. Rev. Fluid Mech.* **1982,** *14,* 153–187.
3. Prasad, A. K.; Koseff, J. R. Combined Forced and Natural Convection Heat Transfer in a Deep Lid-Driven Cavity Flow. *Int. J. Heat Fluid Flow* **1996,** *17,* 460–467.
4. Shankar, P. N.; Deshpande, M. D. Fluid Mechanics in the Driven Cavity. *Annu. Rev. Fluid Mech.* **2000,** *32,* 93–136.
5. Khanafer, K. Comparison of Flow and Heat Transfer Characteristics in a Lid-Driven Cavity between Flexible and Modified Geometry of a Heated Bottom Wall. *Int. J. Heat Mass Transfer* **2014,** *78*(0), 1032–1041.
6. Yapici, K.; Obut, S. Laminar Mixed-Convection Heat Transfer in a Lid-Driven Cavity with Modified Heated Wall. *Heat Transfer Eng.* **2015,** *36*(3), 303–314.
7. Islam, A. W.; Sharif, M. A. R.; Carlson, E. S. Mixed Convection in a Lid Driven Square Cavity with an Isothermally Heated Square Blockage Inside. *Int. J. Heat Mass Transfer* **2012,** *55,* 5244–5255.
8. Khanafer, K.; Aithal, S. M. Laminar Mixed Convection Flow and Heat Transfer Characteristics in a Lid Driven Cavity with a Circular Cylinder. *Int. J. Heat Mass Transfer* **2013,** *66,* 200–209.
9. Ben-Nakhi, A.; Chamkha, A. J. Conjugate Natural Convection in a Square Enclosure with Inclined Thin Fin of Arbitrary Length. *Int. J. Therm. Sci.* **2007,** *46*(5), 467–478.
10. Sun, C.; Yu, B.; Oztop, H. F.; Wang, Y.; Wei, J. Control of Mixed Convection in Lid-Driven Enclosures using Conductive Triangular Fins. *Int. J. Heat Mass Transfer* **2011,** *54*(4), 894–909.
11. Rabienataj Darzi, A. A.; Farhadi, M.; Lavasani, A. M. Two Phase Mixture Model of Nano-enhanced Mixed Convection Heat Transfer in Finned Enclosure. *Chem. Eng. Res. Des.* **2016,** *111,* 294–304.
12. Shi, X.; Khodadadi, J. M. Laminar Fluid Flow and Heat Transfer in a Lid-Drive Cavity due to a Thin Fin. *J. Heat Transfer* **2002,** *124,* 1056–1063.
13. Shi, X.; Khodadadi, J. M. Laminar Natural Convection Heat Transfer in a Differentially Heated Square Cavity due to a Thin Fin on the Hot Wall. *J. Heat Transfer* **2003,** *125,* 624–634.
14. Ben-Nakhi, A.; Chamkha, A. J. Effect of Length and Inclination of a Thin Fin on Natural Convection in a Square Enclosure. *Numer. Heat Transfer A* **2006,** *50*(4), 389–407.
15. Darzi, A. A. R.; Farhadi, M.; Sedighi, K. Numerical Study of the Fin Effect on Mixed Convection Heat Transfer in a Lid-Driven Cavity. *Proc. Inst. Mech. Eng. C* **2011,** *225*(2), 397–406.
16. Xu, F.; Patterson, J. C.; Lei, C. Effect of the Fin Length on Natural Convection Flow Transition in a Cavity. *Int. J. Therm. Sci.* **2013,** *70,* 92–101.
17. Xu, F.; Saha, S. C. Transition to an Unsteady Flow Induced by a Fin on the Sidewall of a Differentially Heated Air-Filled Square Cavity and Heat Transfer. *Int. J. Heat Mass Transfer* **2014,** *71,* 236–244.
18. Roache, P. J. Perspective: A Method for Uniform Reporting of Grid Refinement Studies. *J. Fluids Eng.* **1994,** *116*(3), 405–413.

19. Celik, I. B.; Ghia, U.; Roache, P. J.; Freitas, C. J.; Coleman, H.; Raad, P. E. Procedure for Estimation and Reporting of Uncertainty due to Discretization in CFD Applications. *J. Fluids Eng. 130*(7), 78001.
20. Iwatsu, R.; Hyun, J. M.; Kuwahara, K. Mixed Convection in a Driven Cavity with a Stable Vertical Temperature Gradient. *Int. J. Heat Mass Transfer* **1993,** *36*(6), 1601–1608.
21. Oztop, H. F.; Dagtekin, I. Mixed Convection in Two-Sided Lid-Driven Differentially Heated Square Cavity. *Int. J. Heat Mass Transfer* **2004,** *47,* 1761–1769. (Apr.)
22. Tiwari, R. K.; Das, M. K. Heat Transfer Augmentation in a Two-Sided Lid-Driven Differentially Heated Square Cavity Utilizing Nanofluids. *Int. J. Heat Mass Transfer* **2007,** *50,* 2002–2018.

CHAPTER 40

COMPUTATIONAL AND EXPERIMENTAL STUDIES ON THE FLOW ARRANGEMENT OVER FLAT PLATES OF DIFFERENT CONFIGURATIONS

SAI SARATH KRUTHIVENTI[1,*] M. CHEERANJIVI[2], RAYAPATI SUBBARAO[3], and MD ABID ALI[4]

[1]*Department of Mechanical Engineering, K L University Guntur, Andhra Pradesh, India*

[2]*Department of Mechanical Engineering, Amrita Sai Institute of Technology, Andhra Pradesh, India*

[3]*Department of Mechanical Engineering, NITTTR Kolkata, Kolkata, West Bengal, India*

[4]*Department of Mechanical Engineering, Ramachandra College of Engineering, Eluru, India*

**Corresponding author. E-mail: Satyasai222@gmail.com*

ABSTRACT

Vortex dynamics plays a vital role in the transport of mass, momentum and energy of a flow field. Moreover, vortex manipulation is the key parameter for controlling any device that involves flow mixing. Present work deals with studies on a flat plate with and without chamber edges and when the plate is attached to a triangle with different inclinations and shapes at the top. Mach numbers of 0.05, 0.1, and 0.15 are considered. Parameters such as wake zone dimensions, flow separation and attachment points, drag force, and coefficient of drag are estimated. Experimentation is done using an open channel setup that is fabricated to the designed dimensions with water as the working fluid for the Reynolds numbers ranging from 4000 to 7000. Computational fluid dynamics analysis is also carried out with the same set of parameters and the variations of flow parameters are plotted for various Reynolds numbers and Mach numbers. Flow visualization over steps with different inclinations is made and parameters such as pressure uphill size, separation zone, reverse flow length variation and coefficient of drag with Reynolds numbers are compared. It is observed that the flat plate attached to the triangle shape object has less drag coefficient, when compared with other geometries. A similar phenomenon is observed in both computational and experimental approach. From the results it is observed that the flat plate attached with an equilateral triangle also possesses less wall shear force and less wake zone size in both approaches. The present results showed that the flat plate with triangular geometries has less drag coefficient, when compared with other geometries.

40.1 INTRODUCTION

Reduction of drag over various geometries plays a significant role in aerodynamic studies. Techniques have been developed by earlier researchers for possible drag reduction, considering the variation of geometrical dimensions. By varying the front portion of the object, Bettes[1] reported the reduction of drag by 24%. Yajima and Sano[2] achieved 40% drag reduction by varying the dimensions of the objects. Rathakrishnan[3] reported that objects like forward-facing and backward-facing steps possess less drag, when compared to bluff bodies. Suryanarayana et al.[4] discussed about the reduction in drag for higher Reynolds numbers by varying the object shape. Bansal and Sharma[5] proposed an effective computational fluid dynamics model to obtain the flow structure around a passenger car with different add-on devices. Addition of these devices reduces drag and lift coefficients. Armaly et al.[6] made Laser-Doppler measurements of velocity distribution and re-attachment length in the downstream of a single backward-facing step, mounted in a two-dimensional channel. Such measurements not only yielded the expected primary zone of recirculating flow attached to the backward-facing step, but also showed additional regions of flow separation, downstream of the step and on both sides of the channel test section. There were several computational studies in analyzing the flow structure and drag reduction where the effect of pressure at the back side of the object at which low pressure is created is not well understood.[7] Similarly the effect of positive pressure at the front part of the object that leads to decrease in the drag force is also not clear.

In the present work, an attempt is made to increase the pressure at the backside of the object, where low pressure is created and an attempt is made to reduce the positive pressure, which occurs at the front part of the object that may reduce the drag force. Computational study is also carried out here, using the ANSYS workbench 15.0, for the clear understanding of flow visualization and prediction of flow parameters such as wake zone size, wall shear force, and drag. A flat plate of 25 × 25-mm size with circular, rectangular, and triangular attachments are modeled, and the drag reduction and flow parameters are calculated and presented in plots.

40.2 FLOW STUDIES

40.2.1 EXPERIMENTATION

The flow over various geometries is visualized using a rectangular water flow channel. Schematic view and fabricated experimentation unit are shown in Figure 40.1. Visualization studies are carried out in the test section of 300-mm width and 500-mm length with a uniform water stream of 5-mm depth. Initially, water is allowed to pass through the stagnation chamber and over a wedge that is placed next to the test section. Three wire screens of different hole diameters are placed between the test section and the wedge that leads to uniform velocity. Square blocks of $25 \times 25\ mm^2$ attached with various cross sections such as a circle, fillet or triangle are placed perpendicular to the direction of the flow. Each block is visualized individually and it is ensured that there is no flow taking place at the bottom face of the model. Flow velocity is measured by using the floating particle method. For clear visibility, a die is injected along with water just before the last screen so that it is easy to visualize the flow pattern over the geometries. Entire flow structure is recorded with a video camera by viewing from the measuring scale, drawn within the test section, ensuring no disturbance to the flow. Moreover, it would be easy to measure the distance of reverse flow and the size of the wake zone. Parameters such as drag force, coefficient of drag, flow separation and attachment points, recirculation zone size, and eddies formation are measured and calculated for different Reynolds numbers ranging from 4000 to 7000.

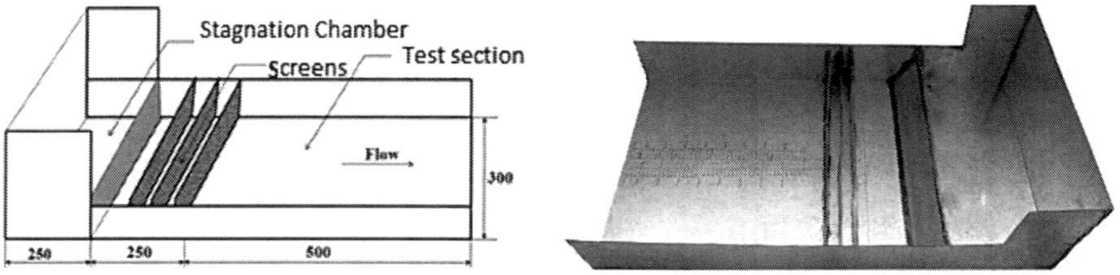

FIGURE 40.1 Schematic view and fabrication of the experimentation unit.

40.2.2 MODELLING

Computational studies are done for better understanding of flow visualization (Fig. 40.2). The effect of forehead geometry of the cube on the flow field is considered. Semicircular and triangular models with an inclination of 45° and 60° with respect to the axis normal to the fluid flow and cube with fillets on both sides are prepared. Material used for the models is aluminum with a density of 2719 kg/m^3, the working fluid used is air with a density of 1225 kg/m^3, and viscosity of 1789×10^{-5} kg/m·s. These models have the same cross-sectional area facing the flow, that is, 25×25 mm^2. The whole setup is prepared in an isolated box (size is extremely large when compared to the model size) to maximize the interaction of air either with the testing model or the inner wall of isolation. All the simulations are carried out at three different Mach numbers of 0.05, 0.1, and 0.15. The pressure at the outlet is taken as atmospheric pressure. K-epsilon turbulence model is used for analysis. Initially, air is passed over the testing geometry assuring uniform flow by setting the velocity of 60 kph and the flow parameters such as wake zone size, coefficient of drag, pressure uphill, and wall shear force are measured. The above procedure is repeated for other velocities and for other geometries.

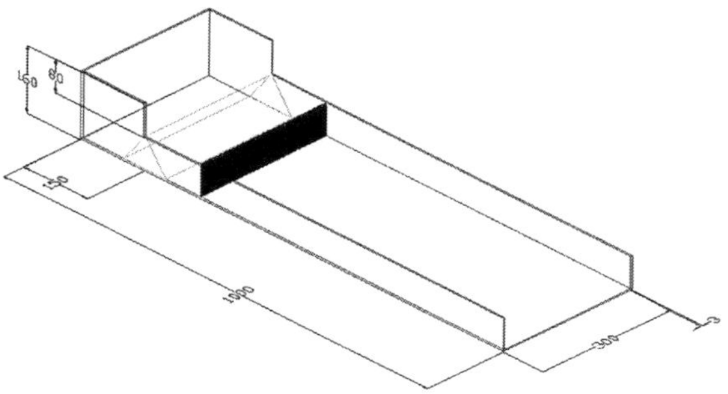

FIGURE 40.2 Computational model.

40.3 RESULTS AND DISCUSSION

40.3.1 COMPUTATIONAL RESULTS

Figure 40.3 shows flow development over the circle and cube cross section without any orientation. As the flow approaches the body, initially it moves smoothly along the walls of the body without any deviation. The streamlines that got separated due to the presence of a bluff body are found to be merging together at a distance of 3 cm downstream. From there, the flow is reversed and moves toward the base of the model due to the low-pressure region created. Here, the reverse flow distance is taken as the distance from the base of the object to the point, where the reverse flow starts. As time progresses, the pressure uphill is observed. The low-pressure region behind the model influences the flow to reverse toward the base. The same phenomenon is observed in Figures 40.4 and 40.5, but the size of the wake zone reduces as the dimensions of the objects are changed. Wake zone size is less for the flat plate attached with a 60° triangle as compared to other objects because the cross section was much steeper when compared to other geometries.

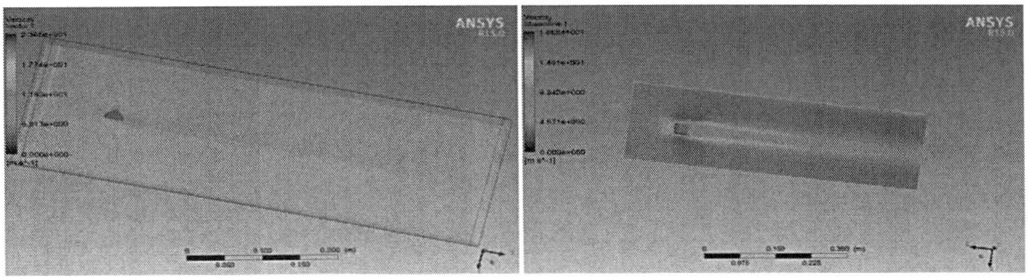

FIGURE 40.3 Flow visualization over circular and cube cross section.

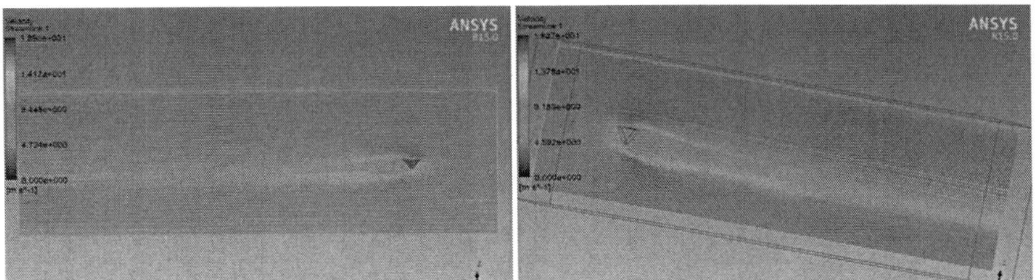

FIGURE 40.4 Flow visualization over triangle with 60° and 45°.

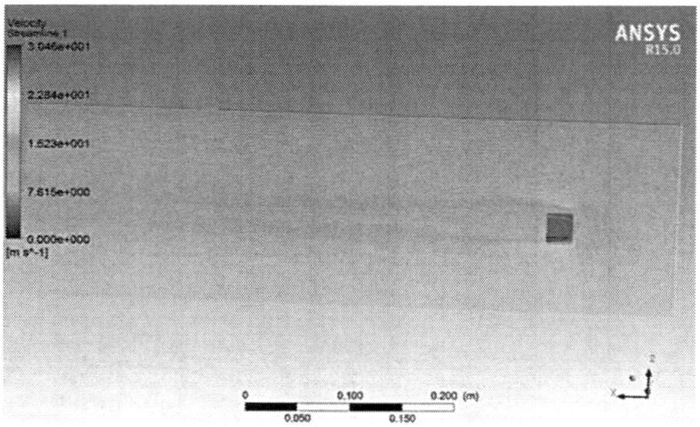

FIGURE 40.5 Flow visualization over filleted cross section.

40.3.2 EXPERIMENTAL ANALYSIS

Experiments are carried out for different model configurations at the specified Mach numbers. The uncertainty for all the measured parameters is about 2%. Figure 40.6 describes the flow visualization over flat plates. The formation of vortices and sizes of recirculation zones are identified and analyzed. At Re=4000 for cube and chamfered cube, it is clearly observed that flow separation takes place ahead of the object, which leads to the formation of wakes behind the object that are indicated with arrows pointing toward the direction of vortices. As Re increases, there will be a drastic change in the formation of vortices and the size of wake zone reduces. This indicates positive impact on flow acceleration. Figure 40.7 shows the variation in the wake zone size with respect to velocity. From the plot, it can be seen that, the wake zone size increases as the Mach number increases for all the models as shown in Figure 40.5. The wake zone size is very less for the flat plate attached to a 60° triangle, followed by a flat plate attached to a 45° triangle. Wake zone size is noted as 0.45 cm for a 60° triangle, which is the minimum when compared to all other geometries studied. This is due to a low suction force and the formation of less number of vortices behind the triangular cross section. These vortices create a low-pressure zone downstream of the object and this low-pressure area increases the drag force over a model. Therefore, for this object drag force is less and this can be clearly seen in Figure 40.8, where the variation of Mach number is shown. From the plot, one can observe that the drag force is also less for the flat plate attached to a 60° triangle, justifying the argument made in the above paragraph regarding the wake region. Indeed, it is true that, a body with a less wake region experiences less drag. This is due to the formation of the rolling effect of the airflow behind the object. Formation of the rolling effect is due to the reverse flow. It is also observed that the influence of geometry on drag is less at lower Mach numbers. As the number increases, the drag force on the body also increases, which is clearly shown in Figure 40.8. The drag force for the fillet and the triangle with 60° seems to be similar while advancing to higher Mach numbers. This is a good indication in the selection of higher Mach numbers to get better flow parameters.

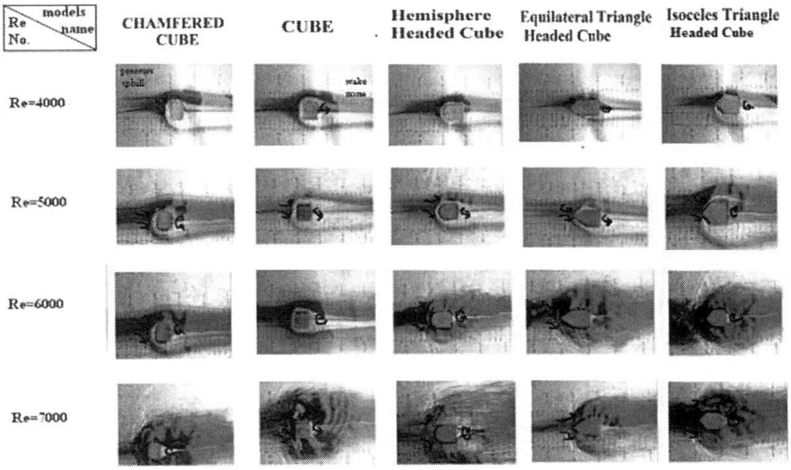

FIGURE 40.6 Flow visualization over various geometries.

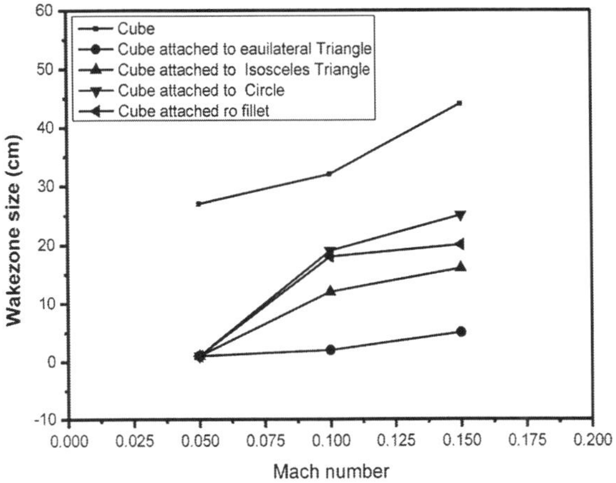

FIGURE 40.7 Variation of wake zone size with Mach number.

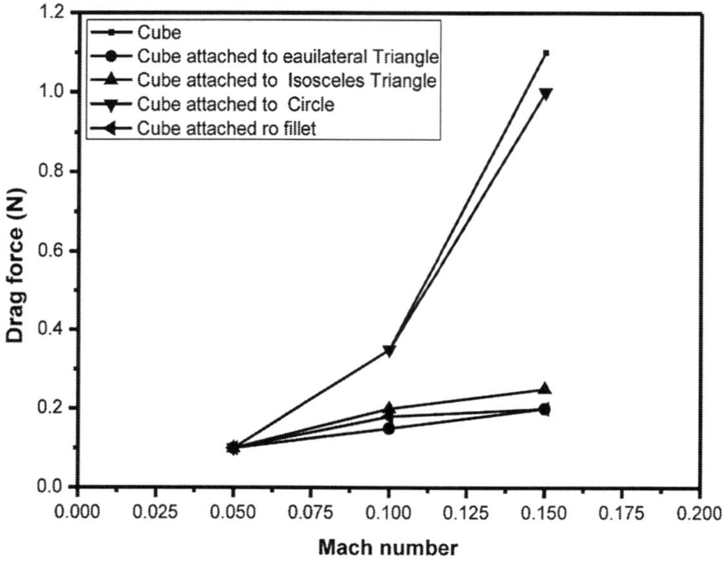

FIGURE 40.8 Variation of drag force with Mach number.

Figure 40.9 shows the variation of wall shear force with respect to the Mach number. From the plot, it is clear that the shear force is dominant for the body with a cube. The wall shear force for the triangle configuration with 60° follows the filleted body pattern for initial velocities and it reduces further as the Mach number is raised. Figure 40.10 shows the variation of flow wake zone size with respect to Re. For Re = 4000, it is observed that there is a wake zone of size 1 cm, and it gradually decreases as Re increases. This happens because of the flow getting accelerated, which in turn reduces the separation of the fluid and the formation of the low pressure zone, which leads to the reduced size of the wake zone. For Re = 4000, reduction in the size of wake zone is observed, which leads to decrease in the positive pressure ahead of the geometry. Figure 40.11 shows the variation of the width of pressure uphill with respect to Re. The width of pressure uphill follows a similar pattern for all the geometries of variable dimensions with respect to Re. The variation of coefficient of drag with respect to Re is shown in Figure 40.12. Coefficient of drag is estimated by using Clift and Gauvin technique. Continuous increment of drag is noticed for all the geometries as Re increases. This is due to creation of some turbulence behind the object. This conveys that one should prefer lower Reynolds numbers, irrespective of the geometry of the material.

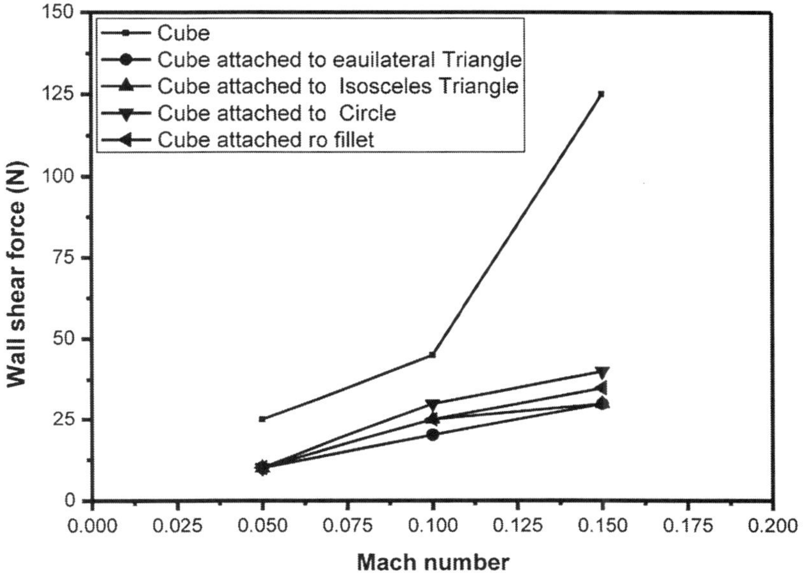

FIGURE 40.9 Variation of wall shear force with Mach number.

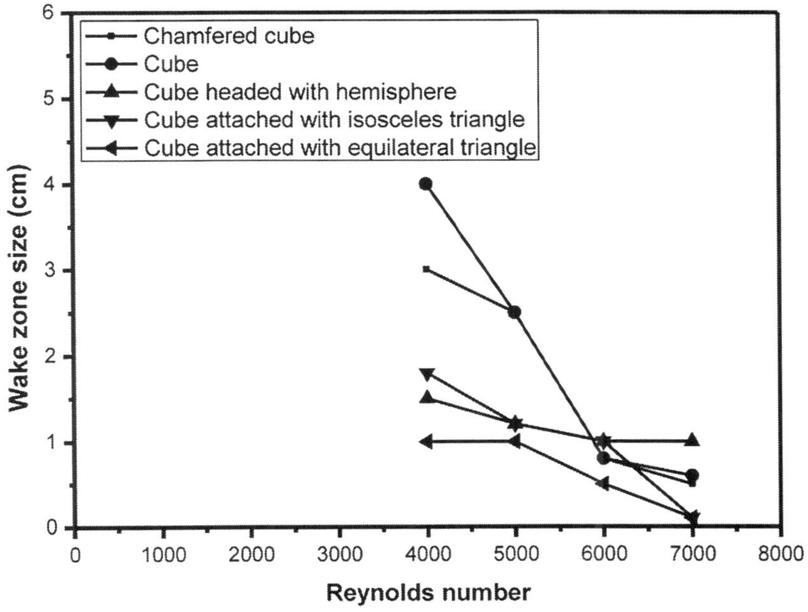

FIGURE 40.10 Variation of wake zone size with Re.

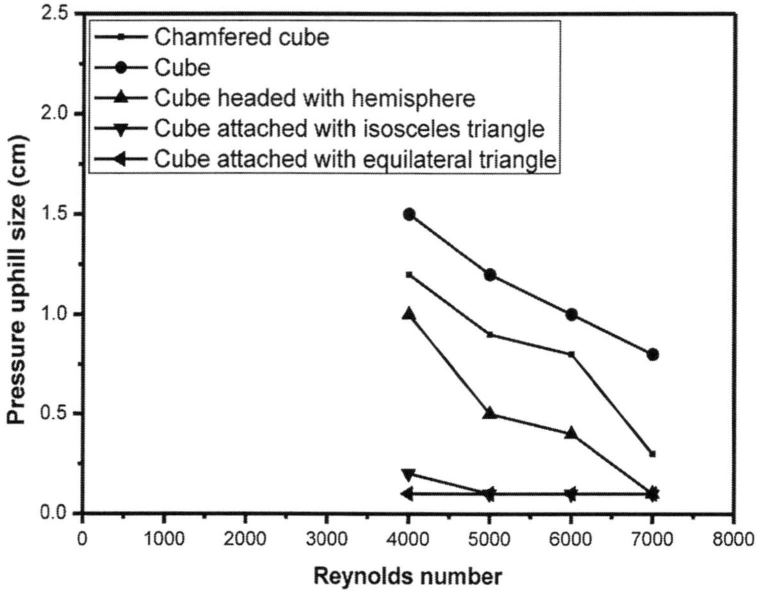

FIGURE 40.11 Variation of pressure uphill size with Re.

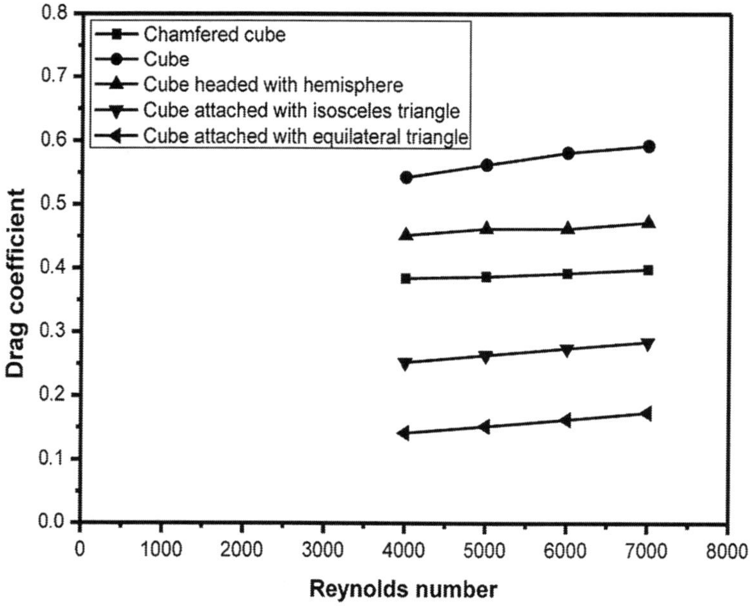

FIGURE 40.12 Variation of drag coefficient with Re.

40.4 COMPARISON

Experiments are conducted in the Re range of 4000–7000. From the results, it is clear that by varying the geometrical dimensions, there is reduction in the wake zone size because of creation of some positive pressure behind the object. This leads to the formation of less drag which is clearly seen in the plots. As shown in Figures 40.3 and 40.6, the merging of flow is at 3 cm downstream. Here, computational study is the means for comparison of the experimental results and flow properties. The comparison shows a good agreement and evidently, these computational studies can be further extended to other geometries and configurations for optimizing the flow parameters.

40.5 CONCLUSIONS

Flow visualization over steps with different inclinations is made and parameters such as pressure uphill size, separation zone, reverse flow length variation, and the coefficient of drag for different Reynolds numbers and Mach numbers are analyzed. Flat plate attached to the triangular shape shows a drag coefficient, when compared with other geometries. Moreover, less wall shear force and less wake zone are observed. Similarly, the wake zone size is less for cube attached with equilateral triangle and hence, it possesses a lower drag. Pressure uphill size is also lowest in the cube attached with an equilateral triangle. Computational studies pave the way for further analysis and for bringing a clear picture of flow parameters with respect to different configurations.

KEYWORDS

- **coefficient of drag**
- **flow separation**
- **flow over flat plate**
- **Mach number**
- **Reynolds number**
- **vortex dynamics**

REFERENCES

1. Bettes, W. H. The Aerodynamic Drag Reduction of Road Vehicles in Earlier. *J. Eng. Sci.* **1982,** *5*, 4–10.
2. Yajima, Y.; Sano, O. A Note on the Drag Reduction of a Circular Cylinder Due to Double Rows of Holes. *J. Fluid Dyn. Res.* **1996,** *18*, 237–243.
3. Rathakrishnan, E. Effect of Splitter Plate on Bluff Body Drag. *AIAA J.* **1999,** *37*(9), 1125–1126.
4. Suryanarayana, G. K.; Pauer, H.; Meier, G. E. A. Bluff-Body Drag Reduction by Passive Ventilation. *J. Exp. Fluids* **1993,** *16*, 73–81.
5. Bansal, R.; Sharma, R. B. Drag Reduction of Passenger Car Using Add-On Devices. *J. Aerodyn.* **2014,** *2014*, 1–13.
6. Armaly, B. F.; Durst, F.; Pereira, J. C. F.; Schönung, B. Experimental and Theoretical Investigation of Backward-Facing Step Flow. *J. Fluid Mech.* **1983,** *127*, 473–496.
7. Mujumdar, A. S.; Wu, Z. Thermal Drying Technologies – Cost-Effective Innovation Aided by Mathematical Modelling Approach. *Drying Technol.* **2008,** *26*, 145–153.

CHAPTER 41

NUMERICAL SIMULATION OF NATURAL CONVECTION IN A NORTH-LIGHT ROOF UNDER WINTER-DAY BOUNDARY CONDITIONS

B. M. PREETHAM[1,*], ASWATHA[2], and K. N. SEETHARAMU[3]

[1]*Department of Mechanical Engineering, Sri Venkateshwara College of Engineering, Bengaluru, Karnataka, India*

[2]*Department of Mechanical Engineering, Bangalore Institute of Technology, Bengaluru, Karnataka, India*

[3]*Department of Mechanical Engineering, PES University, Bengaluru, Karnataka, India*

[*]*Corresponding author. E-mail: preeth.bm@gmail.com*

ABSTRACT

In this chapter, numerical simulation is carried out to study the natural convection flows in a north-light roof under winterday boundary conditions. The winterday boundary conditions such as hot bottom wall due to room heating and cold ceiling due to environmental temperature are adopted for the present study. The steady-state solutions have been obtained for a Rayleigh number ranging from 10^3 to 10^6 and $Pr = 0.7$. In this chapter, natural convection heat transfer in a traditional north-light roof is quantitatively investigated through isotherm pattern, streamline pattern, local and average Nusselt number. The results indicate that magnitude of the stream function is low at lower Rayleigh number due to conduction domination. However, it increases with an increase in Rayleigh number due to the transition from conduction dominant to convection dominated mode. It is observed that the rate of heat transfer is found to be minimum at the center of the bottom wall. It increases further at a greater rate in the right-hand side of the roof than left. This is because of large cooling area at the right side. It is noticed that as the Rayleigh number increases, multiple-cell solution is developed between hot bottom and cold inclined walls.

41.1 INTRODUCTION

The mechanism of convective heat transfer and fluid flow in an enclosure has the center stage for many engineering applications. Some of these applications include floor heating, heat transfer from radiators and roofs, cooling of electronic devices, solar energy collectors, glazing windows, and fire control in buildings. The roofs are the main parts of the building that protect the building from environmental effects such as rain, storm, particle matter in air, and snow. The roofs of the building are built in such a way that it reduces the heat loss from environment to room and vice versa. The roofs of the buildings are constructed based on the climatic conditions, rain, and architectural design. The rate of heat transfer and flow field inside the roof of the building mainly depends on its shape and size. The geometry of roof can be of different shapes such as gable, gambrel, saltbox, and shed roof. Due to the winterday and summerday conditions, the natural convective currents are set up inside the roofs. The buoyancy forces are induced because of the temperature difference between the ceiling of the room and sloping wall of the roof. The investigation of natural convection heat transfer and fluid flow inside the roof is important to analyze energy-saving and to reduce heating or cooling load of buildings. In the present study, natural convection in different roofs of building is thoroughly made. The literature on the transitional and laminar natural convection in closed cavities is reviewed by Fusegi and Hyun.[1] The complexity in fluid flow and heat transfer with realistic boundary conditions was carried out. Asan and Namli[2, 3] studied the energy transfer and fluid flow caused by density difference in a gable roof under winter- and summerday boundary conditions. The comparison has been made between summerday and winterday boundary conditions. It is found that the overall heat transfer mainly depends on aspect ratio (AR) and Rayleigh number for winterday boundary conditions. However, for summerday boundary conditions, Nusselt number is independent of Rayleigh number. Varol et al.[4] carried out the computations on natural convective flows in a saltbox roof for both summer-like and winter-like boundary conditions. The results are compared values of gable roof for same heating length bottom wall. It is found that lower heat transfer is obtained when saltbox-type roofs are used. It is noticed that the energy rate increases in winterday-like boundary conditions than that of summerday boundary conditions. Varol et al.[5] studied steady-state natural convection in a gambrel roof for both summerday and winterday boundary conditions.

It is found that winterday boundary conditions have more influence on heat transfer than that of summer day due to natural convection mechanism. Moreover, it is observed that energy transfer is lower than that of gable roof for the same inclination of sidewall and for same Rayleigh number. The computations on natural convective fluid flow in shed roofs with and without eave of buildings for summer and winter seasons have been carried out by Koca et al.[6,7] The reported results indicate that eave length and AR are the significant parameters on heat transfer for the same Rayleigh number. Recently, a detailed review of the literature on natural convective heat transfer in an attic-shaped space is carried out by Saha and Khan.[8] Their review focused on significant number of recent studies on topics related to energy transfer and flow in attic-shaped space. Sigrid et al.[9] have performed the evaluation and optimization of a traditional north-light roof. Moreover, the industrial plant energy consumption for an optimized north-light roof shape is studied. However, the attempt on heat transfer and fluid flow is not dealt in their study.

In the literature, heat transfer and fluid flow analyses in different shapes such as gable, gambrel, saltbox, and shed roof are carried out. The literature clearly shows that there is a void existing in the analysis of heat transfer and fluid flow in north-light roof. The numerical simulation of north-light roof is much essential than other roofs due to the following reasons. The north-light roof will bring more sunlight inside the working environment from morning to evening. The energy consumption in a building, the one which is having north-light roof is 54% lesser than that of a conventional flat roof.[9]

The chapter deals on the influence of heat transfer with winterday boundary condition in a north-light roof. It is observed in the literature review that the computations are carried out on optimizing the shape of the north-light roof for the thermal management. However, the effects of winterday boundary conditions have not been dealt. Thus, the present study provides the detailed information on natural convection in a north-light roof for a wide range of parameters.

41.2 MATHEMATICAL MODEL

The configuration of the physical domain selected for the analysis is shown in Figure 41.1. It is assumed that roof is filled with Newtonian, incompressible, and viscous fluid.

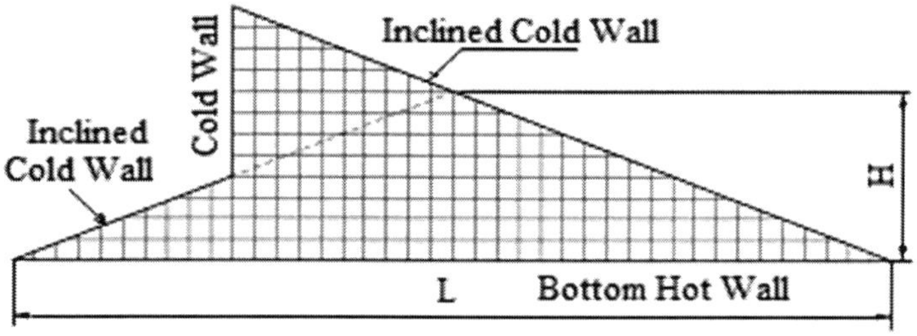

FIGURE 41.1 The physical domain of the system.

During the energy transfer, a variation of thermophysical properties of the working fluid is assumed to be constant. The Boussinesq approximation is invoked for buoyancy term. The gravity forces are acting vertically downward. The differential equations such as conservation of mass, momentum, and energy are governing the natural convection inside the physical domain. The following are the changes of variables used to reduce the governing equation to nondimensional form:

$$X = \frac{x}{L}(1),\ Y = \frac{y}{L},\ U = \frac{\mu L}{\alpha},\ V = \frac{vL}{\alpha}\theta = \frac{T - T_C}{T_h - T_C} \tag{41.1}$$

The following are the nondimensional forms of governing equations:

$$\frac{\partial U}{\partial X} + \frac{\partial V}{\partial Y} = 0 \tag{41.2}$$

$$U\frac{\partial U}{\partial X} + V\frac{\partial U}{\partial Y} = -\frac{\partial P}{\partial X} + Pr\left(\frac{\partial^2 U}{\partial X^2} + \frac{\partial^2 U}{\partial Y^2}\right) \tag{41.3}$$

$$U\frac{\partial V}{\partial X} + V\frac{\partial V}{\partial Y} = -\frac{\partial P}{\partial Y} + Pr\left(\frac{\partial^2 V}{\partial X^2} + \frac{\partial^2 V}{\partial Y^2}\right) + RaPr\theta \tag{41.4}$$

$$U\frac{\partial \theta}{\partial X} + V\frac{\partial \theta}{\partial Y} = \frac{\partial^2 \theta}{\partial X^2} + \frac{\partial^2 \theta}{\partial Y^2} \tag{41.5}$$

The boundary conditions for the physical model, depicted in Figure 41.1, are as follows:

a) In this model, $u=v=0$ for all the walls,
b) On inclined and vertical walls, $T=T_C$
c) At the bottom wall, $T=T_h$.

41.3 EVALUATION OF STREAM FUNCTION AND NUSSELT NUMBER

41.3.1 S\TREAM FUNCTION

The fluid motion is displayed using stream function ψ obtained from the velocity components U and V. The relation between ψ and velocity components[10] for two-dimensional flows is:

$$\frac{\partial^2 \psi}{\partial X^2}+\frac{\partial^2 \psi}{\partial Y^2}=\frac{\partial U}{\partial Y}+\frac{\partial V}{\partial X} \tag{41.6}$$

Using the above definition of the stream function, the motion of the fluid in anticlockwise circulation and clockwise circulation is represented by positive and negative sign, respectively.

41.3.2 NUSSELT NUMBER

The coefficient of heat transfer is representing local Nusselt number (Nu) and is defined as:

$$Nu=-\frac{\partial \theta}{\partial n} \tag{41.7}$$

where n denotes the normal direction on a plane.

The average Nusselt number $\left(\overline{Nu}\right)$ is obtained by integrating local Nusselt number along the wall as follows:

$$\overline{Nu}=\frac{1}{L}\int_0^L Nu_x dX \tag{41.8}$$

41.4 NUMERICAL TESTS

In the present study, a set of governing equations is integrated over the control volume, which produces a set of algebraic equations. The pressure-implicit with splitting of operators algorithm developed by Issa[11] is used to solve the coupled system of governing equations. The set of algebraic equations is solved sequentially using alternating direction implicit (ADI) method.[12] A second-order upwind differencing scheme is used for the formulation of convection contribution to the coefficient in finite-volume equations. The central differencing scheme is used to discrete the diffusion terms. The computation is terminated when all the residuals are less than 10^{-5}.

TABLE 41.1 Comparison of Average Nusselt Number for Gable Roof with Aspect Ratio = 1.

Average Nusselt number ($\overline{Nu}$)	Rayleigh number (Ra)			
	10^3	10^4	10^5	10^6
Nu (ref 3)	4.87	5.12	7.15	12.27
Nu (present)	4.79	5.13	7.08	12.26

The computation domain is discretized into a number of subdomains starting from 31 × 31, 41 × 41, 51 × 51, 61 × 61, 71 × 71, and 121 × 121. It may be noted that the computational grid in north-light roof is generated from normal to the bottom heating wall. The grid tests are made for 31 × 31–121 × 121 to obtain the optimal grid dimensions.

It is observed that grid size 61 × 61 and above shows that there is no appreciable changes in the average Nusselt number in comparison with the literature of Asan and Namli.[3] In the view of this, 61 × 61 grid is selected for further computation. In order to validate the predictive compatibility and accuracy of the present methodology, computations are preferred using the configuration and boundary conditions of Asan and Namli.[3] The finite volume method is preferred to study the numerical simulation of buoyancy flow in a gable roof under winterday boundary conditions. The comparison has been made with mean Nusselt number for AR = 1 including high Rayleigh numbers. The results are presented in Table 41.1. It is observed that there is a good agreement between Asan and Namli[3] and the present study with a maximum discrepancy of 1.6%. The agreement is found to be excellent, which validates the present code.

41.5 RESULTS AND DISCUSSION

41.5.1 EFFECT OF RAYLEIGH NUMBER AND ASPECT RATIO

The numerical simulations are carried out to obtain the heat transfer and flow field inside a north-light roof to cold climate condition due to buoyancy forces. The influence of Rayleigh number and AR, that is, $(H/L)=0.1$ and 0.2, on heat and fluid flow with stream function, temperature contours, local and average Nusselt number is discussed. Figure 41.2a–f illustrate the isotherms and streamlines of the numerical results of various Ra, that is, 10^3–10^6, $H/L=0.1$–0.2, and $Pr=0.7$ for winterday boundary conditions. As expected, the fluid rises up from slightly left side with reference to the midportion of the bottom hot wall. This is because of unsymmetrical geometry and large cooling area on the left side of the north-light roof than the other end. The fluid gets cooled at the top and flows down along the inclined and vertical walls forming two rolls with clockwise and anti-clockwise rotation inside the cavity. It is observed from Figure 41.2a,b that at $Ra=10^3$–10^4, the magnitude of the stream function is very low and the heat transfer is primarily due to conduction. During this conduction-dominant heat transfer, the isotherms are smooth curves, which span the entire enclosure and $\theta \geq 0.7$ is symmetric with respect to vertical centerline. The temperature contours as indicated in Figure 41.2a,b remain invariant up to $Ra=10^4$. For $Ra>10^4$, the circulations near the central regime are stronger and plumelike thermal distribution is obtained. The isotherms $\theta \geq 0.7$ start shifting toward the cold walls as depicted in Figure 41.2c–f for $Ra=10^5$–10^6 and $H/L=0.1$–0.2. The presence of significant convection is also exhibited in temperature contours, which start getting deformed and pushed toward the cold inclined walls.

As Ra increases from 10^4 to 10^6, the buoyancy-driven convective current inside the north-light roof increases as seen from the greater magnitude of stream functions (Fig. 41.2c–f). The circulations are greater at the center and least at the walls due to no-slip boundary conditions. The greater circulations occurred in the left and right side in a north-light roof with respect to central vertical line. The magnitude of the circulations is increased with an increase in Ra and greater at the center of the circulations. The isotherms get compressed to cold inclined walls, which leads to small temperature gradients near the bottom hot and inclined cold walls. It is observed that for $Ra=10^5$ and 10^6 (Fig. 41.2d–f), the thermal boundary layer is developed almost throughout the entire cavity.

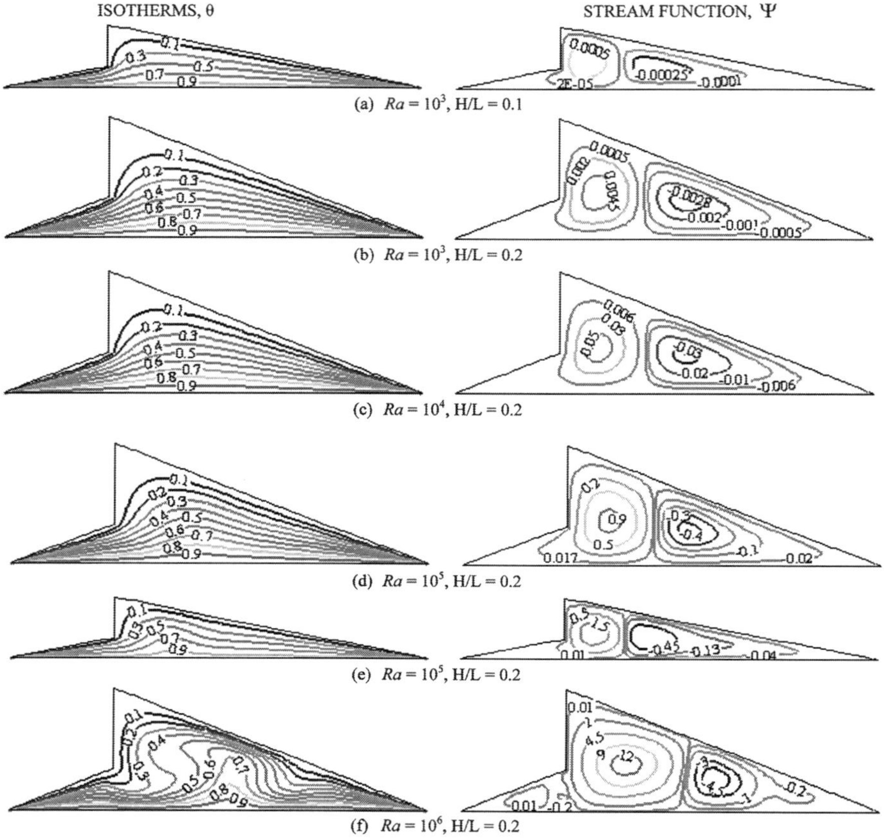

FIGURE 41.2 Isotherms (left) and stream functions (right) for Rayleigh numbers from 10^3 to 10^6. (a) $Ra=10^3$, $H/L=0.1$, (b) $Ra=10^3$, $H/L=0.2$, (c) $Ra=10^4$, $H/L=0.2$, (d) $Ra=10^5$, $H/L=0.2$, (e) $Ra=10^5$, $H/L=0.2$, (f) $Ra=10^6$, $H/L=0.2$.

The comparison between Figure 41.2e,f shows that as AR of north-light roof increases from 0.1 to 0.2, the values of isotherms and stream functions in the core north-light roof increase. The isotherms are concentrated to dragging toward the vertical cold wall due to enlarged space. As the AR increases to 0.2, this trend is increased up to $\theta \leq 0.6$ and becomes dense at the middle of the right inclined cold wall. For $\theta \geq 0.8$, they are symmetric about central vertical line. The magnitudes of the stream functions increase by almost eight times as the AR of the north-light roof increases from 0.1 to 0.2. Multiple cells are set below the left cold vertical walls for AR of 0.2; higher values of stream function are set in the left side than right side. It clearly indicates that the transition from two vortex solution to multiple vortexes depends on Rayleigh number and AR. The convection is more dominated in north-light roof of AR of 0.2 than 0.1.

41.5.2 HEAR TRANSFER RATE AND LOCAL AND AVERAGE NUSSELT NUMBERS

Figure 41.3 shows the effect of *Ra* for AR = 0.2 on the local Nusselt numbers at the bottom hot wall. The heat transfer rate is very high at the right side of the north-light roof for all the *Ra*.

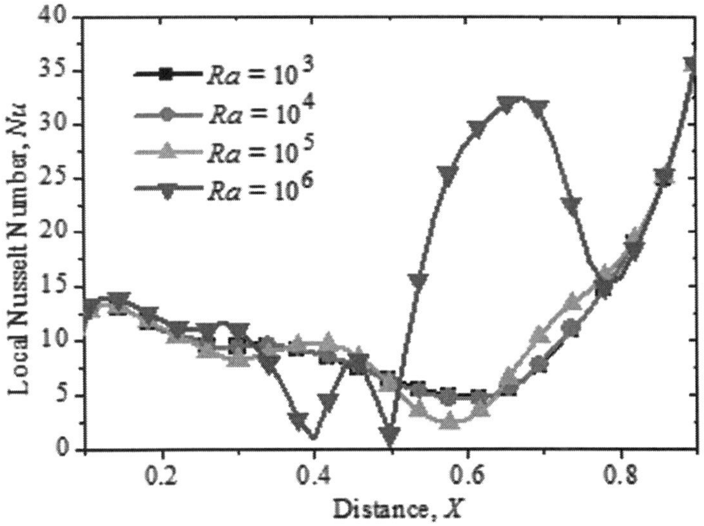

FIGURE 41.3 Variation of local Nusselt number at bottom wall for aspect ratio $(H/L) = 0.2$.

This is because inclined cold wall is very close to hot wall and the span is larger than the left side. The local Nusselt numbers reduces toward the center of the hot wall for the given Rayleigh numbers. As expected, the heat transfer rate is very high at the corners because of the intersection of inclined cold wall and bottom hot wall occurred. The local Nusselt numbers are same for $Ra = 10^3$ and 10^4, due to conduction-dominated heat transfer. For $Ra = 10^5$, the local Nusselt number variations are increasing and decreasing trend and minimum occurred at a distance $X = 0.58$. The heat transfer rate is slightly higher than that for $Ra \leq 10^4$. The physical reason for this is that right central cells drag toward the right corner.

For $Ra = 10^6$, the left bigger cells are extending toward the left side of the bottom hot wall. Hence, the local Nusselt number is increasing and decreasing, and minimum occurred at the center of the hot wall. The Nu steadily increased at $X = 0.5$–0.7, decreased up $X = 0.8$, and then increased. This is because the isotherms are compressed to center of the right inclined wall (Fig. 41.2f). The magnitude of the stream functions at the left side is very high.

The variations of the Nu for different AR and $Ra = 10^5$ are shown in Figure 41.4. It is observed that the Nu is decreasing up to $X \geq 0.5$ for all the AR. However, the minimum value of the Nu is decreased and it shifts to the left side as the AR increased. The heat transfer rate is increased with an increase in AR for $X \geq 0.5$. The overall effect of heat transfer rates is shown in Figure 41.5. The distribution of $\overline{Nu}$ on hot bottom wall is plotted versus logarithmic Ra. The $\overline{Nu}$ is increasing monotonically with an increase in Ra. It is observed that the $\overline{Nu}$ remains constant up to $Ra = 10^4$ due to dominant heat conduction mode for the given AR. The convection-dominated heat transfer occurred for $Ra \geq 10^4$ and the AR is considered for the investigation. It is observed that the changes in $\overline{Nu}$ are more pronounced for higher AR than the lower ones.

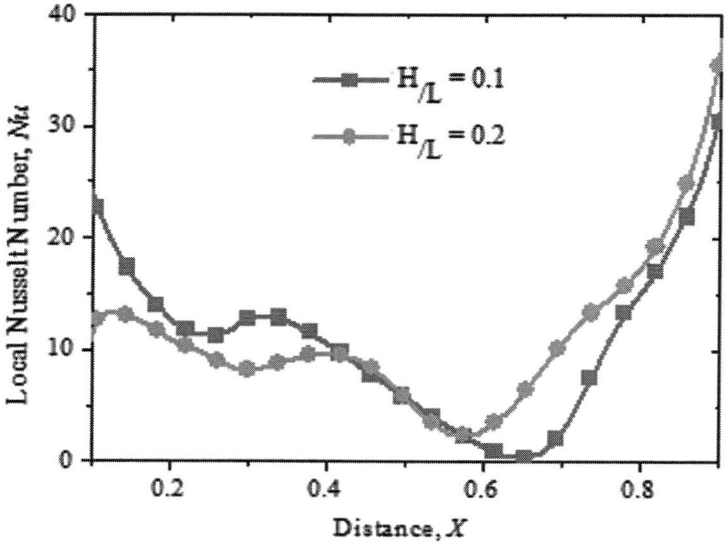

FIGURE 41.4 Variation of local Nusselt number at bottom wall for $Ra = 10^5$.

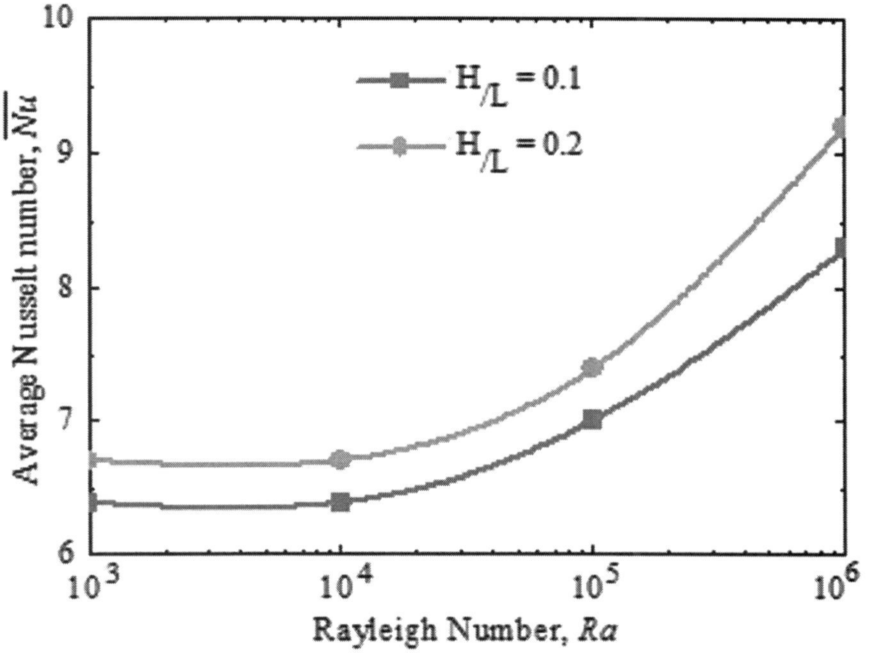

FIGURE 41.5 Variation of average Nusselt number at bottom wall.

41.6 CONCLUSIONS

The chapter has reported the numerical results of laminar, steady, two-dimensional natural convection in a north-light roof for winterday conditions. The following observations have been made:

a) The dominated conductive heat transfer is observed up to $Ra = 10^4$ for all AR under consideration.
b) The $\overline{Nu}$ is increased monotonically with the increase in *Ra* and thermal and flow fields are affected by the shape and AR of the geometry.
c) The heat transfer rate increases with the increase in AR for all *Ra* as computed.
d) It is noticed that as AR increases, minimum value of *Nu* is shifting toward the center of the bottom heating wall.

ACKNOWLEDGMENT

The authors acknowledge all the critical comments of the reviewers which helped to improve the quality of this chapter.

NOMENCLATURE

g—Acceleration due to gravity, $m \cdot s^{-2}$
k—Thermal conductivity, $W \cdot m^{-1} \cdot K^{-1}$
H—Height of the north-light roof with reference to central vertical line of the bottom wall (m)
L—Length of the bottom wall (m)
Nu—Local Nusselt number
$\overline{Nu}$—Average Nusselt number
P—Dimensionless pressure
Pr—Prandtl number
Ra—Rayleigh number
T—Temperature (K)
U—x-component of dimensionless velocity
V—y-component of dimensionless velocity
X—Dimensionless distance along x-coordinate
Y—Dimensionless distance along y-distance

GREEK SYMBOLS

α—Thermal diffusivity ($m^2 \cdot s^{-1}$)
β—Volume expansion coefficient (K^{-1})
θ—Dimensionless temperature
υ—Kinematic viscosity ($m^2 \cdot s^{-1}$)
ρ—Density (kg m^{-3})
ψ—Stream function

SUBSCRIPTS

b—Bottom wall
c—Cold wall
h—Hot wall

KEYWORDS

- **natural convection**
- **steady state**
- **numerical simulation**
- **north light roof truss**
- **winterday boundary conditions**

REFERENCES

1. Fusegi, T; Hyun, J. M., Laminar and Transitional Natural Convection in an Enclosure with Complex and Realistic Conditions. *Int. J. Heat Fluid Flow* **1994,** *15*(4), 258–267.
2. Asan, H.; Namli, L. Laminar Natural Convection in a Pitched Roof of Triangular Cross-Section; Summerday Boundary Conditions. *Energy Build.* **2000,** *33*, 69–73.
3. Asan, H.; Namli, L. Numerical Simulations of Buoyant Flow in a Roof of Triangular Cross-Section Under Winterday Boundary Conditions. *Energy Build.* **2001,** *33*, 753–757.
4. Varol, Y.; Koca, A.; Oztop, H. F. Laminar Natural Convection in Salt Box Roof for Both Summerlike and Winterlike Boundary Conditions. *J. Appl. Sci.* **2006,** *6*(12), 2617–2622.
5. Varol, Y.; Koca, A.; Oztop, H. F. Natural Convection Heat Transfer in Gambrel Roofs. *Build. Environ.* **2007,** *42*, 1291–1297.
6. Oztop, H. F.; Varol, Y.; Koca, A. Laminar Natural Convection Heat Transfer in a Shed Roof with or Without Eave for Summer Season. *Appl. Therm. Eng.* **2007,** *27*, 2252–2265.
7. Koca A; Hakan F; Oztop; Varol Y Numerical Analysis of Natural Convection in Shed Roofs with Eave of Buildings for Cold Climates. *J. Comput. Math. Appl.* **2008,** *56*, 3165–3174.
8. Saha, S. C.; Khan, M. M. K. A Review of Natural Convection and Heat Transfer in Attic-Shaped Space. *Energy Build.* **2011,** *43*, 2564–2571.
9. Sigrid, A.; Liu, H.; Wahed, M.; Zhao, Q. Evaluation and Optimization of a Traditional North-Light Roof on Industrial Plant Energy Consumption. *J. Energ.* **2013,** *6*, 1944–1960.
10. Batchelor, G. K. *An Introduction to Fluid Dynamics*. University Press: Cambridge, 1993.
11. Issa, R. I. Solution of Implicitly Discretised Fluid Flow Equations by Operator Splitting. *J. Comput. Phys.* **1985,** *62*, 40–65.
12. Roache, P. J. *Computational Fluid Dynamics*. Hermosa: Albuquerque, New Mexico, 1982.

CHAPTER 42

NUMERICAL ANALYSIS OF FREE CONVECTION IN A SQUARE ENCLOSURE WITH AN INCLINED HEATED FINNED PLATE

PRAVEEN CHOUDHARY[1,*], HARISHCHANDRA THAKUR[1,2], and SATYENDRA KUMAR PANKAJ[1,3]

[1]*Department of Mechanical Engineering, School of Engineering, Gautam Buddha University, Greater Noida, Uttar Pradesh 201312, India*

[2]*harish@gbu.ac.in*

[3]*satymech92@gmail.com*

[*]*Corresponding author. E-mail: chsmech@gmail.com*

ABSTRACT

Numerical analysis of free convection in the air-filled square enclosure with an inclined finned plate at the angles of 45° and 75° is performed using ANSYS Fluent 15. Temperature contours and flow visualization were investigated. The finned wall is at a constant higher temperature than the sidewalls, while horizontal walls are insulated. Different scenarios were investigated for Rayleigh number 10^4 to 10^6 and value of Prandtl number is 0.71. The study shows the effect of several parameters aspect ratio A, Rayleigh number Ra and L_a on streamlines, temperature contours, and Nusselt number. Thermal buoyancy is responsible for flow inside the cavity or enclosure, which is represented by Ra. Complete domain has been investigated for analysis purpose. The investigation shows the augmentation of heat transfer from the heated surface to the cold surrounding. Numerous engineering applications such as building cooling, heating, ventilation, and flow pattern analysis of electronic cooling devices, and so forth, require an in-depth understanding of convection flow and heat transfer. Therefore, free convection has been extensively used to transfer residual heat from the fins attached at low cost, high durability, and reliability purpose.

42.1 INTRODUCTION

Natural convection heat transfer plays an important role in the cooling application of various electronic hardwares confined in an enclosure. It occurs due to free convection currents. These currents originate when a body force acts on the fluid in which there are density gradients. The net effect results in convection current which is due to buoyancy force. Heat transfer rate is further enhanced by finned surfaces. Heat transfer rate can also be enhanced by changing the orientation of fins and the geometric parameter of fin arrays. Flows due to convection and heat transfer inside all types of enclosures have been the topic of engineering research and studies.

Natural convection helps in establishing temperature distribution within buildings and in determining heat losses in heating, ventilation, and air-conditioning, drying in food processing industries, sterilization for food preservation, distribution of the poisonous product of combustion during fires, cooling of electronic devices, and much more. Convection heat transfer can be forced when the fluid moves by an external agent such as pump or fan. In many cases, natural buoyancy forces are responsible for fluid motion, termed as free or natural convection.

A detailed investigation has been carried out for triangular or conclave enclosure[1–8] under different sets of limiting conditions. Subsequently, the interest focused on a complex model with fins or baffles. De et al.[9] carried out a study of natural convection around the heated cylinder in an enclosure. Sahi et al.[8] studied the effect of free convection in a square enclosure with finned surfaces. Fluid flow and thermal characteristic were also analyzed inside the enclosure with obstruction of varied shape and dimension, several results were obtained for the average Nusselt number and geometric ratios. The observation confirms that the addition of fins or array of the fin at the enclosure surface augments the heat transfer rate between the hot surface and cold surface. Inada et al. carried out the experimental study with fixed fin height and spacing in horizontal fluid layer within the limit.

Thus, the recent development in electronic hardware requires a reliable and efficient cooling system. Different geometries of finned surfaces and limiting conditions markedly alter the flow pattern and hence the transfer of heat inside such an enclosure or cavity. This chapter investigates the fluid flow pattern and heat transfer rate with a heated inclined finned plate (HIFP), inclined at the angles of 45° and 75°. The thin fin is heated and placed at the center of the enclosure. The surrounding sidewalls are at lower temperature. Due to temperature, gradient transfer of heat starts and moves upward. The effect of varying fin lengths and aspect ratios has been discussed and a graph of Nu_{avg} versus Ra has been plotted for two angles and aspect ratio. Graph thus obtained depicts when and where convection is dominant for heat transfer analysis.

42.2 PROBLEM STATEMENT AND GOVERNING EQUATION

The computational domain with related parameters is shown in Figure 42.1. At higher temperature, the temperature at finned plate remains constant, while the sidewalls of the enclosure are at lower temperature. No heat transfers take place from the bottom and the top wall of the enclosure. The thin finned plate is rotated by the angles of 45° and 75°. The assumptions made are flow is 2D, incompressible, steady, and laminar. Viscous dissipation is neglected and thermophysical properties are assumed to be constant, that is, Boussinesq approximation is applied. Several computational parameters in terms of aspect ratio are as under.

$$A = h/H;\ A1 = h1/H;\ A2 = h2/H;\ Ha = ha/H;$$
$$La = la/h1;\ \theta = 45° \text{ and } 75°$$

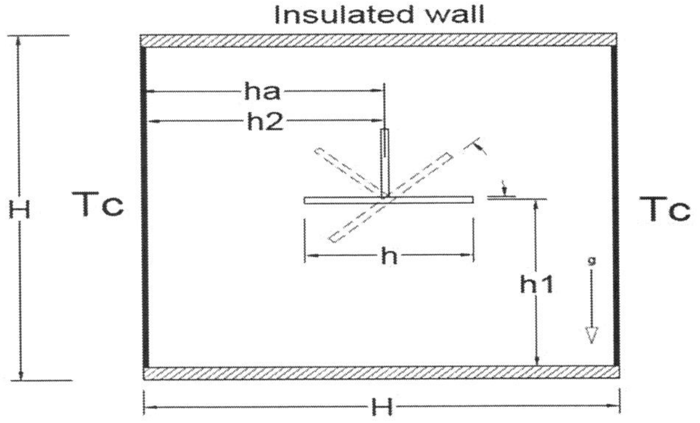

FIGURE 42.1 Computational domain.

42.3 NUMERICAL PROCEDURE

Computational fluid dynamics uses a mathematical model and supercomputers to simulate fluid flow situations for prediction of heat, mass, and momentum transfer and design optimization. It is a proactive analysis and design tool and nowadays widely used for improving reliability, higher product durability, and better performance.

ANSYS Fluent 15 was used to investigate the problem using pressure-based solver and velocity formation taken as absolute. Nodes generated were 30,802 and elements 29,563. Semi-implicit pressure-linked equation scheme and second-order upwind as spatial discretization method were also used. The convergence criterion set for continuity and velocity is 10^{-3} for energy 10^{-6}. Iteration was carried out till a solution is converged. Validation has been performed for $A=0.6$ and $A_1=0.5$ by Oztop et al.[8] and Sahi et al.[8] Table 42.1 shows the result and percentage error, limited to 5% which provides sufficient data to deal with present numerical problem to deal with actual physical problem. Tables 42.2 and 42.3 show the value of mean Nusselt number obtained at the cold wall and hot inclined fin plate, respectively. Variation has been discussed below.

TABLE 42.1 Validation of Results.

Ra	Surface-averaged Nusselt number at cold cylinder wall A=0.6 and A_1=0.5				
	Present study	Oztop[10]	Sahi[8]	Error (%) to Oztop	Error (%) to Sahi
10^4	2.06	2.04	2.035	0.98	1.22
10^5	2.62	2.5	2.54	4.8	3.15
10^6	5.24	5.03	5.04	4.17	3.96

TABLE 42.2 Average Nusselt Numbers for Inclined Fin Plate (Cold Body).

Ra	Aspect ratio (A)	Angle 45°			Angle 75°		
		L_a=0.25	L_a=0.50	L_a=0.75	L_a=0.25	L_a=0.50	L_a=0.75
10^4	0.4	1.697	1.805	1.966	1.657	1.818	2.030
	0.6	2.178	3.030	3.255	2.003	3.398	4.162
10^5	0.4	2.551	2.614	2.702	1.642	1.773	2.001
	0.6	2.970	3.070	3.218	3.211	3.448	4.110
10^6	0.4	2.217	2.299	2.411	2.387	2.451	2.699
	0.6	4.020	4.189	4.290	4.317	4.756	4.940

TABLE 42.3 Average Nusselt Numbers for Inclined Fin (Hot Body).

Ra	Aspect ratio (A)	Angle 45°			Angle 75°		
		L_a=0.25	L_a=0.50	L_a=0.75	L_a=0.25	L_a=0.50	L_a=0.75
10^4	0.4	2.630	2.541	2.519	2.413	2.399	2.325
	0.6	3.234	3.194	3.088	3.477	3.629	3.845
10^5	0.4	5.236	4.987	4.713	3.002	2.915	2.919
	0.6	3.727	3.510	3.416	4.068	5.052	4.316
10^6	0.4	4.516	3.611	3.525	4.073	3.901	3.830
	0.6	5.47	5.32	5.11	5.971	6.111	6.994

The mesh is generated by ANSYS FLUENT 15 software using mesh generation module. The triangular meshing is used (Fig. 42.2).

$$\frac{\partial V_i}{\partial X_i} = 0 \tag{42.1}$$

$$V_j \frac{\partial V_i}{\partial X_j} = -\frac{\partial P}{\partial X_j} + Pr \frac{\partial^2 V_i}{\partial X_j X_j} + Gr\theta\delta_{i2} \tag{42.2}$$

$$V_j \frac{\partial \theta}{\partial X_j} = \frac{1}{Pr} \frac{\partial^2 \theta}{X_j X_j} \tag{42.3}$$

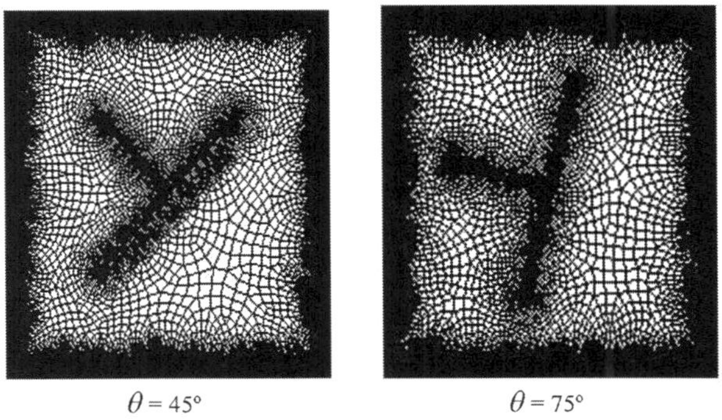

FIGURE 42.2 Mesh generation.

42.4 ISOTHERM AND STREAMLINES

Figures 42.3 and 42.4 depict the flow pattern of isotherms and streamlines against parameters such as Rayleigh number, aspect ratio, and fin length. For Rayleigh number 10^4 and aspect ratio $A = 0.4$, the isotherm plots are smooth and cover the entire enclosure, indicating that conduction is dominant for both angles ($\theta = 45°$ and 75°). For *Ra* 10^5 and 10^6, due to the inclination of the fin, the flow pattern is not identical. For $A = 0.6$, isotherm plum appears at the finned plate except for fin length $L_a = 0.25$ and $\theta = 75°$. When the finned plate rotates, increasing angles tend to squeeze the flow pattern toward the bottom surface and left cold wall; hence, convective heat transfer is more toward bottom rather at the top. Further, with high Rayleigh number 10^6 ($\theta = 45°$ and 75° and $A = 0.4$, 0.6), isotherm pattern changes significantly confirming the mode of heat transfer is convective.

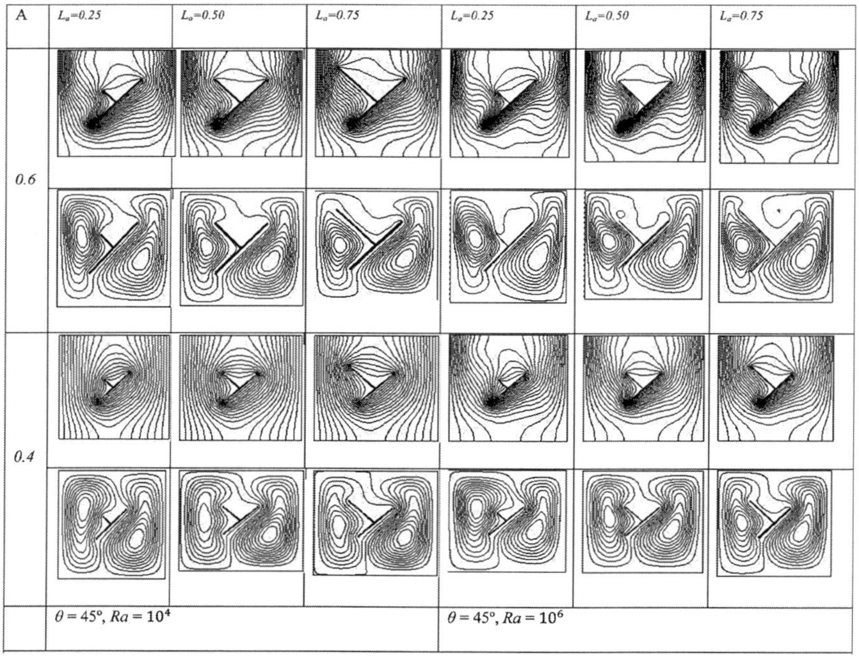

FIGURE 42.3 Streamlines and isotherms for heated inclined finned plate (HIFP) at an angle of 45°, $A_1 = H_a = 0.5$, $L_a = 0.25$, 0.50, and 0.75, and $A = 0.6$ and 0.4.

The flow field represents two vortices rotating in opposite direction for $\theta = 45°$ for $0.25 < L_a < 0.75$. Rotating flow field takes the heat, rises, gets blocked by the adiabatic wall, and then flow down the cooled walls. For $\theta = 75°$ and $A = 0.4$, three vortices are formed out of which flow are interconnected. For $A = 0.6$, three vortices are formed. These vortices enhance the heat transfer rate by the virtue of convective heat transfer and thermal buoyancy.

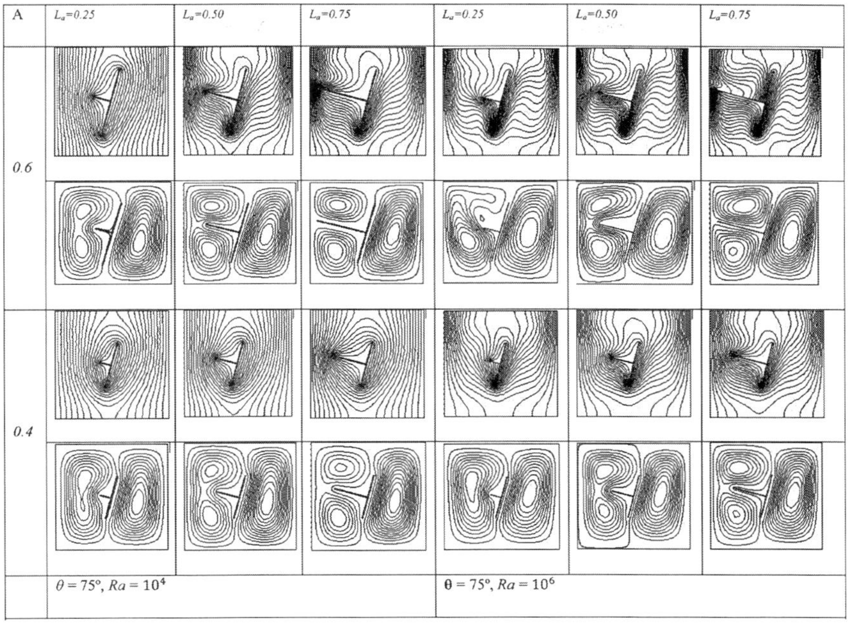

FIGURE 42.4 Streamlines and isotherms for HIFP at an angle of 75°, A_1 = H_a =0.5, L_a=0.25, 0.50, and 0.75, and A=0.6 and 0.4.

42.5 EFFECT OF GEOMETRIC ASPECT RATIO, ANGLE OF DEFLECTION, AND RAYLEIGH NUMBER

Figure 42.5, for aspect ratio A=0.4, Ra 10^6 mean Nusselt number is slightly lower with increasing angle. For A=0.4, θ=75°, and 10^4 and 10^5 Nusselt number is constant for all the finned lengths, isotherm and streamline are identical, clearly indicating that conduction effect is dominant and convection heat transfer is quite weak. For A=0.4, 10^5, and θ=45°, isotherm, and streamline start to distort and hence the convection effect increases resulting in increased value of the mean Nusselt number. For A=0.6, the mean Nusselt number increases with the increase in Rayleigh number. For Ra 10^6, increasing angle tends to squeeze the flow pattern and hence the heat transfer decreases with the increase in finned length. In Figure 42.6, as we move along the finned plate, we observe that, initially, local Nusselt number value is maximum, which decreases to a minimum at 1/5th of a finned length and again attains peak at distance 0.42 and 0.8 of finned length and reduces to a minimum at the tip.

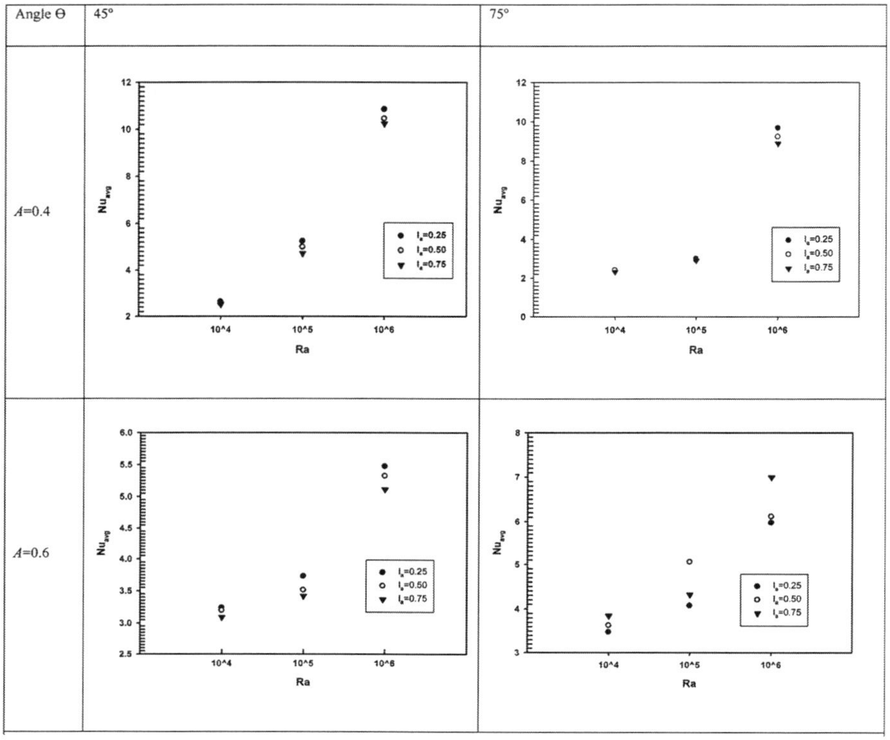

FIGURE 42.5 Average Nusselt number plot at hot plate for different Rayleigh number and different fin lengths.

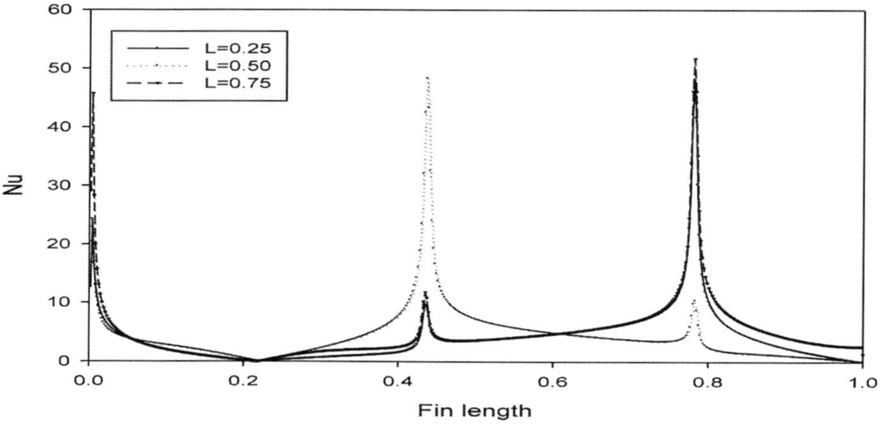

FIGURE 42.6 Variation of average Nusselt number to fin length for $A = 0.4$ and $\theta = 75°$.

42.6 CONCLUSION

Numerical analysis of natural convection in the square enclosure is performed. The study is conducted to see the effect of Rayleigh number, aspect ratio, fin position, and orientation on convection flow field and temperature field.

- For a specified position, the value of Nusselt number increases with the increasing value of Rayleigh number.
- As fin length increases at specified Rayleigh number, the value of Nusselt number also increases.
- Value of Nusselt depends on the orientation of plate.
- Increasing aspect ratio tends to decrease the mean Nusselt number for the HIFP.
- Increasing angle of deflection from horizontal plane tends to slightly decrease the mean Nusselt number for HIFP.

ACKNOWLEDGMENT

We would sincerely like to acknowledge the constant support of our mentor and guide Dr. Harish Chandra Thakur and facility at Gautam Buddha University.

NOMENCLATURE

A—Aspect ratio
g—Acceleration due to gravity (m)
H, (*h*)—Plate enclosure width (m)
k—Thermal conductivity (W)
Fin height (m)
Nu—Average Nusselt number
p—Dimensionless pressure: $P = p/\rho$
Pr—Prandtl number $Pr = v/a$
Ra—Rayleigh number: $Ra = Gr \cdot Pr$
Gr—Grashof number: $Gr = g\beta(T_H - T_C) \cdot H^3/v^2$
Dimensionless velocity components (L/v)
Dimensionless Cartesian coordinate (X_i/L)

GREEK SYMBOLS

β—Thermal expansion coefficient
α—Thermal diffusivity
ν—Kinematic viscosity
θ—Dimensionless temperature $(T-T_c)/(T_h-T_c)$
δ_{ij}—Delta kronecker

KEYWORDS

- **natural convection**
- **square enclosure**
- **plate inside square enclosure**
- **Rayleigh number**

REFERENCES

1. de Vahl Davis, G. Natural Convection of Air in a Square Cavity: A Benchmark Numerical Solution. *Int. J. Numer. Methods Fluids* **1983,** *3*, 249–264.
2. Mahmud, S.; Das, P. K.; Hyder, N.; Sadrul Islam, A. K. M. Free Convection in an Enclosure with Vertical Wavy Walls. *Int. J. Therm. Sci.* **2002,** *41*, 440–446.
3. Ridouane, E. H.; Campo, A. Free Convection Performance of Circular Cavities Having Two Active Curved Vertical Sides and Two Inactive Curved Horizontal Sides. *Appl. Therm. Eng.* **2006,** *26*, 2409–2416.
4. Basak, T.; Roy, S.; Singh, A.; Pandey, B. D. Natural Convection Flow Simulation for Various Angles in a Trapezoidal Enclosure with Linearly Heated Side Wall(s). *Int. J. Heat Mass Transfer* **2009,** *52*, 4413–4425.
5. Kaluri, R. S.; Anandalakshmi, R.; Basak, T. Bejan's Heat Line Analysis of Natural Convection in Right-Angled Triangular Enclosures, Effects of Aspect-Ratio and Thermal Boundary Conditions. *Int. J. Therm. Sci.* **2010,** *49*, 1576–1592.
6. Mahmoodi, M. Numerical Simulation of Free Convection of a Nanofluid in a Square Cavity with an Inside Heater. *Int. J. Therm. Sci.* **2011,** *50*, 2161–2175.
7. Yu, Z.-T.; Xu, X.; Hu, Y.-C.; Fan, L.-W.; Cen, K.-F. Unsteady Natural Convection Heat Transfer from a Heated Horizontal Circular Cylinder to its Air-Filled Coaxial Triangular Enclosure. *Int. J. Heat Mass Transfer* **2011,** *54*, 1563–1571.
8. Sahi, A.; Sadaoui, D.; Meziani, B.; Mansouri, K. Free Convection in a Square Enclosure with a Finned Plate. *Mech. Ind.* **2015,** *16*, 7.

9. De, A. K. Dalal, A. A Numerical Study of Natural Convection Around a Square, Horizontal, Heated Cylinder Placed in an Enclosure. *Int. J. Heat Mass Transfer* **2006,** *49*, 4608–4623.
10. Oztop, H. F.; Dagtekin, I.; Bahloul, A. Comparison of the Position of a Heated Thin Plate Located in a Cavity for Natural Convection. *Int. Commun. Heat Mass Transfer* **2004,** *31*, 121–132.

CHAPTER 43

VISUALIZATION OF THERMAL TRANSPORT IN COMPLEX GEOMETRIES WITH DISCRETE HEAT SOURCES AT BOTTOM AND LEFT WALLS

NARASIMHA SURI TINNALURI[1,*] and JAYA KRISHNA DEVANURI[1,2]

[1]*Department of Mechanical Engineering, National Institute of Technology, Warangal, Telangana 506004, India*

[2]*djayakrishna.iitm@gmail.com*

[*]*Corresponding author. E-mail: tnsuri.suri@gmail.com*

ABSTRACT

The objective of the present work is to investigate the visualization of thermal transport in two-dimensional complex geometries with two discrete heat sources; one at the bottom wall and another at the left wall. A code in C^{++} has been developed to read the mesh from the meshing software and then linked to the in-house developed solver. The governing equations are discretized for the considered non-orthogonal domains by collocated grid-based finite volume method. The present numerical methodology is validated and compared with the commercial software ANSYS Fluent. The results are given in terms of isotherms and heatlines. The contour plots thus obtained are used for the analysis of energy transport in considered geometries.

43.1 INTRODUCTION

Visualization of thermal transport is very much needed for the optimum performance of the equipment. Applications such as heat exchangers, nuclear fuel assemblies, solar energy systems, electronic components, biomedical engineering, and so forth involve intricate geometries where the analysis becomes very complex. To visualize the fluid flow, an energy analog concept, which is based on heat function and heatlines, was first introduced by Kimura and Bejan.[1] Costa[2] presented the heat functions for the laminar natural convection near a vertical wall with isothermal and constant heat flux conditions. The boundary layer problem was solved using the similar method. Deng and Tang[3] studied the numerical visualization of convective heat transfer by streamlines and heatlines. It was mentioned that the transport phenomena was visualized by means of streamlines and heatlines. The concept of heatline visualization observed to provide a more practical and efficient means than the customary ways. Dalal and Das[4] studied two-dimensional (2D) cavity with a wavy right vertical wall. Results were presented by means of streamlines, heatlines, isotherms, local and average Nusselt number for a selected range of Rayleigh number (10^0–10^6). Basak and Roy[5] observed that the food processing requires large temperature for a long time, and the heatline concept was observed to provide insight for the understanding of heat transport phenomena. Basak et al.[6] studied the natural convection and heat flow circulation inside the inverted triangular cavity by considering the heatline concept, the study was carried out with a wide range of Rayleigh number (10^2–10^5) and Prandtl number (0.015, 0.026, 0.7, and 1000). It was found that isotherms are normal to the heatlines during conduction dominant heat transfer and with the increase of Rayleigh number (Ra) the flow patterns get distorted. Kaluri and Basak[7] studied distributed thermal management policy for an energy-efficient method of materials processing applications by natural convection. Analysis was carried out by visualizing the heat flow by heatlines. It was found that the heatline approach is useful in visualizing the heat flow in the cavity with multiple distributed heat sources. Laminar natural convective flow with distributed heat sources in a square cavity was reported by Kaluri et al.[8] Governing equations are solved by using finite element method. Based on heatlines, it was observed that distributed heating was more efficient when compared to that of a conventional bottom heating. Biswal and Basak[9] reported heatline patterns for various types of enclosures with Dirichlet heat function boundary conditions. The governing equations were solved at various ranges of Rayleigh numbers (10^3 and 10^5) and Prandtl numbers (0.015 and 7.2) using Galerkin finite element method. It was concluded that the heatline visualization approach with various heat function boundary conditions can be helpful for the better insight of the energy transport in thermal systems. Krishna et al.[10] studied lid-driven flow in a skewed geometry filled with saturated porous medium. Governing equations were solved numerically by using finite volume method (FVM). A quadrilateral cell which is arranged in a semi-staggered arrangement was employed, and the coordinate transformation was carried out to transform into a square domain. Roychowdhury et al.[11] solved incompressible N–S equations using collocated, non-orthogonal grid. Governing equations were discretized using FVM. The convective formulation was carried out using QUICK scheme. The proposed scheme was validated with bench mark solutions. Suri and Krishna[12] developed a generalized code to read the mesh data from a meshing software GAMBIT. The mesh data thus obtained is then linked with in-house developed collocated grid-based FVM code. The study presented numerical methodology and the comparison was given with ANSYS Fluent.

Based on the above literature, it may be noted that for the better insight of heat transport in complex geometries, several numerical methodologies were adopted and validated. In these studies, grid was generated for a specific domain and later numerical analysis was performed. But in the present study, a generalized code is developed to read mesh data for complex domains and a collocated grid-based FVM has been employed for the analysis of thermal transport in complex domains. It may be noted that visualization of heatlines can play a prominent role for the better insight of thermal transport. Therefore, in the present study the numerical results for energy transport are presented in terms of isotherms and heatlines in 2D complex geometries with a discrete heat source at the bottom and left side walls.

43.2 PROBLEM DEFINITION AND MATHEMATICAL FORMULATION

The schematic diagram of the computational domains with boundary conditions is shown in Figure 43.1a–d. It consists of 2D solid domains of dimensions $L \times H$. The right side of the wall is maintai ned at constant temperature 300 K (T_c) and serves as a heat sink. Two discrete heat sources are located at the bottom and left center portion of the wall which are maintained at constant heat flux 200 W/m^2 (q_b'') and 100 W/m^2 (q_l'') condition and the remaining portion of the geometries are thermally insulated.

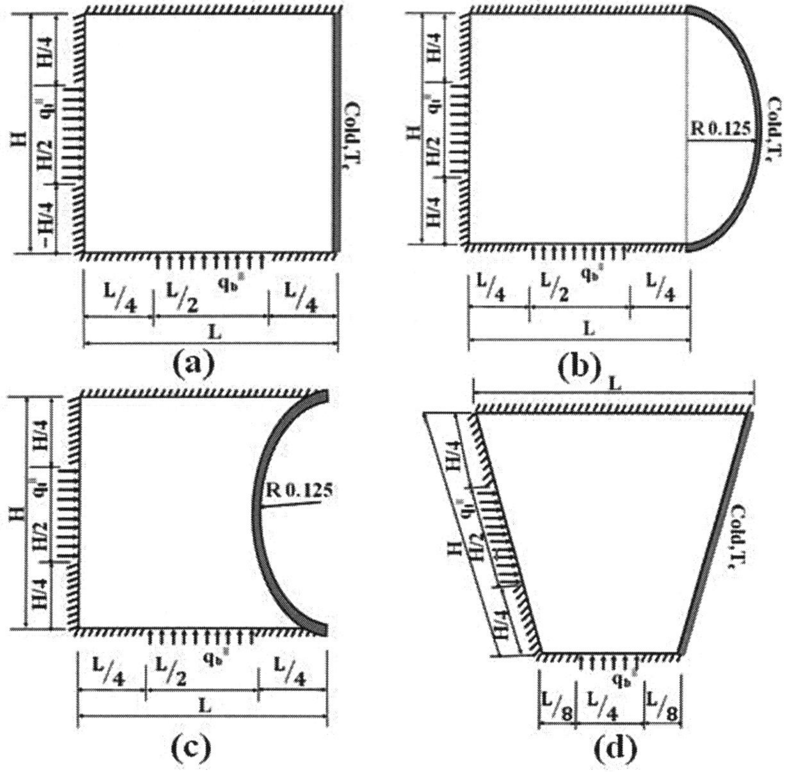

FIGURE 43.1 Schematic diagram of the computational domains with boundary conditions.

The integral form of governing equation employed for the heat transport in complex solid domains can be given as

$$\int_{\Delta V} \frac{\partial}{\partial x}\left(\Gamma \frac{\partial \phi}{\partial x}\right) dx\,dy + \int_{\Delta V} \frac{\partial}{\partial y}\left(\Gamma \frac{\partial \phi}{\partial y}\right) dx\,dy = 0 \tag{43.1}$$

The expression of diffusive flux through the east face can be obtained in terms of the projected areas, and the neighboring node values of ϕ can be evaluated as:

$$J_{De} = \left(\frac{\Gamma_e A_{ex}^1}{V_e}\right)\left[\left\{A_{ex}^1(\phi_E - \phi_P) + A_{ex}^2(\phi_{ne} - \phi_{se})\right\}\right] + \left(\frac{\Gamma_e A_{ey}^1}{V_e}\right)$$
$$\left[\left\{A_{ey}^1(\phi_E - \phi_P) + A_{ey}^2(\phi_{ne} - \phi_{se})\right\}\right] \tag{43.2}$$

Simplification of above Equation 43.2 is as follows

$$J_{De} = d_e^1(\phi_E - \phi_P) + d_e^2(\phi_{ne} - \phi_{se}) \tag{43.3}$$

where $d_e^1 = \frac{\Gamma_e}{V_e}\left[A_{ex}^1.A_{ex}^1 + A_{ey}^1.A_{ey}^1\right]$ and $d_e^2 = \frac{\Gamma_e}{V_e}\left[A_{ex}^1.A_{ex}^2 + A_{ey}^1.A_{ey}^2\right]$

Similarly, J_{Dw}, J_{Dn}, and J_{Ds} can be obtained.

The net diffusive flux contribution for all the four sides can be shown to be of the form

$$J_D = -\left\{d_e^1\phi_E + d_w^1\phi_W + d_n^1\phi_N + d_s^1\phi_S - \left(d_e^1 + d_w^1 + d_n^1 + d_s^1\right)\phi_P + [b_{no}]\right\} \tag{43.4}$$

where

$$b_{no} = \left(d_e^2 + d_w^2\right)\phi_{ne} - \left(d_w^2 + d_n^2\right)\phi_{nw} - \left(d_e^2 + d_s^2\right)\phi_{se} - \left(d_w^2 + d_s^2\right)\phi_{sw}$$

The term b_{no} arises due to non-orthogonality of the grid and it boils down to zero as the grid becomes orthogonal. The corner values are approximated using the four surrounding nodal values. For instance, the northeast corner value is obtained as

$$\phi_{ne} = \frac{1}{4}(\phi_N + \phi_E + \phi_P + \phi_{NE}) \tag{43.5}$$

Similarly, ϕ_{nw}, ϕ_{se}, and ϕ_{sw} can be obtained.

Steady two dimensionless heat function equations can be represented by:

$$\frac{\partial^2\Pi}{\partial X^2} + \frac{\partial^2\Pi}{\partial Y^2} = 0 \tag{43.6}$$

satisfies the heat function Equation 43.6 such that

$$-\frac{\partial\Pi}{\partial X} = -\frac{\partial\phi}{\partial Y}, \frac{\partial\Pi}{\partial Y} = -\frac{\partial\phi}{\partial X} \tag{43.7}$$

The boundary conditions are obtained by integrating Equation 43.7 along with the boundaries at various junction points:

A reference value of $\Pi = 0$ is assumed for the top surface

Top (adiabatic) at $Y=1$; $0<X<1$

$$\Pi(X,1) = \left.\frac{\partial\phi}{\partial y}\right|_{Y=1} = 0 \tag{43.8}$$

Right (cold) at $X=1$; $1>Y>0$

$$\Pi(1,Y)=\Pi(X,1)-\int_0^Y -\left.\frac{\partial\phi}{\partial X}\right|_{X=1} dY \tag{43.9}$$

Left (adiabatic) at $X=0$; $H>Y_c>3H/4$

$$\Pi(0,Y_c)=\Pi(X,1)+\int_{3H/4}^{H} -\left.\frac{\partial\phi}{\partial X}\right|_{X=0} dY \tag{43.10}$$

Left (heat flux) at $X=0$; $3H/4>Y_b>H/4$

$$\Pi(0,Y_b)=\Pi(0,Y_c)+\int_{H/4}^{3H/4} -\left.\frac{\partial\phi}{\partial X}\right|_{X=0} dY \tag{43.11}$$

Left (adiabatic) at $X=0$; $H/4>Y_a>0$

$$\Pi(0,Y_a)=\Pi(0,Y_b)+\int_0^{H/4} -\left.\frac{\partial\phi}{\partial X}\right|_{X=0} dY \tag{43.12}$$

Bottom (adiabatic) at $Y=0$; $0<X_a<L/4$

$$\Pi(X_a,0)=\Pi(0,Y_a)-\int_0^{L/4} -\left.\frac{\partial\phi}{\partial Y}\right|_{Y=0} dX \tag{43.13}$$

Bottom (heat flux) at $Y=0$; $L/4<X_b<3L/4$

$$\Pi(X_b,0)=\Pi(X_a,0)+\int_{L/4}^{3L/4} -\left.\frac{\partial\phi}{\partial Y}\right|_{Y=0} dX \tag{43.14}$$

Bottom (adiabatic) at $Y=0$; $3L/4<X_c<L$

$$\Pi(X_c,0)=\Pi(X_b,0)+\int_{3L/4}^{L} -\left.\frac{\partial\phi}{\partial Y}\right|_{Y=0} dX \tag{43.15}$$

where X and Y are the distances measured along the x- and y-coordinates, respectively; Π is the heat function; ϕ is the primitive variable (temperature); Γ is the diffusion coefficient; d_e^1 and d_e^2 represent orthogonal and non-orthogonal part of diffusive flux for east face of the control volume; and A_{ex}^1 and A_{ey}^2 represent orthogonal and non-orthogonal area in x- and y-directions for the east face of the control volume.

43.3 METHOD OF SOLUTION

The governing equations are discretized using non-orthogonal domains by the finite-volume method. Arbitrary quadrilateral meshes are chosen and numerical work has been carried out for complex domains. Temperature values are defined at the same set of grid points which are located at the center of the cells known as a collocated grid arrangement. Collocated grid arrangement is chosen because the terms in the governing equations are essentially identical between the different balance equations and; hence, the number of coefficients that must be computed and stored is minimized and programming is simplified and exported in neutral file format. A code in C^{++} has been developed to read the mesh from the neutral file of GAMBIT. This code acts as the link between the present in-house finite volume numerical code and meshing software. The set of governing equations obtained is solved using Gauss–Seidel iterative solver, and convergence criterion of 10^{-6} is imposed to terminate the iterations. Contour plots are used for the visualization of isotherms and heatlines for the considered geometries.

43.4 RESULTS AND DISCUSSION

The validity of the present numerical methodology is carried out by comparing with commercial CFD code ANSYS Fluent and is shown in Figure 43.2. The left-hand side of Figure 43.2a specifies isotherms from the present study and the right-hand side gives isotherms from commercial CFD code ANSYS Fluent. Figure 43.2b provides the qualitative comparison with ANSYS Fluent by plotting mid-plane temperature profile. Based on the above two comparisons, it may be noted that the present numerical methodology is in very well agreement with commercial code ANSYS Fluent. Figure 43.3 shows the contours for isotherms and heatlines for the computational domain shown in Figure 43.1d. Basak et al.[6] reported that in conduction dominant regime the isotherms and heatlines are normal to each other. Based on the details provided in Figure 43.2 and 43.3, it may be noted that the validity of the present numerical methodology is justified. Grid generation is a crucial part of the numerical solution. A numerical code has been developed to read the mesh from GAMBIT and linked to the code developed for the visualization of thermal transport. The constraint of the developed code is limited to a quadrilateral grid. A grid size of 80×80 has been considered for the complex domains shown in Figure 43.1a,b.

Figure 43.4 indicates isotherms (left) and heatlines (right) at thermal conductivity (k=0.25 W/m·K) for the complex domains shown in Figure 43.1. As the heatlines provide the direction of heat flow, the visualization for heat transport has been analyzed by plotting heatlines along with the

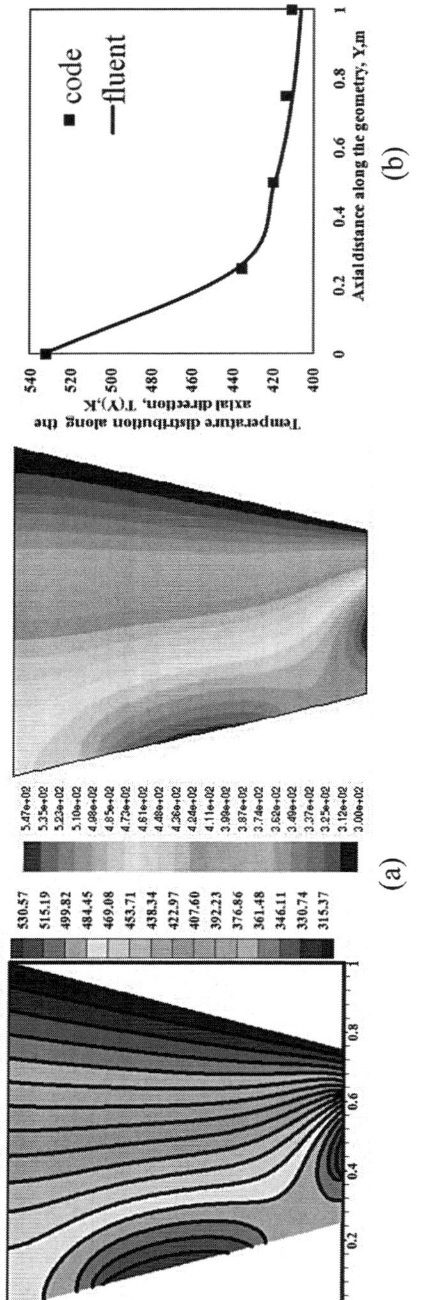

FIGURE 43.2 (a) Comparison of isotherms for the present study (left), ANSYS Fluent (right) and (b) mid plane temperature profile with commercial CFD code.

isotherms. The left column of Figure 43.4 shows the isotherms. Based on the isotherms it may be noted that the maximum temperature can be observed at the center of the bottom wall due to its higher heat flux of 200 W/m^2. It may be noted that due to the presence of discrete heat source which is of higher magnitude when compared to the left wall of 100 W/m^2, large temperature gradients may be noted at the bottom wall. As shown in Figure 43.4a,c, the magnitude of constant temperature lines at the bottom portion is high when compared to Figure 43.4b,d. This difference can be attributed due to the variation in size at the bottom wall of the cavity. Due to the lower magnitude of thermal conductivity, the heat transport is hindered by the increase in the length of the bottom wall. This phenomenon can be supported based on the higher magnitude of heatlines from the bottom wall to constant temperature lines is observed to be maximum at the bottom and left walls and noted to decrease as it is moved toward the right cold wall. Based on heatlines, it can be noted that the heat transport takes place from discrete heat sources of the bottom and left walls to the right cold wall. From Figure 43.4a,c, dense heatlines can be noted toward the middle portion of the right cold wall; whereas from Figure 43.4b,d it is observed at the lower portion. This behavior can be inferred with the size of the discrete heat source. It may be noted that as the length of the heat source increases, the amount of heat to dissipate also increases.

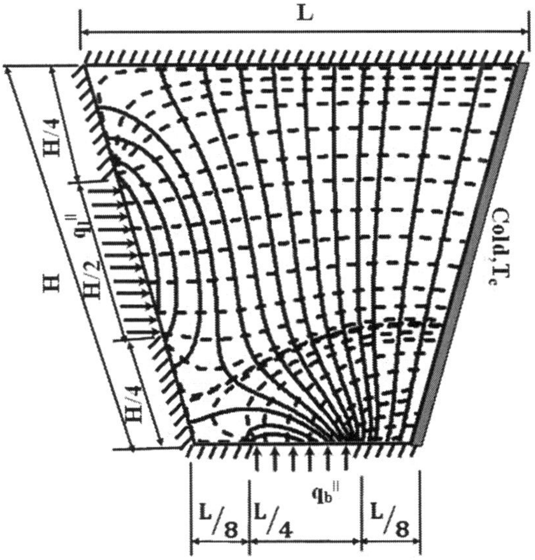

FIGURE 43.3 Comparison of isotherms (__) and heatlines (- - -).

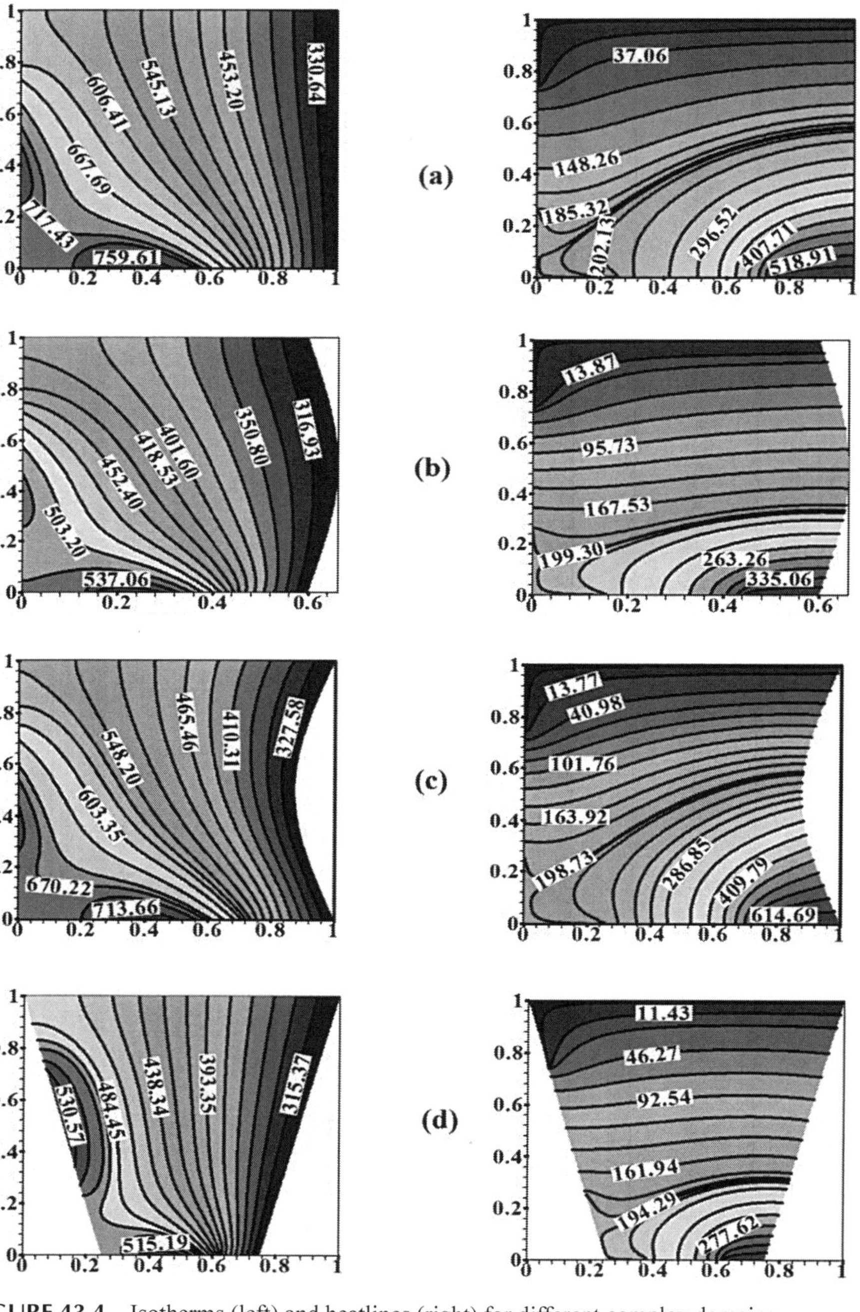

FIGURE 43.4 Isotherms (left) and heatlines (right) for different complex domains.

43.5 CONCLUSIONS

A numerical code has been developed to read the mesh from GAMBIT and linked to the collocated grid-based finite volume solver for the visualization of thermal transport. The governing equations are written in Cartesian coordinate system and are discretized in the physical domain. Results are presented in the terms of isotherms and heatlines for the considered geometries. The results thus obtained are rigorously validated with commercial CFD code ANSYS Fluent. The study could reveal that the heatlines provide an insight for the direction of heat flow which can be inferred from constant heat flux lines from discrete heat sources to the cold wall. Based on the study, it can be concluded that the visualization of heatlines can play a prominent role for the better insight of thermal transport in complex geometries with discrete heat sources. The present numerical methodology can be adopted for the analysis of thermo-hydraulics in complex domains.

NOMENCLATURE

A—projected area, m^2
b_{no}—non-orthogonality of the grid
d^1—orthogonal part of diffusive flux
d^2—non-orthogonal part of diffusive flux
E, N, W, S—east, north, west, south nodes
e, n, w, s—east, north, west, south faces
H—height in vertical direction, m
J_D—diffusive flux
k—thermal conductivity W/m·K
L—length in horizontal direction, m
NE—northeast node
NW—northwest node
q_b''—bottom heater flux strength, W/m^2
q_l''—left heater flux strength, W/m^2
SE—southeast node
SW—southwest node
T—temperature, K
T_c—temperature of cold wall, K
X_a, X_c—distance between bottom adiabatic portion, m
X_b—distance between bottom heat flux, m
Y_a, Y_c—distance between left adiabatic portion, m

Y_b—distance between left heat flux, m
X—distance along x-coordinate, m
x, y—coordinate axis
Y—distance along y-coordinate, m

GREEK SYMBOLS

Γ—diffusion coefficient
Π—heat function
ϕ—primitive variable
ϕ_P—unknown primitive variable (temperature)
ΔV—volume difference, m^3

SUPERSCRIPTS

1—orthogonal terms
2—non-orthogonal terms

KEYWORDS

- **visualization**
- **collocated grid**
- **complex geometry**
- **finite volume method**
- **isotherms**
- **heatlines**
- **discrete heat source**

REFERENCES

1. Kimura, S.; Bejan, A. The Heatline Visualization of Convective Heat Transfer. *ASME: J. Heat Transfer* **1983,** *105*(4), 916–919.
2. Costa, V. A. F. Heatline and Massline Visualization of Laminar Natural Convection Boundary Layers near a Vertical Wall. *Int. J. Heat Mass Transfer* **2000,** *43*(20), 3765–3774.

3. Deng, Q.-H.; Tang, G.-F. Numerical Visualization of Mass and Heat Transport for Conjugate Natural Convection/Heat Conduction by Streamline and Heatline. *Int. J. Heat Mass Transfer* **2002,** *45*(11), 2373–2385.
4. Dalal, A.; Das, M. K. Heatline Method for the Visualization of Natural Convection in a Complicated Cavity. *Int. J. Heat. Mass. Tran,* **2008,** *51*(1–2), 263–272.
5. Basak, T.; Roy, S. Role of Bejan's Heatlines in Heat Flow Visualization and Optimal Thermal Mixing for Differentially Heated Square Enclosures. *Int. J. Heat Mass Transfer* **2008,** *51*(13–14), 3486–3503.
6. Basak, T.; Aravind, G.; Roy, S. Visualization of Heat Flow due to Natural Convection Within Triangular Cavities Using Bejan's Heatline Concept. *Int. J. Heat Mass Transfer* **2009,** *52*(11–12), 2824–2833.
7. Kaluri, R. S.; Basak, T. Analysis of Distributed Thermal Management Policy for Energy-Efficient Processing of Materials by Natural Convection. *Energy* **2010,** *35*(12), 5093–5107.
8. Kaluri, R. S.; Basak, T.; Roy, S. Heatline Approach for Visualization of Heat Flow and Efficient Thermal Mixing with Discrete Heat Sources. *Int. J. Heat Mass Transfer* **2010,** *53*(15–16), 3241–3261.
9. Biswal, P.; Basak, T. Sensitivity of Heat Function Boundary Conditions on Invariance of Bejan's Heatlines for Natural Convection in Enclosures with Various Wall Heatings. *Int. J. Heat Mass Transfer* **2015,** *89,* 1342–1368.
10. Jaya Krishna, D.; Basak, T.; Das, S. K. Numerical Study of Lid-Driven Flow in Orthogonal and Skewed Porous Cavity. *Int. J. Numer. Methods Biomed. Eng.* **2008,** *24,* 815–831.
11. Ghosh Roychowdhury, D.; Das, S. K.; Sundararajan, T. An Efficient Solution Method for Incompressible N-S Equations using Non-Orthogonal Collocated Grid. *Int. J. Numer. Methods Eng.* **1999,** *45,* 741–763.
12. Narasimha Suri, T.; Jaya Krishna, D. A Collocated Grid Based Finite Volume Approach for the Visualization of Heat Transport in 2D Complex Geometries. *Procedia Eng.* **2015,** *127,* 79–86.

CHAPTER 44

NUMERICAL SIMULATION OF HEAT TRANSFER IN LIQUID LITHIUM FLOW THROUGH ELLIPTICAL CYLINDER

NAGARJUNAVARMA GANNA[1,*] and GOVIND NANDIPATI[2,3]

[1]*Department of Mechanical Engineering, Anurag Engineering College, Kodad, Suryapet, Telangana, India*

[2]*Department of Mechanical Engineering, RVR and JC College of Engineering, Guntur, Andhra Pradesh, India*

[3]*govind.nandipati@gmail.com*

[*]*Corresponding author. E-mail: ganna609@gmail.com*

ABSTRACT

In this chapter, work done in open-source computational fluid dynamics (CFD) software OpenFOAM was used to study heat transfer phenomena in liquid lithium flowing through elliptical cylinder. The analysis was done for different aspect ratios and different magnetic inductions with constant wall temperature applied over the geometry. In addition, the magnetic field is imposed orthogonal to the temperature gradient. Thus, numerical results were obtained and results were compared for various aspect ratios and with changing magnitude of magnetic induction as well. The results have shown that heat transfer rate was increased with the increase in aspect ratio and also with the increase in magnetic induction.

44.1 INTRODUCTION

The scope of this chapter is at the crossroads between two important areas of research: magnetohydrodynamics (MHD) on the one hand and computational fluid dynamics (CFD) on the other hand. MHD is the branch of physics which studies the interaction between the flow of electrically conducting fluids and electromagnetic fields. According to Faraday's and Ohm's law, the presence of a magnetic field induces an electric current in a moving conductor.[1] Second, this current distribution is at the origin of an induced magnetic field. Finally, the interaction between the resulting magnetic field and current distribution will cause a body force, which affects the momentum balance of the conducting medium.[4,6,7] The investigation of flow and heat transfer parameters[3] of various fluids have been done by computational methods, which has already shown its strength in solving many flow problems. It has shown its potential in solving the flow problems of both Newtonian and non-Newtonian fluids. MHD applies to a large variety of phenomena. One can think of astrophysical or geophysical processes, like the spontaneous generation of the Earth's magnetic field by the motion of the liquid iron core of the Earth, but MHD flows can also be encountered in applications of more practical interest. The common feature of almost all these industrial and laboratory flows is that the coupling between the flow and the magnetic field is virtually one-way, that is, the magnetic field strongly affects the flow through the generation of a body force, but the flow does not act significantly upon the magnetic field. This regime is known under the name quasi-static MHD. Sometimes, the term liquid metal MHD is also used to refer to this regime. Historically, the first application of MHD concerned flow measurement techniques done by Faraday in 1832. In 1960, the magnetic fields began to be used as tools for flow control and generation. In continuous casting processes, a better quality of the product is achieved by applying a (static) magnetic field during the solidification process; it damps perturbations of the melt flow which may be caused by natural convection or the inflow of the melt. On the other hand, rotating magnetic fields can be used to enhance mixing by inducing nonintrusively a stirring motion in a liquid metal or electrolyte. Such an approach is preferred in configurations where the use of a mechanical mixer is impractical, for example, in high-temperature or corrosion-aggressive environments. Finally, rotating magnetic fields are also applied in heating or levitation processes. Some applications of MHD in materials processing can be found in the article by Davidson[5], where collection and comparison are also done.[8]

Apart from those applications where a magnetic field is imposed on purpose, there are situations in which there is an ambient magnetic field without specific intent for the flow. A notable example is so-called blankets for future thermonuclear fusion devices.[2] Their role is to absorb the energy released in the fusion reaction and transfer it subsequently to a power plant.[12] Moreover, they (blankets) should provide a shield against the neutron irradiation. Liquid metal alloys, like lead–lithium, are primary coolant materials, mainly because of their operability at high temperatures.[13] However, the flow of these media will be heavily affected by the presence of intense magnetic fields (up to 5–10 T) required to confine the plasma in the tokamak reactor.[14] The aim of the chapter is to get a relationship between applied magnetic field, flow and heat transfer parameters which are not estimated yet for wide ranges. Moreover, the information related to heat transfer rate with the applied magnetic field on MHD fluid is relatively very less.[9] It will be appropriate if the study of heat transfer phenomena in liquid lithium flowing through elliptical cylinder for different aspect ratios, uniform magnetic field applied over the geometry and magnetic field is applied alternatively parallel and perpendicular to the temperature gradient.

44.2 PROBLEM DESCRIPTION

Elliptical pipe models (0.15-m length, 0.01-m major axis) have been created for different aspect ratios (ε=major axis/minor axis) of 1, 1.2, and 1.4 (as shown in Figure 44.1). The magnitudes of magnetic field induction values are chosen as $B=0$ T, $B=0.05$ T, $B=0.1$ T, $B=0.15$ T for the examination on elliptical pipe. In the present work, the effect of applied magnetic field which is imposed orthogonal to the flow and aspect ratio on convective heat transfer was calculated under the time-independent flow situation of liquid lithium by maintaining surface wall unchanged on the uniform cross-sectional elliptical pipe for various heat fluxes.

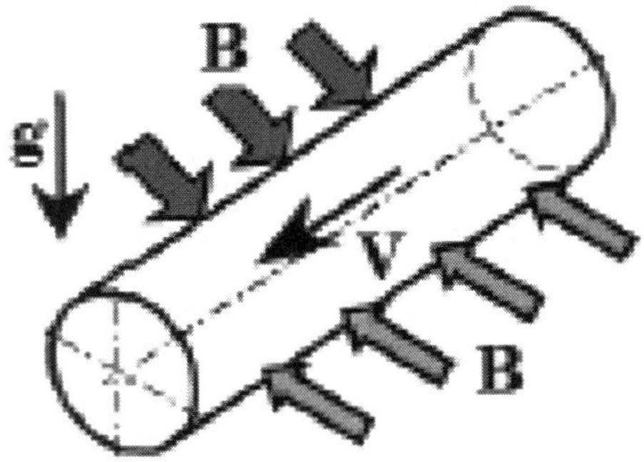

FIGURE 44.1 Physical model.

44.3 PROCEDURE

Salome-Meca 6.6.0 software was used to prepare the geometry and meshing. The abovementioned conditions, that is, time-independent, fully developed liquid lithium MHD flow through the horizontal elliptical pipe under the effect of orthogonally imposed magnetic field is studied. In order to acquire the highest impact, the magnetic field was imposed orthogonally to the flow direction. The applied magnetic field is along the radial direction and liquid metal is flowing along the axial direction. Moreover, the problem was solved and simulated using Open FOAM 2.1.1 open-source CFD software. Then, solved problem was analyzed in post-processing software ParaView.

44.4 GOVERNING EQUATIONS

Applying physical laws (conservation of mass, momentum, and energy) over the incompressible two-dimensional fluid elements, one can get following equations:

$$\frac{\partial u}{\partial x}+\frac{\partial v}{\partial y}=0 \tag{44.1}$$

$$u\frac{\partial u}{\partial x}+v\frac{\partial u}{\partial y}=V_{nf}\frac{\partial^2 u}{\partial y^2}+U_e\frac{dU_e}{dx}-(\frac{\sigma B_o^2}{\rho_{nf}})(u-u_e) \tag{44.2}$$

$$u\frac{\partial T}{\partial x}+v\frac{\partial T}{\partial y}=\alpha_{nf}\frac{\partial^2 T}{\partial y^2}+\frac{v_{nf}}{(\rho c_p)_{nf}}(\frac{\partial u}{\partial y})^2 \tag{44.3}$$

where u and v are the velocity components along the x-axis and y-axis, respectively, T is the temperature, σ is the electrical conductivity, v_{nf} is the effective kinematic viscosity of liquid lithium, and α_{nf} is the effective thermal diffusivity of liquid lithium.[11]

44.4.1 MAGNETIC-FIELD-RELATED EQUATIONS

Maxwell's equations are as follows:

$$\nabla\cdot B=0 \tag{44.4}$$

$$\nabla\times H=0, \tag{44.5}$$

where B is the magnetic induction and H is the magnetic field vector. Further, the magnetic induction, the magnetization vector, M, and the magnetic field vector are related constitutively:

$$B = \mu_0 (M + H) \tag{44.6}$$

where μ_0 is a magnetic permeability in vacuum. Magnetic scalar potential, $\varnothing_m$, is defined as:

$$H = -\nabla\varnothing_m \tag{44.7}$$

Using Maxwell's equations, the flux function for magnetic scalar potential, $\varnothing_m$ may be written as:

$$\nabla\cdot[(1+\frac{\partial M}{\partial H}\nabla\varnothing_m)] = \nabla\cdot[(\frac{\partial M}{\partial T}(T - T_0) + \frac{\partial M}{\partial \alpha_p}(\alpha_p - \alpha_{po}))] \tag{44.8}$$

where α_p, T, and subscript 0 represent the volume fraction of magnetic particles, temperature, and initial conditions, respectively. Within the simulations, $\frac{\partial M}{\partial H} = f$, $\frac{\partial M}{\partial T} = -\beta_m$, M_o, and $\frac{\partial M}{\partial \alpha_p}$ are assumed to be constant, and using Langevin equation, they can be defined.[10]

The dimensions of the elliptical tube are major diameter (a), minor diameter (b), the length (L), and hydraulic diameter (D_h), and aspect ratio=a/b.

Reynolds and Nusselt numbers have been calculated with respect to hydraulic diameter (D_h) as

$$D_h = 4 \times \text{cross-sectional area/wetted perimeter}$$

where D_h=hydraulic diameter of the elliptical tube

$$\text{Area}\ (A = \frac{\pi \times a \times b}{4})\ \text{and perimeter} = \pi\sqrt{\frac{a^2 + b^2}{2}}$$

$$\text{Re} = \frac{u_{av} D_h}{v}$$

$$\text{Nu} = \frac{hD_h}{k}$$

$$\text{Nondimensional average velocity} = V_x / u_{av}$$

$$\text{Nondimensional average temperature} = \frac{T - T_\infty}{T_0 - T_\infty}$$

44.5 RESULTS AND DISCUSSION

Nondimensional average velocities with different aspect ratios were compared and results are shown in Figure 44.2a. Moreover, it is observed that if the aspect ratio is improved, the average velocity is diminished.

Nondimensional average velocity under the influence of magnetic field induction for the values $B=0$ T, $B=0.05$ T, $B=0.1$ T, $B=0.15$ T are compared for the full pipe length at various positions. The results clarified that if the magnetic field induction is raised, the average velocity decreases (Fig. 44.2b).

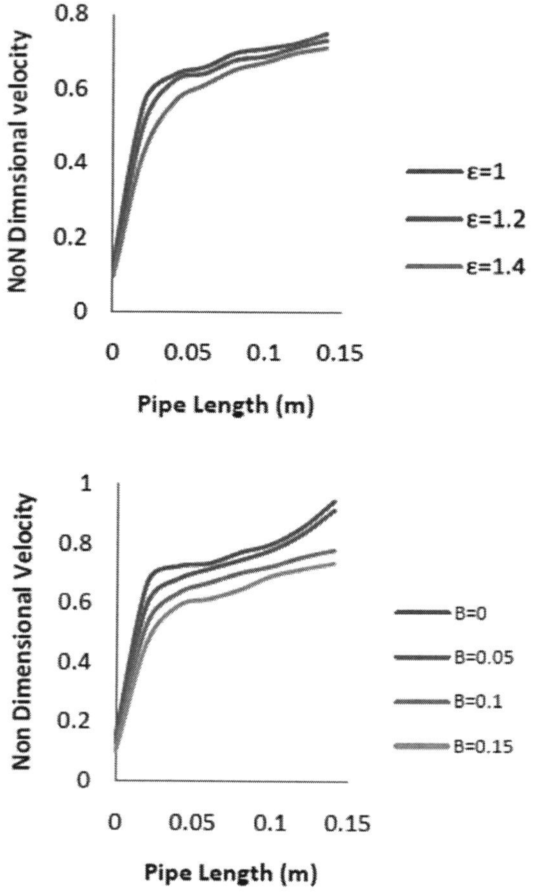

FIGURE 44.2 (a), (b)

Nondimensional temperature with different aspect ratios is compared and results are shown in Figure 44.3a. Moreover, it is observed that if the aspect ratio is increased, the non-dimensional temperature decreased.

Nondimensional temperature under the influence of magnetic field induction for the values $B=0$ T, $B=0.05$ T, $B=0.1$ T, $B=0.15$ T was compared for the full pipe length at various positions. The results showed that if the magnetic field induction is increased, the nondimensional temperature decreased (Fig. 44.3b).

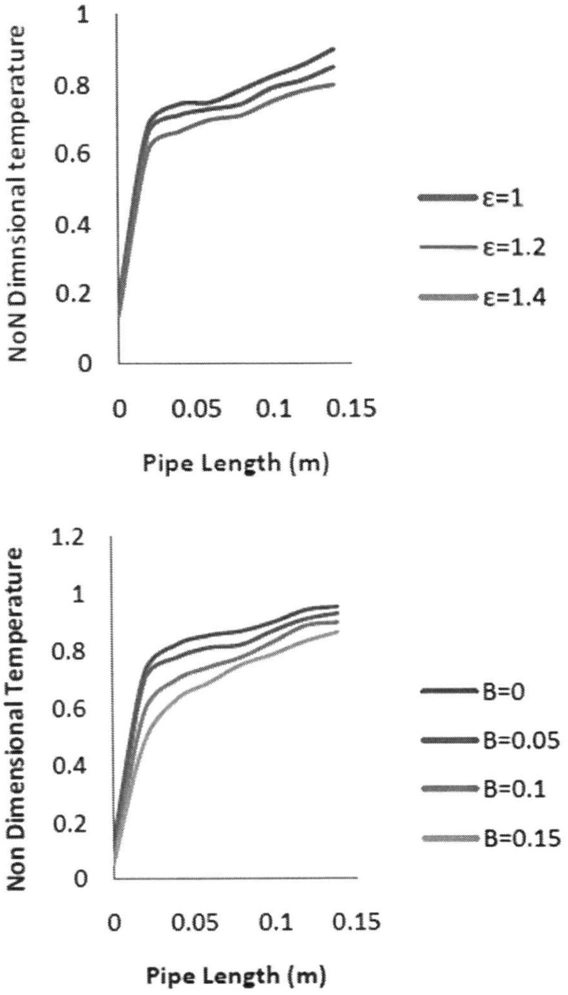

FIGURE 44.3 (a), (b)

In the same manner as explained in the above paragraph, the same type of investigation was preceded for the variation of average Nusselt number for full pipe length was calculated under the influence of orthogonally imposed magnetic field on the elliptical pipe for different aspect ratios. It was observed that the Nusselt number was enhanced by the improved magnetic field induction (as shown in Fig. 44.4).

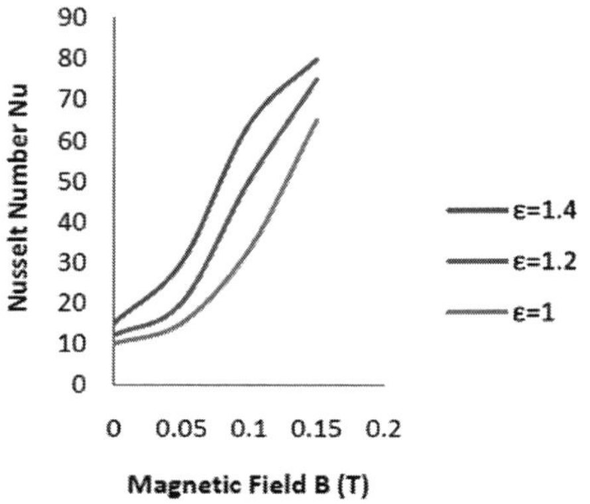

FIGURE 44.4

Same trend was observed for skin friction coefficient as well, that is, skin friction coefficient is improved by the raising of magnitude magnetic field induction for elliptical pipe flow (as shown in Fig. 44.5). This may be because the magnetic field resists the motion of fluid particles as it has the resisting influence on metallic particles.

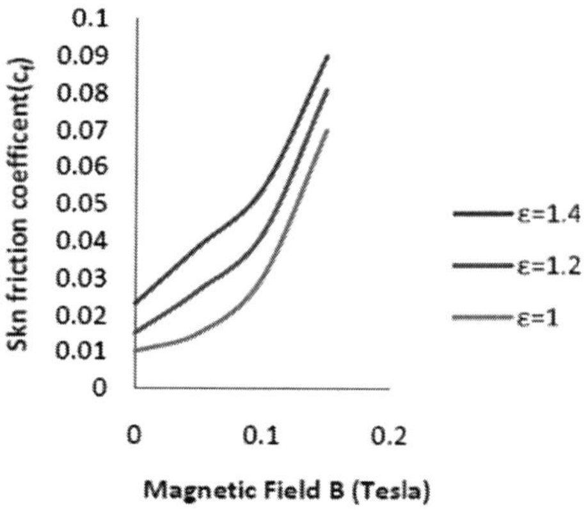

FIGURE 44.5

44.6 CONCLUSION

The dependence of the magnetic field induction on convective heat transfer rate in a uniform cross-sectional horizontal elliptical pipe flow of liquid lithium has been estimated computationally. In this chapter, the analysis is done by Open FOAM open-source CFD software. It is observed that by the raising the magnitude of magnetic field induction on liquid lithium flow, the Nusselt number increases. In every case, the length of major axis is 0.01 m, but the length of minor axis is decreasing with the increase in aspect ratio, which means total cross-sectional area decreases as aspect ratio increases. It is already understood that Nusselt number is increased. From this, it is clear that though the size of cylinder decreases, its Nusselt number is increasing and convective heat transfer rate per unit surface area is also increasing because of magnetic field induction values. When liquid lithium is used, it is preferred to be elliptical cylinder rather than circular cylinder.

KEYWORDS

- **CFD simulation**
- **heat transfer**
- **liquid lithium**
- **elliptical cylinder**

REFERENCES

1. Albets-Chico, X.; Grigoriadis, D. G. E.; Votyakov, E. V.; Kassinos, S. Direct Numerical Simulation of Turbulent Liquid Metal Flow Entering a Magnetic Field. *Fusion Eng. Des.* **2013,** *88*(12), 3108–3124.
2. Buhler, L. Liquid Metal Magneto Hydrodynamics for Fusion Blankets. In *Magneto Hydrodynamics* (Fluid Mechanics and its Applications); Molokov, S., Moreau, R., Moffatt, H. K., Eds.; Springer: USA, 2007; Vol. 80.
3. Cengel, Y. A. *Heat and Mass Transfer*, 3rd ed.; McGraw Hill: USA, 2007.
4. Cowling, T. G. *Magnetohydrodynamics*. Interscience: New York, 1957.
5. Davidson, P. A. Magneto Hydrodynamics in Materials Processing. *Ann. Re. Fluid Mech.* **1999,** *31*, 273–300.
6. Davison, H. W. *Compilation of Thermo Physical Properties of Liquid Lithium*. National Aeronautics and Space Administration: Washington, 1968; p 4.

7. Gedik, E.; Kurt, H.; Recebli, Z.; Balan, C. Two-Dimensional CFD Simulation of Magnetorheological Fluid Between Two Fixed Parallel Plates Applied External Magnetic Field. *Comput. Fluids* **2012,** *63*, 128–134.
8. Hussein, A. M.; Bakar, R. A.; Kadirgama, K; Sharma, K. V. Heat Transfer Enhancement with Elliptical Tube Under Turbulent Flow TiO_2-Water Nanofluid. *Therm. Sci.* **2014,** *Serbia Issue 00,* 0–3.
9. Jafari, A. Simulation of Heat Transfer in a Ferrofluid Using Computational Fluid Dynamics Technique. *Int. J. Heat Fluid Flow* **2008,** *29*, 1197–1202.
10. Jafari, A.; Tynjälä T; Mousavi, S. M.; Sarkomaa, P. CFD Simulation and Evaluation of Controllable Parameters Effect on Thermo Magnetic Convection in Ferrofluids Using Taguchi Technique. *J. Comput. Fluids* **2008.** DOI: 10.1016/j.compfluid (accessed Dec 3, 2007).
11. Khan, Z. H.; Khan, W. A.; Qasim, M.; Ali Shah, I. MHD Stagnation Point Ferrofluid Flow and Heat Transfer Toward a Stretching Sheet. *IEEE Trans. Nanotechnol.* **2014,** *13*(1), 35–40.
12. Listratov, Ya.; Ognerubov, D.; Sviridov, E.; Zikanov, O.; Sviridov, V. Direct Numerical Simulations of Heat Transfer and Convection in MHD Liquid Metal Flow in a Pipe. *Magnetohydrodynamics* **2013,** *49*(1), 87–99.
13. Recebli, Z.; Selimli, S.; Gedik, E. Three Dimensional Numerical Analysis of Magnetic Field Effect on Convective Heat Transfer During the MHD Steady State Laminar Flow of Liquid Lithium in a Cylindrical Pipe. *Comput. Fluids* **2013,** *88*, 410–417.
14. Richardson, L. F. *Weather Prediction by Numerical Process*. University Press: Cambridge, 1922.

CHAPTER 45

PERFORMANCE ANALYSIS OF AN AIR WATER HEATER BY USING HFC REFRIGERANTS

ASHOK BHARATI[1,*] and JAGDEEP KSHIRSAGAR[1,2]

[1]*Department of Mechanical Engineering, Maharashtra Institute of Technology, Aurangabad, Maharashtra 431010, India*

[2]*jagdeep.kshirsagar@mit.asia*

[*]*Corresponding author. E-mail: ashokbharati14@gmail.com*

ABSTRACT

To meet the need of quality of hot water supply and to reduce the power consumption for generating a hot water, there is a requirement of a new system which is highly efficient, energy-saving, and eco-friendly. Again, refrigerant plays an important role in thermal cycle of heat pump. Hence, it is necessary to choose appropriate type of refrigerant with respect to application. It depends on various factors such as working condition, mass flow rate, temperature, and operating pressure of the refrigerant. Moreover, the properties of that particular refrigerant must be coinciding with the requirement. In this chapter, three refrigerants are considered: R22, R134a, and R410A to evaluate the performance of air water heater system working on standard vapor compression cycle. The performance of the refrigerants R22 and non-ozone-depleting R134a and R410A for high-temperature heat pump application is compared. While most reported works were conducted on residential heat pump with evaporating temperature below 0°C, this study considers higher evaporating temperature from 5 to 24°C. The coefficient of performance (COP) of heat pump at condenser temperatures 60 and 65°C is evaluated and compared for refrigerants R22, R134a, and R410A at different evaporating temperatures with different compressors. It is observed that the heating capacity of R410A is 16% higher than that of R134a and 5%

lower than that of R22. Moreover, it is observed that the COP of the system decreases with the increase in condenser temperature and increases with the increase in evaporator temperature for all the three refrigerants.

45.1 INTRODUCTION

The heat pump is not commonly known in a domestic environment, but they are available for domestic purposes. The primary objective of a heat pump is to extract or absorb heat and transfer it from the heat source to heat sink. The advantage of using a heat pump is that useful heat can be extracted from the air water or ground and then transported to heating applications. An air water heater (AWH) is a device which provides both heating and cooling or heating. The heat pump uses electricity to absorb heat from outside air then transfer it to heat sink in heating mode. Refrigeration and allied fields are going through tough time due to the strict norms established due to environmental concern. Refrigerant plays an important role in thermal cycle. It is observed that refrigerants, mainly halogenated refrigerants, are banned due to their harmful results. Research and development in the refrigerant study leads to probable alternatives for traditional refrigerants. Hot water production for low-temperature application using heat pump is the energy-efficient method to meet the need of quality of hot water required. Again to solve energy consumption problems, a use of the new system is required, which is energy-saving, environment-friendly, and highly efficient. An AWH system works on heat pump principle which transfers low-grade thermal energy from air to water by using heat pump cycle. If we compare electric, gas, or solar water heater to air-source water heater, then it is highly efficient, energy-saving, and environment-friendly.

In the present study, heat pump is developed, which is eco-friendly and has more efficiency than the conventional heating system. The work is focused on refrigerants R410A and R134a, which seem to be promising alternatives for R22 for low-temperature heating applications. Again, for comparative study, the experiment is performed on AWH system works on standard vapor compression cycle (VCC). Also, this system generates less CO_2 than conventional heating systems as it uses a renewable natural source of heat from atmospheric air. The performance of the vapor compression heat pump system is calculated on the basis of energy analysis. Energy analysis is based on the first law of thermodynamics. By using this energy analysis, the experimental coefficient of performance (COP) of AWH system is calculated at different temperatures and compared for refrigerants R22, R134a, and R410A.

45.1.1 AIR WATER HEATER (AWH) SYSTEM

- It is an instrument which absorbs the heat from atmospheric air and transfers to water through the mechanism for water heating using thermodynamic principle.
- Like solar water heater works on sunlight, AWH system works with environmental/atmospheric air, independent of sunlight.
- AWH system can extract heat even from low-temperature air, that is, subzero temperature.

45.2 LITERATURE REVIEW

Lee et al.[8] studied the R22 and R407C refrigerants to evaluate the heat transfer capacity of the condenser and to validate the simulation results. Calm[2] studied refrigerants R11, R12, R22, and R500 and observed that most of the current apprehension with refrigerant selections is increased nowadays. Engineers and building owners are involved in traditional chiller specifications. Cabello et al.[1] observed that R134a, R407C, and R22 have been the refrigerants used as working fluids. Saleh and Wendland[10] observed the replacement of the refrigerants R245fa for R11 and R245ca, and RE245 for the intermediate refrigerant R123. Moreover, the refrigerants R141b, RE170, R152a, and RC270 for R12, and propane and propylene were the replacements for R22. Jwo et al.[5] studied the measurements of refrigerating effect for home refrigerator by using R12, R134a, and hydrocarbon refrigerants (R290/R600a). Rocca and Panno[7] observed that the refrigerator used to test the behavior of the R22 substitutive fluids, having a semi-hermetic compressor. Gil and Kasperski[3] observed that the values of maximum COP are 0.422 of n-pentane (R601) and 0.385 of n-hexane (R602) in best simulation condition. Kasperski and Gil[6] observed that there is no hydrocarbon enabling a high value of entrainment ratio in a wide range of generator temperature. Jain et al.[4] studied refrigerants R22, R134a, R410A, R407C, and M20 and concluded that R407C is a potential hydrogen fluoride hydrocarbon (HFC) refrigerant replacement for new and existing systems presently using R22 with minimum investment. Tiwari and Gupta[11] observed that R404a is a refrigerant which provides better cooling capacity than the R134a. Rahhal and Clodic[9] studied refrigerants R290, R32, and R152a and observed that significant COP increase with mixtures mainly composed of R32 and R152a.

45.2.1 POTENTIAL STUDIES OF R407C, R410A, AND R134A

1. *R407C*: It is a blend of three non-ozone-depleting HFC refrigerants (R32, R125, and R134a). It gives a performance as close as to R22.
2. *R410A:* It is a blend of two HFC non-ozone-depleting refrigerants (R32, R125). It provides benefits in efficiency and system size. It exhibits higher pressures and refrigeration capacity than R22.
3. *Freon R134a*: It is a new refrigerant which is a hydrogen fluoride hydrocarbons refrigerant (referred to as HFC).

45.3 EXPERIMENTAL SETUP

The experimental work is carried out to analyze the performance of AWH system by using HFC refrigerant. The detailed analysis of AWH system by using different refrigerants is also carried out. The schematic diagram of the experimental setup used in present study is shown in Figure 45.1 and details are given in Table 45.1. The setup consists of evaporator coil, fan motor assembly, condenser, compressor, thermal expansion valve, switches, and so forth. Figure 45.2 shows the photographic view of experimental setup, consisting of water tank of 100-L capacity, hermetically sealed compressor, evaporator coil, condenser and expansion device, pressure gauges, thermometer, and energy meter.

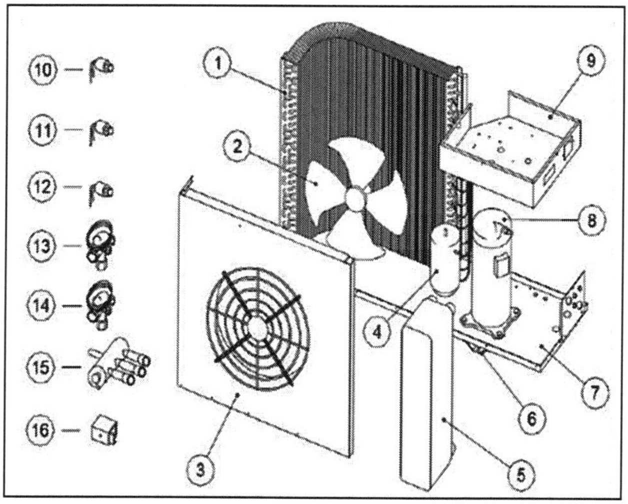

FIGURE 45.1 Schematic diagram of experimental setup.

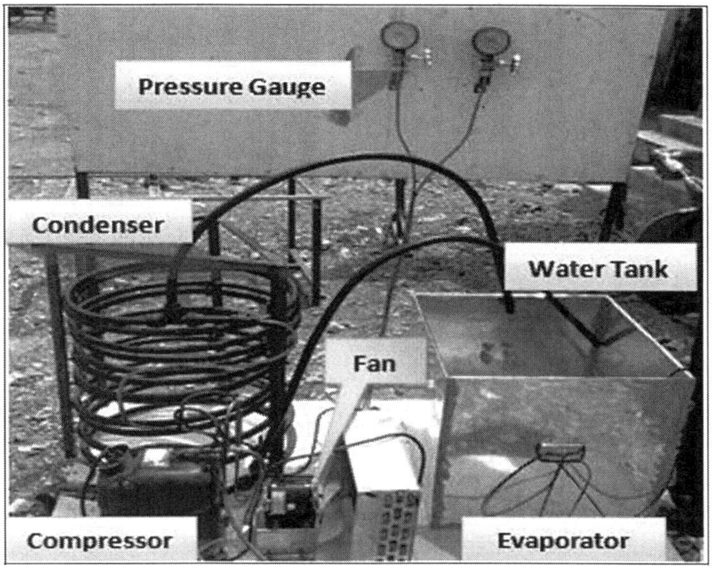

FIGURE 45.2 Photographic view of experimental setup.

TABLE 45.1 Details of Schematic Diagram of Experimental Setup in Figure 45.1.

Item	Description
1	Evaporator/outdoor coil
2	Fan/motor assembly
3	Front panel/fan guard assembly
4	Liquid receiver
5	Condenser/helical coil heat exchanger
6	Foot
7	Base plate
8	Hermetically sealed compressor
9	Electrics and controls panel
10	Refrigerant low-pressure switch
11	Refrigerant high-pressure switch
12	Water low-pressure switch
13	Heating thermal expansion valve
14	Defrost thermal expansion valve
15	Four-way reversing valve
16	Reversing valve solenoid

45.3.1 COMPRESSOR

The compressor is the most important component of AWH. It carries refrigerant in the vapor form at low pressure and temperature to a high pressure and temperature by compressing the refrigerant gas. The driving force used to run compressor is an electric motor, which is integrated into the compressor. In this experiment, the hermetically sealed compressor with model number KCJ511HAEB420 (230 V, 50 Hz) is used.

45.3.2 CONDENSER

A condenser is a device built for efficient heat transfer from one medium to another. In present experiment spiral coil type of condenser, inside copper tube covered by another outside tube of KiTEC material is used. In the condenser, heat transfer takes place from hot refrigerant flowing through the inside copper tube to the water flowing through the outside tube. KiTEC pipes with a pressure range of 13.8 kg/cm^2 at 23°C and 41.0 kg/cm^2 at 60°C are used. It has the thermal conductivity of 0.43 W/m · K, also because of smooth inner surface KiTEC pipe is scale-free and gives higher and consistent flow throughout the service life.

45.3.3 EXPANSION VALVE

Expansion valve is a device used to reduce the pressure of refrigerant which is coming out through condenser. It reduces the pressure from condenser pressure to evaporator pressure by expanding the refrigerant. It also controls the amount of refrigerant flow into the evaporator. In this experiment, the capillary is used as expansion device which is of copper material and having a length of 1.44 m and an inner diameter of 1.2 mm.

45.3.4 EVAPORATOR COIL

An evaporator is a device used to absorb sensible heat and latent heat of the ambient air and transfer them to the refrigerant, which flow inside the evaporator coil. As the liquid form of refrigerant comes in contact with evaporator, its phase change takes place by absorbing heat from air flowing over tubes. As the air passes over the evaporator coil, it becomes cooler than atmospheric air. In this experiment, we used the evaporator coil having 36 copper tubes with aluminum fins.

45.4 DATA REDUCTION

The experimental data of temperature, pressure, water temperature at different time intervals are recorded and used to calculate the performance of AWH system. Using the recorded data of operating pressure and temperatures, the values of enthalpy at inlet and outlet of each component are measured. The performance of AWH system is calculated by using following equations:

$$(COP)_R = \frac{\text{Refrigerating effect}}{\text{Compressor work}} = \frac{h_1 - h_4}{h_2 - h_1} \quad (45.1)$$

$$(COP)_{HP} = \frac{\text{Heating effect}}{\text{Compressor work}} = \frac{h_2 - h_3}{h_2 - h_1} \quad (45.2)$$

$$(COP)_{HP} = (COP)_R + 1 \quad (45.3)$$

where,

h_1 = Enthalpy at compressor inlet (kJ/kg)
h_2 = Enthalpy at compressor outlet (kJ/kg)
h_3 = Enthalpy at condenser outlet (kJ/kg)
$h_4 = h_3$ = Enthalpy at evaporator inlet (kJ/kg)

45.5 RESULTS AND DISCUSSION

The experimental results are shown in this section and the comparison of COP of AWH at different evaporator and condenser temperatures is done for all the refrigerants, R22, R134a, and R410A. When alternative refrigerants are considered to replace a reference refrigerant, first the COP of alternative refrigerants is compared with reference refrigerant. The values of COP of alternative refrigerants should be equal or higher than that of reference refrigerant. In the present study, R22 is used as a reference refrigerant. Refrigerant R134a is a lower capacity and lower pressure alternative refrigerant for R22. Refrigerant R410A is a higher capacity and higher pressure alternative refrigerant for R22. It is used in the market for more than 10 years, and now it is a leading HFC refrigerant for replacing R22 in water heater systems.

Therefore, in the present study, R410A is also used as an alternative refrigerant. This experiment is performed to evaluate and compare the COP of the system at different evaporator and condenser temperature and results are shown below for refrigerants R22, R134a, and R410A. The experiment is performed on experimental test rig of AWH system by using three different refrigerants.

Figure 45.3 shows the effect of evaporator temperature on the COP of AWH system for refrigerants R22, R134a, and R410A. The influences of the evaporator temperature on the performance of the AWH system are predicted by keeping mass of refrigerant and condenser temperature (60°C) constant. It is observed that the COP of AWH system increases as evaporator temperature increases for all the three refrigerants. Simultaneously, it is clear that increase in evaporator temperature leads to decrease in compressor work.

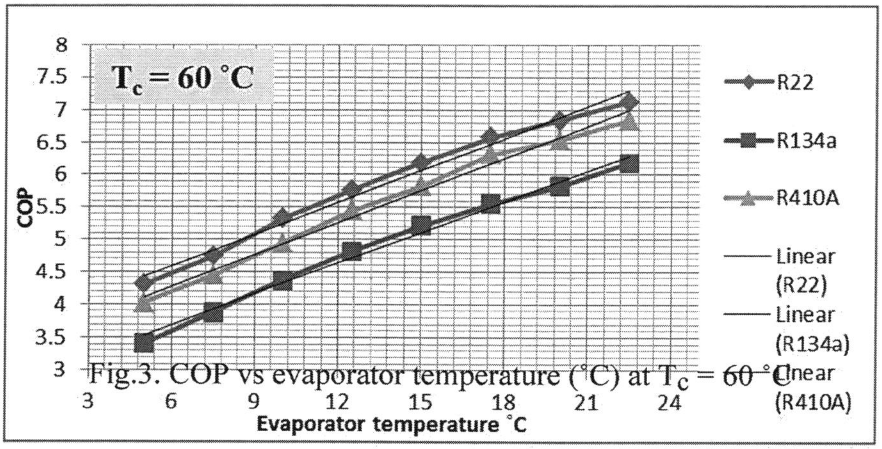

FIGURE 45.3 Graph of coefficient of performance (COP) versus evaporator temperature (°C) at T_c=60°C.

A plot of COP versus evaporator temperature of AWH system is shown in Figure 45.4 when R22, R134a, and R410A are used as the refrigerants. Clearly, the COP increased as evaporator temperature increased at constant condenser temperature of 65°C. Keeping condenser water flow rate and refrigerant flow rate constant, the variations in COP at constant condenser temperature is calculated by considering standard VCC. It is clear that the COP of AWH system decreases as condenser temperature increases for all the three refrigerants.

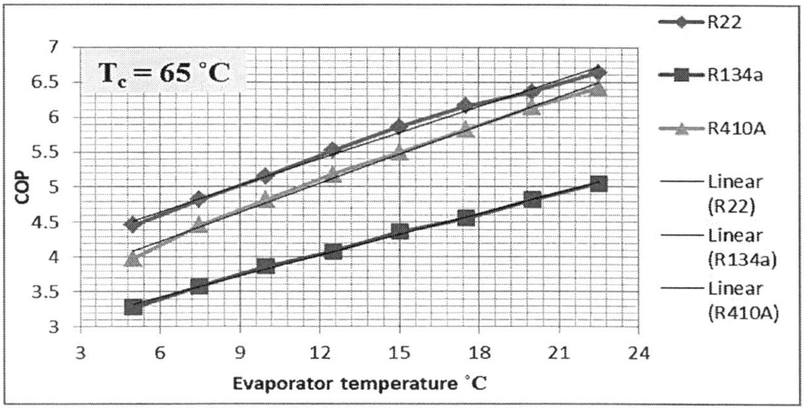

FIGURE 45.4 Graph of COP versus evaporator temperature (°C) at T_c=65°C.

The effect of evaporator temperature on the performance of AWH system by using actual VCC system at 65°C condenser temperature and different refrigerants is shown in Figure 45.5. Clearly, the COP of AWH system increases as the evaporator temperature increased. Moreover, it is clearly seen that the COP of AWH system is calculated by considering subcooling and superheating. Figure 45.5 also shows the combined effect of subcooling and superheating on COP of AWH system at different evaporator temperatures and constant condenser temperatures. However, subcooling increases the COP of AWH system, whereas superheating decreases COP of AWH system.

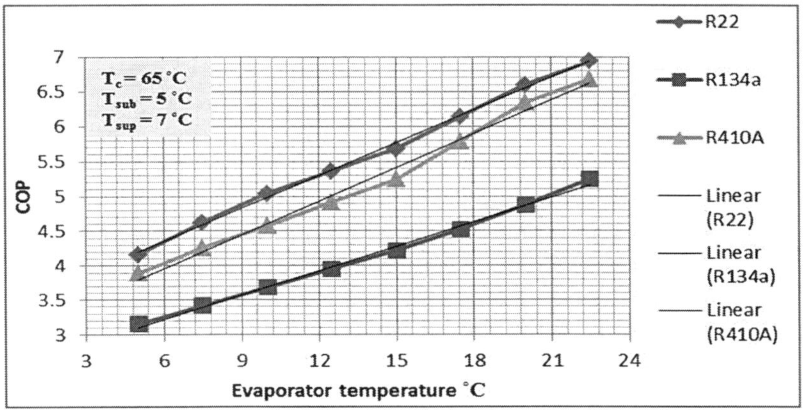

FIGURE 45.5 Graph of COP versus evaporator temperature at T_c = 65°C.

Figure 45.6 shows the variations in COP versus evaporator temperature for refrigerants R22, R134a, and R410A. It shows that COP increases as evaporator temperature increases and condenser temperature decreases after predicting 5°C subcooling and 7°C superheating. It shows that the decrease rate of COP for R22 and R410A is closer as compared to R134a. The subcooling increases the heating effect with the COP without affecting the compressor work. Similarly, the superheating increases the compressor work due to a decrease in vapor volume inside the compressor. In the present study, the overall COP increases with an increase in the evaporator temperature.

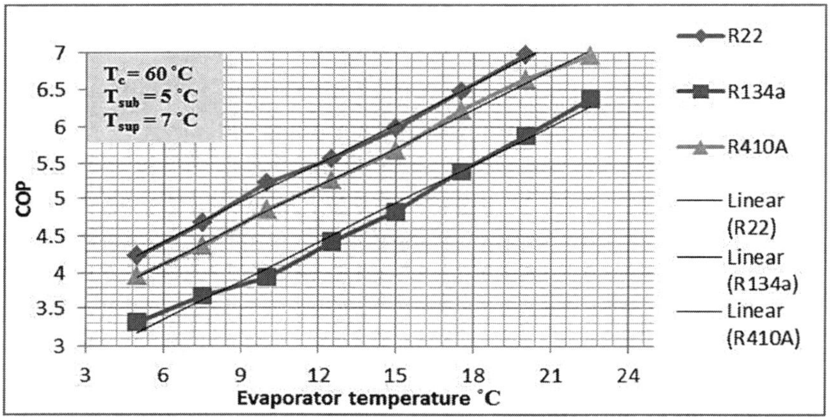

FIGURE 45.6 Graph of COP versus evaporator temperature at T_c = 60°C.

Figure 45.7 shows the effect of condenser water inlet temperature at 60°C condenser temperature on COP of AWH system by using R22, R134a, and R410A as the refrigerants. The results show that COP decreases with increasing condenser water inlet temperature, as the condenser water inlet temperature increases, the heat transfer rate between the refrigerant and hot water decreases. In the present experiment, the variations in condenser water inlet temperature are recorded when heating the 100-L water from initial temperature of water to the set temperature for all three refrigerants by keeping other parameters constant. It is clear that the COP of AWH system decreases as the condenser water inlet temperature increases.

Figure 45.8 COP versus condenser water inlet temperature (°C) for refrigerants R22, R134a, and R410A shows that the COP of AWH system decreases by a rise in condenser water inlet temperature and condenser temperature for all the three refrigerants used in the experimental setup.

The COP for refrigerants R410A and R22 are closer as compared to R134a refrigerant. The increase in condenser water inlet temperature leads to an increase in condenser temperature and the COP of AWH system decreases. It is clear that COP decreases with raise in condenser water inlet temperature as heat transfer rate decreases.

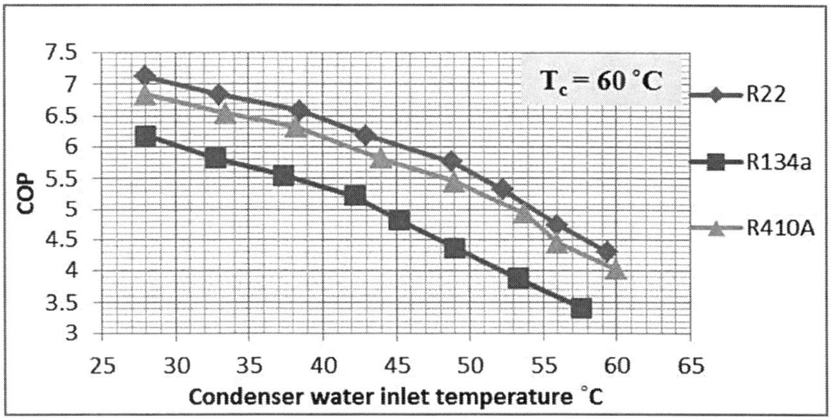

FIGURE 45.7 Graph of COP versus condenser water inlet temperature (°C) at T_c=60°C.

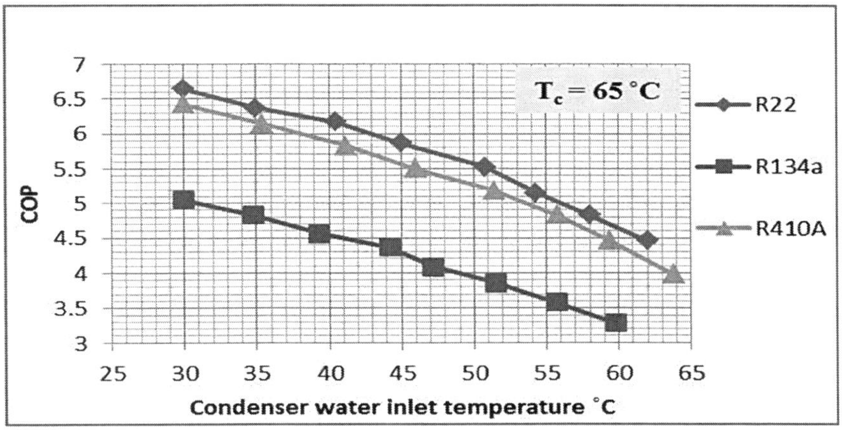

FIGURE 45.8 Graph of COP versus condenser water inlet temperature (°C) at T_c=65°C.

A graph of water temperature inside tank versus time for refrigerants R22, R134a and R410A is shown in Figure 45.9. Clearly, it shows that the time required for heating 60 L of water from initial temperature to 65°C for all the three refrigerants used in the experimental setup.

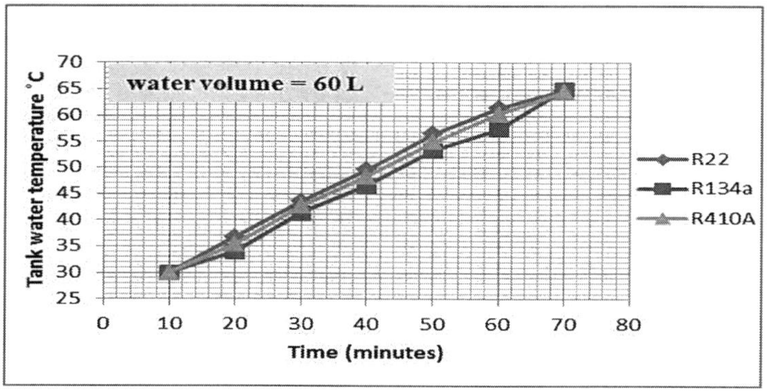

FIGURE 45.9 Tank water temperature (water volume = 60 l) versus time.

Figure 45.10 shows the effect of an increase in water volume on time period for heating 80 L of water from initial temperature to 65°C for refrigerants R22, R134a, and R410A. In the present study, the variation of the time period for making of 80 L hot water as a function of hot water temperature inside a tank. Hence, in this way, we conducted the experiment by using three different refrigerants, R22, R134a, and R410A, and plotted their results on the graph as shown in Figures 45.3–45.10 and their discussion is also given below each and every plot for better understanding of the experimental setup and its results.

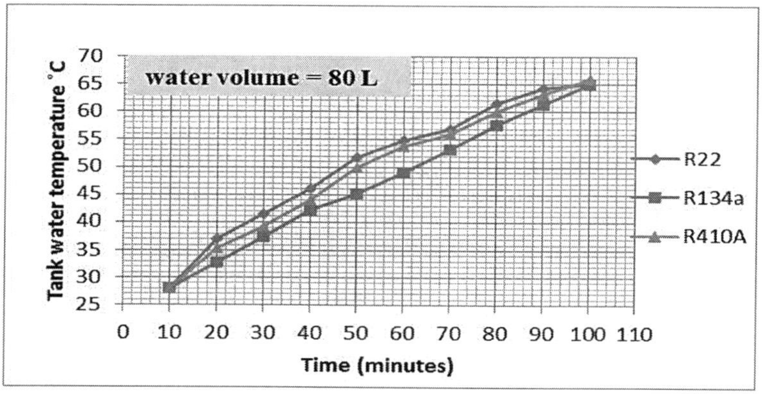

FIGURE 45.10 Tank water temperatures (water volume = 80 l) versus heating time.

45.6 CONCLUSION

The present experiment is performed to compare variations in COP of heat pump using refrigerants R22, R134a, and R410A. From the results, it is concluded that the COP of R22 is 5% greater compared with R410A and 20% greater compared with R134a for low-temperature heat pump applications. However, R22 is environmentally dangerous to use for AWH system due to higher global warming potential and ozone depletion potential values. Again, it is observed that the COP of refrigerant R410A is 16% higher compared with R134a and 5% lower compared with R22. Moreover, the COP decreases with a rise in hot water temperature and condenser temperature. From the above results, it is concluded that R410A is most suitable HFC refrigerant for replacing R22 with minimum investment and efforts. Therefore, the refrigerant R410A having COP 3.92 is more suitable for water heating applications.

NOMENCLATURE

R22—Chlorodifluoromethane ($CHClF_2$)
R134a—Tetrafluoroethane ($C_2H_2F_4$)
COP—Coefficient of performance
AWH—Air water heater
$(COP)_R$—Coefficient of performance of the refrigerator
VCC—Vapor compression cycle
$(COP)_{HP}$—Coefficient of performance of heat pump
h_1—Enthalpy of saturated vapor at evaporator outlet (kJ/kg)
h_2—Enthalpy of superheated vapor at compressor outlet (kJ/kg)
h_3—Enthalpy of saturated liquid at condenser outlet (kJ/kg)
h_4—Enthalpy of liquid–vapor mixture at evaporator Inlet (kJ/kg)
T_c—Condenser temperature (°C)

KEYWORDS

- **refrigerants**
- **COP**
- **condenser temperature**

- **evaporator temperature**
- **energy efficiency**
- **heat pump**

REFERENCES

1. Cabello, R.; Torrella, E.; Navarro-Esbr, J. Experimental Evaluation of a Vapour Compression Plant Performance Using R134a, R407C and R22 as Working Fluids. *Appl. Therm. Eng.* **2004,** *24*, 1905–1917.
2. Calm, J. M. Options and Outlook for Chiller Refrigerants. *Int. J. Refrig.* **2002,** *25*, 705–715.
3. Gila, B.; Kasperskia, J. Performance Analysis of a Solar-Powered Ejector Air-Conditioning Cycle with Heavier Hydrocarbons as Refrigerants. *Energy Procedia* **2014,** *57*, 2619–2628.
4. Jain, V.; Kachhwaha, S. S.; Mishra, R. S. Comparative Performance Study of Vapour Compression Refrigeration System with R22/R134a/R410A/R407C/M20. *Int. J. Energy Environ.* **2011,** *2*, 297–310.
5. Jwo, C.-S.; Ting, C.-C.; Wang, W.-R. Efficiency Analysis of Home Refrigerators by Replacing Hydrocarbon Refrigerants. *Measurement* **2009,** *42*, 697–701.
6. Kasperski, J.; Gil, B. Performance Estimation of Ejector Cycles Using Heavier Hydrocarbon Refrigerants. *Appl. Therm. Eng.* **2014,** *71*, 197–203.
7. La Rocca, V.; Panno, G. Experimental Performance Evaluation of a Vapour Compression Refrigerating Plant when Replacing R22 with Alternative Refrigerant. *Appl. Energy* **2011,** *88*, 2809–2815.
8. Lee, J. H.; Bae, S. W.; Bang, K. H.; Kim, M. H. Experimental and Numerical Research on Condenser Performance for R22 and R407C Refrigerants. *Int. J. Refrig* **2002,** *25*, 372–382.
9. Rahhal, C.; Clodic, D. Method of Choice of Low TEWI Refrigerant Blends. *International Refrigeration and Air Conditioning Conference*, Purdue University, July 17–20, 2006.
10. Saleh, B.; Wendland, M. Screening of Pure Fluids as Alternative Refrigerants. *Int. J. Refrig.* **2006,** *29*, 260–269.
11. Tiwari, A.; Gupta, R. C. Experimental Study of R-404a and R-134a in Domestic Refrigerator. *Int. J. Eng. Sci. Technol. (IJEST)* **2011,** *3*, 6390.

CHAPTER 46

AQUA-AMMONIA-BASED GUI SOFTWARE FOR DESIGN OF A VAPOR ABSORPTION REFRIGERATION SYSTEM UTILIZING AUTOMOBILE WASTE HEAT FOR AIR-CONDITIONING PURPOSES

MONU KUMAR[1,2] and AMIT SHARMA[1,*]

[1]*Department of Mechanical Engineering, DCR University of Science and Technology, Murthal, Sonipat, Haryana 131039, India*

[2]*kumarmonu991@gmail.com*

[*]*Corresponding author. E-mail: amitsharma.me@dcrustm.org*

ABSTRACT

This work presents the thermal design software of vapor absorption refrigeration system (VARS) utilizing the waste heat of an automobile's engine. Aqua-ammonia ($H_2O–NH_3$) is used as working fluids. This is a utility software, which gives the thermal design of all the components of VARS, that is, absorber, generator, heat exchangers, condenser, and evaporator. This has graphical user interface features to make it user-friendly so that practitioners/students/engineers can embark upon the design study of ($H_2O–NH_3$) VARS for air-conditioning purposes utilizing waste heat of the vehicle/ automobile. It is web-based software to increase the usage.

46.1 INTRODUCTION

The commonly utilized method of cooling which is used in vehicles is the vapor compression (VC) cycle, but the refrigerants used in VC refrigeration systems are usually hydrocarbons such as hydrofluorocarbons and hydrochlorofluorocarbons, which are not environmentally friendly, resulting in unwanted changes in the atmosphere and environment such as ozone layer depletion, global warming, and so forth; also the air-conditioning (AC) system applies more load, thus drawing more power from the engine shaft to operate the compressor of the VC system. This also leads to adverse environmental changes resulting from more energy consumption (fossil fuel) and pollutant emission.[1]

The vapor absorption refrigeration system (VARS) for automobiles addresses this concern by lowering the temperature of a small space inside the vehicle by utilizing the waste heat of engine exhaust gases.[2] It is known that an internal combustion (IC) engine has an efficiency of about 35–40%,[3] which means that only one-third of the energy produced by the combustion of the fuel is converted into useful work, that is, into mechanical output and about 60–65% of the energy in the form of heat is lost in the environment. In which about 28–30% is lost by lubrication losses and coolant, around 30–32% is lost through exhaust gases from the exhaust pipes and the remainder of the energy is lost by radiation and convection. In a VARS, the heat required for running the system can be obtained from the energy which is wasted into the atmosphere from the IC engine.[4,5,6,7] Hence, to utilize the waste heat (exhaust gasses heat and coolant heat) from an engine, the vapor absorption refrigerant system can be used, which increases the overall efficiency of a car.

This work presents the working operation of a graphical user interface (GUI)-based software which designs the various components as shown in Figure 46.1 of a VAR cycle. This chapter depicts/explains the working of software developed by the authors, which tries to provide a utility helpful for designers/new entrants/students in the field of design of VARS for automobile AC purposes utilizing the waste heat of an engine exhaust.

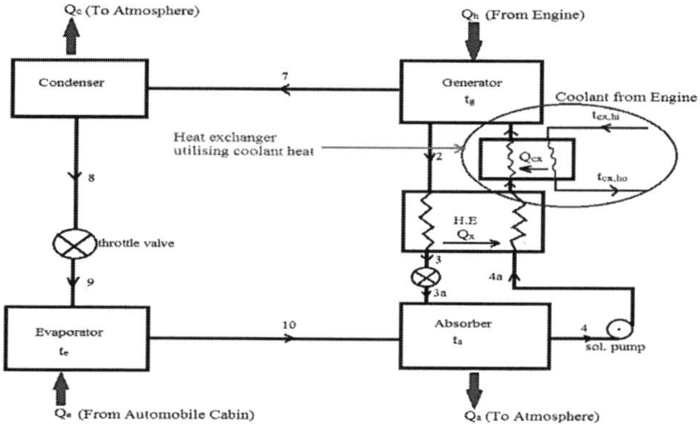

FIGURE 46.1 System utilizing coolant and exhaust heat from engine.

46.2 GRAPHICAL USER INTERFACE-BASED SOFTWARE

The software has been developed as per the design modeling of a VARS-based automobile AC, utilizing waste heat of the engine exhaust and coolant with the help of two softwares, that is, Microsoft visual studio and Microsoft structured query language (SQL) server. Programming has been done in C-Sharp language having ASP.NET environment and the database (properties tables, etc.) has been saved in Microsoft SQL server with a connectivity of ADO.NET.

The software is web-based to increase its usage. Hence, there are no requirements for minimum computer specifications; anyone can access it online from anywhere and for the link and accessing the software, contact the author of this chapter.

46.2.1 ASSUMPTIONS

Common assumptions observed are as follows:

- Thermophysical properties have been taken at mean temperature.[8,9]
- Single- and double-phase fluid (i.e., vapor and liquid refrigerant).
- Tube material is carbon steel (because ammonia reacts with copper).[5,10]
- No water vapor formation takes place before condenser (for design purpose).
- Effectiveness of every heat exchanger has been taken as 0.6 (as per requirement of the system).[9,11]
- Specific heat capacity taken at various temperature ranges from properties tables.[12]

46.2.2 WORKING OF THE SOFTWARE

This is a utility software, giving the thermal design of all the components of VARS, that is, absorber, generator, heat exchangers, condenser, and evaporator. This software provides facility to the user to work in the International System of Units (SI) or foot–pound–second units. In this work, illustrations are in SI units.

46.2.2.1 INPUT VALUES WINDOW

First, various requirements for the system need to be fed or input. Figure 46.2 is an input value window in which we are required to input the value of the generator temperature (T_g), exhaust gasses temperature at the inlet to the generator (T_{ei}), evaporator temperature (T_e) which we want to maintain, absorber temperature (T_a) corresponding to cooling fluid temperature, high pressure (p1), low pressure (p2), and tonnage of AC (TR) as shown in Figure 46.2.

FIGURE 46.2 Input values window.

This window also shows design pattern options, required for the number of tubes passes and the number of tubes (N) as a requirement for various components desired; in flowchart option, one can upload the overall flow diagram of this system utilizing the waste heat of an engine and this diagram shows the full algorithm for software programming; in table values option, one can get all the properties tables used in the software for aqua-ammonia system.

Working of the software is explained by various conditions required as input to the input window given in Table 46.1. The number of tubes passes (N) as 10 for all components except the generator where it is 16. Temperature of exhaust gasses (T_{ei}) is taken as 536°C, internal conditioned air temperature ($T_{e,ho}$) set as 22°C, internal diameter of generator as 0.074 m, internal diameter of generator's tubes as 0.0096 m, lateral pitch as 0.017 m, mass flow rate of exhaust gasses as 0.03386 kg/s, and coolant temperature as 110 C.[13]

TABLE 46.1 Various Input Conditions.

Parameters	Values
Generator temperature, T_g	130°C
High side pressure, p_1	20 bar
Low side pressure, p_2	2 bar
Absorber temperature,[1] T_a	35°C
Air-conditioner (AC) capacity (tonnage)	1.5 TR
Evaporator temperature, T_e	5°C

Source: Adapted from ref 10.

By using these input conditions in the input window and submitting it, one can get the design results for all the components of VARS utilizing waste heat of an automobile engine as summarized in Table 46.2. Table 46.2 results are the compilation of results of design pattern window or output window of the software as shown in Figures 46.3 and 46.4.

TABLE 46.2 Design Parameters at Given Input Conditions.

Design parameters	Condenser	Evaporator	Absorber	Liquid–liquid heat exchanger	Coolant heat exchanger	Generator
Q (kJ/kg)	1680	1340	1947	4340	567	2287
LMTD (°C)	99.7	20.7	11.2	39.62	11.6	264
L (m)	0.343	0.285	0.815	0.323	0.341	0.092
D (m)	0.004	0.007	0.096	0.216	0.228	0.074

46.2.2.2 OUTPUT WINDOW

In Figure 46.3, all tabs shown are the components of the VARS. One can get the design parameters of these components by opening these tabs, as shown in Figure 46.4, which shows an open tab of the generator.

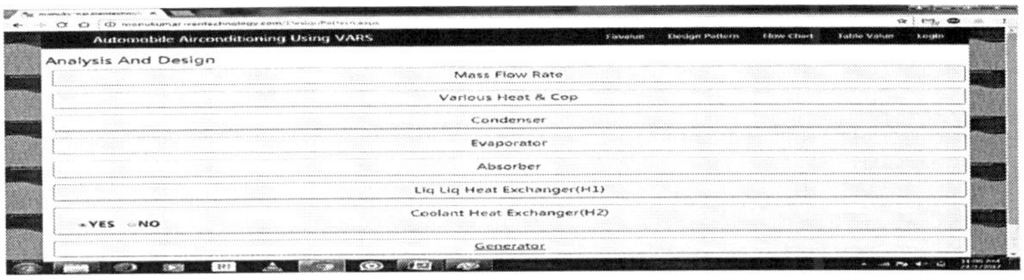

FIGURE 46.3 Output window or design pattern.

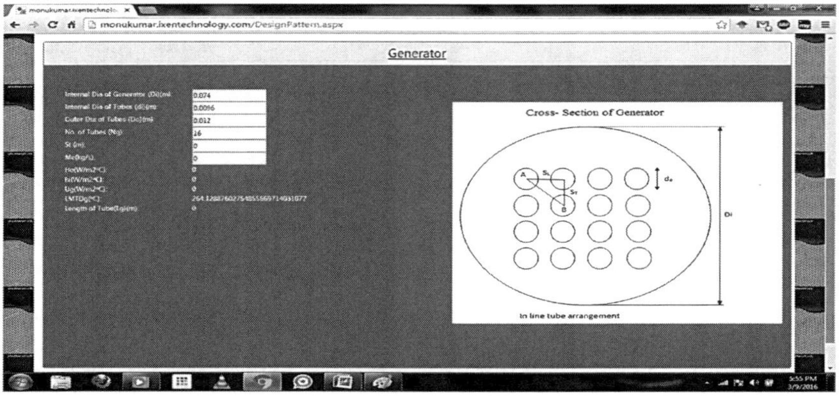

FIGURE 46.4 Output window with open generator tab.

46.3 CONCLUSIONS

The development of design software for a VARS-based car AC system by utilizing waste heat of an engine has been presented in this work. The following conclusions can be drawn from the results presented in this study:

- The system finds application in a variety of fields for providing the most economical and efficient cooling as it uses the waste heat energy.
- As ammonia is an eco-friendly refrigerant that has zero ozone-depletion potential and global warming potential, it has no environmental impact.
- Low-grade energy or waste energy of an engine for automobile AC has been utilized, thus improving the efficiency of the automobiles' engine.
- Sizing of an automobile AC can be determined based on the aqua-ammonia refrigeration system utilizing waste heat of an engine.
- Design parameters can be calculated with the help of GUI-based software easily.

KEYWORDS

- **design software**
- **vapor absorption refrigeration**
- **air-conditioning**

REFERENCES

1. Horuz, I. An Alternative Road Transport Refrigeration. *Int. J. Eng. Environ. Sci.* **1998,** *22*, 211–222.
2. Horuz, I. Vapor Absorption Refrigeration in Road Transport Vehicles. *J. Energy Eng.* **1999,** *125*, 48–58.
3. Sai Lavanya, R.; Murthy, B. S. R. Design of Solar Water Cooler Using Aqua Ammonia Absorption Refrigeration System. *Int. J. Adv. Eng. Res. Stud.* **2013,** *2*, 20–24.
4. http://4mechtech.Blogspot.In/2013/12/Practical-Absorption-System.Html (n.d.) (accessed April 21, 2016).
5. Ballaney, P. L. *Refrigeration and Air Conditioning*, 16th ed.; Khanna Publishers: Delhi, 2015; pp 483–528.
6. Stockers, W. F. *Design of Thermal Systems;* 3rd ed. McGraw Hill: New York, 1989.
7. http://kids.britannica.com/elementary/art-87059/Many-parts-work-together-to-make-an-automobile-run (accessed April 22, 2016).
8. Horuz, I.; Callander, T. M. S. Experimental Investigation of a Vapor Absorption Refrigeration System. *Int. J. Refrig.* **2004**, *27*, 10–16.
9. Balaji, K.; Senthil Kumar, R. Study of Vapour Absorption System Using Waste Heat in Sugar Industry. *IOSR J. Eng.* **2012,** *2*, 34–39.
10. Arora, C. P. *Refrigeration and Air Conditioning*, 3rd ed.; McGraw Hill Education (India) Private Limited: India, 2017; pp 402–420.
11. AlQdah, K.; Alsaqoor, S.; Al-Jarrah, A. Design and Fabrication of Auto Air Conditioner Generator Utilizing Exhaust Waste Energy from a Diesel Engine. *Int. J. Thermal Environ. Eng.* **2011**, *3*(2), 87–93.
12. Mathur, M. L.; Mehta, F. S. *Refrigerant and Psychrometric Properties;* Jain Brothers: New Delhi, 2014.
13. Amrutkar, P. S.; Patil, S. R. Automotive Radiator Sizing and Rating—Simulation Approach. *IOSR J. Mec. Civ. Eng.* 01–05. ISSN(e): 2278–1684, ISSN(p): 2320–334X.

CHAPTER 47

EXPERIMENTAL STUDY OF THE EFFECT OF TWO-PHASE (AIR–WATER) FLOW CHARACTERISTICS ON PIPE VIBRATION AT ATMOSPHERIC CONDITIONS

S. R. TODKAR[1,*], T. R. ANIL[2,*], U. C. KAPALE[3], and A. N. CHAPGAON[4]

[1]*Department of Mechanical Engineering, D. Y. Patil College of Engineering and Technology, Kolhapur, Maharashtra 416006, India*

[2]*Department of Mechanical Engineering, Gogte Institute of Technology, Belgaum, Karnataka 591236, India*

[3]*Department of Mechanical Engineering, S.G. Balekundri Institute of Technology, Belgaum, Karnataka 591236, India, E-mail: uday_kapale@hotmail.com*

[4]*Department of Mechanical Engineering, Ashokrao Mane Group of Institute, Vathar, Wadgaon, Kolhapur, Maharashtra, India, E-mail: ashok.chapgaon@gmail.com*

[*]*Corresponding author. E-mail: srtodkar@gmail.com, aniltr@gmail.com*

ABSTRACT

Flow-induced vibrations due to the two-phase flow (air–water) have been largely investigated earlier by many researchers. In this work, an experimental set up has been fabricated to study the effect of two-phase flow (air–water) characteristics on pipe vibration for the different support condition, that is, cantilever end condition and both end fixed condition. To carry out the experiments, the test section is made of steel material of 0.0239 m in diameter and 1.2 m in length. Experiments are carried out for different flow patterns or flow characteristics and void fraction. The flow characteristics are obtained by varying the quantity of air and water flowing through test section. At the same time, the vibrations induced due to two-phase flow (air–water) for each pattern are measured in terms of acceleration (rms) with the help of an accelerometer from fast Fourier transform analyzer. The results of experimental analysis are compared with those obtained from computational fluid dynamics (CFD) analysis (Fluent 15.0) software. Moreover, these results are found to be in good agreement and satisfactory with 4–10% error between experimental analysis and CFD analysis.

47.1 INTRODUCTION

The flow-induced vibration in pipe due to two phases of liquid and gas commonly occurs in piping systems of the nuclear and thermal power plant, chemical, and petroleum industries. These vibrations are of complex nature because these depend upon two-phase flow regime, that is, characteristics of two-phase flow and the void fraction.[8,2] Recently, flow-induced vibration due to two-phase flow causes accidents considerably resulting huge economic loss to enterprises Therefore, to ensure longer life and operate piping systems with the best efficiency, it is necessary to study the damage caused by flow-induced vibration and the criteria for assessing and preventing vibration.

From the literature review, it is clear that there are no quantitative measures to decide upon the existence of distinguished flow patterns. Therefore, it is decided to study the different flow patterns or regimes and the effect of these flow patterns on pipe vibration for different support conditions, that is, cantilever end and both end fixed for the first time.

47.2 METHODOLOGIES

The methodology is developed to obtain different flow patterns or flow characteristics by passing two-phase flow (air–water) through a test section.[3,7,9,15] Therefore, an experimental setup has been fabricated to obtain different flow regimes by controlling flow velocity of water and air simultaneously The vibrations induced due to two-phase flow are recorded using fast Fourier transform (FFT) analyzer.[4,6]

47.3 EXPERIMENTAL SETUP

Figure 47.1 shows the schematic diagram of the experimental setup to obtain different flow regimes.[2,9] It consists of a centrifugal pump (1.5 hp, 1440 rpm) which is used to pump the water through a test section from a water reservoir. The flow rate of water from the pump is measured by calibrated Venturi meter with the help of manometer. To have two phases flow (air–water), an air compressor (0–7 kg/cm^2, 2 hp, 2850 rpm) is used to provide compressed air at regulated outlet pressure using flow control valve. The flow rate of air is measured by calibrated orifice meter with the help of manometer. Both air and water are passed through a mixing section to form different flow patterns.[10,11] The viewing section is provided with glass view port of 0.5 m in length and 0.0256 m in diameter for observing different flow characteristics. The test section is made of steel material of 0.0239 m in diameter and 1.2 m length fixed at one end for cantilever condition and both end fixed condition. To measure induced vibrations in the test section, the accelerometer is mounted at the end of test section for cantilever end condition and output of accelerometer which is connected to single channel FFT analyzer (Adash 4400-VA4Pro). And similarly for both end fixed condition, the accelerometer is mounted at the center of test section and output of accelerometer is connected to single channel FFT analyzer. From

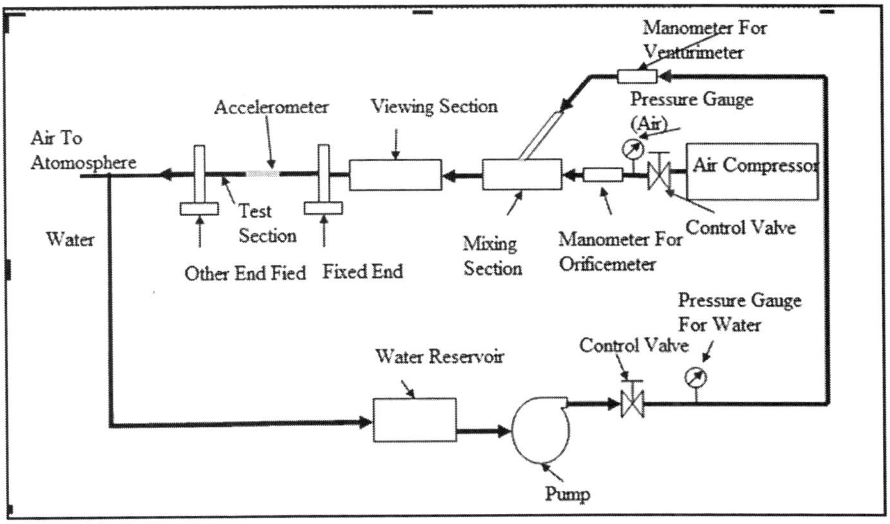

FIGURE 47.1 The schematic diagram of the experimental setup.

the velocity of water and air, corresponding volume flow rate of water and air, mass flow rate of water, and air and void fraction are calculated from the formula.[13] A similar procedure is followed for other flow patterns such as bubbly, elongated bubbly, stratified smooth, and stratified wavy which are obtained by varying simultaneously velocity of water and air as shown in Table 47.1. The corresponding flow patterns are observed and recorded using high-speed high-resolution camera.[2,5,15,18]

47.4 PARAMETERS

47.4.1 MEASUREMENTS OF INPUT PARAMETERS

Volume flow rate of air, Q_A, **m³/s:** An orifice meter is used to measure the volume flow rate of air. This device determines the superficial air velocity by measuring the manometer reading of water.

$$Q_A = C_{do} \times a_A \times \sqrt{2gh_A},\ m^3 / s,\ \text{where } a_A = \text{Area of orifice meter}^{13}$$

Volume flow rate of water Q_W, **m³/s:** A Venturi meter is used to measure the volume flow rate of water. This device determines the superficial velocity of water by measuring the manometer reading of mercury.

$$Q_W = C_{dv} \times a_w \times \sqrt{2gh_w},\ m^3 / s,\ \text{where } a_W = \text{Area of Venturi meter}^{13}$$

Measurements of output parameters: The output two-phase flow parameters to be measured are the mixture velocity of water and air and the void fraction.

Mixture velocity, $V_m = (V_A + V_W)$ (m/s): Mixture velocity is the sum of the superficial velocity of air and water.[7]

Measurement of vibrations: For measurement of flow-induced vibration due to two-phase flow (air–water), the accelerometer is mounted at the end of test section for cantilever end and at the center of test section for both end fixed end conditions. The output of accelerometer is connected to single channel FFT analyzer to measure displacement, velocity, and acceleration (RMS) induced in the pipe due to each flow patterns obtained by varying velocity of water and air, simultaneously.[4] For analysis, the graphs are plotted for mixture velocity of air and water against accelerations for cantilever end and both ends fixed conditions of test section to observe the effect of mixture velocity on pipe vibrations.

TABLE 47.1 Experimental and CFD Analysis.

Type of flow by visual inspection		Cantilever end condition						Both ends fixed		
Sr. no		**Mixture velocity Vm (m/s) expt**	**Mixture velocity, Vm (m/s) (CFD software)**	**% deviation mixture velocity**	**Accelertion a (m/s^2) CFD Software**	**Acceleration a (m/s^2) CFD Software**	**% deviation acc.**	**Acceleration a (m/s^2) expt (rms)**	**Acceleration (m/s^2) (rms) CFD Software**	**% deviation acc.**
1	Annular flow	9.46	**7.75**	18.08	1.47	**1.38**	6.12	4.86	**4.46**	8.23
2		17.79	16.30	8.39	1.30	1.28	1.54	2.33	2.46	5.58
3		26.55	25.30	4.69	2.20	2.18	0.91	3.56	3.43	3.65
4		28.77	27.50	4.40	2.50	2.38	4 80	2.21	2.29	3.62
5		28.90	27.63	4.39	2.48	2.18	12.10	3.19	3.28	2.82
6	Elongated bubble flow	7.94	**7.30**	8.06	1.80	**1.78**	1.11	0.70	**0.79**	13.29
7		8.10	6.57	18.92	0.60	0.57	5.00	0.79	0.90	13.92
8		8.62	8.44	2 05	0.45	0.41	8.39	0.62	0.70	12.90
9		8.76	8.26	5 69	0.46	0.42	8.70	0.91	1.02	12.09
10		8 76	8.26	5.69	0.46	0.42	8.70	0.62	0 70	12.90
11	Bubble flow	8.97	**7.35**	18.03	0.37	**0.39**	5.12	0.82	**0.85**	3.66
12		9.38	7.81	16.77	0.64	0.59	7.81	0.61	0.69	13.11
13		9.86	8.19	16.91	0.42	0.4	4 76	0.84	0.90	7.14
14		10.23	8.55	16.43	0.42	0.4	4.76	0.62	0.68	9.68
15		10.81	9.40	13.01	0.35	0.33	5.98	0 74	0.80	8.11

TABLE 47.1 *(Continued)*

Type of flow by visual inspection					Cantilever end condition			Both ends fixed		
Sr. no		**Mixture velocity Vm (m/s) expt**	**Mixture velocity, Vm (m/s) (CFD software)**	**% deviation mixture velocity**	**Accelertion a (m/s²) CFD Software**	**Acceleration a (m/s²) CFD Software**	**% deviation acc.**	**Acceleration a (m/s²) expt (rms)**	**Acceleration (m/s²) (rms) CFD Software**	**% deviation acc.**
16	Stratified wavy flow	11.62	**10.66**	8.28	0.65	**0.59**	9.23	1.49	**1.58**	6.04
17		12.42	12.22	1.58	1.13	1.05	7 08	1.65	1.71	3.64
18		13.71	13.18	3.86	1.12	1.05	6.25	1.97	2.05	4.06
19		15.62	15 29	2 10	0.62	0.59	4.22	1.15	1.17	1.74
20		16.46	16.18	1.70	1.39	1.21	12.95	0.98	1.04	6.12
21	Stratified smooth flow	5.68	**5.12**	9.87	0.46	**0.49**	6.52	1.41	**1.46**	3.55
22		8.12	8.01	1.36	0.42	0.4	4.08	0.36	0.40	11.11
23		10.34	10.44	0.99	0.57	0.53	7.34	1.15	1.20	4.35
24		11.06	10.44	5.62	0.59	0.55	6.78	1.27	1.33	4.72
25		12.58	12.46	0.94	0.30	0.27	10.00	1.38	1.45	5.07

*Comparison between the results of vibration (acceleration, c: CFD analyst's experimental analysis of two phase flow).

*The frequency of readmes is taken at the interval of 10 s.

47.5 COMPUTATIONAL FLUID DYNAMICS ANALYSIS

To validate the experimental results obtained, computational fluid dynamics (CFD) analysis is carried out using ANSYS Fluent 15.0 software. The working fluid considered is two-phase (air–water) at atmospheric condition. The 3D tube is modeled in ANSYS 15.0 meshing and exported to Fluent. The optimum mesh size is selected having nodes 30,804 and elements 20,779. The solver conditions are set for transient flow. The pressure-based solver with mixture model and transient structural model is used for fluid flow and structure interaction, respectively. Transient 3D planar simulation is activated along with absolute velocity formulation. For the entire range of flow rates, shear stress transport (SST) k-ω model is considered[1] using these solver conditions set, for each quality (void fraction), average properties are calculated for air and water. Thus, for every void fraction and flow patterns at each flow rate, the model is simulated to get the mixture velocity, void fraction, and acceleration data.

47.6 OBSERVATIONS

The results obtained from experimental test are those compared with CFD analysis software as shown in Table 47.1 and Figure 47.2.

The reading of experimental and CFD analysis is as shown in Table 47.1.

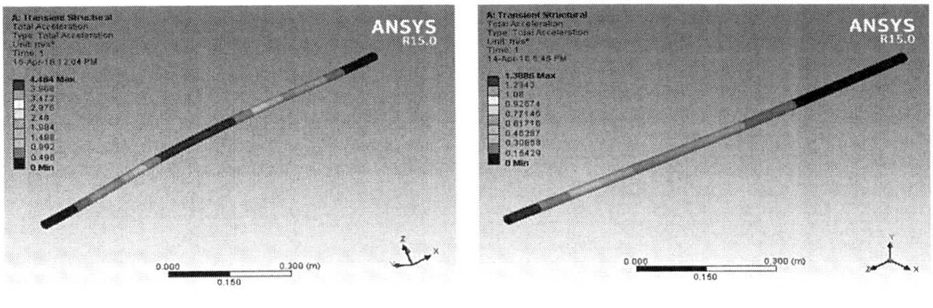

Max.Acceleration-4.464 m/s^2 Max.Acceleration-1.3886 m/s^2

FIGURE 47.2 CFD analysis of annular flow in terms of acceleration for both ends fixed (maximum acceleration 4.464 m/s^2) and cantilever end (maximum acceleration 1.3886 m/s^2).

The graphs of comparison between the experimental and CFD analysis are as shown in Graph 47.1.

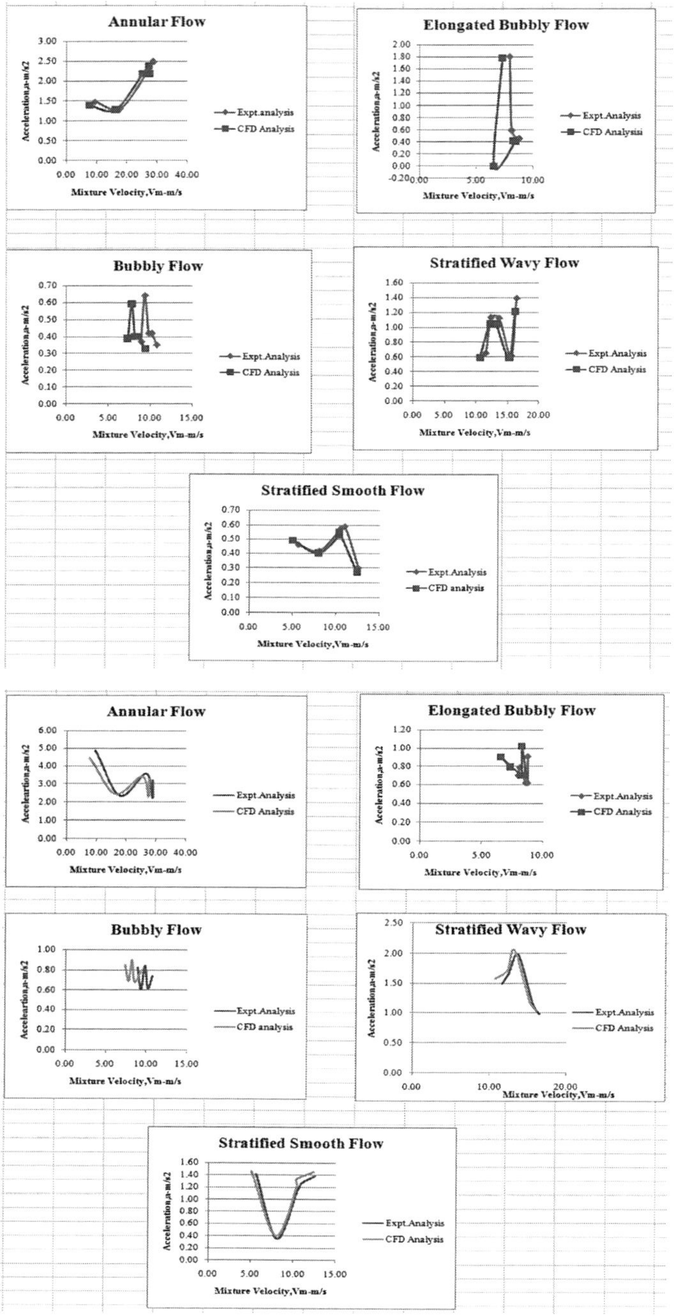

GRAPH 47.1 Mixture velocity versus acceleration for cantilever end.

47.7 CONCLUSION

From the results presented, there is good agreement between experiment results obtained and CFD analysis found to be satisfactory with 4–10% error for different end condition, that is, cantilever end and both ends fixed. It is seen that as the mixture velocity increases, induced accelerations (vibration) increase for all the patterns for different end condition. However, there is an increase in acceleration for both end fixed condition than the cantilever end condition which might be due to the additional constraint. And the acceleration reaches to maximum value for annular pattern due to turbulent nature of flow as compared to stratified smooth flow. In addition, it is observed that the vibration amplitude first decreased as the void fraction increases. However, further increase of void fraction, the pipe vibration amplitude starts increasing signaling instability. Moreover, it is seen that when the two-phase flow is closer to bubble flow, the vibration excitation is more random. When the flow pattern approaches slug flow, a more periodic excitation is observed.

ACKNOWLEDGMENT

This work is supported by Department of Mechanical Engineering, D. Y. Patil College of Engineering and Technology, Kolhapur, Maharashtra, India.

KEYWORDS

- **two-phase flow**
- **flow regimes**
- **void fraction**
- **flow-induced vibration (FIV)**
- **computational fluid dynamic (CFD)**

REFERENCES

1. Versteeg, H. K.; Malalasekera, W. *An Introduction to Computational Fluid Dynamics;* Longman Scientific & Technical, Longman Group Limited: USA, 1995, ISBN 470-23515-2(USA).

2. Bhagwat, S. M. Study of Flow Patterns and Void Fraction in Vertical Downward Two Phase Flow. Thesis, Amravati University, Maharashtra, 2008.
3. Fan, Z.; Lueeryran, F.; Hanratty, T. J. *Initiation of Slugs in Horizontal Gas-Liquid Flows. Department of Chemical Engineering, Fluid Mechanics and Transport Phenomena;* University of Illinois: Urbana, November 1993.
4. Gama, A. L.; Ferreira, L. R. dos S.; Walter Filho, P. H. A. Experimental Study on the Measurement of Two-Phase Flow Rate Using Pipe Vibration. *Proceedings of COBEM, 20th International Congress of Mechanical Engineering,* Gramado, RS, Brazil, November 15–20, 2009.
5. Ghajar, A. J. Non-Boiling Heat Transfer in Gas–Liquid Flow in Pipes – a Tutorial School of Mechanical and Aerospace Engineering Oklahoma State University, Stillwater, OK 74078, USA; ghajar@ceat.okstate.edu.
6. Kim, D., Chang, S. Flow-Induced Vibration in Two-Phase Flow with Wire Coil Inserts. *Int. J. Multiphase Flow* **2008,** *34*, 325–332.
7. Lamari, M. L. An Experimental Investigation of Two-Phase (Air-Water) Flow Regimes in a Horizontal Tube at Near Atmospheric Conditions. Thesis, Department of Mechanical and Aerospace Engineering, Carleton University, Ottawa, Ontario, Canada, 2000.
8. Mitra, D. R. Fluid-Elastic Instability in Tube Arrays Subjected to Air-Water and Steam-Water Cross-Flow. Thesis, Amravati University of California, Los Angeles, 2005.
9. Mohammadi, M. An Investigation on the Effect of Gas-Liquid Flow Characteristics on Pipe Vibration, Faculty of Engineering, Mechanical and Manufacturing Department, University Putra Malaysia, March 2009.
10. Oshinowo, O. Two Phase Flow in a Vertical Tube Coil. Ph.D. Thesis, Department of Chemical Engineering and Applied Chemistry, University of Toronto, Toronto, 1971.
11. Paras, G. Characterization of Downward Two Phase Flow by Neutron Noise Analysis, Nuclear Engineering. M.S. Thesis, University of Washington, 1982.
12. Sasakawa, T.; Serizawa, A.; Kawara, Z. Fluid-Elastic Vibration in Two-Phase Cross Flow. *Exp. Therm. Fluid Sci.* **2005,** *29*, 403–413.
13. Schobeiri, M. T. *Fluid Mechanics for Engineers: a Graduate Textbook.* Springer Science and Business Media: Verlag Berlin Heidelberg, 2010, ISBN : 978-3-642-11593-6.
14. Sim, W. G.; Bae, B. M.; Mureithi, N. W. An Experimental Study on Characteristics of Two-Phase Flows in Vertical Pipe. *J. Mech. Sci. Technol.* **2010,** *24*, 1981–1988.
15. Subramanian, R. S. Elementary Aspects of Two-Phase Flow in Pipes Two-Phase Flow. *Exp. Therm. Fluid Sci.* **2007,** *9*(1), 34–38.
16. Thomson, W. T. *Theory of Vibration with Applications*, 4th ed.; Taylor & Francis, Milton Park: Cheltenham, U.K., 2003.
17. Ujang, P. M.; Lawrence, C. J.; Hale, C. P.; Hewitt, G. F. Slug Initiation and Evolution in Two-Phase Horizontal Flow. *Int. J. Multiphase Flow* **2006,** *32*, 527–552 (Department of Chemical Engineering, Imperial College London).
18. Yamazaki, Y.; Yamaguchi, K. Characteristics of Two Phase Downflow in Tubes: Flow Pattern, Void Fraction and Pressure Drop. *J. Nucl. Sci. Technol.* **1979,** *16*(4), 245–255.
19. Zhang, M. M.; Katz, J.; Prosperetti, A. Enhancement of Channel Wall Vibration Due to Acoustic Excitation of an Internal Bubble Flow. *J. Fluids Struct.* **2010,** *26*, 994–1017 (Department of Mechanical Engineering, The Johns Hopkins University, Baltimore).

CHAPTER 48

NUMERICAL STUDY OF EQUIVALENCE RATIO AND SWIRL EFFECT ON THE PERFORMANCE AND EMISSIONS OF AN HCCI ENGINE

T. KARTHIKEYA SHARMA[1,*], DHANARAJ SAVARY NASAN[1,3], G. AMBA PRASAD RAO[2,*], and K. MADHU MURTHY[2,4]

[1]*Department of Mechanical Engineering National Institute of Technology, Andhra Pradesh, Tadepalligudem, AP, India*

[2]*Department of Mechanical Engineering, National Institute of Technology, Warangal, Telangana-506004*

[3]*savarynasan@gmail.com*

[4]*madhumurthyk@gmail.com*

[*]*Corresponding author. E-mail: karthikeya.sharma3@gmail.com; ambaprasadrao@gmail.com*

ABSTRACT

A promising combustion strategy that combines the advantages of both spark ignition (SI) and compression ignition (CI) combustion modes is homogeneous charge compression ignition (HCCI) combustion mode. A volumetric combustion of lean mixture of charge is the beauty of HCCI combustion leading to low NO_x emissions and soot. In the present work, a CI engine with HCCI combustion is analyzed to study the effect of swirl motion of intake charge at different equivalence ratios on performance and emissions using three-zone extended coherent flame combustion model (ECFM-3Z, CI). The analysis is done considering equivalence ratios ranging from 0.26 to 0.86 and swirl ratios from 1 to 4. HCCI engine with each equivalence ratio was analyzed at all the four swirl ratios using ECFM-3Z CI model. First, validation of the model is done with the existing literature. The present study revealed that ECFM-3Z of STAR-CD has well predicted the performance and emissions of CI engine in HCCI mode. The simulation results show that the in-cylinder pressures, temperatures, NO_x emissions and piston work were reduced with an increase in swirl ratio irrespective of the equivalence ratio. However, wall heat transfer losses, CO_2 emissions, and CO emissions increased with increase in swirl ratio irrespective of the equivalence ratio. The combined results of the equivalence ratio and swirl ratio show that higher in-cylinder pressures, higher temperatures, higher CO_2 emissions are observed at lower swirl ratios and higher equivalence ratios. The lower CO and NO_x results are observed at higher swirl ratios and lower compressions ratios. Low wall heat transfer losses are observed with high swirl ratios and low equivalence ratios. Swirl ratio and equivalence ratio are optimized for the chosen engine geometry. It is observed that there is a trade-off between the emissions and piston work.

48.1 INTRODUCTION

Internal combustion (IC) engines have become indispensable prime movers over the past one and half century. Though the performance of conventional spark ignition (SI) and compression ignition (CI) engines is satisfactory, SI engine suffers from poor part load efficiency and high CO emissions. The CI engine yields high particulate and NO_x emissions. These effects may be attributed to their conventional combustion process. Of late, a hybrid combustion process called homogeneous charge compression ignition (HCCI) equipped with advanced low-temperature combustion technology is gaining attention by researchers. In principle, HCCI involves the volumetric auto combustion of a premixed fuel, air, and diluents at low to moderate temperatures and at high compression ratios. The other associated advantages with HCCI mode of combustion have been well documented and presented as a potentially promising combustion mode for internal combustion engines.[1–3]

Swirl helps in homogeneous mixture formation of the fuel and air.[4] It also helps in NO_x emission reduction.[5] The increase in swirl ratio reduces the peak temperatures by increasing the heat transfer to the combustion chamber parts. This leads to a low-temperature combustion process resulting in low NO_x emissions.[6]

Performing these explorations (under different operating parameters with induction induced swirl) solely in the laboratory would be inefficient, expensive, and impractical since there are many variables that exhibit complex interaction. Because of this reason, a computational fluid dynamics (CFD) tool, STAR-CD, is chosen for the analysis. Several modifications were made to STAR-CD and es-ice module so that these could be used for HCCI engine modeling. The different combustion models which are well developed for predicting engine processes are transient interactive flamelet (TIF) model, digital analysis of reaction system–transient interactive flamelets model (DARS-TIF), G-equation model,[7] extended coherent flame combustion model-3 zones[8](ECFM-3Z), and the equilibrium-limited extended coherent flame model-combustion limited by equilibrium enthalpy (ECFM-CLEH).[9,10] Each model has its own limitations and is suitable for a specific set of problems. Generally speaking; ECFM-3Z and ECFM-CLEH can be used for all types of combustion regime whereas ECFM-3Z is mostly suitable for homogeneous turbulent premixed combustion with SI and CI. Various combustion models' applicabilities are shown in Table 48.1. Owing to its wide applicability, in the present work ECFM-3Z has been used to study the effect of swirl motion of intake charge on emissions and performance of HCCI engine. Figure 48.1 depicts the schematic representation of the three zones of the ECFM-3Z model. This model is capable of simulating the complex mechanisms such as turbulent mixing, flame propagation, diffusion combustion, and pollutant emission that characterize modern IC engines.

TABLE 48.1 Combustion Models Capabilities.

Model	Applicability
G-equation	Partially premixed SI and CI
DARS-TIF	Compression ignition
ECFM	Nonhomogeneous premixed SI
ECFM-3Z	Premixed and Non-premixed SI and CI

SI: spark ignition; CI: compression ignition; DARS-TIF: digital analysis of reaction system–transient interactive flamelets model; ECFM: extended coherent flame model.

Induction-induced swirl has a predominant effect on mixture formation and rapid spreading of the flame front in the conventional combustion process of a CI engine. This has been well documented in the literature. However, it is observed that no work has been done on the effect of swirl in HCCI mode.`

The main objective of the present study is to analyze the effect equivalence ratio under varying swirl ratios on the performance and emissions of an HCCI engine. As HCCI engines are known for their low specific fuel consumption (SFC), an attempt is made to find a suitable equivalence ratio at which the maximum performance and low emissions could be obtained. In this regard, a computational attempt is made under four induction-induced swirl ratios varying from 1 to 4 and equivalence ratios from 0.26 to 0.46.

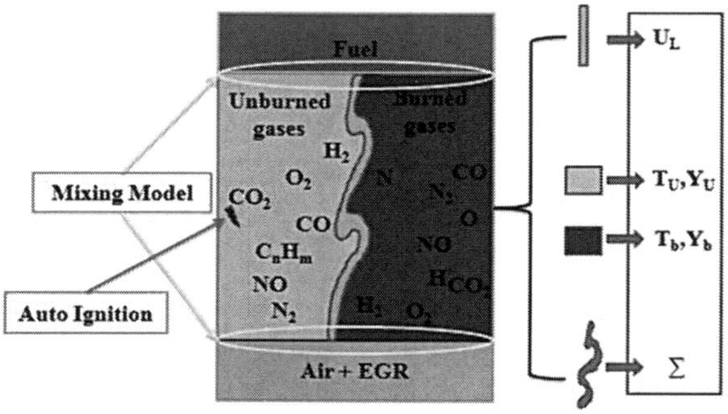

FIGURE 48.1 Schematic representation of three zones of extended coherent flame model (ECFM-3Z) model.

48.2 METHODOLOGY

The software used in the present work makes use of CFD with finite volume approach. The respective governing equations for conservation of mass, momentum, energy, and species are solved consecutively in a solver, es-ice, an expert system developed for internal combustion engines. Near the wall region to estimate the flow, heat and mass transfer, the standard k-ω model was used. The ECFM-3Z model is a general-purpose combustion model capable of simulating the complex mechanisms of turbulent mixing, flame propagation, diffusion combustion, and pollutant emission that characterize modern internal combustion engines. "3Z" stands for three zones of mixing: the unmixed fuel zone; the mixed gases zone; and unmixed air plus exhaust gas recirculation (EGR) zone. The three zones are too small to be resolved by the mesh and are; therefore, modeled as sub-grid quantities. The mixed zone is the result of turbulent and molecular mixing between gases in the other two zones and is where combustion takes place. The flame propagation phase is modeled by the flame surface density transport equation incorporating the theoretical flame speed. The engine specifications considered for the analysis are shown in Table 48.2. The analysis was done from the second cycle after the engine has started.

TABLE 48.2 Engine Specifications.

Engine specifications	**Value**
Displacement volume	1600 cm^3
Bore	12.065 cm
Stroke	14 cm
Connecting rod length	26 cm
Compression ratio	21:1
Fuel	n-dodecane
Operating conditions	
Engine speed	1000 rpm
Equivalence ratio	0.26
Inlet temperature air (T_{air})	353 K
Inlet air pressure (P_{air})	0.1 MPa
Cylinder wall temperature (T_{wall})	450 K
EGR	0%

EGR: exhaust gas recirculation.

48.3 COMPUTATIONAL FLUID DYNAMICS MODEL SETUP

The piston bowl shape and 3D mesh of the piston bowl sector are shown in Figure 48.2. The computational mesh consists of 0.312×10^6 cells. The entire mesh consists of cylinder and one-sixth of piston bowl created in HyperMesh—a mesh generation utility and is imported into STAR-CD for solutions. A spline has been developed based on the imported model; 2D template was cut by the spline to cut the 3D mesh with 80 radial cells, 140 axial cells, 5 top dead center layers, and 40 axial block cells.

Energy efficiency of the engine is analyzed by gross indicated work per cycle *(W)* calculated from the cylinder pressure and piston displacement using Equation 48.1:

$$W(Nm) = \frac{\pi a B_2}{8} \int_{\theta_1}^{\theta_2} p(\theta) \left[2\sin(\theta) \frac{\text{asin}(2\theta)}{\sqrt{l^2 - a^2 sin^2(\theta)}} \right] d\theta \quad (48.1)$$

where *a, l,* and *B* are the crank radius, connecting rod length and cylinder bore, respectively, and θ_1 and θ_2 are the beginning and the end of the valve-closing period.

The indicated power per cylinder (P) is related to the indicated work per cycle by using Equation 48.2:

$$P(\text{kW}) = \frac{WN}{60000 n_R} \tag{48.2}$$

where n_R=2 is the number of crank revolutions for each power stroke per cylinder and N is the engine speed (rpm). The indicated SFC (ISFC) is shown in Equation 48.3:

$$\text{FC}(\text{g} / \text{kWh}) = \frac{30\text{m}_{\text{fuel}} N}{P} \tag{48.3}$$

In Equation 48.1, the power and ISFC analyses can be viewed as being only qualitative rather than quantitative in this study.

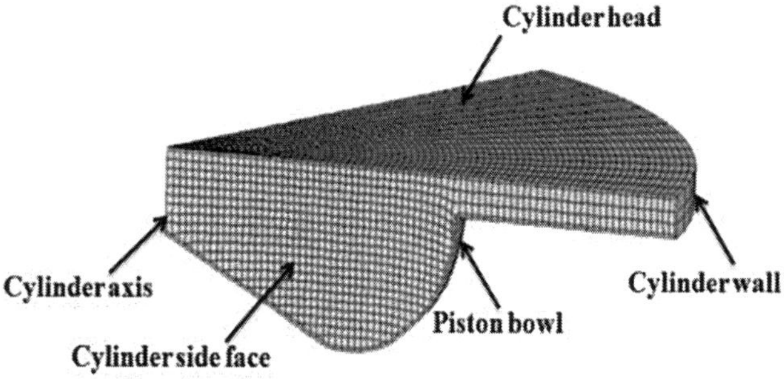

FIGURE 48.2 Schematic representation of 3D piston bowl shape at top dead center.

48.4 MODELING STRATEGY

The STAR-CD used in the present study has integrated several sub models such as turbulence, fuel spray and atomization, wall function, ignition, combustion, NO_x, and soot models for various types of combustion modes in CI as well as SI engine computations. As initial values of k and ε are not known prior, the turbulence initialization is done using *I-L* model. For this purpose, local turbulence intensity, *I*, and length scale, *L*, are related as

$$k_\infty = 3/2\, I^2 V^2_{\ \infty} \tag{48.4}$$

$$\varepsilon_\infty = C^{3/4}_{\ \mu}(^{3/2}/L) \tag{48.5}$$

This practice will ensure that k and ε and the turbulent viscosity μ_t, will all scale correctly with V_∞, which is desirable from both the physical realism and numerical stability point of view. Moreover, the turbulent intensity is defined using the same velocity vector magnitude as that of stagnation quantities.

The combustion is modeled using ECFM-3Z. As far as fluid properties are concerned, ideal gas law and temperature dependent constant pressure specific heat (C_p) are chosen.

The ECFM-3Z incorporates the following models in its operation.

48.5 INITIAL AND BOUNDARY CONDITIONS

To begin with, an absolute pressure 1.02 bar, with 0% EGR, temperature to 353 K, equivalence ratio as 0.26 are taken as initial values. Fixed boundary wall temperatures are taken with combustion dome regions as 450 K, piston crown regions as 450 K, and cylinder wall regions as 400 K; the Angleberger wall function mode is considered.[11] The "two-layer" and low Reynolds number approaches, were applied at the boundary layers by solving the mass, momentum, and turbulence equations (the latter in their "low Reynolds number" form) within them.[11] The hybrid wall boundary condition, which is a combination of two layered and low-Reynolds number wall boundary conditions, is considered in this analysis. This hybrid wall boundary condition removes the burden of having to ensure a small enough near-wall value for y^+ (by creating a sufficiently fine mesh next to the wall). The y^+ independency of the hybrid wall condition is achieved using either an asymptotic expression valid for $0.1 < y^+ < 100$ or by blending low-Reynolds and high-Reynolds number expressions for shear stress, thermal energy, and chemical species wall fluxes. This treatment provides valid boundary conditions for momentum, turbulence, energy, and species variables for a wide range of near-wall mesh densities. Standard wall functions are used to calculate the variables at the near wall cells and the corresponding quantities on the wall. The initial conditions were specified at initial value compensation, consisting of a quiescent flow field at pressure and temperature for full load condition.

48.6 VALIDATION OF THREE-ZONE EXTENDED COHERENT FLAME COMBUSTION MODEL

STAR-CD is a well-known commercial CFD package, being adopted by many renowned researchers and well-established research organizations in the field of automotive IC engines. The results obtained through this package are validated with the experimental results by many authors like Pasupathy Venkateswaran et al.[5], Zellat et al.[12], and Bakhshan et al.[13] A comparison of the CI engine in HCCI mode is done in this chapter by considering the ECFM-3Z compression ignition model for combustion analysis. The present chapter deals with the simulation of CI engine in HCCI mode, using a fuel vaporizer to achieve excellent HCCI combustion in a single cylinder air-cooled direct injection diesel engine. No modifications were made to the combustion system. Ganesh et al.[14] conducted experiments with diesel vapor induction without EGR and diesel vapor induction with 0, 10, and 20% EGR. Validation of the present model with the experimental results of Ganesh et al.[14] was done by considering all the engine specifications.

Ganesh et al.[14] considered a vaporized diesel fuel with air to form a homogeneous mixture and inducted into the cylinder during the intake stroke. To control the early ignition of diesel vapor–air mixture, cooled (30°C) EGR technique was adopted. For the validation purpose, the results are compared with respect to engine performance and emissions in the following figures. It is observed that the simulated results are in good agreement with the experimental results. The comparison of the plots between simulation and experimental results are shown in Figure 48.3. In the figures, EDVI represents the experimental diesel vapor injection and SDVI represents simulated diesel vapor induction at respective EGR concentrations.

48.7 RESULTS AND DISCUSSION

In the present chapter, the effect of equivalence ratio on the performance of HCCI engine has been studied under varying swirl ratios ranging from 1 to 4. The analysis has been carried out on a single cylinder HCCI engine, using a three-zone extended coherent flame (ECFM-3Z) CFD model. The results are plotted and discussed below.

48.7.1 IN-CYLINDER PRESSURES

The variation of in-cylinder pressures with equivalence ratio using swirl ratio 1 is plotted in Figure 48.4. The in-cylinder pressures increase with increase in equivalence ratio irrespective of the swirl ratio. The increase in in-cylinder pressures is due to rich mixture participating in combustion spontaneously liberating high temperatures.

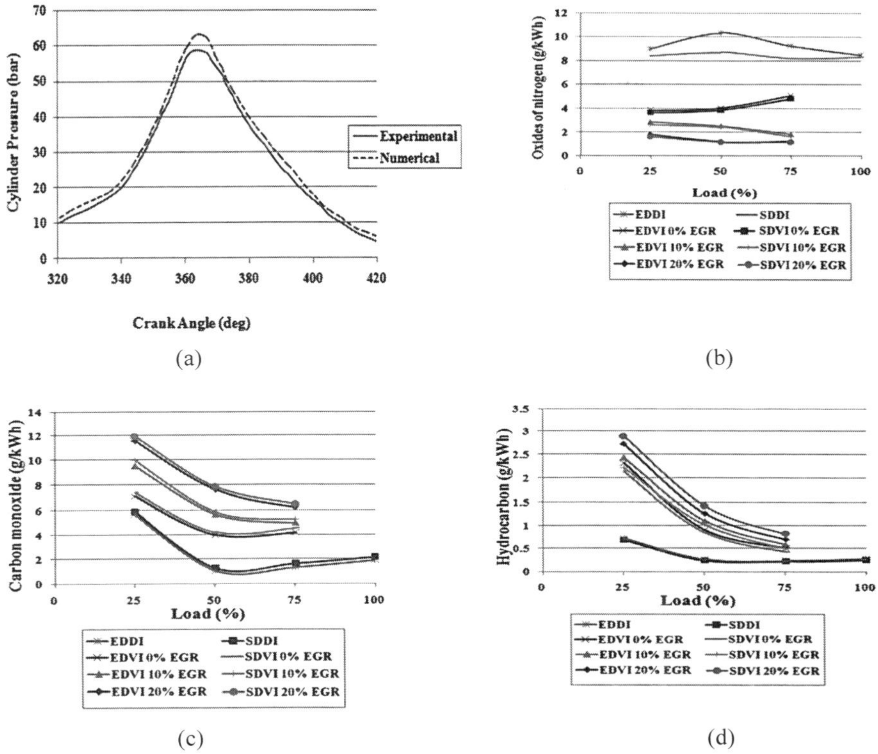

FIGURE 48.3 Validation of the ECFM-3Z compression ignition model with the experimental results of external mixture formation of HCCI engine.

However, with an increase in swirl ratio, the in-cylinder peak pressures are getting decreased which can be attributed to increase in in-cylinder turbulence that enhances the wall heat transfer losses. The variation of wall heat transfer losses with swirl ratio is illustrated in Figure 48.5. It can be observed clearly that wall heat transfer losses increase with increase in equivalence ratio at all swirl ratios.[15] From Figure 48.4, it can be seen that in-cylinder pressures at 0.86 equivalence ratio are higher compared to other equivalence ratios for a swirl ratio of 1. A maximum in-cylinder pressure of 13.7143 MPa at 719.775° crank angle (CA) and 30.145 MPa at 718.9725° CA was obtained at an equivalence ratio of 0.26 and 0.86, respectively. A total increase of 119.80% in in-cylinder pressures was obtained when the equivalence ratio was increased from 0.26 to 0.86. The peak values of in-cylinder pressures obtained for various equivalence ratios and swirl ratios are summarized in Table 48.3. From Table 48.4, it is observed that irrespective of swirl ratios, the in-cylinder pressures increased with increase in equivalence ratio. A maximum percentage increase of 119.98 in in-cylinder pressures was found at swirl ratio 3. Similarly, maximum in-cylinder pressures decrease with increase in swirl ratio irrespective of the equivalence ratio. The percentage decrease in peak pressures is low for equivalence ratio of 0.46, while there is not much variation in other equivalence ratios.

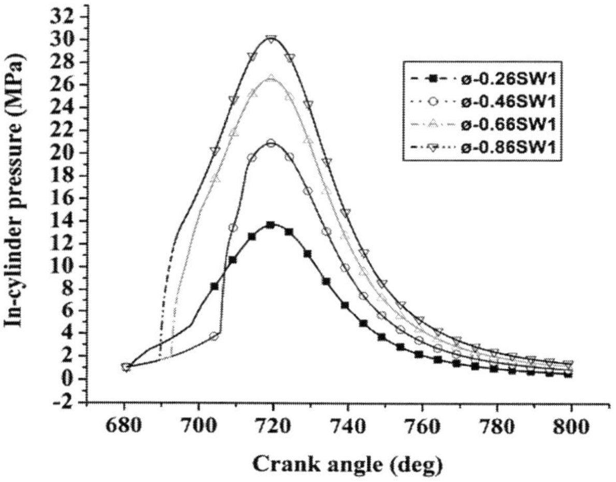

FIGURE 48.4 In-cylinder pressure versus crank angle (CA).

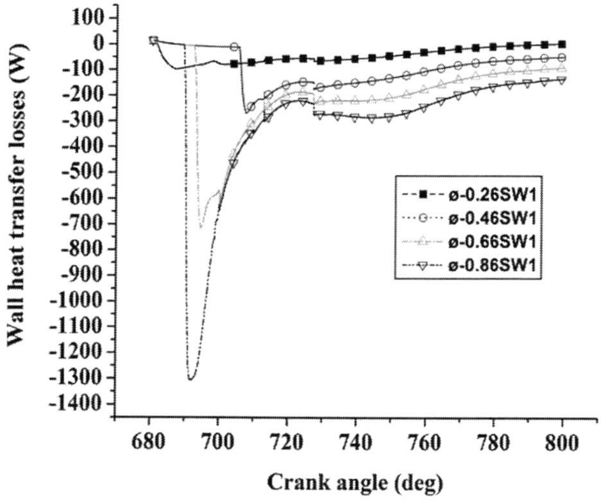

FIGURE 48.5 Wall heat transfer loss versus CA.

TABLE 48.3 Variation of Maximum In-Cylinder Pressures and Their Percentage Variation with Respect to Equivalence Ratio and Swirl Ratio.

In-cylinder pressures	Ø=0.26	Ø=0.46	Ø=0.66	Ø=0.86	Percentage increase
SW1	13.7	20.8	26.5	30.1	119.80
SW2	13.6	20.8	26.5	29.8	119.75
SW3	13.4	20.7	26.0	29.6	119.98
SW4	13.3	20.6	25.8	29.3	119.78
Percentage decrease	2.7	1.2	2.7	2.7	

48.7.2 IN-CYLINDER TEMPERATURES

Figure 48.6 shows the variation of in-cylinder temperatures with increase in equivalence ratio at swirl ratio 4. Though the numerical experiments are performed for different swirl ratios, low in-cylinder temperatures are realized only at higher swirl ratios (SW=4). This is in tune with the HCCI combustion which signifies the low-temperature combustion process. However, in-cylinder temperatures are observed to be higher with increase in equivalence ratio at all swirl ratios.

This is due to the fact that with increase in equivalence ratio more fuel gets into the combustion chamber; thus, rich mixture participation in combustion increases the magnitude of heat release leading to increase in in-cylinder temperatures.[16] Moreover, increased swirl ratios have enhanced the wall heat transfer losses with increased turbulence. The increased turbulence in the combustion chamber with swirl ratio facilitates toward the well homogeneous mixture formation, and this leads toward the ideal volumetric combustion causing low-temperature combustion. Increase in equivalence ratio resulted in higher wall heat transfer losses are the second reason for the reduced in-cylinder temperatures.

A maximum in-cylinder temperature of 1177.306 K at 717.075° CA and 2571.708 K at 710.05° CA was obtained at 0.26 and 0.86 equivalence ratios, respectively. A total increase of 118.44% in in-cylinder temperatures was obtained when the equivalence ratio was increased from 0.26 to 0.86. Table 48.4 shows the maximum in-cylinder temperatures with respect to equivalence ratio and swirl ratio. From Table 48.5, it is observed that irrespective of swirl ratios, the in-cylinder temperatures increased with increase in equivalence ratio. Moreover, peak cycle temperatures decrease with increase in swirl ratio irrespective of the equivalence ratio. The percentage decrease in in-cylinder temperatures is low for equivalence ratio of 0.46, when compared to other equivalence ratios.

TABLE 48.4 Variation of Maximum In-Cylinder Temperatures and Their Percentage Variation with Respect to Equivalence Ratio and Swirl Ratio.

In-cylinder temperatures	Ø=0.26	Ø=0.46	Ø=0.66	Ø=0.86	Percentage increase
SW1	1206.4	1850.8	2342.5	2603.2	115.76
SW2	1197.3	1850.8	2342.5	2591.6	116.45
SW3	1187.7	1847.8	2318.5	2580.7	117.27
SW4	1177.3	1837.9	2309.6	2571.7	118.44
Percentage decrease	2.4	0.69	1.40	1.20	

48.7.3 CO AND CO_2 EMISSIONS

In-cylinder temperatures play a major role in the formation of emissions. Higher in-cylinder temperatures facilitate the conversion of CO to CO_2 and lower temperatures reduce the formation of NO_x. A clear reduction from 1177.306 K at 717.075° CA and 2571.708 K at 710.05° CA was obtained at 0.26 and 0.86 equivalence ratios, respectively, and it can be seen from Figure 48.6. Usually, CO emissions increase with increase in swirl ratio; but with increase in equivalence ratio, CO emissions got reduced. The reason for decrease in CO emissions is due to increase in in-cylinder temperatures with increase in equivalence ratio that facilitates the reaction between CO and O_2 to form CO_2.[17] Figure 48.7 represents the variation of CO emissions with equivalence ratio for a swirl ratio 1. CO emissions of 0.138 and 0.00313 g/kg fuel are obtained when the engine runs under equivalence ratios of 0.26 and 0.46 for swirl ratio 1. Greater drop in CO emissions was found at equivalence ratio of 0.46. A total decrease of 97.73% in CO emissions was obtained between the equivalence ratios of 0.26 and 0.46. These emission values are observed to be low when compared with the CO emissions at swirl ratio 4 at all equivalence ratios.

The variation of CO_2 emissions with equivalence ratio at swirl ratio 4 is illustrated in Figure 48.8. The increase in CO_2 emissions with increase in equivalence ratio at higher swirl ratio is due to increase in-cylinder temperatures facilitating the conversion of CO to CO_2 by reacting with the O_2 available. CO_2 emissions of 0.0699 and 0.902 g/kg fuel are obtained when the equivalence ratio increased from 0.26 to 0.86 at swirl ratio 1. The CO_2 emissions obtained at swirl 4 are lower when compared to swirl ratio1.

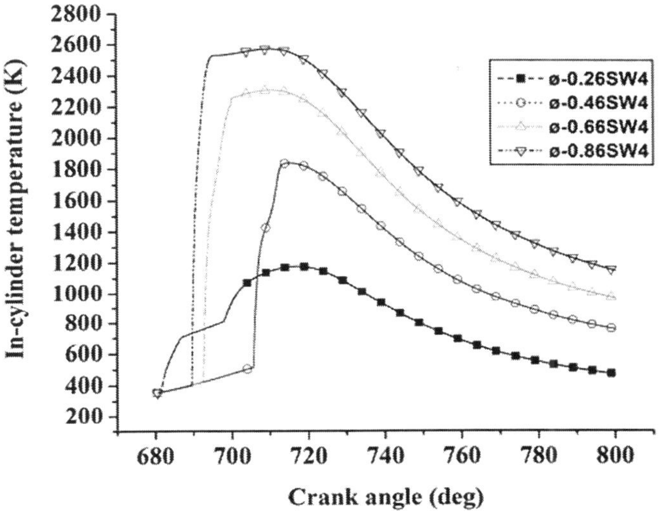

FIGURE 48.6 Temperature versus CA.

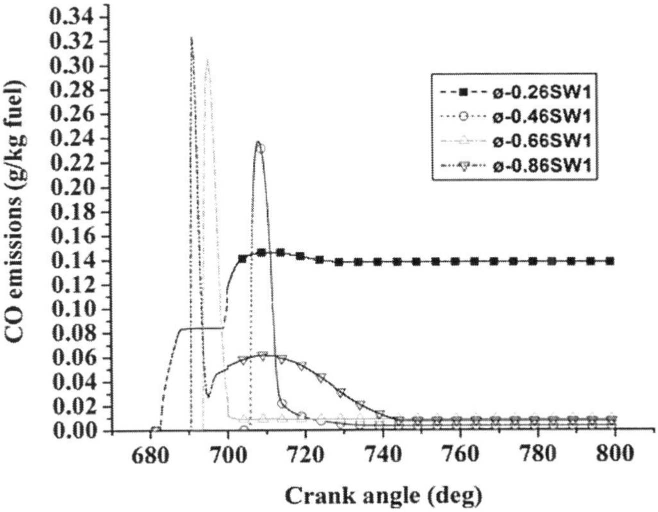

FIGURE 48.7 CO emissions versus CA.

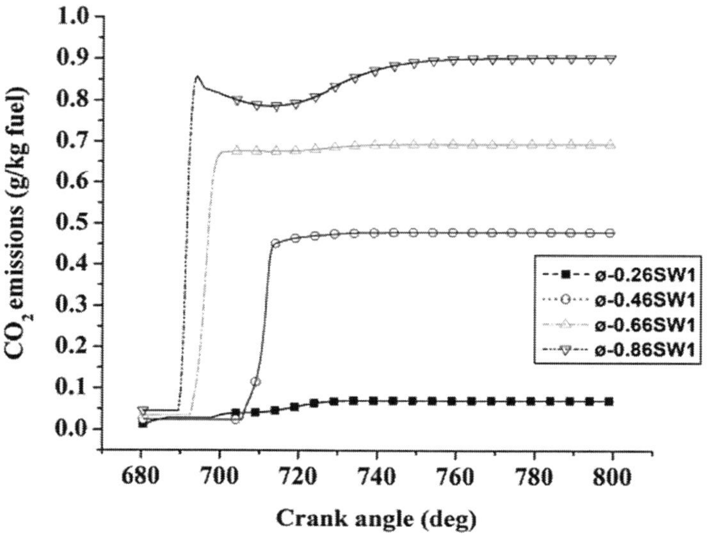

FIGURE 48.8 CO_2 emissions versus CA.

48.7.4 NO_X EMISSIONS

The formation of NO_X is highly dependent on the in-cylinder temperatures, oxygen concentration, and residence time for the reaction to take place. From Figure 48.9, with increase in equivalence ratio increase in NO_x emissions can be observed. Increase in in-cylinder temperatures is the reason for increase in NO_x emissions with equivalence ratio.[18] With increase in swirl ratio, the NO_x emissions decrease irrespective of equivalence ratio; increased wall heat transfer losses at higher swirl ratios cause this effect. This concludes that higher swirl ratios and lower equivalence ratios are favorable for low NO_x emissions.

Out of all the equivalence ratios analyzed, 0.46 gives the lowest NO_x emissions, proving it to be the optimum equivalence ratio for this particular engine. Lowest NO_x emissions of 0.00893 and 0.00558 g/kg fuel are obtained when the engine run with 0.26 and 0.46 equivalence ratios at swirl ratio 4. Thus, a total decrease of 37.51% in NO_x emissions was obtained when the equivalence ratio increased from 0.26 to 0.46 at swirl ratio 4. The NO_x emissions obtained at swirl ratio 4 are low when compared to swirl ratio 1.

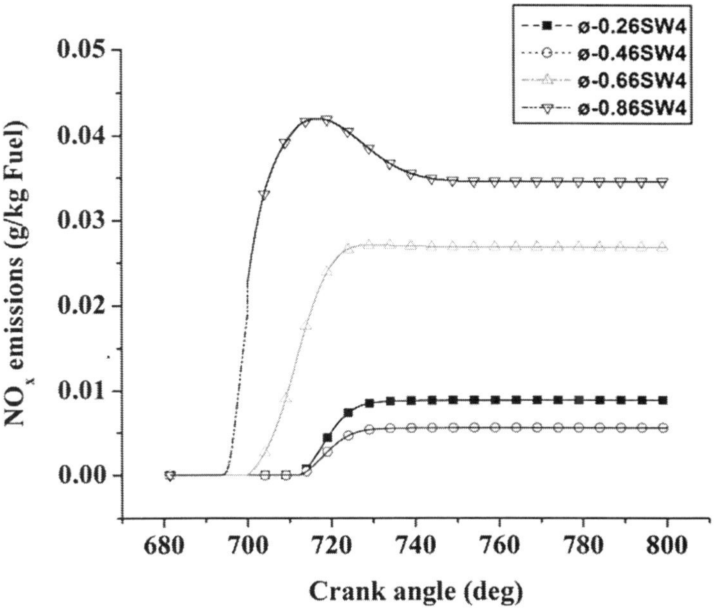

FIGURE 48.9 NO_x emissions versus CA.

48.8 CONCLUSIONS

In the present numerical study, prediction of HCCI engine performance under four swirl ratios with increase in equivalence ratio has been done using extended coherent flame combustion analysis considering three zones. Based on the study, the following conclusions are arrived.

- ECFM-3Z of STAR-CD has well predicted the performance and emissions of CI engine in HCCI mode.
- It is observed that swirl ratio is a major factor in achieving HCCI mode of combustion with a significant reduction in harmful NO_x emissions.
- Increase in in-cylinder pressures is observed with increase in equivalence ratio.
- Reduction in in-cylinder pressures and temperatures are observed with increase in swirl ratio.
- Equivalence ratio 0.46 is found to be the optimum equivalence ratio for the chosen engine delivering low NO_x and CO emissions and high piston work.

- The study revealed that there is a trade-off between emissions and piston work at higher equivalence ratios and lower swirl ratios.
- Equivalence ratio 0.46 with low swirl ratios is favorable to get low CO emissions and high piston work.
- Equivalence ratio 0.46 with high swirl ratios is favorable to get low NOx emissions.
- HCCI combustion can be regarded as low-temperature combustion as there is a significant decrease in in-cylinder temperatures and pressures at lower equivalence ratios and higher swirl ratios.

ACKNOWLEDGMENTS

The authors thank Dr. Raja Banerjee, Associate Professor, IIT Hyderabad, for allowing to use computational facility, Mr. B. Siva Nageswara Rao from CD-adapco, Bengaluru and Mr. P. Madhu, computer lab supervisor IIT Hyderabad, for their support during the simulation work.

KEYWORDS

- **HCCI**
- **ECFM**
- **swirl ratio**
- **CO and NOx emissions**
- **spark ignition**
- **internal combustion**
- **homogeneous charge compression ignition**
- **Reynolds number**

REFERENCES

1. Izadi Najafabadi, M.; Abdul Aziz, N. Homogeneous Charge Compression Ignition Combustion: Challenges and Proposed Solutions. *J. Combust.* **2013,** *2013*, Article ID 783789, 14. DOI: 10.1155/2013/783789.
2. Sharma, T. K.; Rao, G. A. P.; Madhumurthy, K. Combustion Analysis of Ethanol in HCCI Engine. *Trends Mech. Eng.* **2012,** *3*(1), 1–13.
3. Sharma, T. K.; Rao, G. A. P.; Murthy, K. M. Effect of Swirl on Performance and Emissions of CI Engine in HCCI Mode. *J. Braz. Soc. Mech. Sci. Eng.* **2014,** 1–12. DOI: 10.1007/s40430-014-0247-7.

4. Manimaran, R.; Thundil Karuppa Raj, R. Computational Studies of Swirl Ratio and Injection Timing on Atomization in a Direct Injection Diesel Engine. *Front. Heat Mass Transfer (FHMT)* **2014,** *5*(1), 15–30.
5. Pasupathy Venkateswaran, S.; Nagarajan, G. Effects of the Re-Entrant Bowl Geometry on a DI Turbocharged Diesel Engine Performance and Emissions—A CFD Approach. *J. Eng. Gas Turbines Power* **2010,** *132*(12), 122803.
6. Maroteaux, F., L. Noel, and A. Ahmed. Numerical Investigations on Methods to Control the Rate of Heat Release of HCCI Combustion Using Reduced Mechanism of n-Heptane with a Multidimensional CFD Code. *Combust. Theory and Modell.* **2007,** *11*(4), 501–525.
7. Cipolla, G.; Vassallo, A.; Catania, A. E.; Spessa, E.; Stan, C.; Drischmann, L. Combined Application of CFD Modeling and Pressure-Based Combustion Diagnostics for the Development of a Low Compression Ratio High-Performance Diesel Engine. No. 2007-24-0034. SAE Technical Paper, 2007.
8. Colin, O.; Benkenida, A. The 3-Zone Extended Coherent Flame Model (ECFM3Z) for Computing Premixed/Diffusion Combustion. *Oil Gas Sci. Technol. Rev. IFP* **2004,** *59*(6), 593–609.
9. Ravet, F.; Abouri, D.; Zellat, M.; Duranti, S. Advances in Combustion Modeling in STAR-CD: Validation of ECFM CLE-H Model to Engine Analysis. 18th Int. Multidimensional Engine Users' Meeting at the SAE Congress, Detroit, April 13, 2008.
10. Subramanian, G.; Vervish, L.; Ravet, F, New Developments in Turbulent Combustion Modeling for Engine Design: ECFM-CLEH Combustion Submodel. SAE International 2007-01-0154.
11. Moureau, V.; Lartigue, G.; Sommerer, Y.; Angelberger, C.; Colin, O.; Poinsot, T. Numerical Methods for Unsteady Compressible Multi-Component Reacting Flows on Fixed and Moving Grids. *J. Comput. Phys.* **2005,** *202*(2), 710–736.
12. Zellat, M; et al. Towards a Universal Combustion Model in STAR-CD for IC Engines: from GDI to HCCI and Application to DI Diesel Combustion Optimization. Proc. 14th Int. Multidimensional Engine User's Meeting, SAE Cong, 2005.
13. Bakhshan, Y.; et al. Multi-dimensional Simulation of n-Heptane Combustion under HCCI Engine Condition Using Detailed Chemical Kinetics. *J. Eng. Res.* **2011,** *22*(22), 3–12.
14. Ganesh, D.; Nagarajan, G. Homogeneous Charge Compression Ignition (HCCI) Combustion of Diesel Fuel with External Mixture Formation. *Energy* **2010,** *35*(1), 148–157.
15. Maurya, R. K.; Agarwal, A. K. Experimental Investigation on the Effect of Intake Air Temperature and Air–Fuel Ratio on Cycle-to-Cycle Variations of HCCI Combustion and Performance Parameters. *App. Energy* **2011,** *88*(4), 1153–1163.
16. Christensen, M.; Johansson, B. Influence of Mixture Quality on Homogeneous Charge Compression Ignition. *SAE Trans. J. Fuels Lubr.* **1998,** *107,* 1–13.
17. Aceves, S. M.; Smith, J. R.; Westbrook, C. K.; Pitz, W. J. Compression Ratio Effect on Methane HCCI Combustion. *J. Eng. Gas Turbines Power* **1999,** *121*(3), 569–574.
18. Flowers, D.; Aceves, S.; Westbrook, C. K.; Smith, J. R.; Dibble, R. Detailed Chemical Kinetic Simulation of Natural Gas HCCI Combustion: Gas Composition Effects and Investigation of Control Strategies. *J. Eng. Gas Turbines Power* **2001,** *123*(2), 433–439.

CHAPTER 49

COMPUTATIONAL FLUID DYNAMICS STUDY OF SERPENTINE FLOW FIELD PROTON EXCHANGE MEMBRANE FUEL CELL PERFORMANCE

VENKATESWARLU VELISALA* and G. NAGA SRINIVASULU

Department of Mechanical Engineering, National Institute of Technology Warangal, Telangana, India

**Corresponding author. E-mail: 2venkee@gmail.com*

ABSTRACT

Fuel cell (FC) is an electrochemical device which converts the chemical energy of fuel directly into electrical energy with high efficiency and without any toxic by-products. Flow field plates are the backbone of the FC. It is an essential component of the FC with multifunctional character. Modeling the transport phenomena in an FC plays a vital role in the development of FCs. Numerical modeling saves the operational time and cost during testing and development of FC systems. In the present work, a complete three-dimensional steady-state, single-phase isothermal computational fluid dynamics (CFD) model of a proton exchange membrane FC with a single serpentine flow channel is proposed to analyze the species transport phenomenon. The numerical model was developed using the commercial CFD code (ANSYS Fluent 15.0) and simulations were carried out with species concentration on anode side as H_2-0.8, O_2-0, and H_2O-0.2 and on the cathode side H_2-0, O_2-0.2, and H_2O-0.1 to get important parameters such as variation of hydrogen, oxygen, water mass fraction, and liquid water activity in the flow channels, membrane water content and membrane protonic conductivity. The numerical results show that at lower cell voltage (0.4 V) which corresponds to higher current density, the hydrogen and oxygen consumption and water production are high. Finally, the polarization curve obtained was compared with the available experimental data in the literature. It is found that the numerical results are in good agreement with the experimental results.

49.1 INTRODUCTION

The ever-increasing energy demand, emission-free energy generation, and other ecological issues have encouraged many researchers to look for advanced efficient energy conversion technologies.[1] Within such a perspective, fuel cell (FC) systems may be considered as a good alternative due to practical merits such as the high-energy density, superb dynamic response, low hostility to the environment, and lightweight as well as easy and fast recharging through a replacement or a refilled fuel cartridge.[2] FCs are categorized based on the type of electrolyte materials.[3] The commonly available FC technologies include polymer electrolyte membrane (PEMFC), alkaline (AFC), phosphoric acid (PAFC), molten carbonate (MCFC), and solid oxide (SOFC)-based FCs.

PEMFCs have numerous distinctive characteristics as compared with other FC types, such as low operating temperature, silent operation, high power density, quick start, faster response, and high modularity makes them the most encouraging contender for future power-generating devices in applications such as automobiles, distributed power generation, and compact electronic devices.[4] Designing and building of FC system are laborious and expensive.[5] The alternative is modeling and simulation of the FC system; this can permit the assessment of the FC performance, lowering the cost and time along the design stage and tests.[6]

Berning et al.[7] developed three-dimensional (3D) FC models taking into account computational liquid elements approach. Barreras et al.[8] conducted both experimental and numerical work to analyze the flow distribution in a bipolar plate of a PEMFC. Lum and McGuirk[9] developed a 3D computational fluid dynamics (CFD) model of PEMFC to carry out some parametric studies. Carcadea et al.[10] presented a 3D CFD model to understand the physical and chemical processes taking place inside PEM FC. To examine the performance and transport phenomena in the PEMFCs, Jang et al.[11] also developed 3D numerical models with conventional flow field designs (parallel flow field, Z-type flow field, and serpentine flow field). Suresh et al.[12] proposed an improved serpentine flow field with an enhanced cross-flow and carried out both numerical and experimental work with the proposed design. Guo et al.[13] developed a 3D simulation PEMFC model with bioinspired flow field designs and carried out both numerical and experimental studies to examine the performance. Khazaee and Sabadbafan[14] developed 3D single-phase PEMFC models with rectangular, triangular, and elliptical cross section geometries and investigated the effect of increasing the number of channels and cross-sectional area of the channels on the FC performance.

This chapter presents a complete 3D steady-state single-phase isothermal CFD model of PEMFC with a single serpentine flow channel to study the performance. Using this model, it is possible to understand the transport phenomena taking place inside the cell which cannot be studied experimentally. The model is well adapted for the PEMFC and incorporates the key physical and electrochemical processes that happen in the cell. This study will be useful for the design and operation of practical PEMFC.

49.2 MODEL DESCRIPTION

The present work presents a 3D CFD model of the PEMFC shown in Figure 49.1a that incorporates the significant physical processes and the important parameters affecting FC performance. The flow of reactants takes place in the flow channels and a carbon paper, which serves as a gas diffusion layer (GDL). The flow channel contains a number of bends and/or dead-ends to facilitate diffusion of the reactants through the GDL while the gas passes through the channel. The single FC containing the membrane electrode assembly with a single serpentine flow channel was simulated using Fluent software.

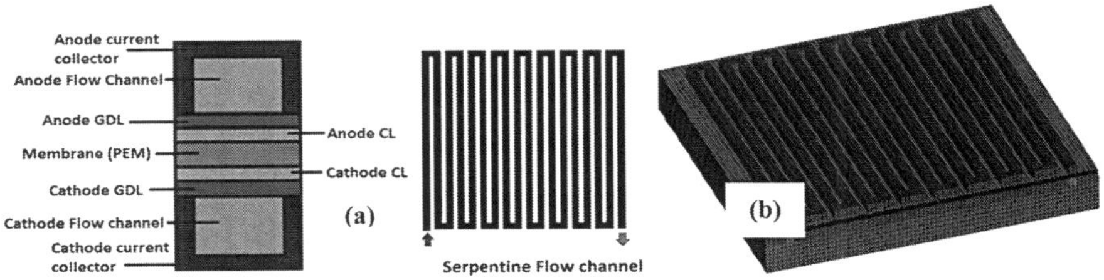

FIGURE 49.1 (a) Proton exchange membrane (PEM) fuel cell (FC) assembly with serpentine flow field and (b) computational mesh.

The whole computational domain consists of serpentine channels, gas diffusion layers (GDLs), catalyst layers (CLs) for anode and cathode, and a membrane. All the components are modeled, assembled, and meshed with SolidWorks© CAD software and ICEM CFD, respectively. The 3D meshed geometry of the PEMFC created in the ICEM CFD was shown in Figure 49.1b.

The special add-on module for the FC embedded in Fluent software numerically solves the governing equations such as continuity, momentum, energy, species transport, and charge conservation by finite volume method in the form of SIMPLE algorithm and by pressure correction method that were given in Table 49.1.

TABLE 49.1 Governing Equations.[16]

Continuity equation	$\frac{\partial(\rho u)}{\partial x}+\frac{\partial(\rho v)}{\partial y}+\frac{\partial(\rho w)}{\partial z}=\frac{\partial \rho}{\partial t}$
Momentum equation	$u\frac{\partial(\rho u)}{\partial x}+v\frac{\partial(\rho u)}{\partial y}+w\frac{\partial(\rho u)}{\partial z} = \frac{\partial \tilde{n}}{\partial x}+\frac{\partial}{\partial x}\left(\mu\frac{\partial u}{\partial x}\right)+\frac{\partial}{\partial y}\left(\mu\frac{\partial u}{\partial y}\right)+\frac{\partial}{\partial z}\left(\mu\frac{\partial u}{\partial z}\right)+S_{mom,x}$
Species transport equation	$u\frac{\partial(\rho y_i)}{\partial x}+v\frac{\partial(\rho y_i)}{\partial y}+w\frac{\partial(\rho y_i)}{\partial z}=\frac{\partial(j_{x,i})}{\partial x}+\frac{\partial(j_{y,i})}{\partial y}+\frac{\partial(j_{z,i})}{\partial z}+S_i$
Energy equation	$\nabla(\vec{v}(\rho E+P))=\nabla.\left(K_{eff}\nabla T-\sum_j h_j\vec{j}_j+(\tau_{eff}.\vec{v})\right)+S_h$
Charge conservation	$\nabla.(\sigma_{sol}\nabla\varphi_{sol})+R_{sol}=0$ $\nabla.(\sigma_{mem}\nabla\varphi_{mem})+R_{mem}=0$

49.2.1 ASSUMPTIONS

a) The fuel and gas mixtures are considered as perfect gases.
b) The flow is steady, laminar, and incompressible.
c) The system is isothermal.
d) The electrochemical reactions take place only in the catalyst layer.
e) Porous structures such as GDLs, the CLs, and the membrane (PEM) were considered as isotropic.
f) Neglect the gravity effect.[15]

The generated mesh file then exported to the fluent solver, and boundary conditions as well as parameters shown in Table 49.2 were applied before solution initialization. The pressure drop, reactants mass fraction in the flow channels, membrane water content, and current density distributions were calculated numerically and presented.

TABLE 49.2 Cell Design Parameters, Transport Properties and Key Operating Parameters Used in the Simulation.

Property	Value
Flow channel length	7 cm
Flow channel height	0.1 cm
Flow channel width	0.1 cm
Rib width	0.1 cm
Gas diffusion layer (GDL) thickness	0.025 cm
Catalyst layer(CL) thickness	0.006 cm
Membrane thickness	0.015 cm
GDL porosity	0.5
CL porosity	0.5
Membrane porosity	0.5
Oxygen diffusion coefficient	3e-5
Hydrogen diffusion coefficient	3e-5
Operating parameters	
Cell operating temperature	333 K
Cell operating pressure	1 bar
Fuel	H_2
Oxidant	O_2
Fuel flow rate	2.05e-7 kg/s
Oxidant flow rate	1.67e-6 kg/s

49.3 RESULTS AND DISCUSSION

CFD study of the serpentine flow field PEMFC performance has been explored. The polarization and performance characteristics of the PEMFC are shown below. Other parameters such as pressure drop, concentration distribution of hydrogen (H_2), oxygen (O_2), liquid water activity along the channel and the membrane water content for peak power performance of the serpentine flow field PEMFC are shown below.

The pressure drop takes place in the flow channel and it is one of the key factors to take into account in order to achieve an efficient flow channel design. If the pressure drops are less, then the power required to drive the reactants will be less given that more energy is required to drive the reactants effectively which results in improvement in the net performance obtained of the PEMFC.

Figure 49.2 shows the pressure drop in the anode and cathode flow channel, respectively at 0.4 V cell potential. This pressure drop in the flow channel is an important consideration in the design of PEMFC as it will help in deciding the type of blowers or compressors to be used to maintain sufficient pressure in the channel. This pressure drop also influences the electrochemistry of the FC.

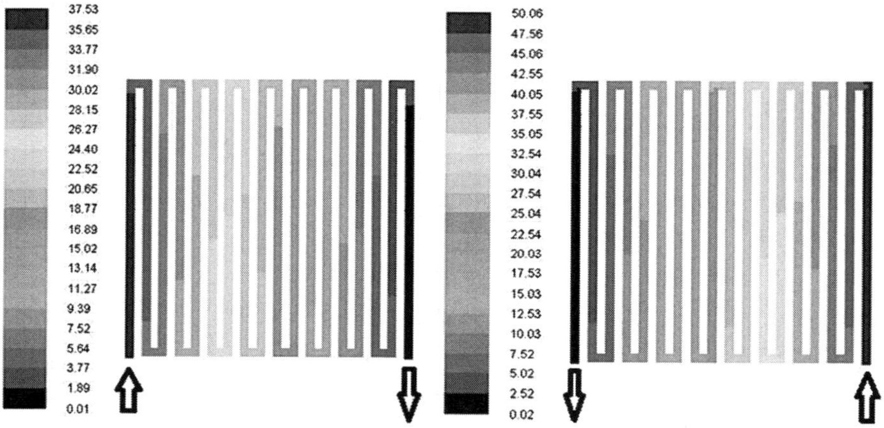

FIGURE 49.2 Contours of pressure variation along the flow channel of anode (left) and cathode (right).

Figure 49.3 shows the hydrogen and oxygen mass fraction distribution along the anode and cathode flow channel at cell potential of 0.4 V. Both H_2 and O_2 concentrations were more at the inlet and reduce gradually toward the outlet of the channel. The reduction of species concentration in the channels is due to the consumption of reactants in the reaction.

To achieve optimum performance from the cell and to prevent larger temperature gradients inside the cell, it is good to have a uniform distribution of current density over catalyst layer.

Figure 49.4 shows the current flux density distribution over the cathode catalyst layer at a cell potential of 0.4 V. Current flux density distribution is uniform over the cathode catalyst layer.

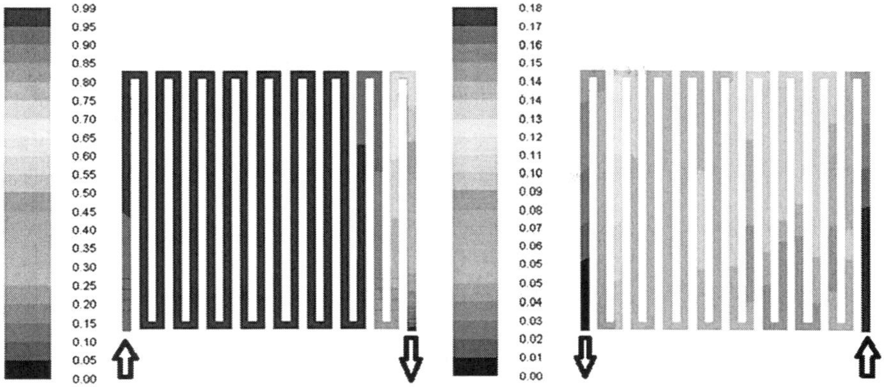

FIGURE 49.3 Contours of H_2 and O_2 mass fraction along the anode (left) and cathode (right) channel.

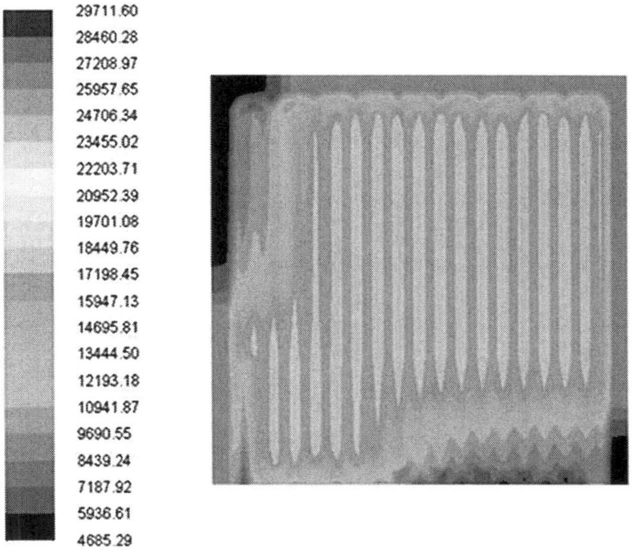

FIGURE 49.4 Current flux density distribution over cathode catalyst layer at 0.4 V.

Figure 49.5 shows the liquid water activity in the cathode channel at 0.4 V cell potential. This water is produced on the cathode because of the oxygen reduction reaction. The liquid water activity is less at the inlet and more at outlet of the cathode flow channel and also the water production rate is more at higher current densities.

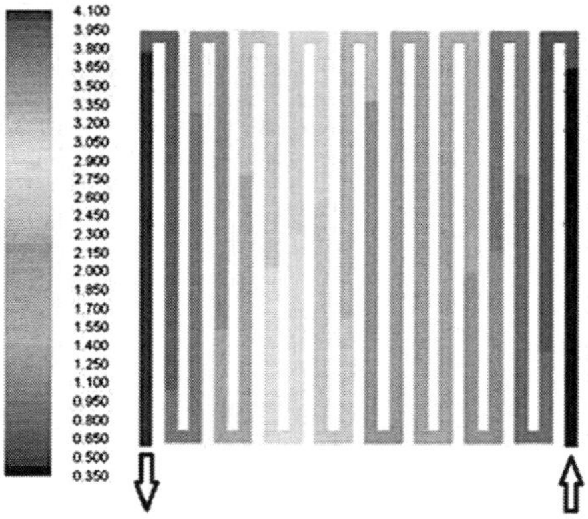

FIGURE 49.5 Liquid water activity in the cathode channel.

Both dry-out and flooding membrane will negatively affect the cell behavior. So in order to avoid dry-out and flooding of the membrane, the water content inside the membrane should be distributed uniformly. Figure 49.6 shows the membrane water content distribution at cell potential of 0.4. Water content distribution in the membrane is comparable to the current density distribution in Figure 49.4.

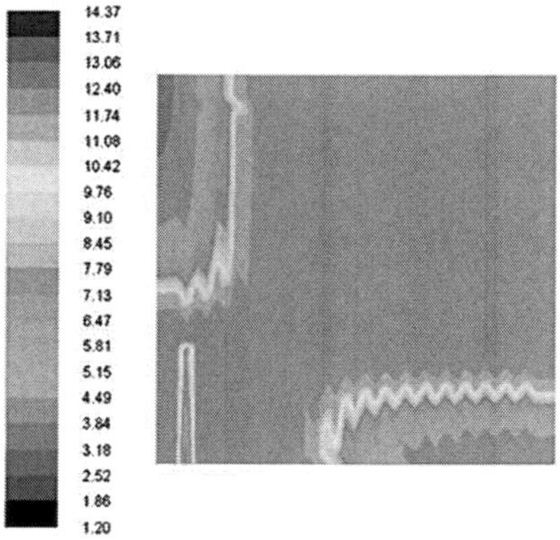

FIGURE 49.6 Membrane water content at 0.4 V.

Figure 49.7 shows the polarization and the power curve of the present PEMFC model and it is observed that peak power performance (power density of 0.32 W/cm^2) is obtained at 0.4 V cell potential at 0.8 A/cm^2 of current density.

Finally, the proposed model is validated by comparing the polarization and performance curves of simulation data with the experimental data[17] shown in Figure 49.7. The simulation results slightly overpredict the experimental results.

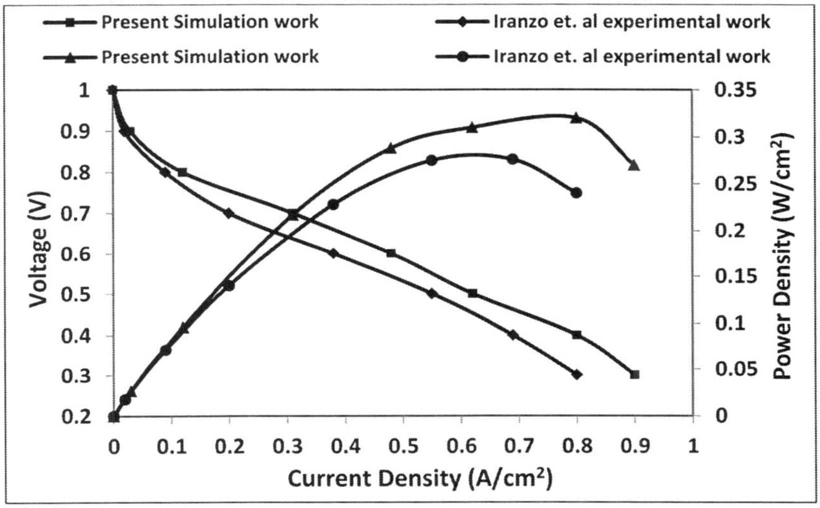

FIGURE 49.7 Polarization and performance curve.

49.4 CONCLUSIONS

A CFD study on serpentine flow field PEMFC was carried out and key parameters such as pressure drop, reactants mass fraction, liquid water activity, the membrane water content, polarization, and performance of the PEMFC have been analyzed. It is concluded that the cell delivered a peak power density of 0.32 W/cm^2 at 0.4 V cell potential at 0.8 A/cm^2 of current density. The simulation results of the proposed model are compared with the experimental data that showed good agreement.

ACKNOWLEDGMENTS

The authors acknowledge the financial support provided by Technical Education Quality Improvement Programme (TEQIP)-II/CoE—Centre for Sustainable Energy Studies, National Institute of Technology, Warangal.

KEYWORDS

- **fuel cell**
- **flow field**
- **modeling**
- **species**
- **performance**

REFERENCES

1. Beicha, A.; Zaamouche, R. Electrochemical Model for Proton Exchange Membrane Fuel Cells Systems. *J. Power Technol.* **2013,** *93*(1), 27–36.
2. Venkateswarlu Velisala; Srinivasulu, G. N.; Srinivasa, B.; Rao, K. V. K. Review on Challenges of Direct Liquid Fuel Cells for Portable Application. *World J. Eng.* **2015,** *12*(6), 591–606.
3. Larminie, J.; Dicks, A. *Fuel Cell Systems Explained,* 2001, 2nd ed. John Wiley & Sons: West Sussex, p 93.
4. Boettner, D. D.; Paganelli, G.; Guezennec, Y. G.; Rizzoni, G.; Moran, M. J. Proton Exchange Membrane Fuel Cell System Model for Automotive Vehicle Simulation and Control. *J. Energy Resour. Technol.* **2002,** *124*(21), 20–22.
5. Mammar, K.; Chaker, A. Neural Network-Based Modeling of PEM fuel cell and Controller Synthesis of a Stand-Alone System for Residential Application. *IJCSI Int. J. Comput. Sci.* **2012,** *9*(6), 244–253.
6. Corrêa, J. M.; Member, S.; Farret, F. A.; Canha, L. N.; Simões, M. G.; Member, S. An Electrochemical-Based Fuel-Cell Model Suitable for Electrical Engineering Automation Approach. *IEEE Trans. Ind. Electron.* **2004,** *51*(5), 1103–1112.
7. Berning, T.; Lu, D. M.; Djilali, N. Three-Dimensional Computational Analysis of Transport Phenomena in a PEM Fuel Cell. *J. Power Sources* **2002,** *106*(1), 284–294.
8. Barreras, F., Lozano, A.; Valiño, L.; Marín, C.; Pascau, A. Flow Distribution in a Bipolar Plate of a Proton Exchange Membrane Fuel Cell: Experiments and Numerical Simulation Studies. *J. Power Sources* **2005,** *144*(1), 54–66.
9. Lum, K. W.; McGuirk, J. J. Three-Dimensional Model of a Complete Polymer Electrolyte Membrane Fuel Cell–Model Formulation, Validation and Parametric Studies. *J. Power Sources* **2005,** *143*(1–2), 103–124.
10. Carcadea, E.; Ene, H.; Ingham, D. B.; Lazar, R.; Ma, L.; Pourkashanian, M.; Stefanescu, I. A Computational Fluid Dynamics Analysis of a PEM Fuel Cell System for Power Generation. *Int. J. Numer. Methods Heat Fluid Flow* **2007,** *17*(3), 302–312.
11. Jang, J.-H.; Yan, W.-M.; Li, H.-Y.; Tsai, W.-C. Three-Dimensional Numerical Study on Cell Performance and Transport Phenomena of PEM Fuel Cells with Conventional Flow Fields. *Int. J. Hydrogen Energy* **2008,** *33*(1), 156–164.

12. Suresh, P. V.; Jayanti, S.; Deshpande, A. P.; Haridoss, P. An Improved Serpentine Flow Field with Enhanced Cross-Flow for Fuel Cell Applications. *Int. J. Hydrogen Energy* **2011,** *36*(10), 6067–6072.
13. Guo, N.; Leu, M. C.; Koylu, U. O. Bio-Inspired Flow Field Designs for Polymer Electrolyte Membrane Fuel Cells. *Int. J. Hydrogen Energy* **2014,** *39*(36), 21185–21195.
14. Khazaee, I.; Sabadbafan, H. Numerical Study of Changing the Geometry of the Flow Field of a PEM Fuel Cell. *Heat Mass Transf.* **2016,** *52*(5), 993–1003.
15. Yu, L.; Ren, G.; Qin, M.; Jiang, X. Simulation of Performance Influencing Factors of Proton Exchange Membrane Fuel Cells. *Eng. Appl. Comput. Fluid Mech.* **2008,** *2*(3), 344–353.
16. Hashemi, F.; Rowshanzamir, S.; Rezakazemi, M. CFD Simulation of PEM Fuel Cell Performance: Effect of Straight and Serpentine Flow Fields. *Math. Comput. Model.* **2012,** *55*(3–4), 1540–1557.
17. Iranzo, A.; Muñoz, M.; López, E.; Pino, J.; Rosa, F. Experimental Fuel Cell Performance Analysis Under Different Operating Conditions and Bipolar Plate Designs. *Int. J. Hydrogen Energy* **2010,** *35*(20), 11437–11447.

INDEX

D

E

F

G

T

U

V

W

X

Y

Z

Printed and bound by PG in the USA
USA2018PGIL